Natural Remediation of Environmental Contaminants:

Its Role in Ecological Risk Assessment and Risk Management

Other titles from the Society of Environmental Toxicology and Chemistry (SETAC):

Environmental Contaminants and Terrestrial Vertebrates: Effects on Populations, Communities, and Ecosystems
Albers, Heinz, Ohlendorf, editors
2000

Ecotoxicology of Amphibians and Reptiles
Sparling, Linder, Bishop, editors
2000

Evaluation of Persistence and Long-range Transport of Organic Chemicals in the Environment
Klečka, Boethling, Franklin, Grady, Graham, Howard, Kannan, Larson, Mackay, Muir, van de Meent, editors
2000

Multiple Stressors in Ecological Risk and Impact Assessment: Approaches to Risk Estimation
Ferenc and Foran, editors
2000

Evaluating and Communicating Subsistence Seafood Safety in a Cross-cultural Context: Lessons Learned from the Exxon Valdez Oil Spill
Field, Fall, Nightwander, Peacock, Varanasi, editors
2000

Ecotoxicology and Risk Assessment for Wetlands
Lewis, Mayer, Powell, Nelson, Klaine, Henry, Dickson
1999

Reproductive and Developmental Effects of Contaminants in Oviparous Vertebrates
Di Giulio and Tillitt, editors
1999

Linkage of Effects to Tissue Residues: Development of a Comprehensive Database for Aquatic Organisms Exposed to Inorganic and Organic Chemicals
Jarvinen and Ankley, editors
1999

Sustainable Environmental Management
Barnthouse, Biddinger, Cooper, Fava, Gillett, Holland, Yosie, editors
1998

Chemical Ranking & Scoring: Guidelines for Relative Assessments of Chemicals
Swanson and Socha, editors
1997

For information about any SETAC publication, including SETAC's international journal, *Environmental Toxicology and Chemistry*, contact the SETAC Office nearest you.

1010 N. 12th Avenue
Pensacola, Florida, USA 32501-3367
T 850 469 1500 F 850 469 9778
E setac@setac.org

Av de la Toison d'Or
B-1060 Brussels, Belgium
T 32 2 772 72 81 F 32 2 770 53 83
E setaceu@setac.org

http://www.setac.org

Environmental Quality Through Science®

Natural Remediation of Environmental Contaminants:

Its Role in Ecological Risk Assessment and Risk Management

Edited by

Michael Swindoll
ExxonMobil Biomedical Sciences, Inc.
Annandale, New Jersey

Ralph G. Stahl, Jr.
E.I. duPont de Nemours
Corporate Remediation
Wilmington, Delaware

Stephen J. Ells
U.S. Environmental Protection Agency
Office of Emergency and Remedial Response
Washington, D.C.

SETAC General Publications Series

Current Coordinating Editor of SETAC Books
C. G. Ingersoll
U.S. Geological Survey

Publication sponsored by the Society of Environmental Toxicology and Chemistry (SETAC) and the SETAC Foundation for Environmental Education

SETAC PRESS

Cover by Michael Kenney Graphic Design and Advertising
Copyediting and typesetting by Wordsmiths Unlimited
Indexing by Celia McCoy

Library of Congress Cataloging-in-Publication Data

Natural remediation of environmental contaminants : its role in ecological risk assessment and risk management / edited by Charles Michael Swindoll, Ralph G. Stahl, Jr., Stephen J. Ells.

p. cm. -- (SETAC general publication series)

Includes bibliographical references and index.

ISBN 1-880611-33-3 (alk. paper)

1. Hazardous wastes--Natural attenuation. 2. Environmental risk assessment. I. Swindoll, Charles Michael, 1954–. II. Stahl, Ralph G., 1953– III. Ells, Stephen J. (Stephen John), 1950– IV. Series.

TD1060.N387 2000
628.5--dc21 00-058800

International Standard Book Number 1-880611-33-3
Printed in the United States of America
07 06 05 04 03 02 01 00 10 9 8 7 6 5 4 3 2 1

∞ The paper used in this publication meets the minimum requirements of the American National Standard for Information Sciences—Permanence of Paper for Printed Library Materials, ANSI Z39.48-1984.

Reference Listing: Swindoll M, Stahl Jr RG, Ells SJ. 2000. Natural remediation of environmental contaminants : Its role in ecological risk assessment and risk management. Published by the Society of Environmental Toxicology and Chemistry (SETAC). 472 p.

The SETAC General Publications Series

The SETAC General Publications Series was established by the Society of Environmental Toxicology and Chemistry (SETAC) to provide in-depth reviews and critical appraisals on scientific subjects relevant to understanding the impacts of chemicals and technology on the environment. The series consists of books on topics reviewed and recommended by the SETAC Board of Directors and approved by the Publications Advisory Council for their importance, timeliness, and contribution to multidisciplinary approaches to solving environmental problems. The diversity and breadth of subjects covered in the series reflect the wide range of disciplines encompassed by environmental toxicology, environmental chemistry, and hazard and risk assessment. These volumes attempt to present the reader with authoritative coverage of the literature, as well as paradigms, methodologies, and controversies; research needs; and new developments specific to the featured topics. This volume underwent peer review.

The SETAC General Publications are useful to environmental scientists in research, research management, chemical manufacturing, chemical regulation, and education, as well as to students considering careers in these areas. The series provides information for keeping abreast of recent developments in familiar subject areas and for rapid introduction to principles and approaches in new subject areas.

Contents

List of Figures

List of Tables

Acknowledgments

The editors wish to acknowledge the many contributors who made this book possible. We thank the chapter authors for contributing their time and dedication to this effort. We also thank the many reviewers who provided important input, including peer reviewers Gene Mancini and Nigel Blakely, who offered valuable suggestions. We thank Stacey Hagman and others at SETAC who provided valuable support and time to bring this effort to completion, as did Mimi Meredith at Wordsmiths Unlimited. Last but not least, we acknowledge the continued support of our families.

The content of this publication does not necessarily reflect the position or the policy of the United States government, and an official endorsement should not be inferred.

Contributors

C. Marjorie Aelion
Department of Environmental Health Sciences
University of South Carolina
Columbia, South Carolina

Herbert E. Allen
Department of Civil and Environmental Engineering
University of Delaware
Newark, Delaware

Robert C. Barrick
Exponent
Bellevue, Washington

Charles A. Bleckmann
Department of Systems and Engineering Management
Air Force Institute of Technology
Wright-Patterson Air Force Base, Ohio

D. Frederick Bodishbaugh
Exponent
Bellevue, Washington

T. Prabhakar Clement
Pacific Northwest National Laboratory
Battelle
Richland, Washington

John P. Connolly
Quantitative Environmental Analysis, LLC
Montvale, New Jersey

John W. Davis
The Dow Chemical Company
Midland, Michigan

Stephen J. Ells
USEPA Office of Emergency and Remedial Response
Washington DC

David Glaser
Quantitative Environmental Analysis, LLC
Montvale, New Jersey

Mark N. Goltz
Department of Systems and Engineering Management
Air Force Institute of Technology
Wright-Patterson Air Force Base, Ohio

Duane Graves
IT Corporation
Knoxville, Tennessee

Mary Ann Hawke
University of Washington
Seattle, Washington

Edward C. Heyse
Environmental Restoration
Air Force Center for Environmental Excellence
Brooks Air Force Base, Texas

James P. Hickey
U.S. Geological Survey
Ann Arbor, Michigan

Timothy J. Iannuzzi
Exponent
Landover, Maryland

Steven M. Jones
Exponent
Exton, Pennsylvania

Michael A. Kamrin
Institute for Environmental Toxicology
Michigan State University
East Lansing, Michigan

James R. Karr
University of Washington
Seattle, Washington

Michael D. Lee
Terra Systems, Inc.
Wilmington, Delaware

Greg Linder
HeronWorks Farm
Brooks, Oregon

David F. Ludwig
Exponent
Landover, Maryland

Teresa Michelsen
Avocet Consulting
Kenmore, Washington

Christina Noftsker
Exponent
Bellevue, Washington

Gary A. Pascoe
EA Engineering Science and Technology
Bellevue, Washington

Dora R. Passino-Reader
U.S. Geological Survey
Ann Arbor, Michigan

Robert A. Pastorok
Exponent
Bellevue, Washington

Brent M. Peyton
Chemical Engineering Department
Washington State University
Pullman, Washington

Douglas P. Reagan
URS Greiner Woodward Clyde
Denver, Colorado

Michael V. Ruby
Exponent
Boulder, Colorado

Daniel W. Smith
Exponent
Exton, Pennsylvania

Ralph G. Stahl, Jr.
Corporate Remediation
E.I. duPont de Nemours
Wilmington, Delaware

Michael Swindoll
ExxonMobil Biomedical Sciences, Inc.
Annandale, New Jersey

Alan F. Weston
Conestoga-Rovers & Associates
Niagara Falls, New York

Yujun Yin
University of Delaware
Department of Civil and Environmental Engineering
Newark, Delaware

Introduction

ECOSYSTEMS ARE DYNAMIC and possess innate, yet finite, abilities to adapt to environmental conditions and changes. For many environmental contaminants, the potential risks they pose to ecological receptors decrease over time because of changes in chemical properties and the natural assimilative abilities of the environment. A variety of terms describes these processes, including "weathering," "natural attenuation," "passive remediation," "intrinsic bioremediation," "natural restoration," and "bioattenuation." In this book, the term "natural remediation" (NR) is used to characterize these processes. Specifically, NR is defined as the sum of the nonanthropogenic chemical, physical, and biological processes that reduce potential risks to ecological receptors by altering the concentration, mobility, toxicity, and/or bioavailability of chemical contaminants. In addition, biological and ecological responses to contamination may alter the potential impact of the contaminants and thus also contribute to the NR phenomenon.

Natural remediation can be effective at some sites; however, for a variety of reasons, NR is not an appropriate risk management alternative for all sites. This book is intended to raise awareness of NR and its role in ecological risk assessment (ERA) and risk management among environmental scientists, regulatory personnel, risk assessors, risk managers, and other remediation practitioners. While our primary focus is on the technical aspects of NR, we also discuss other components of risk management, including societal and regulatory considerations, costs, feasibility of engineered remediation, and potential impacts and risks of remediation to humans and ecological receptors.

The reader should be aware that regulators, risk assessors, and other environmental practitioners hold diverse points of view concerning the appropriateness of NR as a risk management alternative. That diversity is also reflected in this book. For some, the typical first reaction is to equate NR with a "no action" or a "no further action" decision. But NR is not a no-action decision; it can result in risk reduction. Some in the regulatory community can be expected to remain skeptical and resistant to implementing NR. This skepticism can be addressed by including a monitoring program as part of the NR plan, by securing the commitment of the responsible parties to undertake such a monitoring effort, or when necessary, by implementing a contingency plan if conditions change such that NR is no longer viable.

The risk reduction that NR affords humans is not specifically addressed in this book. Although the focus of this book is on risk reduction to ecological receptors, the reader should understand that risk management decisions typically consider baseline risks to both human health and the environment as well as the risks from implementing various alternatives. From a scientific standpoint, it is expected that risk reduction to humans occurs in a similar manner as for ecological receptors and that a reduction in risk to either receptor normally reduces risk to the other. There are exceptions to this, and no unilateral conclusions should be drawn from this statement. For example, the use of institutional controls such as fish consumption advisories may reduce exposures and risks to humans, but by themselves, they are not protective of ecological receptors.

A major focus of this book is to describe the precepts of NR and to provide examples of where it has been or could be used in site cleanups. Under appropriate conditions, the risks posed by most contaminants will be reduced to some degree by NR. However, the magnitude of these risk reductions varies depending on site-specific factors, including the type of contaminant and the site conditions. For some contaminants, the risks they pose will not be reduced to acceptable levels within an acceptable time frame, because of either intrinsic properties of the contaminant or site-specific conditions. This may be the case for many of the persistent, bioaccumulative, and toxic (PBT) chemicals. The challenge for risk assessors, risk managers, scientists, and other environmental professionals is to determine the appropriate remedial action for a site.

Sometimes these professionals act to remove contaminants from a site without clearly understanding the impact of the selected remedial action or the role that NR might play in restoration. Implementing such remedial action may result in long-term detrimental impacts to the ecosystems, whereas NR may be sufficiently protective such that additional, often intrusive remediation is unnecessary or able to be minimized. In situations where remedial action may not significantly reduce ecological risks, an understanding of NR can be used to assess the potential long-term impact of the contamination and to develop appropriate management actions. Appropriate assessment of NR should allow for the net environmental benefit to be factored into the selection of remedial alternatives.

The book is organized into 16 chapters. Chapters 1 through 6 provide an overview of NR and describe NR processes and their assessment, ecological responses to contamination, modeling of NR, and the potential role of NR in risk assessment and management. Chapters 7 through 10 address the NR of some of the chemicals that are often of greatest environmental interest, specifically chlorinated compounds, polycyclic aromatic hydrocarbons, and other petroleum hydrocarbons, metals, and radionuclides. Chapters 11 through 15 discuss NR in a variety of ecosystems and habitats and include case histories of contaminated sites where NR played a role in risk-management decisions. Chapter 16 addresses NR in the context of risk management, provides insight into the perspectives of various stakeholders, and identifies technical challenges and research needs.

Ultimately, NR is one of a suite of potential actions that should be considered when risk mitigation is addressed. We hope that this book will encourage scientists and policy-makers to include it routinely in the assessment of contaminated sites. However, only when a sufficient experience base has developed will NR find more routine application owing to the confidence gained from understanding the benefits and lessons derived from its use.

Michael Swindoll
ExxonMobil Biomedical Sciences, Inc.

Ralph G. Stahl, Jr.
E.I. duPont de Nemours

Stephen J. Ells
U.S. Environmental Protection Agency
Office of Emergency and Remedial Response

1

Natural Remediation of Environmental Contaminants: Its Role in Ecological Risk Assessment and Risk Management

Michael Swindoll, Ralph G. Stahl Jr., Stephen J. Ells

CHEMICAL CONTAMINANTS, including metals, organic compounds, and radionuclides, occur in the environment as the result of human activities and natural phenomena. In instances where wildlife or other biota are exposed to these chemicals, there is the potential for adverse ecological impacts. As stewards of Earth's ecological resources, we must balance the responsibility to protect and maintain these resources with the need to provide for continued societal growth and utilization of these resources. The recognition of this responsibility and of the potential adverse impact of contamination on ecological resources led to the enactment of environmental regulatory programs in the U.S. and other countries and to an increased scientific examination of the fate and effects of environmental contaminants.

Environmental regulations and reforms have contributed to significant improvements in environmental quality in many countries. Changes in many industrial and societal practices have reduced or eliminated sources of contaminants to the environment. With these changes, improvements to environmental quality are expected to continue. However, with these improvements has come the realization that contamination cannot be completely eliminated, nor is complete contaminant elimination necessary in order to protect and maintain healthy, sustainable ecosystems.

Natural Remediation of Environmental Contaminants. Michael Swindoll et al., editors.
 ISBN 1-880611-33-3.

This chapter provides an overview of the processes of natural remediation (NR), of how NR can be evaluated during ecological risk assessments (ERAs), and of how NR can be selected appropriately as a response to contamination. Natural remediation will not be appropriate to all sites, all contaminants, or all receptors. In those instances where chemical contamination poses an immediate threat to ecological resources, corrective actions may be required to mitigate the threat. In instances where NR is selected as the appropriate risk management alternative, considerable future effort and resources may be required to ensure that it was an appropriate choice, that is, that unacceptable risks have been or are being mitigated.

REGULATORY CONSTRUCTS

Since the 1970s, numerous regulatory programs have been implemented in the U.S. and other developed nations to restore and maintain environmental quality. These programs often require actions that reduce or mitigate the exposure of humans and ecological receptors to harmful levels of chemical contaminants. Under the 2 primary legislative drivers in the U.S. for addressing contaminated sites, the Comprehensive Environmental Response, Compensation and Liability Act (CERCLA) and the Resource Conservation and Recovery Act (RCRA), assessing potential ecological risks posed by chemical contaminants is a key element of determining appropriate response actions. Where chemical contaminants pose potentially unacceptable risk to ecological receptors because of the magnitude, duration, or frequency of exposure, appropriate response actions are normally required to protect ecological receptors.

Historically, decisions for undertaking response actions for contaminated sites were typically based on human health–based regulatory standards or criteria that defined contaminant cleanup levels. Until recently, the use and understanding of risk assessment for humans and ecological receptors were not fully developed, and most regulatory cleanup levels were not based on rigorous, site-specific risk information. As a result, the processes for addressing appropriate response actions were often generic, rigid, costly, and in some cases, overly conservative. At the same time, there was not a wide understanding of natural processes that affect environmental contaminants or of how these processes affect the fate and effects of contaminants.

Today, risk-based approaches for establishing response actions and making risk management decisions are being used because they establish cleanup levels appropriate for site-specific conditions and, as such, are not under- or overprotective of ecological resources. Guidances for ERAs (U.S. Environmental Protection Agency [USEPA] 1992, 1997, 1998a; Wentsel et al. 1996), as well as for risk-based corrective action (RBCA; ASTM 1994, 2000), use a tiered or phased approach for assessing potential risks to human and ecological receptors. Tiered assessment approaches provide the benefit of allowing decisions to be made at the earliest possible time and therefore may contribute to more streamlined and cost-effective evaluations and corrective actions.

LINKAGE WITH ECOLOGICAL RISK ASSESSMENT

Assessments of potential risk posed by environmental contaminants and by potential response actions are important to the risk management process. To be most useful, assessments of ecological risk should include evaluations of risks associated with response action alternatives that may meet management goals. A properly performed ERA, coupled with other information, provides decision-makers and risk managers with details helpful in selecting the response alternative that best meets management objectives. During the ERA, sufficient information should be derived to allow for a technically defensible comparison of all response alternatives. In most instances, both potential impacts of not implementing a response action and potential impacts of implementing a response action should be evaluated. Historically, risk evaluations of both the ecological impacts of the proposed action and the NR alternative were seldom undertaken. Therefore, informed selection of the alternative that meets management goals and provides the greatest net environmental benefit was not always possible. However, in recent years, the greater utilization of risk-based decision-making has meant that assessments increasingly include consideration of the NR alternative.

NATURAL REMEDIATION AND USEPA ECOLOGICAL RISK ASSESSMENT GUIDELINES

It is important to understand how NR fits, or should fit, into the agency-wide guidelines for ERA (USEPA 1998a) and into other guidance such as the Superfund guidance on ERA (USEPA 1997, 1999). The USEPA guidelines consist of 3 steps:

1) Problem formulation
2) Analysis (exposure and effects)
3) Risk characterization.

Ways in which NR can be integrated into the ERA process are discussed in the following sections.

Problem formulation

Problem formulation, the first step in the USEPA guidelines for ERA, consists of a number of very important activities. Actions include gathering past and current information, selecting assessment endpoints, measuring effects and exposure, building the conceptual model, and making an analysis plan. One of the most important actions is the discussion that occurs between the risk assessor and the risk manager. The interface between these 2 individuals, or groups of individuals, is crucial to the successful outcome of the ERA. It is here, perhaps for the first time, that the concepts of NR are discussed. Management goals, types of potential remedial action that might be feasible, and constraints on the timing, resources, and scope that will apply to the effort — all affect what the ERA will ultimately provide. During these discussions, it is important to resolve whether NR, and its potential to reduce the estimated ecological risk over time, will or will not be evaluated as an appropriate option during the ERA (Figure 1-1). It is also important at this step to begin to understand which receptors might be at risk and what types of ecosystems are involved. Some receptors and ecosystems will be much more adaptable than others at compensating for the effects from chemical stressors. For these, NR will be more applicable and valuable than will other types of remediation. Of course, NR will not be an option at all sites and under all circumstances because some ecological risks could be large enough that they require more immediate attention.

- ✓ There is no evidence of an imminent threat to ecological resources.
- ✓ The environmental contaminants are amenable to natural remediation (NR).
- ✓ Ecosystem characteristics that support NR are present and expected to continue.
- ✓ Sufficient time is available for NR.
- ✓ Protecting existing ecosystem integrity is important.
- ✓ It is feasible to measure and track how NR reduces ecological risks.

Figure 1-1 Situations in which natural remediation may be appropriate

When NR is to be considered in the ERA, this should be agreed to no later than the problem formulation step. This is important because the analysis plan will be developed during problem formulation, and this plan will provide the framework under which data important to evaluate NR will be collected. Furthermore, the conceptual model is developed during problem formulation, making it crucial that NR be shown as part of this conceptual vision. Omitting NR from the conceptual model and the analysis plan can cause problems later in the ERA when it becomes increasingly difficult to secure the resources and time to obtain additional data. And, in some cases, data to support whether or not NR is occurring and is sufficiently protective are already outside the "mainstream" dataset that risk assessor and risk manager are accustomed to dealing with. This may present a barrier to obtaining agreement on collecting the relevant NR data.

Analysis

The analysis step of the ERA consists of a technical evaluation of potential exposure and characterization of ecological effects that may result from exposure to a stressor. There are 3 main activities in the analysis step:

1) Assess ecological effects.
2) Assess exposure.
3) Characterize the ecosystem and receptors at risk.

Assess ecological effects

It is important during the effects analysis activity to understand which chemical stressors are present and how those stressors are likely to impact exposed receptors. The toxicity of some chemical stressors is known to decline with time. For example, polycyclic aromatic hydrocarbons (PAHs) and petroleum hydrocarbons (PHs) can be attenuated by organic materials in soils and sediments (Swartz and Di Toro 1997), reducing the toxicity of these chemicals. Natural remediation can be linked to the ERA during the effects analysis by collecting data that will shed light on whether the attenuation and reduction processes are present and, if so, whether they are likely to have reduced the toxicity of the chemical stressor. Without collecting the appropriate data, however, there will continue to be high uncertainty as to whether or not the NR process is occurring or could occur.

Assess exposure

One of the most important aspects of the exposure assessment activity is to understand what exposure pathways exist, which of the existing pathways are complete or potentially complete, and the circumstances under which the exposure occurs or may be mitigated. It is during this activity that the chemical stressor is further evaluated for conditions that might alter the exposure pathway or the magnitude, duration, and frequency of exposure. For example, humic acids and other organic matter in soils and water have the ability to bind the chemical stressor to some degree and, in so doing, to reduce its bioavailability. These attenuating substances, and their ability to bind toxic chemicals, may ultimately provide one of the most important processes in NR. The ability of the receptors in question to metabolize the chemical stressor to a less toxic form, and ultimately to excrete it, are important features to consider. In some cases, information on the receptor's metabolic ability is readily available in the published literature; in others, this may not be true. As a result, obtaining site-specific information on the receptor's ability to metabolize contaminants might prove highly valuable to the risk assessor when he or she estimates the ecological risk.

Characterize the ecosystem and receptors at risk

Along with the assessment of exposure and effects, characterizing the ecosystem potentially at risk also provides important information that can be used to understand NR. This can be important when assessing whether NR will be feasible, given the type of ecosystem at risk. In instances where NR is not feasible, this understanding may facilitate the selection of response actions (e.g., soil amendments that bind the chemical stressor) that enable the ecosystem to mitigate the effects of chemical stressors.

Other important aspects of NR that can be determined during the analysis step include whether or not the ecosystem has intrinsic capabilities of degrading the chemical stressor, such as may be the case where the ecosystem has a well-established and metabolically diverse microbial or fungal community. Again, because of the diversity of microbial and fungal communities, it is important to obtain site-specific information.

When the ecosystem at risk is being characterized, "indicator parameters" can be analyzed to help determine whether a particular ecosystem is capable of mitigating the effects of the chemical stressors. Indicator parameters are those analytes that directly or indirectly indicate NR processes. For example, it may be useful to collect information on the presence of oxygen in surficial aquatic systems, levels of nutrients that would support microbial and other degradative systems, and concentrations of "natural attenuators" in the media of interest. Natural attenuators can include humic acids, clay and clay-like substances in soils, organic matter, and other materials that are known to retard the bioavailability, toxicity, or mobility of chemical stressors or to enhance their degradation. These analytical results would provide valuable insight into the ability of NR processes to affect contaminants in an ecosystem.

Risk characterization

During the risk characterization step, the risk assessor is faced with describing the ecological risk, estimating its magnitude, and describing the uncertainties in the ERA (Browner 1995). Natural remediation comes into play throughout this step, particularly when the risk assessor is communicating the results of the ERA to the risk manager. For example, the risk posed by a chemical stressor may be mitigated by the presence of natural attenuators or other factors that reduce exposure or hazard. Knowing this, the risk assessor can explain that the ecological risk may be less than that represented by the numerical estimate because the estimate may not reflect these mitigating factors. It is also important that the risk assessor communicates what information, were it to be obtained, would help reduce the uncertainty in the ERA.

RISK MANAGEMENT

The ERA is ultimately used for making risk management decisions, although these decisions are not wholly dependent upon the results of the ERA. Other kinds of information, including policy, cost, feasibility, timing, engineering constraints, and public interest, all play into the final risk management decision. However, it is at this point in the overall effort that the role of NR is prominent. In general, the risk manager is faced with 3 options:

1) some type of active remediation,
2) NR, or
3) a combination of the two.

If NR is selected as the risk management action, it is likely that this decision will include a requirement for a long-term chemical and/or biological monitoring plan to ensure that the natural processes are adequately protective or are continuing to reduce the ecological risk to levels that will be protective.

Thus, selecting NR may involve a substantial amount of effort, and in the end, it may not be the least costly alternative. One obvious advantage to selecting NR as the risk management option is that it tends to have the smallest environmental impact from habitat destruction or alteration: Soils or sediments are generally not excavated, and thus, the cleanup area is not physically disrupted. By comparison, active remediation, in which soils, sediments, or other materials may be removed from an area and disposed of, can have a much larger and more immediate environmental impact and may take much longer for the ecosystem to restore. Of course, it will be the risk manager's job to balance the need for active remediation against the NR option. Where the risk to ecological receptors is deemed unacceptable, even in the short term, active remediation may be the most appropriate risk management action.

THE CHALLENGES AHEAD

Today, there are 2 major hurdles that limit the application of NR in ERA: regulatory approval and experience. Regulatory approval is becoming less of an issue, at least in some situations. In its strategy for managing contaminated sediments, USEPA discusses NR and the benefits derived from its application (USEPA 1998b); however, applying NR is not without potential side effects. For example, in situations where natural resource damages may be assessed, the selection of the NR option may increase the liability of a company seeking to remediate a contaminated site or spill.

The other hurdle to overcome is the lack of experience in applying NR at contaminated sites. At this time, there are few data to help determine the difference between risk reduction provided by NR and that provided by active remediation. If NR becomes more widely applied, there will be a greater experience base on which to draw.

REFERENCES

[ASTM] American Society for Testing and Materials. 1994. Guide for risk-based corrective action applied at petroleum release sites. Volume 11.04, Annual book of ASTM standards. Philadelphia PA: ASTM. E1739.

[ASTM] American Society for Testing and Materials. 2000. Standard guide for risk-based corrective action; 2000 April 10. Philadelphia PA: ASTM. E208100.

Browner C. 1995. Policy for risk characterization at the U.S. Environmental Protection Agency. Washington DC: USEPA.

Swartz RC, Di Toro MD. 1997. Sediments as complex mixtures: An overview of the methods to assess ecotoxicological significance. In: Ingersoll CG, Dillon T, Biddinger GR, editors. Ecological risk assessment of contaminated sediments. Pensacola FL: Society of Environmental Toxicology and Chemistry (SETAC).

[USEPA] U.S. Environmental Protection Agency. 1992. Framework for ecological risk assessment. Washington DC: USEPA. EPA/630/R-92/001.

[USEPA] U.S. Environmental Protection Agency. 1997. Ecological risk assessment guidance for Superfund: Process for designing and conducting ecological risk assessments. Interim final report. Washington DC: USEPA. EPA–540-R-97-006.

[USEPA] U.S. Environmental Protection Agency. 1998a. Guidelines for ecological risk assessment. Washington DC: USEPA. EPA/630/R-95/001F.

[USEPA] U.S. Environmental Protection Agency. 1998b. EPA's contaminated sediment management strategy. Washington DC: USEPA. EPA-823-F-98-001.

[USEPA] U.S. Environmental Protection Agency. 1999. Ecological risk assessment and risk management principles for Superfund sites. Washington DC: Office of Solid Waste and Emergency Response. Directive 9285.7. 28 p.

Wentsel RS, La Point TW, Simini M, Checkai RT, Ludwig D, Brewer LW. 1996. Tri-service procedural guidelines for ecological risk assessments. Aberdeen Proving Ground MD: U.S. Army Edgewood Research, Development, and Engineering Center.

2

Natural Remediation in the Risk Management Process: Goals, Options, and Monitoring

Douglas P. Reagan

DECISIONS REGARDING implementation of the natural remediation (NR) option should be based on protection of both human and environmental health. In general, NR will not be appropriate at sites or in portions of sites where human health risks exceed criteria for acceptability. However, at many sites, cleanups are driven primarily by risks to ecological resources. This chapter therefore considers NR primarily in the context of ecological risk management.

Reluctance to consider NR in risk management decisions has been based largely on a general perception that this is a "do nothing" option that "lets the polluter off the hook" without adequately addressing the potential risk. Part of the pressure to act comes from a general perception by the public that assessment without subsequent action is a waste of money. Unfortunately, the resulting pressure to "do something" can lead to remedial actions such as removing sediment from a wetland or dredging a stream channel to remove sediments containing contaminants at concentrations that may pose only marginal ecological risk. In the larger context, more ecological values may have been protected if the wetland had been left to remediate naturally. While an approach that causes more harm than the original problem may seem inappropriate, cleanup decisions at many contaminated sites have been based on the narrow objective of reducing potential ecological risk from the contaminants in question (e.g., effects on benthic invertebrate populations) without appropriate consideration of the effects that cleanup actions may have on the overall ecosystem.

Natural Remediation of Environmental Contaminants. Michael Swindoll et al., editors.

Such decisions often put risk managers at odds with natural resource managers. One such resource manager at a site where I worked a few years ago eloquently expressed the dilemma in the following way: "We don't see any effects of contamination, but we sure know what bulldozers will do to the ecosystem!" Such an approach may also result in additional liabilities, as adverse effects from contaminant cleanup actions on natural resources and the services they provide are considered natural resources injuries under Natural Resource Damage Assessment (NRDA) regulations (e.g., 43 Code of Federal Regulations [CFR] 11.73).

The processes that are broadly classified as NR have been operating in nature for eons. These processes continue to operate in the natural environment at locations such as outcrops of highly mineralized rock or hot springs and seeps where metals and other chemicals are released at concentrations that produce adverse effects in surrounding ecosystems. It is therefore logical to consider natural processes in making remedial decisions. If an objective of ecological risk assessment (ERA) and risk management is the overall protection of ecological values at a site, NR is a particularly important tool to consider.

If NR is implemented at a site, the major remedial activity will be monitoring remediation. Because NR processes are by definition natural, the monitoring program will focus on measuring reductions of stressor effects or of the exposure of ecological endpoints to environmental stressors (e.g., chemical contamination, physical or biological stressors).

This chapter addresses issues relevant to selecting and implementing the NR option in the risk management process. Risk management consists of 3 elements (Pittinger et al. 1998):

1) Identifying issues
2) Setting goals
3) Identifying options.

This chapter is organized to generally parallel this process. The following section, "Considering the Natural Remediation Option," discusses some of the issues involved in determining the suitability of the NR option to meet remedial goals. Each site is unique with respect to its suite of stressors, ecological characteristics, and geographic location; therefore, NR may not be appropriate at many sites. Because the NR option focuses on monitoring, the development of monitoring goals and the design of a monitoring program are discussed in the sections titled "Risk Management Goals" and "Designing a Monitoring Program for Natural Remediation" respectively.

CONSIDERING THE NATURAL REMEDIATION OPTION

The ultimate goal of the ecological risk management process is selecting the most reasonable and acceptable remedial option (Pittinger et al. 1998), and the National

Contingency Plan (NCP) requires consideration of a "no action" alternative (i.e., NR) at the feasibility stage. To properly evaluate the NR option, risk managers need to understand total ecosystem values (see the section titled "Risk Management Goals") to avoid the situation where remediation causes more harm than would have occurred had the contamination been left in place and the risk reduced to acceptable levels by natural processes. This does not mean that NR is appropriate in all circumstances. It does, however, suggest that NR should be considered as an alternative to engineered solutions that may cause additional ecological impacts. The decisions to implement the NR option are based on a number of considerations:

1) What are the valued ecological resources at this site?
2) Which NR processes are operating?
3) How long will recovery take?
4) What are the other remedial alternatives?
5) What are the costs associated with each remedial alternative?

What are the valued ecological resources at the site?

Knowing the ecological values at a site is basic to any consideration of NR. These values generally equate to the ecological functions that are necessary for maintaining a healthy ecosystem and the services the ecosystem provides to humans. If a site is highly disturbed or in an industrial setting, ecological values may be limited, and cleanup decisions may be based largely on other criteria. At sites with high ecological value (e.g., wetlands, mature forest habitat, prairie grasslands, areas supporting endangered or threatened species), ecological values should be considered in making remedial decisions.

Which natural remediation processes are operating?

To seriously consider the NR option, we first need to know which, if any, NR processes are operating in the affected area. Are there site data to indicate that these processes are working? How good is the evidence? What do we know about the mechanisms of these processes under the environmental conditions at the site? What remedial alternatives can be implemented if NR fails to solve the problem?

How long will recovery take?

How long do we reasonably expect natural processes to take to reduce the risk from in situ contamination to acceptable levels? Are trend data available for the site? What are the results from similar sites? Such data may be available from measurements taken during the site assessment of contaminant concentrations in exposure media or from the tissues of exposed organisms to determine whether there is a declining trend in concentrations that is attributable to NR processes.

The U.S. Department of Interior (USDOI) NRDA regulations (43 CFR 11.73) list a number of factors to consider when estimating natural recovery times that are appropriate for estimating NR rates, including the following:

- Patterns of ecological succession in the area
- Growth or reproductive patterns, life cycle, and ecological requirements of species involved, including their reaction to or tolerance of the contaminants (stressors)
- Bioaccumulation of contaminants in the food web
- Chemical, physical, or biological removal rates of the contaminants from the media involved, especially as they relate to local conditions.

What are the other remedial alternatives?

Natural remediation should be considered along with other alternatives for reducing risk at the affected site. The 2 most important questions to be addressed are these:

1) What are the risk reduction benefits of each alternative?
2) What are the relative effects of implementing each alternative on overall ecological values at the site?

While contamination may adversely affect one ecosystem component, it may (or may not) produce adverse effects at higher levels of ecological organization. For example, contamination may affect soil invertebrate populations in a forest ecosystem, but at levels that do not affect the food web of the ecosystem. However, active removal of the contaminated soil to which the soil invertebrates are exposed may result in the removal of all plant and animal communities at the site, resulting in loss of habitat, primary productivity, biological diversity, and other ecological values.

Remedial decisions should weigh the relative costs and benefits of using NR against those for other remedial alternatives (e.g., dredging sediments) that may reduce the risk from stressors more rapidly but that could cause large-scale adverse effects on overall ecosystem values. These values should be evaluated systematically using an approach that considers both ecological and societal values and that considers all levels of ecological organization. The general assessment endpoint (GAE) approach described in the section titled "Ecologically Relevant Values" provides an appropriate framework, and one that is consistent with the approach for selecting assessment endpoints for risk assessment and monitoring purposes.

What are the costs associated with each remedial alternative?

Although cost is not a consideration in the selection of remedial alternatives, cost effectiveness should be considered. In the U.S., the NCP requires that a detailed analysis be performed on a limited number of viable approaches to risk abatement.

Cost is also a required evaluation criterion in the analysis of alternatives for the feasibility study (of remediation) under the Comprehensive Response Compensation and Liability Act (CERCLA). Costs associated with active remediation include planning, implementation (e.g., dredging, soil excavation, capping and the associated long-term operation and maintenance of options such as groundwater pump and treat), and monitoring.

Monitoring is the major direct cost of implementing the NR option, and this cost can be greater than the cost of implementing other remedial alternatives. Potential costs associated with ongoing natural resource injuries or service losses that create liabilities under the NRDA regulations should also be considered (see the section titled "Special Monitoring Considerations"). The costs of permitting ongoing contaminant effects to continue throughout NR should be weighed against the additional natural resource injuries or service losses that could occur as the result of remedial activities.

RISK MANAGEMENT GOALS

Goals are needed in the risk management process for these reasons: 1) to select the appropriate remedy and 2) to focus the subsequent monitoring program. Identifying goals for the monitoring program is particularly important when NR is the remedial option selected.

Assessment endpoint identification should systematically identify all ecological values before determining which ones may be affected by environmental stressors. While this process should have been conducted as part of the risk assessment, it may not have provided a value framework adequate for making risk management decisions. For example, assessment endpoints affected by site-related stressors (e.g., chemical contaminants) may have been identified, but the larger context of ecological values necessary for evaluating effects of remedial options may not.

Selecting values to be protected in the risk management process is all too often an informal activity that focuses on species and populations and ignores stressor effects on ecological values at higher levels of ecological organization. Because all levels of ecological organization should be considered, the process of determining ecological values can be complex, particularly at sites that support a variety of natural ecosystems. The identification process is complicated by the need to address the large numbers of species, the different scales of ecological organization, and the number of endpoints and criteria that may be important (Pittinger et al. 1998). Documenting the rationale for selecting particular values and not others is generally brief and superficial at best.

Establishing general goals for risk management

Vogt et al. (1997) point out the importance of indexing species into functional groups based on their sensitivities to environmental variables as a means of understanding the effects of stressors on ecosystems. This section briefly describes a comprehensive and systematic process for identifying ecological values to be protected for the ecosystem potentially affected by environmental stressors and by remedial activities. The process of identifying those values, or GAEs, thoroughly documents the rationale for determining values for each potentially affected ecosystem. These GAEs encompass ecological and human use values at all levels of ecological organization, thus providing a systematic basis for determining goals for remedy selection and monitoring.

The process of identifying GAEs occurs in 2 parts. Part 1 identifies ecological values from a strictly ecological perspective. Part 2 considers the human values associated with the ecological resources of the ecosystems under evaluation.

Ecologically relevant values

The ultimate ecological value to be protected is a healthy, sustainable ecosystem. This concept is consistent with current principles of ecosystem management and with our current understanding of ecosystems (Kaufmann et al. 1994; Reichman and Pulliam 1996; Boyce and Haney 1997).

The process of identifying ecological GAEs consists of 5 steps:

1) Identify values common to all ecosystems.
2) Identify functional components of potentially affected (regional) ecosystems.
3) Develop a functional food web of potentially affected ecosystems.
4) Determine ecologically relevant attributes of the functional components of these ecosystems.
5) State the ecologically relevant GAEs.

This progression from general values to ecosystem-specific values is hierarchical and scientifically based.

Once these ecological GAEs have been determined, societal values for the ecosystem are developed. A detailed description of the overall approach is provided in Reagan (2001). A brief example of this process follows, based on a terrestrial site in western North America.

Identify values common to all ecosystems. GAE identification proceeds from the fundamental value of preserving a healthy and sustainable ecosystem to the identification of valued characteristics common to all ecosystems:

- Biological diversity (biodiversity). This property describes ecological structure, including species, community, habitat, and genetic diversity.

Biological diversity usually is considered most relevant in the context of species diversity (i.e., the variety and abundance of species) (Norton 1987; Magurran 1988; Kellert 1996) that is typical of the particular ecosystem under consideration.

- Functional integrity. An ecosystem with functional integrity is composed of living organisms within populations of different taxa, interacting to sustain normal ecosystem processes. Not only are all of the ecosystem components present, but they interact according to organizational principles typical of that ecosystem.
- Nutrient and energy dynamics. Appropriate nutrient and energy dynamics operate within the ecosystem to maintain balanced populations of indigenous species at levels typical of a healthy ecosystem. For an ecosystem to function normally, the flow rates of energy and nutrients (e.g., primary productivity and decomposition) should fall in a range typical of that type of ecosystem.

These values express ecosystem structure and function in terms that can be evaluated in the risk assessment and monitoring. Because they are values relevant to all ecosystems (Kormondy 1969; Brewer 1979; Odum 1993), they are appropriate values to be protected in the risk management process and to be evaluated in remedial monitoring. All other ecologically relevant issues are subsets of these valued characteristics.

Identify functional components. Values of regional ecosystems present at the site are the next level to be determined. Because food webs provide essential structural organization in ecosystems (Gallopin 1972) and because all organisms in an ecosystem are part of the food web, the food web is used as a framework to identify basic functional components of potentially affected ecosystems.

Food webs are typically composed of 3 basic trophic categories:

1) Producers: Organisms that manufacture food from inorganic compounds by photosynthesis or chemosynthesis (e.g., green plants)
2) Consumers: Organisms that ingest other organisms (e.g., animals that consume plants and/or other animals)
3) Decomposers: Organisms that derive their nourishment from dead organic matter (e.g., fungi and bacteria).

These categories are based on the broad interrelationships among groups of organisms but do not describe the many ways in which these interactions may occur. The 3 fundamental food web categories are, therefore, further categorized into functional components, based on a general knowledge of the taxa present in the regional ecosystems at the site. By using functional rather than taxonomic groups, we are able to determine broad functional aspects of the ecosystem and important interrelationships among components, thus avoiding the impossible task of determining the ecological relevance of each species or merely pursuing traditional goals and targets that miss many important ecological values (Slocombe 1998).

To illustrate this approach, consider the food web of a typical site in western North America that has a mosaic of grassland and woodland habitats. Functional components are developed on the basis of food, foraging location, and feeding habits of species present in the terrestrial ecosystem. Table 2-1 lists the functional components for the site. Three categories of producers are identified. Based on food type and location, 4 ecological components are identified within the herbivorous consumers trophic category. Some feeding guilds are arbitrarily grouped (e.g., fruit- and seedeaters [frugivores and granivores], leafeaters [folivores] and browsers) because of the similarity of food type for many species in this particular ecosystem. Similar functional groups are identified for the other intermediate consumers and top predators. All species can be classified into one or more of the functional components.

Table 2-1 Functional components of a food web at a site in western North America

Basic trophic category	Functional component
Producers	Herbaceous plants
	Woody shrubs
	Trees
	Mycorrhizal fungi (enhancers)
Consumers	Herbivores
	Terrestrial frugivores and granivores
	Arboreal frugivores and granivores
	Folivores and browsers
	Nectarivores
	Carnivores and omnivores
	Intermediate and small predators
	Omnivores
	Top terrestrial predators
	Top arboreal predators
	Detritivores and scavengers
Decomposers	Chemical decomposers

Develop a functional food web. The functional components described in the preceding section define the general range of feeding preferences and location (strata) in the ecosystem. Figure 2-1 presents a food web based on these functional

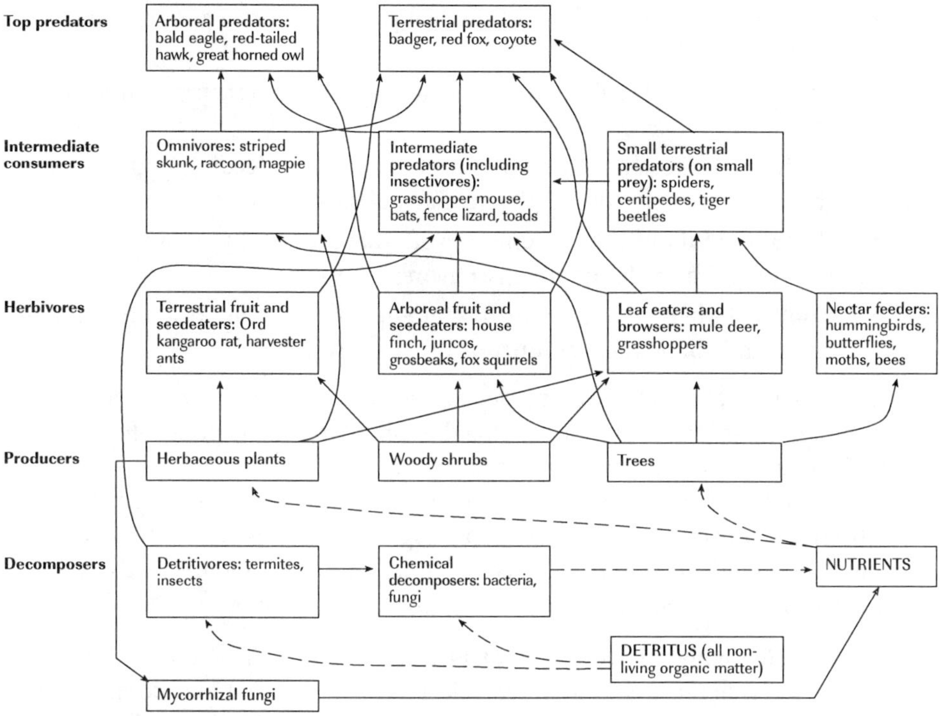

Figure 2-1 Food web based on functional components for a terrestrial ecosystem in western North America (modified from Reagan 2001). Arrows indicate flow of energy and nutrients; dashed lines indicate recycling and flux of energy and nutrients from the decomposition processes.

components. Although this food web is a gross oversimplification of reality, it is sufficient to permit the qualitative evaluation of interactions among functional components of the ecosystem and to predict possible effects at higher levels of ecological organization.

Determine ecologically relevant attributes. Each functional component of a food web may have additional ecologically relevant attributes that define its overall ecological value. For many functional components, the nontrophic attributes may be more important than their role in the food web. Common attributes for terrestrial ecosystems include these:

- Food: Source of energy and nutrients
- Habitat: Shelter or structural support for other organisms
- Primary production: Processes of converting inorganic chemicals to organic compounds that can provide energy and nutrients to other living components of the ecosystem

- Decomposition: Breakdown of nonliving organic matter and recycling nutrients
- Seed dispersal: Distribution of seeds and spores from their origin to other locations
- Pollination: Cross-fertilization by nectar and pollen-feeding animals that is the sole means of reproduction in many plant species
- Predation: Killing and consumption of other animals by carnivores, including top predators and insectivores
- Control: Either "top-down" or "bottom-up" effect on the structure (composition and abundance) or function of the ecosystem.

Other attributes may be relevant, depending on the ecosystem. For example, some species act as "engineers," modifying the habitat for other species (Jones et al. 1994). Still other attributes are appropriate for aquatic ecosystems.

The various attributes are then determined for each functional component of the ecosystem and arranged in a table displaying the attributes of each functional component (Table 2-2). The arrangement of these components in the landscape and the scale of the ecosystem at the site are important considerations.

State the ecologically relevant general assessment endpoints. USEPA defines an "endpoint" as "a characteristic of an ecological component...that may be affected by exposure to a stressor," and defines an "assessment endpoint" as "an explicit expression of the environmental value that is to be protected" (USEPA 1992). Therefore, GAEs based on functional components can be developed directly from Table 2-2, which provides a comprehensive summary of the attributes of all functional components of the ecosystem. General assessment endpoints that clearly define the values to be protected in this form are understandable to risk managers and other stakeholders in the risk management process.

Globally relevant assessment endpoints (i.e., biodiversity, functional integrity, and nutrient and energy dynamics) define values at the ecosystem and community levels of ecological organization. Ecosystem-specific endpoints are then defined, based on the significant attributes of the ecological components of the ecosystem. Only a few of these statements are provided below to illustrate the point; however, in practice, statements for all ecosystem values would be identified as the basis for making risk management decisions.

- Trees are valued components of the forest ecosystem because of their role in primary production, providing habitat for arboreal species, and as food for a variety of herbivorous animals.
- Terrestrial fruit and seedeaters are valued components of the ecosystem because of their importance as a food source to higher-level carnivores and their role as seed and propagule dispersers.

Table 2-2 Ecologically relevant attributes of functional components identified for a terrestrial ecosystem in western North America

	Ecologically relevant attributes						
Functional components	Food	Habitat	Primary production	Decomposition	Seed dispersal	Pollination	Control
Herbaceous plants	X	X	X				
Woody shrubs	X	X	X				
Trees	X	X	X				
Mycorrhizal fungi			X				X
Terrestrial frugivores and granivores	X				X		
Arboreal frugivores and granivores	X				X		
Folivores and browsers	X						
Nectarivores	X					X	
Intermediate and small predators	X						
Omnivores	X						
Top terrestrial and arboreal predators							X
Detritivores and scavengers				X			
Chemical decomposers				X			

- Nectar and pollen eaters are valued components of the ecosystem because of their importance in plant reproduction (i.e., pollination).
- Mycorrhizal fungi (fungi associated with plant roots) are valued components of the ecosystem because of their importance in nutrient recycling and regeneration of plants.
- Chemical decomposers are valued components of the ecosystem because of their importance in decomposition, in nutrient recycling, and as a food source.

Many sites have ecosystems composed of a variety of habitat and vegetation types (e.g., forest, woodland, grassland, freshwater streams). From a food web perspective, these habitats usually form a mosaic with many functional components feeding at the interface between ecosystems or, as in the case of many top predators, foraging throughout all vegetation types. Because many exotic (nonnative) species have invaded ecosystems (e.g., weeds), not all species present may be considered valued components, even though they function within the ecosystem. Spatial architecture (i.e., how ecosystem components are arranged in the environment) is also an important consideration.

Societally relevant values

Management goals are inextricably related to the societal values of ecological resources. These values include formally recognized and protected ecological values such as threatened and endangered species and game species. Where members of indigenous communities (e.g., Native American tribes) are present, the values identified may include a broad range of ecological components used for food, medicine, and spiritual purposes.

Determining societal values involves a systematic search of all pertinent data sources. These include existing information on rare and threatened species, interviews with regional experts, information on commercially and recreationally important resources, and input from the public (including indigenous communities) on natural resource values. Other ecological values for the ecosystem may be identified, based on a review of regional management goals and plans for potentially affected areas.

Representative GAEs based on societal values of ecological components can be stated in a format similar to that used for ecological values, as illustrated by the following examples:

- Game species are valued components of the ecosystem and are to be protected because of their importance for consumptive uses such as hunting and fishing for recreation and for food.
- Threatened and endangered species and their habitats are valued components of the ecosystem to be protected because of their regulatory stature and global value.
- Watershed health (water quality and quantity, erosion control, wetlands, etc.) is a valued characteristic of the ecosystem and is to be protected because of its importance to local communities.

Using general assessment endpoints in remedy selection

The primary aim of remedy selection is to reduce risk to acceptable levels. Remedies generally address removing the source or interrupting exposure pathways to sensitive receptors or ecological endpoints. At sites with significant ecological resources, remedy selection should also consider the adverse effects of alternative remedies on the ecological values (i.e., the GAEs) at the site. GAEs thus provide a basis for evaluating how various remedial options might affect ecological services and societal values of resources, thereby avoiding situations where the remedy causes more environmental injury than do the site-related stressors.

Selecting assessment endpoints for monitoring

Assuming that NR is a selected option, a monitoring program will be needed to determine whether NR is occurring as predicted. The monitoring program should establish clear goals. Monitoring should also consider a variety of factors, such as the types of ecosystems involved, the types of stressors present and their exposure pathways, and the parameters that can and should be measured.

Monitoring NR has the same general goals as monitoring other types of remediation, that is, to determine whether the risks posed by stressors (e.g., contamination) have been reduced to acceptable levels. The assessment of ecological risk is based on the identification of assessment endpoints.

Whereas the evaluation of remedial alternatives should consider overall ecosystem values, a monitoring program should be based on the assessment endpoints. Assessment endpoints are explicit expressions of values to be protected (Suter 1989; USEPA 1992). These should have been developed as part of the ERA. If, however, they were not developed as part of the ERA, they should be developed and reviewed at this point in the process. If they were developed as part of the ERA, the assessment endpoints should be reviewed at this point.

The USEPA (1998) provides 3 criteria for selecting assessment endpoints:

1) Ecological relevance
2) Relevance to management goals (societal relevance)
3) Susceptibility to known or potential stressors.

Ecologically relevant endpoints are those that reflect important ecological characteristics and are functionally related to other endpoints. Endpoints based on ecological values that people care about have management relevance and are therefore more likely to be used in risk management decisions (USEPA 1992, 1998). The first 2 criteria encompass both the ecologically relevant values and the human values derived from the ecosystem.

The third criterion, susceptibility to environmental stressors, addresses both sensitivity and exposure to stressors. These stressors can be any physical, chemical, or biological entities capable of inducing adverse responses (USEPA 1992). Constrained by current regulations, most risk assessments focus only on chemical contamination and disregard other related stressors. For example, mine sites produce waste rock and tailings that can create direct physical stress by burying surface vegetation and depositing sediment in streams, in addition to producing environmental stress by the release of chemical contaminants. The direct physical effects of this discharge can disrupt both terrestrial and aquatic ecosystems, but these effects are sometimes disregarded in contaminant-focused risk assessments.

For chemical stressors (i.e., contaminants), properties such as the ability to biomagnify within food chains or to affect certain receptor groups are important considerations. Identifying assessment endpoints based on ecological and societal relevance and subsequently evaluating susceptibility to environmental stressors can remove the "most sensitive species" bias that could lead to trivial conclusions in ERA and to inappropriate endpoints in a subsequent monitoring program.

Consistent with USEPA (1992, 1998) guidance, the additional criteria applied to the selection of site-specific assessment endpoints for risk from site-related stressors involve the susceptibility to known or potential environmental stressors.

Site-specific assessment endpoints identification for risk management requires answers to the following:

- Which functional components are potentially affected by environmental stressors?
- Which effects are potentially significant enough to require remedial action?

Measurement endpoints related to the assessment endpoints will subsequently need to be considered for monitoring purposes. Measurement endpoint selection criteria should include the answers to these questions:

- Which plant and animal taxa potentially inhabiting the site are sensitive to which contaminants?
- What exposure pathways exist between contaminant sources and sensitive taxa?
- Which aspects of ecosystem organization (e.g., biodiversity) could be affected by contamination effects on functional components?

Assessment endpoint identification at this stage is based on ecological function. Once assessment endpoints have been determined, selecting measurement endpoints to evaluate assessment endpoints will require identification of the appropriate taxa for potentially affected functional groups. Identification of appropriate taxa should be based on the dominant taxa within the exposed functional component that is most sensitive to the environmental stressors of concern. Higher taxa may have members belonging to more than one functional component. For example, among birds, some species may be seedeaters (e.g., finches), but others will be omnivorous (e.g., robin) or strictly carnivorous (e.g., hawks and eagles). Site-specific assessment endpoint selection, therefore, depends more on the exposure pathway for the functional component than on the characteristic of any particular species.

Once the relevant taxa for each of the potentially exposed functional components has been determined, the potential effect on higher levels of ecosystem organization can be assessed by examining the functional food web and the attributes of each functional component of the web.

A number of factors are involved in selecting site-specific assessment endpoints. The GAEs provide a comprehensive summary of the ecological and societal values for the ecosystem and provide the context for ensuring that the ecological consequences of contamination are adequately considered in the selection of site-specific assessment endpoints. Identifying GAEs also provides a means of communicating the complex process to the public and documenting the thought processes involved in assessment endpoint selection.

DESIGNING A MONITORING PROGRAM FOR NATURAL REMEDIATION

Once a decision has been made to implement the NR option, a monitoring program must be established. Monitoring can involve a variety of measurements, such as these:

- Contaminant concentrations in exposure media (e.g., soils or sediments)
- Contaminant concentrations in the tissues of receptor species
- Processes that reduce exposure pathways (e.g., sedimentation)
- Ecological effects of stressor exposure (e.g., reproductive success in birds, benthic biodiversity, primary productivity).

Data may also be collected for other regulatory purposes (e.g., quantifying natural resource injuries and service losses during the period of recovery).

Remediation goals and assessment endpoints that the risk assessment has identified as adversely affected will dictate the types of measurements to be used in the monitoring program. Because one of the fundamental goals of monitoring is to determine and document that the risk from environmental stressors (e.g., contamination) has been reduced to acceptable levels, application of a decision-making process will be fundamental to the design of the NR monitoring program. Monitoring should collect all (and only) data necessary to determine whether the remediation goals have been achieved, and it should adequately document the remediation process.

Establishing the framework

The data quality objective (DQO) approach (USEPA 1993) provides an appropriate framework for designing a monitoring program. The process consists of the following 7 steps:

1) State the problem.
2) Identify the decision to be made.
3) Identify inputs to the decision.
4) Define the study boundaries.
5) Develop decision rules.
6) Specify limits on decision errors.
7) Optimize the monitoring study design.

State the problem

For monitoring purposes, the problem can be stated in terms of the risk to ecological values identified in the risk assessment and in the subsequent risk management decision process. A short summary statement referencing appropriate documents and summarizing the remedial issues should be sufficient.

Identify the decision to be made

There can be several levels of decision to be made in the monitoring program, including these:

- Is NR occurring, and if so, is the rate within acceptable limits?
- When can NR be considered successfully complete?

All such decisions should be identified early in the development of a monitoring program.

Identify inputs to the decision

Inputs to the decisions should come primarily from data collected as part of the monitoring program. Information needed to support decisions is identified in order to determine what new environmental measurements are required. Potential sampling approaches to measure effects and/or stressor levels and stressor thresholds (e.g., contaminant-specific action levels) should be defined.

Define the study boundaries

Boundary definition includes the spatial and temporal aspects of the environmental media involved. The geographic areas of interest, characteristics of affected populations, scale of decision-making, and time frame to which the decision applies should be specified. Monitoring boundaries should generally be restricted to the affected locations identified in the risk assessment, locations at which the NR remedial option will be implemented.

Develop decision rules

The development of decision rules is critical to monitoring program development. Logical "if...then" statements define the conditions that allow the risk manager to choose among alternative actions (e.g., continue monitoring or declare the remediation successfully complete). Decision rules can be simple or complex, depending on the variety of stressors present, their spatial distribution, potentially affected receptors, and other criteria. In some situations, it may be appropriate to employ community-level analyses of ecotoxicological effects in biological monitoring (Kedwards et al. 1999).

Specify limits on decision errors

Specifying the limits of decision errors is an important step (and one that will be used in the last step) in optimizing the design of the monitoring program. The purpose of this step is to design a monitoring program that reduces the chances of making decision errors to an acceptable level. Both sampling errors and measurement errors are combined into a total study error that is directly related to decision error.

Optimize the monitoring study design

In the final step, the monitoring study design is optimized based on a review of the DQO outputs. The purpose is to identify the most resource-effective sampling design to generate data that satisfy the specified DQOs. Following these steps ensures that the resulting monitoring program will collect all (and only) data necessary to make the various decisions specified in the monitoring program. Knowing in advance what decisions are to be made, what data are necessary to make the decisions, and how the outcomes of data analysis will be used to make decisions is beneficial to all stakeholders. The substantial savings of effort and costs over the life of the program more than justify the application of the DQO process.

Special monitoring considerations

Once the ecological and human use values have been established, other considerations may be important for establishing monitoring goals. Although it may appear obvious, the type of ecosystem affected by the contamination that will be allowed to remediate naturally must be considered. The composition of these communities will dictate what species or aspects of ecological organization will be selected for monitoring. In general, it will be appropriate to select at least one measure for each of the affected assessment endpoints identified in the risk assessment.

In some instances, monitoring may focus on indicators (McKenzie et al. 1992). These may be selected because of their societal value (e.g., charismatic megafauna) or because the indicator species has been identified as the species most sensitive to the adverse effects of stressors in the ecosystem. In the latter instance, it is assumed that if the effects or exposure to this species is reduced to acceptable levels, risks to the ecosystem are also within acceptable levels.

Other important considerations in designing a monitoring program are the types of stressors involved, the exposure pathways, and the target receptors. For example, at relatively low concentrations, organochlorine pesticides (OCPs) bioaccumulate through food chains and produce adverse effects on birds. Monitoring could address concentrations of OCPs either in exposure media or in bird tissues. The abundance and regulatory status of potentially affected birds would also have to be considered before such tissue analysis could be incorporated into a monitoring program, to ensure that monitoring itself does not cause an adverse effect.

Integration with natural resource damage assessments

Monitoring also provides an opportunity to collect additional data for other regulatory purposes. Data from ERAs frequently can be used by trustees to determine natural resource injuries and service losses resulting from contamination. Pertinent data to address NRDA concerns could be incorporated into monitoring for NR. Quantitative data on the severity, spatial extent, and duration of injuries or service losses is not required for remediation monitoring but could provide a basis for resolving natural resource issues. Such data could easily be incorporated into the monitoring program.

Natural resource injuries or service losses may have occurred at a site, whether or not the risk assessment indicates a current unacceptable risk. Under the NRDA regulations, losses from remedial activities (e.g., habitat removal) count as indirect natural resource injuries. At sites where NR is selected as a remedial option, monitoring will address risk reduction but may not satisfactorily document resource recovery for NRDA purposes. This could be due to differences between risk management goals and NRDA restoration goals.

The overall goal of risk management is to reduce risk to acceptable levels in a timely and cost-effective manner; therefore, monitoring goals and data collection will focus on determining when risk has been reduced to acceptable levels. The general goals of restoration under NRDA rules are to restore natural resources and their services to baseline levels and to make the public whole for the interim lost use of injured resources. Under the current NRDA rule (USDOI 1997), there are strict definitions of natural resource injury, which do not necessarily correspond exactly to the adverse effects identified in risk assessments.

Because the goals of the ERA and NRDA processes are similar, and because the restorations for natural resource injuries and services losses are similar, it may be possible to address NRDA issues by making only minor modifications to the monitoring program. These issues can be addressed informally as part of the risk management efforts and not necessarily in the context of a formal NRDA.

SUMMARY AND CONCLUSIONS

Natural remediation is a viable option at many sites where environmental stressors (e.g., chemical contaminants) have created unacceptable risk to ecological endpoints. Several factors should be considered in selecting the NR option. These include answers to the following:

- What are the valued ecological resources at the site?
- Which NR processes are operating?

- What are the other remedial alternatives?
- What are the costs associated with each alternative?

Risk management goals must also be established as part of the process of selecting the appropriate remedial options.

Goals for selecting remedial options and for subsequent monitoring should be based on the ecological and societal values of the affected ecosystems. A systematic and comprehensive process that considers all species and all levels of ecological organization can be used for both remediation and monitoring. The GAE process identifies ecological values for each potentially affected ecosystem. The ultimate goal is to protect a healthy, functional ecosystem. Values common to all ecosystems (e.g., biodiversity) are then identified, and values particular to the potentially affected regional ecosystems are addressed. This hierarchical approach identifies values at all levels of ecological organization. Societal values (e.g., recreation, resource extraction) are then identified. Remedial options are then screened on the basis of the net benefit of remedial options by evaluating both the effects of environmental stressors and the effects of the remedial alternatives (e.g., dredging sediments). Selecting a remedial option involves other considerations: cost effectiveness and community acceptance, for example.

At sites where the remedial option has been selected, a monitoring program will be established. The DQO process provides a suitable framework for designing a monitoring program. This process identifies all decisions to be made, all data needs for making decisions, and decision criteria for analyzing data to make the decisions. It ensures that all (and only) necessary data will be collected in the monitoring program. In some instances, other regulatory considerations, such as data collection for NRDA purposes, can also be incorporated into the monitoring program.

Awareness that many site activities (e.g., mining) produce significant physical and chemical stressors is becoming more widespread. This is particularly true in developing countries, where regulations permit flexible approaches to addressing total site-related risks. Discussion in this chapter has focused on sites where chemical contaminants are the identified stressors of concern; however, the approach for addressing sites with physical and biological stressors would follow much the same process. For example, at sites where the discharge of mine-related wastes into streams causes sedimentation and flooding, physical stressors are identified by establishing natural thresholds for sedimentation and flooding rates as screening benchmarks and determining no-effects and lowest-adverse-effects thresholds, much as is done for naturally occurring chemical stressors such as metals. Natural remediation in such situations might consist of the physical processes that return the streambed to its original levels and reduce flooding to acceptable levels.

As the risk assessment and risk management processes become more integrated, NR is more likely to be implemented as an appropriate remedial option. Our understanding of NR processes has advanced substantially in the past decade, as have approaches to establishing effective monitoring programs. The successful application of NR at appropriately chosen sites will provide a basis for greater confidence in applying the NR option.

ACKNOWLEDGMENTS

Many of the ideas presented in this chapter were based on my experiences with the risk assessment and risk management processes at a variety of industrial and federal sites in the United States and at sites in Australia and Papua New Guinea. This diverse experience has allowed me to see how NR processes can and should work in a risk management context. I thank Jim Doenges and Laurel Pye for their thoughtful comments and suggestions, which have substantially improved the quality and content of this chapter. Finally, I thank my wife, Julie, for her continued patience, understanding, and support throughout the preparation of this chapter.

REFERENCES

Boyce MS, Haney A. 1997. Ecosystem management: Applications for sustainable forest and wildlife resources. New Haven CT: Yale Univ Pr. 342 p.

Brewer R. 1979. Principles of ecology. Toronto ON: WB Saunders. 299 p.

Gallopin CG. 1972. Structural properties of food webs. Systems analysis and simulation in ecology. New York NY: Academic Pr. p 241–282.

Jones GC, Lawton JH, Shachak M. 1994. Organisms as ecosystem engineers. *Oikos* 69:373–386.

Kaufmann RM., Graham RT, Boyce Jr DA, Moir WH, Perry L, Reynolds RT, Bassett RL, Mehlhop P, Edminster CB, Block WH, Corn PS. 1994. An ecological basis for ecosystem management. Washington DC: U.S. Department of Agriculture (USDA) Forest Service. General Technical Report RM-246.

Kedwards TJ, Maund SJ, Chapman PF. Community level analysis of ecotoxicological field studies: I. Biological monitoring. *Environ Toxicol Chem* 18(2):149–157.

Kellert RS. 1996. The value of life: Biological diversity and human society. Washington DC: Island Pr/Shearwater. 263 p.

Kormondy JE. 1969. Concepts of ecology. Englewood Cliffs NJ: Prentice Hall. 209 p.

Magurran AE. 1988. Ecological diversity and its measurement. Princeton NJ: Princeton Univ Pr. 179 p.

McKenzie DH, Hyatt DE, McDonald VJ, editors. 1992. Ecological indicators. London UK: Elsevier Applied Science. 1567 p.

Norton BG. 1987. Why preserve natural variety? Princeton NJ: Princeton Univ Pr. 281 p.

Odum EP. 1993. Ecology and our endangered life support systems. 2nd ed. Sunderland MA: Sinauer Assoc. 301 p.

Pittinger CA, Bachman R, Barton AL, Clark JR, deFur PL, Ells SJ, Slimak MW, Stahl RG, Wentsel RS. 1998. A multi-stakeholder framework for ecological risk management:

Summary from a SETAC technical workshop. SETAC Workshop on Framework for Ecological Risk Management; 1997 June 23–25; Williamsburg VA. Pensacola FL: Society of Environmental Toxicology and Chemistry (SETAC).

Reagan DP. 2001. Determining ecological values: A critical step in assessing ecological risk. In: D. Paustenbach, editor. Human and ecological risk assessment. New York NY: J Wiley. In press.

Reichman OJ, Pulliam HR. 1996. The scientific basis for ecosystem management. *Ecol Appl* 6:694–696.

Slocombe DS. 1998. Defining goals and criteria for ecosystem-based management. *Environ Manage* 22(4):483–493.

Suter II GW. 1989. Ecological endpoints. In: Warren-Hicks W, Parkhurst BR, Baker Jr SS, editors. Ecological assessment of hazardous waste sites: A field and laboratory reference document. Corvallis OR: Environmental Research Laboratory. EPA 600/3-89/013. p 2-1–2-8.

[USDOI] U.S. Department of Interior. 1997. Natural resource damage assessments. *Code of Federal Regulations* 43 CFR Part 11.

[USEPA] U.S. Environmental Protection Agency. 1992. Framework for ecological risk assessment. Washington DC: Risk Assessment Forum, USEPA. EPA/630/R-92/001.

[USEPA] U.S. Environmental Protection Agency. 1993. Data quality objectives process for Superfund: Interim final guidance. Washington DC: USEPA. EPA 540-R-93-071.

[USEPA] U.S. Environmental Protection Agency. 1998. Guidelines for ecological risk assessment. *Federal Register* 63(93).

Vogt KA, Gordon JC, Wargo JP, Vogt DJ, Absjornsen H, Palmiotto PA, Clark HJ, O'Hara JL, Keeton WS, Patel-Wynand T, Witten W. 1997. Ecosystems: Balancing nature with management. New York NY: Springer. 470 p.

3

Ecological Responses and Compensatory Mechanisms: Using Ecological Knowledge to Improve Remediation Efforts

Mary Ann Hawke and James R. Karr

CHANGE IN NATURE is ubiquitous and occurs over a broad range of spatial and temporal scales (Inouye 1995). Living organisms are constantly challenged by changes; those unable to cope with change perish. Generally, the pace of change is slow, and organisms are able to adapt to their changing environment. Several major periods of extinction in Earth's history are exceptions to this pattern. Whether massive and rapid or of smaller dimensions and gradual, change comes when environments move outside the optimum range of conditions (the range of tolerance) that permits a species to survive. Tolerance limits—minimum and maximum—for physical, chemical, and biological conditions determine not only the presence of an organism but also its abundance and distribution.

Gradual changes in a region's biota form a normal process that reflects an ongoing struggle by each species to sustain healthy local and regional populations. Changes in local ecosystems can be subtle or extensive, but the obvious result of human activities over the past 200 years is a dramatic change in biotic communities over much of the earth (Hannah et al. 1994). Many stresses introduced by humans are too sudden, violent, arrhythmic, or incessant for adaptation to occur. These stresses can result in drastic changes in living systems, including severe oscillations, rather than relative stability (Odum 1969). Fortunately, the natural resilience of living systems has often kept society out of serious trouble (Anderson 1998), although local changes have sometimes been enough to exceed the capacity of regional ecological systems to sustain themselves (Catton 1993; Diamond 1994, 1997).

Natural Remediation of Environmental Contaminants. Michael Swindoll et al., editors.
 ISBN 1-880611-33-3.

Humans face the contradiction that change is critical and constant in living systems, yet changes of the magnitude created by human actions threaten the ability of living systems to sustain themselves. This poses numerous challenges to society. Can we rely upon biological systems to recover naturally from human-introduced stressors? Can we capitalize on the adaptive responses of ecosystems to help speed and improve engineered remediation efforts? How do we decide when and where to avoid actions that may be detrimental? Can knowledge of ecological principles help us to answer these questions? By applying existing and rapidly expanding knowledge of living systems and their properties, we can indeed improve environmental policy (Crosby 1998). For example, McLendon and Redente (1997) argue that our failure to recognize the effects of ecological factors and processes limits our ability to develop and manage successful restoration projects. Hence, we miss the opportunity to use that knowledge to accomplish our goals in a timely and cost-effective manner. They predict that ignoring ecological principles and natural processes will increase the probability of restoration failure.

Use of ecological knowledge may make management efforts more effective, less expensive, more easily and rapidly implemented, and more likely to yield desirable results (MacMahon 1987). If attention to ecological theory can achieve all of these laudable goals, adds MacMahon, it deserves our scrutiny.

A rich literature reflects the evolution of thought about the responses of ecosystems to disturbance (Cairns et al. 1975; Cairns 1980; Barrett 1981; Odum 1985; Pickett and White 1985; Kelly and Harwell 1990; Cech et al. 1998) and the concept of ecological health (Rapport et al. 1985, 1995; Costanza et al. 1992; Karr 1993, 1996, 1999; Suter 1993; Wicklum and Davies 1995; Cech et al. 1998). Important issues range from the focus on elements (structure) and processes (function) in living systems to concepts such as succession (primary and secondary), biodiversity, and resilience. Some argue that the concept of health should not be applied to ecological systems, but few argue with the importance of measurement of ecological, especially biological, change stimulated by the activities of humans. Moreover, these discussions reflect the trend away from the traditional toxicological emphasis on single species and single stressors, toward multiple stressors and multiple levels of biological organization (Levin et al. 1984; Karr and Chu 1997).

Ecologists now think more about how to improve decision-making with direct measures of the status of ecological systems, including identification of biological endpoints to evaluate remediation success. For this reason, the science of restoration ecology has blossomed in recent years, highlighting the relevance of ecology to remedial efforts and refining the often obscure differences between mitigation, reclamation, rehabilitation, and restoration (Jordan et al. 1987; Cairns 1991, 1993). Coupled with this is the development of broader contexts for thinking about ecological risk—a focus that extends to a broader range of stressors (not just chemical stressors) and to what is at risk before, during, and after remediation activities (Karr 1995; Karr and Chu 1997). We need to recognize that ecological health (the condition of a living system) represents a continuum, varying from near pristine to severely degraded. It is not binary or, as regulatory thinking often suggests, "impaired" or "unimpaired" (Karr and Chu 1999). Those decisions are context specific rather than absolute. Similarly, the choice of chemical remediation methods is dictated by the desired outcome; outcomes lie along a continuum from no action to sequestration of pollutants on site, to natural remediation (NR), to complete removal of all contaminants.

Ecologists are often asked by those charged with remediation of damaged and contaminated sites to answer a variety of questions: Can an ecological system be restored once it is degraded? If so, how long will it take and will it be restored to the same condition? If change is natural, and ecosystems have mechanisms to compensate for perturbations, how much change is too much? Is it possible to measure the condition of an ecological system? Here, we address these questions, review basic concepts about ecological systems, especially their living components, and examine the ways in which these systems respond to stress. Topics such as biodiversity, succession, ecological health, biological integrity, and ecological restoration are covered. Our goal is to clarify how they relate to NR. We also discuss how understanding ecological responses and compensatory mechanisms and applying that knowledge in planning and decision-making can help improve the efficiency of NR.

POINT 1: It is counterproductive to focus narrowly on the effects of chemicals

Two dimensions of ecological risk must be incorporated in remediation activities. First, existing chemical and radiological contamination often poses human and ecological health risks that must be reduced. Second, the reduction of those risks should not result in new ecological risk from physical or other disruptions. Regulatory and legal drivers of remedial processes often dictate the reduction of chemicals to below some threshold concentration. Yet ecological risks in the "during" and "after" phases of remediation must also be considered carefully. The impact of the proposed action must be presented clearly and must be part of the management decision process. We should not inadvertently trade one costly problem for another. Remediation that adopts a narrow chemical focus may result in massive cleanup expenditures, expenditures that go beyond the needed level of risk reduction. Too often, expensive engineered remediation continues beyond a level that yields significant environmental benefit. In these circumstances, prudent remediation efforts will treat other primary factors that limit environmental quality.

How can we avoid causing extensive ecological damage during remediation? We can begin by examining and expanding the definition of NR. Currently, mention of NR usually brings to mind the idea of using existing organisms (such as plants or microbes) to break down or sequester environmental contaminants. Natural remediation as defined in this book involves "those nonanthropogenic chemical, physical, and/or biological processes that reduce potential risks to ecological receptors by altering the concentration, mobility, toxicity, and/or bioavailability of chemical contaminants." We must recognize that adverse biological effects could also arise from physical disturbance or biological stressors (like invasive species or pathogenic organisms). We therefore adopt a broad definition of "stressor" (Suter 1993; USEPA 1997a): "any physical, chemical, or biological entity that induces an adverse response." Ecologically speaking, mitigation or remediation of chemical contaminants is not the end but only the beginning of the remediation and restoration process.

Further, the scope of remedial efforts derived from narrow conceptions, such as restoring only endangered or charismatic species, may do more to address symptoms than causes. Thus, remedial efforts, and assessments to evaluate project success, must transcend the level at which the problem is detected. By considering all levels, one is more likely to understand the system and thus develop comprehensive and successful approaches to remediation.

POINT 2: Understanding evolutionary history, especially biological responses to natural and human-induced disturbances, is crucial to remediation

Stress may be defined as "a perturbation that is applied to a system by a stressor which is foreign to that system or which may be natural to it but, in the instance concerned, is applied at an excessive level" (Barrett 1981). Naturally occurring stressors include abiotic (extreme temperatures, fire, drought, floods, storms, earthquakes, and volcanoes) and biotic factors (predators, grazers, pests, pathogens, and invasive nonindigenous species). Anthropogenic stressors may include chemical or radiological contaminants, physical disturbance, pathogens, and nonindigenous species introduced by humans. Human actions may also cause extreme or incessant stress from natural events such as altered frequency of fire or flooding. Various authors have classified stresses on natural ecosystems from human activities. Rapport et al. (1985) defined 4 major stresses:

1) Harvesting renewable resources
2) Discharging pollutants
3) Restructuring physical environments
4) Introducing nonindigenous species.

Based on work in aquatic systems, Karr and colleagues (Karr 1991; Angermeier and Karr 1994) defined the primary organizing processes for living systems and how they are influenced by human actions. They defined 5 classes of interacting factors and thus 5 major ways that human actions degrade living systems:

1) Alteration of chemical environment (e.g., chemical pollution)
2) Alteration of physical environments (e.g., channelization)
3) Alteration of trophic web (e.g., destruction of riparian vegetation, organic enrichment)
4) Altered temporal variation (e.g., increased flooding)
5) Altered biotic interactions (e.g., changes in disease frequency, overharvest of valued species, introduction of nonindigenous taxa).

Regardless of which factor is influenced by human actions, chronic stress may have a different effect than does an acute stress that is quickly followed by recovery and return to an unstressed state (Odum 1985). Some ecologists (see Yount and Niemi 1990) propose replacing "acute" and "chronic" with "pulse" and "press" because the latter are inclusive of stressors that are not necessarily chemical, whereas the former emphasize toxicology and chemical stressors.

From the conception of "land sickness" and its causes (Leopold 1941) to modern efforts to identify and confirm the responses of living systems to diverse stresses (Rapport et al. 1985; Karr and Chu 1999), the identification and use of biological or ecological indicators is critical to successful remediation. Those efforts must include both species and assemblage levels of focus and analysis.

Many species have evolved adaptive responses to both abiotic and biotic natural stressors. Unlike acclimation, which is a physiological adjustment to slowly changing conditions, adaptation is a population-level phenomenon based in genetic change driven by natural selection. Examples of adaptations of plants to naturally occurring abiotic stressors include the seeds of some conifer species, which germinate only after being exposed to fire, and desert plants with physical (reduced or sunken stomata, thick cuticle, reduced or absent leaves) and physiological (crassulacean acid metabolism [CAM]) adaptations to arid conditions. In response to naturally occurring biotic stressors, some plants produce phytoalexins, making them resistant to pathogens; others regenerate quickly when burned or browsed. Stream fish exhibit behavioral adaptations such as downstream or lateral (onto the floodplain) movements (and subsequent recolonization) in response to flooding (Yount and Niemi 1990). Viceroy butterflies avoid predators by mimicking the coloration of the foul-tasting Monarch butterfly. A classic example of adaptation in response to the effects of human actions is the peppered moths (*Biston betularia*) in England. After industrial air pollutants darkened the trunks of trees, the black form of moth became more abundant than the light form.

At the level of assemblages, stressors can change the biology of places in many ways. In grassland systems, physical disturbances such as grazing or agriculture can lead to the replacement of native perennial grasses with annual or perennial nonindigenous grasses, a shift that persists even decades after the stress is removed (Clampitt 1993; Brandt and Rickard 1994). These nonindigenous grasses and other "opportunistic species" are capable of increasing their populations suddenly in response to changed conditions (Ricklefs 1979; Cairns and Pratt 1990). An increase in the relative abundance of opportunists, often nonindigenous species, is common.

These opportunists can often be distinguished from "pioneer species," native taxa characteristic of early successional areas. Pioneers are usually replaced through succession within a few years after their arrival. The earliest pioneers (often wind-dispersed microbes, mosses, or lichens) capitalize on high dispersal ability to rapidly colonize newly created or disturbed areas. The characteristics of both pioneers and opportunists are these: numerous small seeds, wind or animal dispersion, long and latent seed viability in soil, low root to shoot ratio, rapid growth rate, small size at maturity, and low shade tolerance (Ricklefs 1979). Later successional species that replace pioneers have limited dispersal ability, slower growth rates, larger size at maturity, and seeds that are relatively large to provide nutrients for seedlings in a highly competitive environment (Ricklefs 1979). In general, nonindigenous species tend to displace native abundant species, and the invaders are more short-lived and opportunistic (Rapport et al. 1985). This may lead to "retrogression," or the apparent reversion to an earlier stage of succession. In this case, changes tend to be the reverse of those that occur during succession: simplification of the communities, opening of niches, loss of the nutrient inventory accumulated during succession, and reduced stability.

In vertebrates, disturbance may stimulate individuals to change physiologically (e.g., acclimatize), behaviorally (e.g., adopt a different resting posture), or ecologically (e.g., feed in a different place or alter reproductive output), and it may change the relative fitnesses of phenotypes (Karr and Freemark 1985). At the population level, attributes such as density, spatial distribution, mortality rate, reproductive rate, genetic structure, and extinction probabilities may change. At the community level, changes may occur in species composition, relative abundance, or trophic structure because of differences in the responses of species to disturbances. For ecosystems, functional rate processes such as energy flow and nutrient cycling may also be altered as a result of the responses of component species (Karr and Freemark 1985).

Stream biotas evolve life-history characteristics that favor flexibility or adaptability as a result of the variety of disturbances to which they are subjected (Yount and Niemi 1990). For example, many lotic (running water) species have short generation times, high fecundity, and good dispersal ability. Variable conditions in lotic systems, especially small streams, favor species that are flexible (opportunistic) in their requirements. Hypereutrophic systems too characteristically have shorter food chains with a lower degree of organization, lower diversity, and higher productivity at lower trophic levels (Haskell et al. 1992). Schindler (1987) presents a careful review of aquatic ecosystem responses to anthropogenic stress.

Awareness of these natural characteristics of species and ecosystems is critical to insightful remediation or restoration. By taking advantage of inherent characteristics of the biota, especially the responses of organisms to stressors, more effective remediation approaches can be defined. Plants, for example, have evolved a number of adaptations to deal with toxic levels of metals and other pollutants. Knowledge of these attributes has been used to treat wastewater for more than 300 years (USEPA 1997b). Because the exudates from the roots of plants influence the soil–root interface, they can directly or indirectly alter the mobilization of elements in the rhizosphere. Metal tolerant species can stabilize contaminated soil by limiting the mobility and bioavailability of metals, as well as by minimizing erosion or sediment migration. Certain plant species hyperaccumulate metals from the soil and concentrate them (a process known as "phytoextraction"). Plants also affect soil–water dynamics, and the establishment of vegetation cover over engineered soil "caps" can influence the infiltration rate and potentially alter the movement of contaminants in the soil (Link et al. 1994, 1995).

The response of organisms to environmental conditions is an integrated result of the direct and indirect impacts of contaminants, natural stressors, or a combination of both natural and human-induced perturbations (Adams et al. 1998). Biological systems under stress commonly exhibit a certain sequence of responses (Haskell et al. 1992). Because species and ecosystems exhibit considerable uniformity in their responses to stress (Rapport et al. 1985), a core set of indicators of ecological distress can be identified. This fundamental principle is the foundation of multimetric evaluations of biological condition such as the index of biological integrity (IBI) (Karr 1981, 1991; Karr and Chu 1999). Cairns and Niederlehner (1993) review what is known about ecological responses to stress (see also Woodwell 1970, 1990; Odum 1985; Rapport et al. 1985; Schindler 1987; Resh et al. 1988; Schaeffer et al. 1988; Pratt 1990; Karr and Chu 1999). General patterns include a tendency toward smaller dominant life forms, reduction in species diversity, rapid shifts in species composition, and an increase in the amplitude of population fluctuations. Substitution of sensitive taxa by tolerant ones, further simplification of community structure, an increase in respiration, nutrient loss and turnover, and an imbalance of production:respiration or maintenance:biomass ratios often follows.

The most widespread signs of ecosystem response to stress include reduction in species diversity, reversal of normal successional trends, changes in primary productivity, changes in nutrient cycling, and temporary reduction in average size of dominant biota (Rapport et al. 1985). A species by itself that has small ecosystem effects (e.g., a pollinator or seed disperser) can have large indirect effects if it influences the abundance of another species with large direct ecosystem effects (e.g., an important canopy species; Chapin et al. 1997). Two mechanisms whereby ecosystems resist or cope with stress are 1) the replacement of the stress-sensitive species with functionally similar but more resistant species and 2) the use of external biogeochemical feedback loops (Rapport et al. 1985).

An issue of *Science* (August 1997) addressed the ongoing debate among scientists regarding the role that species diversity plays in maintaining ecological processes. While some believe there is evidence to show that higher levels of species diversity benefit ecosystem processes, others provide results that indicate it is the kinds of organisms (i.e., their biological characteristics or functional groupings) present in a system that are important, rather than their numbers (Grime 1997). The idea is that the greater the diversity within a functional group, the greater its resilience in responding to environmental change because any decrease in the abundance of one species will be compensated for by increases in other functionally similar species (Chapin et al. 1997). In other words, the current debate centers around the question of the relative roles of species diversity and species composition in the functioning of an ecosystem.

When anthropogenic disturbances are similar in magnitude and type to natural disturbances, living systems respond in much the same way to both. All too often, anthropogenic disturbances do not mimic natural events because they are either unique or too sudden, violent, arrhythmic, or incessant for organisms to evolve adaptive responses. Instead of relative stability, drastic change results, including severe oscillations in living systems (Odum 1969). When human reproductive and consumptive behavior alters the natural disturbance pattern in an environment, the changes that result outpace the capacity of biological systems to evolve.

Often the result is massive change in living systems. Many of those changes may threaten human welfare. Keeping human-induced stresses or disturbances within the bounds of natural events is one important strategy to protect human interests. But to do so, we must understand the dynamics of natural disturbance. Further, we must design resource management and remediation projects within the context of local ecological circumstances. Our record on that score is abysmal (Ludwig et al. 1993), while our hubris goes unbounded (Karr 1994).

POINT 3: Remediation must be grounded in the natural history of the regional biota

Too often, remediation projects focus on physical and chemical endpoints, with little direct reference to living systems as endpoints. When organisms are used as endpoints, the large, more obvious organisms (e.g., vascular plants) are commonly the target. This approach fails to recognize the key roles played by smaller organisms—"the little things that run the world" (Wilson 1987; Alper 1998)—and the larger framework of biological condition. Microbes, lichens, fungi, and invertebrates are critical for their contributions to develop and stabilize soil, to fix nitrogen, to pollinate plants, even to act as disease vectors that may keep populations of other organisms in check. Without these smaller organisms, remediation projects are likely to be narrowly focused or superficial at best and failures at worst. Neglecting to recognize that a lichen was responsible for nitrogen production led to the failure of one remediation project (Diamond 1987). In fact, "the resilience of many systems may well be determined by the ability of microbes to reestablish the conditions upon which the larger organisms depend" (Cairns and Niederlehner 1993). But it is not only the small organisms that play critical roles in living systems. In tropical forests, for example, a substantial proportion of tree species depends on vertebrates for pollination, seed dispersal, or both (Leigh 1999). Even top carnivores play a keystone role as stabilizers of the trophic structure of terrestrial ecosystems, illustrating the importance of intact faunas to the ability of ecological processes to operate unhindered (Terborgh et al. 1999). The key message is that remediation projects, wherever they are attempted, must be grounded in detailed knowledge of the natural history of the regional biota. That natural history must span a range of spatial and temporal scales relevant to the taxa in the region.

Conceptually, biological systems are organized hierarchically from genes to individuals, populations, communities, ecosystems, landscapes, and finally to the biosphere. Organizational hierarchies can be useful tools for understanding complex biological phenomena. For example, taxonomic, genetic, and ecological hierarchies may be recognized, and biotic units at any level within the hierarchy may be termed "elements" (Angermeier and Karr 1994). Thus, species, families, and orders are taxonomic elements, genes and chromosomes are genetic elements, and populations and biomes are ecological elements. The effects of stressors must be explored and understood at multiple levels of organization; they may even be beneficial at one level and detrimental at another (Odum 1985). In the Everglades of South Florida, for example, seasonally dry periods concentrate fish in small areas, where birds remove 76% of the fish (Karr and Freemark 1985). Fish abundances are low, but species richness is high. In periods when bird populations are low, fish mortality increases to 96%; local extinctions are common in these years. Predation by birds (a biological disturbance) reduces overall mortality by a physical disturbance (drought) in normal years, but those interactions are out of line in unusual years. Because of spatial, temporal, and trophic level complexities, narrowly focused remediation in this kind of situation is unlikely to succeed.

Because of the complexity of ecological systems, no single scale of study is correct, and no scale should be ignored. Generally, the richest evaluations derive from integration of data from diverse spatial and temporal scales and levels of biological organization. An important level of integration heretofore ignored is the landscape, a scale with critical levels of integration that cannot be understood at finer levels of resolution (MacMahon 1987; Forman 1995). Ecotoxicologists have studied biological systems at this and other levels (Anderson 1998), but too often their focus is more narrow and the application of their results fraught with uncertainty.

In the temporal dimension, a critical first step is to establish that the duration of a study adequately permits a full display of natural variability, cyclic phenomena, and normal successional processes (Cairns 1991). It is important to incorporate multiple scales when restoration projects are being planned. "Restoration efforts should begin with an explicit statement of the spatial and temporal scale of the project, since the ability to understand a phenomenon depends upon the extent to which it is studied at the level of organization that most influences it" (Cairns 1991).

The elements of biological systems (also known as biological "structures") extend over a broad range of organizational scales from genes to populations, species, assemblages, and landscapes. Biodiversity is essentially described by the composition, relative frequencies, and spatiotemporal distribution of biological elements. The complements of elements are biological processes (also known as biological "functions") on which those elements depend (Angermeier and Karr 1994). Those processes are integral to the maintenance of biological diversity across all levels of biological organization (Table 3-1). Elements are usually measured as the number or abundance of biota (e.g., individuals, species) present, whereas processes are usually measured as rates (Karr 1993). Elements and processes occur at all levels of organization, not just at the community or ecosystem level. Both can be influenced by stressors, and changes in either can have negative affects on human goals.

Consideration of both temporal and spatial scales is also important in determining relative risk and measuring the success of bioremediation. A large part of ecosystem heterogeneity arises from the consequences of disturbances operating at various temporal and spatial scales (Karr and Freemark 1985). Ecological elements are typically defined by spatial extent (e.g., a pond community or a desert landscape), yet most levels can correctly encompass a wide range of spatial scales (Angermeier and Karr 1994).

POINT 4: Remediation and restoration must go beyond introducing a few species or constructing a few components of physical habitat

The term "biological diversity" (biodiversity) emerged with recognition that species extinction rates were exceeding natural extinction rates by approximately 1,000 to 10,000 times (Noss 1990; Pimentel et al. 1997). The annual economic and environmental benefits of biodiversity in the U.S. is estimated at $300 billion (Pimentel et al. 1997). Costanza et al. (1997) estimate the annual value of goods and services by Earth's living systems at $33 trillion. Higher diversity may increase system stability in these ways (Chapin et al. 1997):

1) Providing alternative pathways of energy flow among trophic levels
2) Reducing the susceptibility of ecosystems to invasion after disturbances
3) Reducing the spread of pathogens by increasing the average distance between individuals of a species.

Table 3-1 Elements, processes, and potential indicators of biological condition for 6 levels of organization within 3 biological categories

Biological category	Elements (levels)	Processes	Indicators
Taxonomic	Species	Range expansion or contraction Extinction Evolution	Range size Number of populations Population size Isolating mechanisms
Genetic	Gene	Mutation Recombination Selection	Number of alleles Degree of linkage Inbreeding or outbreeding depression
Ecological	Individual	Health	Disease Deformities Individual size and condition index Growth rates
	Population	Changes in abundance Colonization or extinction Evolution Migration	Age or size structure Dispersal behavior Presence of particular taxa (e.g., intolerants) Gene flow
	Assemblage	Competitive exclusion Predation or parasitism Energy flow Nutrient cycling	Number of species Dominance Number of trophic links Stream distance for 1 carbon molecule to complete passage through food chain (spiraling length)
	Landscape	Disturbance Succession Soil formation Metapopulation dynamics	Fragmentation Percentage of disturbed land Number of communities Sources and sinks Number and character of metapopulations

Source: Karr and Chu 1999.

Biological diversity as a concept can be vague, and it is often poorly defined. Too often, it is considered equal to the simpler concept of "species diversity" (the number of species in a place) (Angermeier and Karr 1994). Biodiversity is not simply the number of genes, species, ecosystems, or any other group of things in a defined area (Noss 1990). Early definitions emphasized the elements of living systems—genes, species, and ecosystems—as they ignored crucial processes, such as interspecific interactions, natural disturbances, and nutrient cycles (Noss 1990). Another misconception is to equate a loss in biodiversity with species extinction. Local and regional loss of forests, or declines in salmon stocks, for example, are still significant losses of biodiversity regardless of whether any species become extinct (Angermeier and Karr 1994).

Understanding biodiversity in a broader context has implications when it comes to clearly defining the goals of any effort to restore an ecosystem, because simply replacing a laundry list of species will not necessarily ensure a healthy, self-sustaining ecological system. Generally, the system to be restored will not operate efficiently and in a sustained way unless critical elements are present, including those that nature normally provides but that may be easily overlooked (Diamond 1987). Neither is it sufficient simply to have a functioning ecosystem. Even the most disturbed system still functions, but its level of function is inappropriate to the region or to the expectations of human society (e.g., streams dominated by sewage-sludge worms [Tubificidae] due to organic enrichment). As Cairns and Niederlehner (1993) argue, "Environmental protection must assure that organisms are robust and healthy rather than merely just surviving. A census merely affirms that they are or are not present." Censuses do not convey information on the vitality of organisms or on the vitality of the life-support systems they depend on.

"Restoration" is defined as the return of an ecosystem to a close approximation of its predisturbance condition (National Academy of Sciences [NAS] 1992) and includes both the elements of a damaged ecosystem as well as the processes that support them (Cairns 1991). The NAS (1992) defines restoration as the reestablishment of predisturbance functions and related physical, chemical, and biological characteristics. They describe the goal of restoration as "the emulation of a natural, functioning self-regulating system that is integrated with the ecological landscape in which it occurs." In many regulations and public laws, the terms "habitat creation," "reclamation," and "rehabilitation" are often used. Habitat creation (bringing into being a new ecosystem that previously did not exist on the site), reclamation (adapting a wild or natural resource to serve a utilitarian human purpose), and rehabilitation (visibly improving a natural resource, or returning it to good condition or working order) all imply the creation of a new or altered use to serve a particular human purpose (NAS 1992). Restoration is different from habitat creation, reclamation, and rehabilitation; it is a holistic process not achieved through the isolated manipulation of individual elements. Merely recreating a form without the functions, or recreating the functions in an artificial configuration that bears little resemblance to a natural form, does not constitute restoration (NAS 1992). "Mitigation" is defined as the alleviation of any or all detrimental effects arising from a given action, or the actions taken to avoid, reduce, or compensate for the effects of environmental damage (NAS 1992). Mitigation often does not involve any in-kind restoration or creation (Table 3-2).

Bradshaw (1987) argues that there are 2 stages to restoration: the first involves identifying and understanding what is wrong, and the second involves reversing those effects. That reversal should come from an alleviation of the causes, not from a masking of the symptoms.

Table 3-2 Definitions of common terms

Term	Definition
Ecological or biological integrity	Condition at sites with little or no influence from human actions. Organisms living at sites are products of the evolutionary and biogeographic processes influencing that site. Implies correspondence with some original condition and refers to the capacity to support and maintain a balanced, integrated, adaptive biological system having the full range of elements (genes, species, assemblages) and processes (mutation, demography, biotic interactions, nutrient and energy dynamics, and metapopulation processes) expected in the natural habitat of a region (Karr and Dudley 1981; Karr 1996).
Ecological health	Preferred state in terms of biological condition of sites modified by human activity and supportive of use for that activity indefinitely (i.e., that land use is sustainable).
Ecosystem health	Being stable, sustainable, active; maintaining organization and autonomy over time; resilient to stress (Haskell et al. 1992).
Habitat creation	Bringing into being a new ecosystem that previously did not exist on the site (National Academy of Science [NAS] 1992).
Homeostasis (stability)	Maintaining favorable internal conditions despite fluctuations in external conditions (Willard and Klarquist 1995).
Mitigation	Alleviating any or all detrimental effects arising from a given action, which need not involve in-kind restoration or creation (NAS 1992).
Persistence or inertia	Continuing to exist and resisting the effects of disturbance (Willard and Klarquist 1995).
Reclamation	Adapting a wild or natural resource to serve a utilitarian human purpose, or the amelioration of the effects of disturbance at a site and subsequent revegetation (McLendon and Redente 1997).
Rehabilitation	Improving a natural resource, or returning it to good condition or working order (NAS 1992).
Resilience	Ability to recover or return to the previous state following disturbance (Cairns and Niederlehner 1993).
Restoration	Returning an ecosystem to a close approximation of its condition prior to disturbance; a holistic process not achieved through the isolated manipulation of individual elements (NAS 1992).
Revegetation	Establishing some vegetative cover at a site without first ameliorating the effects of the disturbance (McLendon and Redente 1997).
Sustainability	Ability to maintain structure (organization) and function (vigor) over time in the face of external stress (resilience) (Costanza 1992).

Cairns (1991) describes several different types of restoration:

Class A: Complete restoration to a self-maintaining system whose biological condition is comparable to the predisturbance state. In this case, the goal is restoration of biological integrity, the living system characteristic of that site that is the product of evolutionary and biogeographic processes in the relative absence of humans (Karr 1999).

Class B: Reestablishment of only selected attributes to a system that likely is not self-maintaining.

Class C: Production of an alternative ecosystem unlike the predisturbance ecosystem, which may or may not be self-sustaining.

Class D: Recovery left primarily to natural processes.

Class E: Restoration of hazardous waste sites with organisms that stabilize the site and minimize offsite movement of contaminants.

Class F: Restoration in an area with a high risk of ecological accidents.

Class G: Restoration after contamination by genetically engineered organisms.

Class H: Restoration designed primarily to protect the health of adjacent ecosystems.

The goals of any restoration effort must be clearly defined, and they may include returning the system to the equivalent of minimally disturbed places (no pristine or undisturbed places remain). It is impractical (even impossible) for at least 4 reasons to return a system to predisturbance condition (Diamond 1987):

1) Local and regional extinctions prevent complete recovery.
2) Introduced or nonindigenous taxa are present and often impossible to eradicate.
3) Restoration is attempted on small areas that will inevitably be influenced by their surroundings.
4) Restored areas often are not self-sustaining, and thus they must be managed by humans.

In many circumstances, then, restoration should not be judged by the criterion of complete recovery of biological integrity, the predisturbance condition (Cairns 1989, 1991). Cairns (1991) proposes to judge restoration success based on 6 criteria:

1) Ecological damage is contained, and contaminants or other negative effects are not exported to adjacent ecosystems.
2) The restored ecosystem interacts with adjacent ecosystems.
3) The restored system is self-sustaining with all major functional groups represented.
4) Successional processes are comparable to a similar but undisturbed ecosystem.
5) Ecosystem services provided are comparable to the predisturbance condition.
6) Natural variability can be accommodated.

To date, most restoration efforts do not fit Cairn's Class A criterion. Generally, goals are more modest, such as revegetation of mined lands, reestablishment of a fishery, or reconstruction of wetlands (Cairns 1991). In the case of revegetation, the goal may be to achieve only a Class E restoration: stabilize the soil surface. Bratton (1992) believes that our adherence to the idea that a site must be returned to a predisturbance state may be impractical. Often, the area to be restored is isolated and is no longer integrated with surrounding ecosystems, so certain elements (e.g., large predators) may be missing and difficult to replace. We may not know what the predisturbance state was or what it might have evolved into had it not been disturbed. Acceptable signs of a successful restoration effort may therefore include the return of all species originally present on the site, a high species diversity, the presence of rare or unique species, or the return of natural disturbance cycles such as fire regimes (Bratton 1992).

Early restoration efforts often involved an attempt to reestablish the process of succession by simply letting nature take over after the removal of the stressor (Cairns 1991). These early attempts used either an agricultural model or a climax community model, emphasizing productivity and stability as indicators of ecosystem health (Bratton 1992). One of the problems with letting nature take its course is the speed with which recovery occurs. Sometimes the lack of a seed bank or immigrants from nearby communities are limiting factors. If the only action taken has been to stop the stressor from acting at the site, the soil itself is still damaged and may need to be treated either physically or chemically before vegetation can take hold. Often, plants may need to be reintroduced using agricultural or forestry techniques (Bradshaw 1987).

Where soil needs to be rebuilt, nonindigenous species may exhibit desirable physiological mechanisms (like nitrogen fixation or cation pumping; Aber 1987). In these circumstances, restoration may be a 2-stage process, with nonindigenous species being planted first to protect and improve the soil. They can eventually be removed in favor of native species, although the reality is that this is often extremely difficult or impossible. Weeds are adapted to grow on richer agricultural soils; therefore, weed problems in restoration are more common in richer soils than in poorer soils.

Most past studies of disturbed ecosystems have emphasized analysis of population fluctuations, an approach that all too often fails to provide an understanding of the complexity of ecological systems (Miller 1987). Miller notes, for example, that ecologists were slow to recognize and appreciate the importance of mycorrhizal associations and other soil microbiological processes. Soil microbiology and plant–microbe associations have likely been overlooked in the past because they are less conspicuous and are deemed more difficult to work with. Yet they are critical to nutrient cycling, water relations, and competition.

POINT 5: Diverse conceptual frameworks in ecology can mislead remediation efforts if goals and protocols are based on theories without empirical validation or on outdated language and ideas

Our mental models and ideals of naturalness (including an emphasis on natural homeostasis and a return to past ecosystem states) may constrain the ways in which we use ecological restoration to solve environmental problems (Bratton 1992). For decades, ecologists have viewed ecological systems through glasses tinted by concepts such as equilibrium and homogeneity. Disturbances were considered rare, typically stochastic events that upset the status quo in largely static systems. In effect, ecological science marginalized disturbance as having little consequence to empirical, experimental, and theoretical advances (Botkin 1990; Karr 1994).

The focus on old language and theoretical concepts has contributed to a constriction in our thinking. For example, the concept of succession was developed and fine-tuned to accommodate the dynamic nature of presumed static systems (Karr 1994). Ecological succession occurs when the composition and function of communities change over time. It has been historically described as reasonably orderly and directional, passing through several transitional or "seral" stages and culminating in a stabilized system or "climax" (Odum 1971). Clements (1916), who originally proposed many of the concepts of succession in his monograph on vegetation, believed that a region had only one true climax (the monoclimax theory) and recognized just 14 climax communities in North America. We now recognize that a continuum of possible outcomes exists, instead of this canalized movement to an inevitable endpoint. The composition of vegetation varies continuously over environmental gradients of nutrients, moisture, slope, exposure, and other factors (Ricklefs 1979). While classical succession and the concepts of equilibrium, stability, and homeostasis dominated ecological thought in the 1960s and 1970s, there has been much debate in recent years about the empirical evidence available to support these views (Wu and Loucks 1995). A recent review of the competing perspectives on ecological stability outlined some of the criticisms of the "balance of nature" concept and presented other paradigms proposed to explain the behavior of ecological systems (Wu and Loucks 1995).

Adherence to the notion that biological systems will return to a state of equilibrium on their own may result in the belief that reclamation of disturbed land is solely a technical problem and may lead to the setting of only a few simple objectives, including stabilization, pollution control, visual improvement, and general amenity (Bratton 1992). Much has been published in ecological literature about the recovery and responses of ecosystems to stress, yet there still has been a lack of success in many remedial efforts. Attempts to replace or mitigate specific structures or functions did not retain or replace the complex, especially biological, roles played by ecological systems. Many dimensions of living systems have been lost, with little or no effort to track that loss. It is perhaps the most pervasive but largely unseen consequence of modern regulatory activities.

We suggest that the reason the rhetoric does not match the accomplishments is the failure to recognize the importance of biological endpoints and the need to measure and communicate system status using those endpoints. Ecological engineering that performs functions such as the sequestration and removal of contaminants or soil stabilization should mark only the beginning of the process. For example, reducing the level of a contaminant to below a legally mandated threshold is not an improvement unless our accomplishments are evaluated against explicit biological goals. There is also too much emphasis on the ecological risks posed by contaminants present before the remediation process begins and not enough emphasis on the risks posed by activities that occur during or after remedial activities. Currently there is a lack of integrative and easily understood biological endpoints—endpoints that define the biological condition of existing systems or the condition of restored (remediated) systems. We propose that careful planning and analysis would at least define more integrative biological endpoints and avoid one of the most dangerous scenarios: damage to things we do not recognize as important. The lack of a common biological currency to express the condition of ecological systems, to define the biological health of a site or landscape, results in a narrow focus that highlights selected and narrowly conceived goals (Karr 1998).

POINT 6: Environmental policy should focus more on well-defined patterns in the real world, especially integrative biological endpoints, and less on intermediate or bureaucratic measures of success such as chemical characterization, permits issued, or habitat area

In Point 2, we discussed selected responses of biological systems to disturbance, including adaptations made by organisms. While disturbance is responsible for many of the heterogeneities in biological systems, cataloging disturbance patterns and biological processes or theorizing about general responses of ecosystems to disturbance is not enough for real understanding (Karr 1994). To quote Rapport (1992), "This is less a question for theory than for brute empiricism." Dissatisfaction with assessment methods and their consequences is widespread and growing (e.g., among wetland biologists Kusler and Niering 1998; Larson 1998), and improved alternatives are now available (Karr and Chu 1999).

Despite all our knowledge, many remediation and restoration efforts are not succeeding. For example, remediating areas by seeding nonindigenous grasses may stabilize the soil but does little to support the native flora and fauna. One remediation project failed because it ignored a lichen responsible for nitrogen production

(Diamond 1987). Wetland protection projects have failed to protect natural wetlands, and little evidence can be mustered to show that mitigation efforts are successful (Zedler 1996; Karr 1998).

A critical problem is a lack of integrative and easily understood biological endpoints, endpoints that define the biological condition of existing systems or the desired biological condition of restored or created systems. Biological endpoints focus our attention on what is at risk, rather than relying on chemical surrogates to infer the condition of biological systems. Direct assessment of biological condition, or health, is critical to the protection of the benefits that society derives from living systems. How can we talk about repairing the damage caused by human activity if we do not know what is normal or healthy for a given ecosystem? How do we measure a deviation from normal? How do we evaluate whether a system can be restored to a healthy state? These questions are implied when we ask of remediation activities, "How clean is clean enough?" Such concepts and questions generate much discussion among ecologists and philosophers, as well as among political leaders, decision-makers, and citizens.

Application of concepts such as health and integrity to ecological contexts is commonplace and common sense (Karr 1999). Integrity, for example, appears in the law (U.S. Clean Water Act, Canada National Park Act, European Union Water Policy) and in international treaties (U.S.–Canada Treaty on the Great Lakes), as well as in the scientific and philosophy literature (Westra and Lemons 1995; Westra 1998; Pimentel et al. 2000).

Although the terms "biological or ecological integrity" and "ecological health" are often used synonymously, we believe that they should be reserved for different contexts (Karr 1996, 1999). "Ecological health" describes the preferred state of sites modified by human activity (e.g., cultivated areas, plantations, forests, industrial parks, and cities). These sites may be considered healthy when present use neither degrades them in ways that preclude that use in the future nor degrades areas beyond their borders. In contrast, "ecological integrity" most appropriately refers to the condition at sites with little or no influence from human actions; the organisms living there are products of the evolutionary and biogeographic processes influencing that site. Ecological integrity implies correspondence with some original condition and refers to the capacity to support and maintain a balanced, integrated, adaptive biological system that has the full range of elements (genes, species, assemblages) and processes (mutation, demography, biotic interactions, nutrient and energy dynamics, and metapopulation processes) expected in the natural habitat of a region (Karr and Dudley 1981; Karr 1991, 1996; Angermeier and Karr 1994). Degradation of integrity results in biotic impoverishment, the systematic reduction in the capacity of the earth to support living systems (Karr 1993, 1996). A biological system is healthy and has ecological integrity when its inherent potential is realized, its condition is stable, its capacity for self-repair is maintained, and external support for maintenance is minimal (Karr 1993). Another definition states that: "An ecological system is healthy and free from 'distress syndrome' if it is stable and sustainable—that is, if it is active and maintains its organization and autonomy over time and is resilient to stress" (Haskell et al. 1992). Here, the concept of sustainability is key, since it implies that the system can maintain its elements and processes over time. For example, even an agricultural community can be healthy, although many of its native species have been replaced with cultivated species, provided that the overall system maintains sufficient complexity to protect its self-organizing qualities (Haskell et al. 1992).

There has been much debate about the appropriateness of using the term "health" to describe ecological systems (Rapport et al. 1985; Costanza 1992; Ehrenfeld 1992; Karr 1993; Suter 1993; Wicklum and Davies 1995; Fairbrother 1998). The terms "health" and "integrity" have been used in long-range policy documents as goals of protection efforts, although some scholars maintain that the terms have never been defined clearly enough to make them useful in practice (Haskell et al. 1992). Ehrenfeld (1992) argues that "Health is an idea that transcends scientific definition. Although it contains values that are not amenable to scientific methods of exploration, they are no less important or necessary because of that. Health is a bridging concept connecting two worlds: it is not operational in science if you try to pin it down, yet it can be helpful in communicating with nonscientists." Rapport (1992) proposes that a healthy ecosystem is whatever ecologists, environmentalists, and the public at large deem it to be, and points out that what is desired or healthy must take into account social and cultural as well as ecological values. Wilson (1998) thinks the argument comes down to the basic question of whether ecosystem health is a subjective concept (a metaphor for those things the inquirer deems important) or whether it possesses objective, determinable properties. One argument supporting the concept of ecological health (Haskell et al. 1992) states: "An ecosystem health approach is needed because complex ecosystems must be assessed in ways that reflect their overall performance and not just the production of desired species. Single species management has led to many disasters in the past, but before there can be true multicriteria management there must be an adequate definition of health."

The first step toward effective biological assessment is to realize that the goal is to measure and evaluate the consequences of human actions on biological systems. The relevant measurement endpoint is biological condition; detecting change in that endpoint, comparing the change with a minimally disturbed baseline condition, identifying the causes of the change, and communicating these findings to decision-makers and citizens should be our goal (Karr and Chu 1999). Understanding baseline, or reference, conditions in different places is analogous to veterinarians' learning what indicates health in different kinds of animals. "Healthy" for a lizard is not the same as "healthy" for a dog. Knowing geophysical setting and undisturbed biological condition—in other words, knowing what produces and constitutes biological integrity—must underpin any effort to restore biological systems (Karr and Chu 1999).

For several decades, ecotoxicology has been the primary discipline invoked to guide remediation. But toxicology as a discipline has not dealt with many important causes of degradation; has not dealt adequately with the cumulative, antagonistic, and other effects of complex mixes of contaminants; and has not adequately integrated population, assemblage, and landscape consequences of human actions. Renewed interest among ecologists in measuring and understanding the consequences of human actions to living systems (Davis and Simon 1995; Karr and Chu 1999; Simon 1999) has taken the strengths of toxicology (e.g., the concept of dose–response curves) and applied them in the broader context needed to assess the effects of human actions. Teamwork and joint efforts that cross disciplinary boundaries are required. For too long, chemists, toxicologists, and ecologists have talked primarily to one another and in their own tongues (Crosby 1998). They need to develop a common language that will resonate with a concerned public and with policy-makers responsible for decisions about our collective future. The evolution and integration of these 2 approaches can make environmental policy more cost effective and efficient (Karr 1998).

Remediation efforts, in short, must be grounded in a systematic and comprehensive approach to the study of disturbance. Disturbance and organismal responses to disturbance play a primary role in the evolution of organismal natural histories and the biology of multispecies assemblages (Karr 1994). Understanding of disturbance and its consequences has major theoretical interest and becomes integral to controlling the influence of humans on natural resources. The ultimate challenge of the 21st century is to develop a more informed approach to managing anthropogenic disturbances in natural and human landscapes (Karr 1994; Chu and Karr 2001).

CONCLUSIONS

Because ecological systems are complex and variable, both defining the approach and measuring the effects of remediation can be a challenge. Remediation activity itself presents additional challenges because of disconnects between policy goals and legal drivers, between scientific foundations and technological capabilities. But one thing is clear: Leaving contaminated sites to purge themselves by natural methods is often socially and scientifically unacceptable. In short, the "no action" alternative is commonly unacceptable. The extreme alternative ("muck and truck") is equally unacceptable in many circumstances because of the secondary damage often caused by such approaches. Remediation, then, must be guided by a more insightful analysis of ecological context, with an eye toward reducing ecological risks and using the power of ecological processes to advance remediation goals.

We recommend keeping the following scientific, technical, and policy issues in mind.

- Define remediation goals clearly.
- Go beyond simple reduction of chemicals or the achievement of mandated thresholds to accomplish defined biological goals.
- Use biological rather than bureaucratic or chemical measures as endpoints.
- Ensure that biological endpoints include multiple dimensions of living systems, from the health of individuals to, for example, the status of populations and the dynamics of nutrient cycles across landscapes.
- Avoid the tendency to focus on a few species (e.g., selected microorganisms, endangered species, or selected, easily established plants) at remediation sites.
- Treat the causes of degradation rather than the symptoms.
- Ensure that remedial actions do not stimulate secondary effects, such as those from physical or biological disturbances associated with remediation.
- Understand the difference between remediation and restoration.
- Design remediation projects with natural system capabilities in mind. Take advantage of natural processes to enhance remediation success and to save both money and time.
- Create planning and implementation strategies that explicitly incorporate the relevant temporal and spatial scales for critical chemical, physical, and biological processes in the system to be remediated.
- Think in terms of the cumulative dose of human activities and their biological (human and nonhuman) consequences when you design remediation projects. Begin first with an analysis of past events that have created the challenge, and then consider the consequences of remediation projects during and after remediation.

REFERENCES

Aber JD. 1987. Restored forests and the identification of critical factors in species-site interactions. In: Jordan III WR, Gilpin ME, Aber JD, editors. Restoration ecology: A synthetic approach to ecological research. New York NY: Cambridge Univ Pr. p 241–250.

Adams SM, Ham KD, LeHew RF. 1998. A framework for evaluating organism responses to multiple stressors: Mechanisms of effect and importance of modifying ecological factors. In: Cech Jr JJ, Wilson BW, Crosby DG, editors. Multiple stresses in ecosystems. Boca Raton FL: Lewis. p 13–22.

Alper J. 1998. Ecosystem "engineers" shape habitats for other species. *Science* 280:1195–1196.

Anderson DW. 1998. Evaluation and impact of multiple stressors on ecosystems: Four classic case histories. In: Cech Jr JJ, Wilson BW, Crosby DG, editors. Multiple stresses in ecosystems. Boca Raton FL: Lewis. p 3–7.

Angermeier PL, Karr JR. 1994. Biological integrity versus biological diversity as policy directives. *Bioscience* 44:690–697.

Barrett GW. 1981. Stress ecology: An integrative approach. In: Barrett GW, Rosenberg R, editors. Stress effects on natural ecosystems. New York NY: Wiley. p 3–12.

Botkin, DB. 1990. Discordant harmonies: A new ecology for the twenty-first century. New York NY: Oxford Univ Pr.

Bradshaw AD. 1987. The reclamation of derelict land and the ecology of ecosystems. In: Jordan III WR, Gilpin ME, Aber JD, editors. Restoration ecology: A synthetic approach to ecological research. New York NY: Cambridge Univ Pr. p 53–74.

Brandt CA, Rickard WH. 1994. Alien taxa in the North American shrub-steppe four decades after cessation of livestock grazing and cultivation agriculture. *Biol Conserv* 68:95–105.

Bratton SP. 1992. Alternative models of ecosystem restoration. In: Costanza R, Norton BG, Haskell BD, editors. Ecosystem health: New goals for environmental management. Washington DC: Island Pr. p 170–189.

Cairns Jr J, editor. 1980. The recovery process in damaged ecosystems. Ann Arbor MI: Ann Arbor Science. 167 p.

Cairns Jr J. 1989. Restoring damaged ecosystems: Is predisturbance condition a viable option? *Environ Professional* 11:152–159.

Cairns Jr J. 1991. The status of the theoretical and applied science of restoration ecology. *Environ Professional* 13:186–194.

Cairns Jr J. 1993. Is restoration ecology practical? *Restor Ecol* 1:3–7.

Cairns Jr J, Dickson KL, Herricks EE. 1975. Recovery and restoration of damaged ecosystems. International Symposium on the Recovery of Damaged Ecosystems; 1975 March 16–25; Blacksburg VA. Univ Pr Virginia. 531 p.

Cairns Jr J, Niederlehner BR. 1993. Ecological function and resilience: Neglected criteria for environmental impact assessment and ecological risk analysis. *Environ Professional* 15:116–124.

Cairns Jr J, Pratt JR. 1990. Biotic impoverishment: Effects of anthropogenic stress. In: Woodwell GM, editor. The earth in transition: Patterns and processes of biotic impoverishment. New York NY: Cambridge Univ Pr. p 495–505.

Catton WR, Jr. 1993. Carrying capacity and the death of a culture: A tale of two autopsies. *Sociol Inquiry* 63:202–223.

Cech Jr J, Wilson BW, Crosby DG, editors. 1998. Multiple stresses in ecosystems. Boca Raton FL: Lewis. 202 p.

Chapin III FS, Walker BH, Hobbs RJ, Hooper DU, Lawton JH, Sala OE, Tilman D. 1997. Biotic control over the functioning of ecosystems. *Science* 277:500–504.

Chu EW, Karr JR. 2001. Environmental impact: Concept and measurement. In: Levin SA, editor. Encyclopedia of biodiversity. San Diego CA: Academic Pr. In press.

Clampitt CA. 1993. Effects of human disturbances on prairies and the regional endemic *Aster curtus* in western Washington. *Northwest Sci* 67:163–169.

Clements FE. 1916. Plant succession: Analysis of the development of vegetation. *Publ Carnegie Institut Wash* 242:1–512.

Costanza R. 1992. Toward an operational definition of ecosystem health. In: Costanza R, Norton BG, Haskell BD, editors. Ecosystem health: New goals for environmental management. Washington DC: Island Pr. p 239–256.

Costanza R, d'Arge R, de Groot R, Farber R, Grasso M, Hannon B, Limburg K, Naeem S, O' Neill RV, Paruelo J, Raskin RG, Sutton P, van den Belt M. 1997. The value of the world's ecosystem services and natural capital. *Nature* 387:253–260.

Costanza R, Norton BG, Haskell BD, editors. 1992. Ecosystem health: New goals for environmental management. Washington DC: Island Pr. 269 p.

Crosby DG. 1998. Epilogue: Multiple stresses in ecosystems. In: Cech Jr J, Wilson BW, Crosby DG, editors. Multiple stresses in ecosystems. Boca Raton FL: Lewis. p 189–191.

Davis WS, Simon TP, editors. 1995. Biological assessment and criteria: Tools for water resource planning and decision making. Boca Raton FL: Lewis. 415 p.

Diamond JM. 1987. Reflections on goals and on the relationship between theory and practice In: Jordan III WR, Gilpin ME, Aber JD, editors. Restoration ecology: A synthetic approach to ecological research. New York NY: Cambridge Univ Pr. p 329–336.

Diamond JM. 1994. Ecological collapses of past civilizations. *Proceedings Am Philosoph Soc* 138:363–370.

Diamond JM. 1997. Guns, germs, and steel: The fates of human societies. New York NY: WW Norton. 480 p.

Ehrenfeld D. 1992. Ecosystem health and ecological theories. In: Costanza R, Norton BG, Haskell BD, editors. Ecosystem health: New goals for environmental management. Washington DC: Island Pr. p 135–143.

Fairbrother A. 1998. Establishing the health of ecosystems. In: Cech Jr JJ, Wilson BW, Crosby DG, editors. Multiple stresses in ecosystems. Boca Raton FL: Lewis. p 101–107.

Forman RTT. 1995. Land mosaics: The ecology of landscapes and regions. New York NY: Cambridge Univ Pr. 632 p.

Grime JP. 1997. Biodiversity and ecosystem function: The debate deepens. *Science* 277:1260–1261.

Hannah L, Lohse D, Hutchinson C, Carr JL, Lankerani A. 1994. A preliminary inventory of human disturbance of world ecosystems. *Ambio* 23:246–250.

Haskell BD, Norton BG, Costanza R. 1992. What is ecosystem health and why should we worry about it? In: Costanza R, Norton BG, Haskell BD, editors. Ecosystem health: New goals for environmental management. Washington DC: Island Pr. p 3–20.

Inouye DW. 1995. Variation in undisturbed plant and animal populations and its implications for studies of recovering ecosystems. In: Cairns Jr J, editor. Rehabilitating damaged ecosystems. 2nd ed. Boca Raton FL: Lewis. p 357–371.

Jordan III WR, Gilpin ME, Aber JD, editors. 1987. Restoration ecology. A synthetic approach to ecological research. New York NY: Cambridge. 342 p.

Karr JR. 1981. Assessment of biotic integrity using fish communities. *Fisheries* (Bethesda) 6(6):21–27.

Karr JR. 1991. Biological integrity: A long-neglected aspect of water resource management. *Ecol Appl* 1:66–84.

Karr JR. 1993. Defining and assessing ecological integrity: Beyond water quality. *Environ Toxicol Chem* 12:1521–1531.

Karr JR. 1994. Defining disturbance. In: Stouder DJ, Fresh KL, Feller RJ, editors. Theory and application in fish feeding ecology. The Belle W. Baruch Library in Marine Science, No. 19. Columbia SC: Univ South Carolina Pr. p 285–291.

Karr JR. 1995. Risk assessment: We need more than an ecological veneer. *Human Ecol Risk Assess* 1:436–442.

Karr JR. 1996. Ecological integrity and ecological health are not the same. In: Schulze P, editor. Engineering within ecological constraints. Washington DC: National Academy Pr. p 100–113.

Karr JR. 1998. Seeking suitable endpoints: Biological monitoring for wetland assessment. Report to USEPA Region 10, Seattle, WA.

Karr JR. 1999. Defining and measuring river health. *Freshwater Biol* 41:1–14.

Karr JR, Chu EW. 1997. Biological monitoring: Essential foundation for ecological risk assessment. *Human Ecol Risk Assess* 3:993–1004.

Karr JR, Chu EW. 1999. Restoring life in running waters: Better biological monitoring. Washington DC: Island Pr. 206 p.

Karr JR, Dudley DR. 1981. Ecological perspective on water quality goals. *Environ Manage* 5:55–68.

Karr JR, Freemark KE. 1985. Disturbance and vertebrates: An integrative perspective. In: Pickett STA, White PS, editors. The ecology of natural disturbance and patch dynamics. New York NY: Academic Pr. p 153–168.

Kelly JR, Harwell MA. 1990. Indicators of ecosystem recovery. *Environ Manage* 14:527–545.

Kusler J, Niering W. 1998. Wetland assessment: Have we lost our way? *Natl Wetlands Newsletter* 20:8–14.

Larson JS. 1998. Wetland landscape assessment: A way out of the dilemma. Presentation at Society for State Wetland Managers meeting; 1998 April 15; Washington DC.

Leigh Jr EG. 1999. Tropical forest ecology: A view from Barro Colorado Island. New York NY: Oxford Univ Pr. 245 p.

Leopold A. 1941. Wilderness as a land laboratory. *Living Wilderness* 6:3.

Levin SA, Kimball KD. 1984. New perspectives in ecotoxicology. *Environ Manage* 8:375–442.

Link SO, Waugh WJ, Downs JL, Thiede ME, Chatters JC, Gee GW. 1994. Effects of coppice dune topography and vegetation on soil water dynamics in a cold-desert ecosystem. *J Arid Environ* 27:265–278.

Link SO, Wing NR, Gee GW. 1995. The development of permanent isolation barriers for buried wastes in cool deserts: Hanford, Washington. *J Arid Land Studies* 4:215–224.

Ludwig D, Hilborn R, Walters C. 1993. Uncertainty, resource exploitation, and conservation: Lessons from history. *Science* 260:17–18.

MacMahon JA. 1987. Disturbed lands and ecological theory: An essay about a mutualistic association. In: Jordan III WR, Gilpin ME, Aber JD, editors. Restoration ecology: A synthetic approach to ecological research. New York NY: Cambridge Univ Pr. p 221–237.

McLendon T, Redente EF. 1997. Revegetation manual for the environmental restoration contractor. Richland WA: Bechtel Hanford. BHI-00971 Rev. 0.

Miller RM. 1987. Mycorrhizae and succession. In: Jordan III WR, Gilpin ME, Aber JD, editors. Restoration ecology: A synthetic approach to ecological research. New York NY: Cambridge Univ Pr. p 205–219.

[NAS] National Academy of Sciences. 1992. Restoration of aquatic ecosystems. Science, technology, and public policy. Washington DC: National Academy Pr.

Noss RF. 1990. Indicators for monitoring biodiversity: A hierarchical approach. *Conserv Biol* 4:355–364.

Odum EP. 1969. The strategy of ecosystem development. *Science* 164:262–270.

Odum EP. 1971. Fundamentals of ecology. Philadelphia PA: WB Saunders. 574 p.

Odum EP. 1985. Trends expected in stressed ecosystems. *Bioscience* 35:419–422.

Pickett STA, White PS, editors. 1985. The ecology of natural disturbance and patch dynamics. Orlando FL: Academic Pr. 472 p.

Pimentel D, Westra L, Noss RF. 2000. Ecological integrity: Integrating environment, conservation, and health. Washington DC: Island Pr. 375 p.

Pimentel D, Wilson C, McCullum C, Huang R, Dwen P, Flack J, Tran Q, Saltman T, Cliff B. 1997. Economic and environmental benefits of biodiversity. *Bioscience* 47:747–757.

Pratt JR. 1990. Aquatic community response to stress: Prediction and detection of adverse effects. In: Landis WG, van der Schalie WH, editors. Volume 13, Aquatic toxicology and risk assessment. Philadelphia PA: American Society for Testing and Materials (ASTM). STP 1096. p 16–26.

Rapport DJ. 1992. What is clinical ecology? In: Costanza R, Norton BG, Haskell BD, editors. Ecosystem health: New goals for environmental management. Washington DC: Island Pr. p 144–156.

Rapport DJ, Gaudet CL, Calow P, editors. 1995. Evaluating and monitoring the health of large-scale ecosystems. NATO Advanced Research Workshop; 1993 Oct 10–15; Montebello, QC, Canada. New York NY: Springer-Verlag. 454 p.

Rapport DJ, Regier HA, Hutchinson TC. 1985. Ecosystem behavior under stress. *Am Natur* 125:617–640.

Resh VH, Brown AV, Covich AP, Gurtz ME, Li HW, Minshall GW, Reice SR, Sheldon AL, Wallace JB, Wissmar RC. 1988. The role of disturbance in stream ecology. *J N Am Benthol Soc* 7:433–455.

Ricklefs RE. 1979. Ecology. 2nd ed. New York NY: Chiron Pr. 966 p.

Schaeffer DJ, Herricks EE, Kerster HW. 1988. Ecosystem health: I. Measuring ecosystem health. *Environ Manage* 12:445–455.

Schindler DW. 1987. Detecting ecosystem responses to anthropogenic stress. *Can J Fish Aquat Sci* 44:6–25.

Simon TP. 1999. Assessing the sustainability and biological integrity of water resources using fish communities. Boca Raton FL: CRC Pr. 671 p.

Suter II GW. 1993. Ecological risk assessment. Boca Raton FL: Lewis. 538 p.

Terborgh J, Estes JA, Paquet P, Ralls K, Boyd-Heger D, Miller BJ, Noss RF. 1999. The role of top carnivores in regulating terrestrial ecosystems. In: Soule ME, Terborgh J, editors. Continental conservation: Scientific foundations of regional reserve networks. Washington DC: Island Pr. p 39–64.

[USEPA] U.S. Environmental Protection Agency. 1997a. Ecological risk assessment guidance for Superfund: Process for designing and conducting ecological risk assessments. Interim final. Washington DC: Office of Solid Waste and Emergency Response (OSWER 9285.7-25). EPA-540-R-97-006.

[USEPA] U.S. Environmental Protection Agency. 1997b. Recent developments for in situ treatment of metal contaminated soils. Washington DC: Office of Solid Waste and Emergency Response. EPA-542-R-97-004.

Westra L. 1998. Living in integrity: A global ethic to restore a fragmented earth. Lanham MD: Rowman and Littlefield. 271 p.

Westra L, Lemons J, editors. 1995. Perspective on ecological integrity. Boston MA: Kluwer Academic. 279 p.

Wicklum D, Davies RW. 1995. Ecosystem health and integrity? *Can J Botany* 73:997–1000.

Willard DE, Klarquist J. 1995. Mitigation banks: A strategy for sustainable ecosystem function. In: Cairns Jr J, editor. Rehabilitating damaged ecosystems. Boca Raton FL: Lewis. p 97–132.

Wilson BW. 1998. Ecosystem health: Some perspectives. In: Cech Jr JJ, Wilson BW, Crosby DG, editors. Multiple stresses in ecosystems. Boca Raton FL: Lewis. p 53–57.

Wilson EO. 1987. The little things that run the world. *Conserv Biol* 1:344–346.

Woodwell GM. 1970. Effects of pollution on the structure and physiology of ecosystems. *Science* 168:429–433.

Woodwell GM. 1990. The earth in transition: Patterns and processes of biotic impoverishment. New York NY: Cambridge Univ Pr. 530 p.

Wu J, Loucks OL. 1995. From balance of nature to hierarchical patch dynamics: A paradigm shift in ecology. *Quart Rev Biol* 70:439–466.

Yount DJ, Niemi, GJ. 1990. Recovery of lotic communities and ecosystems from disturbance: A narrative review of case studies. *Environ Manage* 14:547–569.

Zedler JB, editor. 1996. Ecological issues in wetland mitigation: A forum. *Ecol Appl* 6:33–101.

4

Natural Remediation Processes

Charles A. Bleckmann, Edward C. Heyse, Mark N. Goltz

NATURAL REMEDIATION (NR) of environmental contaminants occurs as the result of physical, chemical, and biological processes that reduce the potential for contaminants to adversely affect ecological resources. NR processes lower these risks by altering chemical properties, by reducing contaminant concentrations, or by modifying the toxicity, bioavailability, or mobility of the contaminants. These processes, long known to contribute to the cleansing of natural environments such as rivers and lakes, have been employed to great benefit by society for waste processing and disposal. Historically, the treatment of sewage has relied almost exclusively on natural processes. Although the assimilative properties of the environment have long been recognized, only recently have the regulatory and scientific communities begun to consider NR as a potentially effective and appropriate remedial alternative for contaminated soils, sediments, or groundwater. Increasingly, NR processes are being embraced as valuable tools in the arsenal of environmental remediation alternatives (Hart 1996).

Contaminants released into the environment enter compartments such as the atmosphere, surface water, groundwater, and soil, where they may occur at concentrations that can produce adverse impacts (Figure 4-1). Typically, contaminants do not remain in a single compartment, but instead move among them. The movement among compartments may either enhance or retard NR processes and significantly change the contaminant's degradation, bioavailability, or reactivity. The interfaces between compartments (e.g., air–water, water–solid, and solid–air) are particularly important in promoting NR processes. Although gas and liquid compartments are largely homogeneous, each may contain particulates that provide important

Natural Remediation of Environmental Contaminants. Michael Swindoll et al., editors.

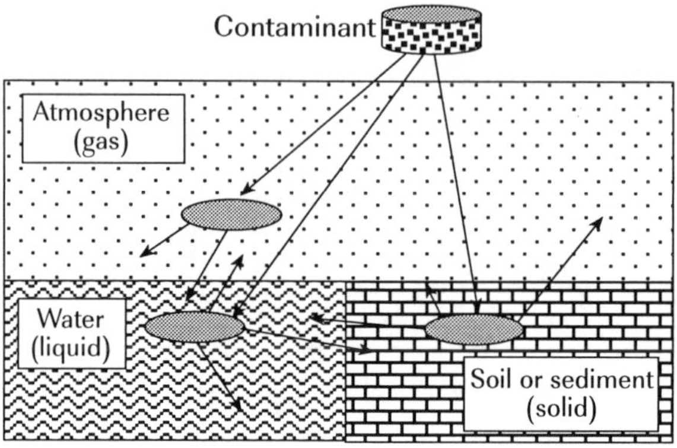

Figure 4-1 Contaminants released into the environment move between liquid, gas, and solid phases (Materials in solid phase may partition between organic and inorganic fractions.)

reaction sites. Soil consists of organic and inorganic fractions, with movement of materials occurring between the two. Soil pores or interstitial spaces (as much as 50% of bulk soil volume) are the primary sites of contaminant interactions with the solids, liquids, and gases that make up a soil system.

The processes that contribute to reductions in risks posed by environmental contaminants vary in significance based on contaminant and environmental setting. This chapter identifies the variety of naturally occurring processes (Table 4-1) that are important to environmentally significant organic and inorganic chemicals, and it describes the potential impact of these processes on the fate and transport of these chemicals. Overviews of dominant NR processes, including transfer, transport, and transformations, are presented along with information needed to understand and quantify NR of specific environmental contaminants. This information can be applied to predict the potential of NR to meet site remediation objectives. The technical literature germane to NR processes is immense and spans topics ranging from environmental chemistry, microbiology, and engineering to the ecological sciences. Recent reviews and primary literature are referenced extensively in this chapter to facilitate access to the technical foundation of this topic of emerging environmental significance.

NATURAL REMEDIATION PROCESSES

No matter which scale is used to bound environmental contamination—from that of a microscopic particle to an entire ecosystem—the same phases, gas (air), liquid (water), and solid (soil), are present and the same physical, chemical, and biological processes contribute to the fate and transport of the contaminant. Processes that contribute to the NR of contaminants can be divided into 3 groups:

Table 4-1 Natural remediation processes

		Transformation	
Transport	Transfer	Abiotic	Biotic[a]
Advection	Volatilization	Oxidation	Biodegradation
Dispersion	Condensation	Reduction	Cometabolism
Sedimentation	Sorption and desorption	Hydrolysis	Transpiration
Diffusion	Ion exchange	Photolysis	Biotransformation
	Partitioning	Precipitation	Sequestration
		Dissolution	
		Complexation	
		Ionization	
		Polymerization	

[a] Biotic processes are not inherently different from the other categories; however, organisms may enhance the rates of the other processes.

1) Transport processes are responsible for the movement of contaminants within a compartment, and they include advection, dispersion, diffusion, and sedimentation.
2) Phase transfer processes, such as precipitation, solubilization, volatilization, and sorption, can result in the transition or movement of a contaminant from one compartment to another.
3) Transformation processes alter the chemical structure of the contaminant molecules. Microbially mediated degradation of an organic compound is an example of perhaps the most common of the environmental transformation processes.

Table 4-2 Relative importance of biotic and abiotic reactions in natural remediation processes

Process	Biotic	Abiotic
Phase transfer	+	+++
Transport	+	+++
Transformation	+++	+

+++ = Very important
\+ = Important

In this chapter, various transport, phase transfer, and transformation processes that can contribute to the fate and transport of environmental contaminants are identified and discussed in the context of their role in NR.

Although transport and phase transfer processes are primarily abiotic (Table 4-2) and do not directly reduce contaminant mass, they play an important role in determining the ultimate fate and effect of environmental contaminants. Transport and phase transfer processes are crucial to NR, in that they may serve to reduce contaminant concentrations to harmless levels, sequester contaminants and thus reduce potential exposure to organisms, or increase the reactivity of contaminants (either making them bioavailable or chemically reactive) so that the contaminants may undergo transformation. Transformation processes are largely biotic, although abiotic mechanisms (e.g., photolysis, hydrolysis) and conditions (e.g., temperature, pH, oxygen concentrations) may be involved.

Transport processes

Transport occurs as the result of either fluid movement, gravity acting upon solid particles, or chemical diffusion. The transport processes typically most important to the NR of contaminants are advection and dispersion, sedimentation, and diffusion.

Advection and dispersion occur as the result of fluid movement. Advection is the transport of dissolved or suspended contaminant mass by the movement of the fluid (e.g., air, water) in which the molecules reside (Weber and DiGiano 1996). As the fluid moves, variations in fluid velocity in space and time lead to spreading of the contaminant molecules. As different fluids with varying velocities merge, mixing of molecules occurs. This spreading and mixing, due to variations in advective velocity, is termed "dispersion." Weber and DiGiano (1996) identify 3 types of dispersion:

1) Turbulent (or eddy) dispersion is due to random velocity fluctuations in fluid flow and is efficient at 3-dimensional dispersing (i.e., mixing and spreading) of contaminants. Turbulent dispersion is important in the atmosphere and in fast-flowing rivers and channels.
2) Taylor dispersion occurs in narrow channels and pore spaces, such as aquifers and other porous media, where fluid flow is relatively slow (i.e., laminar) and is due to the parabolic distribution of fluid velocities resulting from drag at the fluid–solid boundary.
3) Hydrodynamic (or mechanical) dispersion also occurs in aquifers and other porous media and is due to media heterogeneities that cause variations in velocity and flow paths (Bedient et al. 1994).

Sedimentation can be an important mass transport mechanism that removes particulates, and chemicals associated with them, from mobile phases. Environmental contaminants sorbed onto particulates may be removed from stream flow (either air or water) and deposited due to gravity, the process of sedimentation. Likewise, resuspension into the fluid phase by scouring may occur because of shear stress as the moving fluid passes over the sediment. However, unlike advection and dispersion, sedimentation does not act on dissolved substances. Dissolved chemicals must first go through a phase transfer process (e.g., sorption onto a suspended particulate) before sedimentation plays a role. Coagulation or consolidation of fine suspended particulates serves to enhance sedimentation.

Diffusion is the spreading and mixing of contaminant from areas of high contaminant concentration to areas of low concentration because of random molecular motion (Cussler 1984). Although mixing from diffusion and mixing from dispersion are the results of completely different phenomena, both may be described by Fick's law of diffusion, which states that the flux (mass transfer per unit area) of contaminant is proportional to the contaminant concentration gradient.

Transport processes (e.g., advection, dispersion, sedimentation, and diffusion) can be extremely important to NR. Transport processes can reduce biological exposure to potentially toxic concentrations of contaminants. Advection may transport a contaminant to a location where it is less likely to present a hazard to sensitive organisms. Advection acting alone does not decrease contaminant mass or concentration. However, in concert with dispersion, advection can decrease contaminant concentrations. In addition, dispersion and diffusion processes serve to mix contaminant with other chemicals and/or organisms that can transform it. Sedimentation reduces contaminant mass in the mobile phase and thus may serve to reduce exposures and the bioavailability of contaminants to biota.

Phase transfer processes

Phase transfer processes are movements that distribute a chemical among different ecosystem compartments (e.g., air, water, soils or sediment, organic liquids, colloids, or organisms). Some of the phase transfer processes important to environmental contaminants are identified in Table 4-3. These processes result in a change to the compartment in which the chemical resides but do not alter its molecular structure. Phase transfers between compartments are critical to the ultimate fate of contaminants. Mackay et al. (1985) describe ecosystem-wide chemical distribution using the concept of "fugacity," the tendency of a substance to escape the environmental compartment in which it exists. Similarly, equilibrium partition or distribution coefficients are used to describe chemical distribution among phases.

Equilibrium distribution coefficients are used to quantify how a contaminant distributes itself, at equilibrium, among environmental compartments. These coefficients are typically defined as the contaminant concentration in one phase or compartment divided by the concentration in a second phase or compartment. Some important equilibrium distribution coefficients are defined in Table 4-3. These coefficients describe the extent to which a contaminant is distributed between compartments at equilibrium but do not consider the rate at which contaminants move between compartments. Although environmental systems are rarely at equilibrium, equilibrium coefficients are useful for predicting the direction of environmental transfers. Such processes may be rate-limited by micro-scale transport or chemical reaction kinetics (Brusseau and Rao 1989). In the first case, a contaminant molecule cannot move from one compartment to the other until it reaches the interface between the compartments, and the transport to the interface

Table 4-3 Transfer processes and equilibrium distribution coefficients

Phase	Process	Equilibrium distribution coefficient
Air–water	Volatilization	Henry' s law constant, K_H mole fraction in air per mole fraction in water
Soil–water	Sorption	Distribution coefficient, K_d mg/kg soil per mg/L water
	Ion exchange	Ion exchange constant, K_{ie} (mg ion 1 per kg soil) × (mg ion 2 per L water) per (mg ion 2 per kg soil) × (mg ion 1 per L water)
Organic–water	Partitioning	Organic C normalized partition coefficient, K_{oc} mg/kg organic carbon per mg/L water
Lipid–water	Partitioning	Octanol–water partition coefficient, K_{ow} mg/L octanol per mg/L water

may be a relatively slow, diffusive process. The second case is applicable only when the transfer involves some type of chemical reaction.

The most common of the phase transfer processes is partitioning of a chemical between environmental compartments. The partitioning of chemicals between water and air (volatilization) can be described by Henry's law constants (Ashworth et al. 1988). Partitioning between an organic phase and water may be described by partition coefficients that are derived using Raoult's law (Cline et al. 1991; Lee, Hagwall et al. 1992; Lee, Rao, and Okuda 1992; Chen et al. 1994).

If all soil organic matter is assumed to be similar, distribution coefficients between water and soil (or sediment) normalized to the organic C content of the soil will be constant for a given contaminant. This organic C partition coefficient has been predicted by the octanol–water partition coefficient (K_{OW}; Karickhoff et al. 1979; Means et al. 1980; Karickhoff 1981). A recent challenge to this theory is work demonstrating that not all organic matter is the same. Sorption by organic matter with a more rigid, crystalline structure may be due to adsorption (Weber et al. 1992; Pignatello and Xing 1996; Weber and Huang 1996; Huang et al. 1997), while sorption of hydrophobic organic compounds by dissolved organic matter and organic matter colloids is a partitioning process (Chiou et al. 1979; Liu and Amy 1993; Chin et al. 1997).

Adsorption processes are surface phenomena involving chemical reactions between a solid and a chemical (Garnier et al. 1997; Roy and Dzomback 1997). Ion exchange is an important adsorption process, particularly for cations reacting with clay mineral surfaces and negatively charged functional groups on organic matter. Soils and sediments have so many varied surfaces (mineral surfaces, oxide coatings, organic matter, etc.) that the phase transfer process may be difficult to define. The effects of all of these various sorptive interactions are often lumped together using a sorption distribution coefficient.

Phase transfer processes can have important effects on NR. For instance, contaminants from a surfacewater ecosystem can be removed from the water column either by volatilization to the atmosphere or by sorption to particles and subsequent burial through sedimentation. Phase transfer processes can either retard or enhance environmental transport of contaminants. For example, contaminants adsorbed to soil surfaces will be transported with the groundwater at a rate slower than the mean groundwater velocity. Transfer mechanisms can also lead to enhanced transport with contaminant moving slightly faster than the water phase, such as soil–gas partitioning and gas-phase diffusive transport to areas upgradient of water flow (Hinchee and Reisinger 1987). In addition, low solubility contaminants, which could be expected to travel slowly because of partitioning into soils, may experience facilitated transport because of colloid partitioning (Schwarzenbach et al. 1993). Finally, sorption processes that remove contaminants from the water phase may reduce contaminant bioavailability (Scow and Hutson 1992).

Transformation processes

In contrast to physical or chemical processes that transport materials, transformation processes produce a change in molecular structure. Transformation processes can break down organic materials into innocuous compounds or can immobilize inorganics and reduce their availability or reactivity to acceptable levels. However, in some instances, the transformation products represent a greater environmental threat than the original material. Examples of such undesirable end products are the formation of vinyl chloride (VC) during degradation of chlorinated hydrocarbons, the methylation of metallic Hg in sediments, or the production of nitrosamines from secondary or tertiary amines or quaternary N compounds (Alexander 1981).

Transformations most often occur as the result of biotic reactions; however, abiotic reactions or combinations of biotic and abiotic reactions can also be important. Both biotic and abiotic transformation rates may be strongly dependent upon transport processes.

Abiotic transformations

Abiotic transformations result from physical or chemical processes (Table 4-1) and may accomplish contaminant destruction directly. These same processes can also modify materials so destruction can be completed by biotic processes. Chemical transformation processes include oxidation, reduction, hydrolysis, photolysis, precipitation or dissolution of salts, or complexation with other solutes. The rate and extent of abiotic reactions depends upon environmental conditions (e.g., sunlight, oxygen, water availability, temperature) and the contaminant chemical structure. Weber (1998), Crosby (1998), and Lyman (1995) have compiled recent, extensive reviews on the broad topic of abiotic processes.

Oxidation and reduction (redox) reactions involve the movement of electrons between compounds. Both organic and inorganic contaminants are subject to redox reactions. Oxygen most often serves as an electron acceptor in oxidation reactions, although halogens and other electron acceptors, such as nitrogen oxides, may also be important. In the atmosphere and in aerated soils and water environments with plentiful oxygen, contaminants are usually oxidized. In anoxic environments, such as saturated soils and sediments, reductive reactions are typically involved in the transformation of contaminants.

Oxidation may result in the addition of oxygen to contaminant molecules and in a change in the contaminant's solubility and availability. Oxidation of iron from ferrous (Fe^{+2}) to ferric (Fe^{+3}) forms decreases solubility, causing precipitation of compounds such as ferric hydroxide. Oxidation of reactive organics, such as phenols or anilines, may make them more available or accessible for subsequent microbial transformations. In aquatic environments, the presence of metal ions and mineral oxides may catalyze oxidation reactions.

Abiotic reduction reactions often result in the addition of H to reactive contaminants such as the nitroaromatics in explosives and dyes, pesticides, and some halogenated organics. Reduction reactions may be the primary route of degradation for some of these compounds. Minerals containing Fe and S may be the primary source of reducing potential in soil and sediment environments.

Hydrolysis is the reaction of a compound with water. During hydrolysis, an existing covalent bond in the contaminant molecule is broken, and a new bond forms with a hydroxyl (OH^-) group. The reactions are usually nonreversible and are often catalyzed by acids or bases. Daily changes in the pH of surface waters occur because of photosynthesis in sunlight (raises pH) and respiration at night (lowers pH). This cycle ensures the availability and variability of reactants (H^+ and OH^-) to enhance hydrolysis. Many classes of contaminants are susceptible to hydrolytic reactions, including carbonates, carbamates, ureas, halogenated aliphatics, and organophosphates.

Photolysis is a chemical reaction resulting from the addition of light energy. Photolytic reactions are classified as direct or indirect. Direct reactions are the result of the absorption of ultraviolet (UV) energy by chemical bonds and the subsequent breaking of those bonds. Indirect reactions may result from the formation of reactive species by light and subsequent reactions such as oxidation or hydrolysis. Atmospheric reactions are most often the result of photolysis, but photolysis may also be important in surface waters.

Contaminant precipitation, the formation of solids from materials in solution, and contaminant dissolution, movement from a solid into solution, may act in both phase change and transformation processes. In particular, metal ions are susceptible to transformations, from available, dissolved forms to more stable, solid forms. Environmental factors, such as temperature, pH, and the concentrations of other dissolved species (both solids and gases), increase or decrease the solubility of contaminants.

Complexation is the result of ionic reactions of metal ions and ligands such as humic compounds and humic and fulvic acids. Complexation reactions can change solubility and toxicity and can result in attenuation of the metal ions. The reactions are related to pH, temperature, the presence of organic ligands, metal ions, and the ionic strength of the solution. Complexation reaction rates may vary over several orders of magnitude.

Ionization or dissociation, the separation of an electrolyte into ions of opposite charge, occurs in aqueous systems, is pH dependent, and is reversible. Both organic and inorganic contaminants may undergo ionization. Metal salts, organic acids ($^-$COOH), and N-containing organic bases, such as amines or anilines, are the most common contaminants influenced by ionization.

Polymerization is the bonding of 2 or more monomers to form a polymer. Both organic and inorganic contaminants can polymerize. For example, phenoxy radicals, formed by oxidations, may polymerize in the environment. The reaction of metal ions and water may produce metal hydroxo polymers that may persist in aqueous systems for years.

In summary, although a number of abiotic reactions that occur in the environment are directly related to NR, oxidations, reductions, and hydrolysis reactions are by far the most important. Most of these reactions reduce the risk associated with environmental contaminants.

Biotic transformations

Microorganisms (e.g., bacteria and fungi) are typically the most important organisms contributing to the transformation of organic contaminants. The importance of microorganisms to NR stems from their ubiquitous distribution in soil, sludge, and water, and from their role as the primary agents responsible for the cycling of elements in nature (Alexander 1994). Microorganisms are the dominant biota in soil systems in terms of mass and activity. Bacteria are often found in very high numbers in most soils (2 × 109 cells/gm), in the deeper subsurface (106 to 107 cells/cm^3), and in both marine and freshwater ecosystems (105 to 106 cells/cm^3) (Whitman et al. 1998). Microorganisms use both natural and anthropogenic organics as nutrient and energy sources. Their metabolic activity may alter the target molecules through mineralization, through conversion of the compound to CO_2 and H_2O, and through transformation or conversion into simpler compounds (Alexander 1981). Microbial metabolic processes are typically the most important mechanisms contributing to the degradation of organic molecules.

Terminal electron acceptors (TEAs) are required for the final step in the redox reactions involved in the ultimate mineralization of organics (Table 4-4). The most prevalent TEA available in the microbes' environment usually determines which of the various microbial populations will be the most metabolically active. Aerobic

Table 4-4 Terminal electron acceptors, microorganisms, and energy

Reduction potential of half-reaction (Eh) (mV / electron equivalent)	Terminal electron acceptor (TEA)	Common organisms	Oxidation state, marginal conditions, organisms, or activities	Process or reduced end product	H_2 concentration (nM)	Free energy change for oxidation of CH_2O (kJ / electron equivalent)
+810	Oxygen	Aerobes	Aerobic	H_2O		−125
+750	Nitrate (NO_3^-)	Nitrate reducers	Aerotolerant anaerobes	N_2, nitrite (NO_2)	< 0.05	−119
+580	Manganese (MnIV)	Various	Anoxic conditions	Mn(II)	< 0.05	−103
+420 to +560	Halogenated organics (RX)		Facultative anaerobes	Dehalogenated carbons (RH, X^-)		−87.4 to −101
+60	Iron (FeIII)	Iron reducers		Fe(II)	0.2	−52.6
−180	Organic molecules (CH_2O)	Fermenters	Anaerobic reductive dehalogenation	Alcohols, fatty acids, ketones, H_2		−29.7
−210	Sulfate (SO_4^{2-})	Sulfate reducers	Strict anaerobes	H_2S, HS^-, S, S^{2-}	1 to 1.5	−26.4
−240	Carbon dioxide (CO_2)	Methanogens	Simultaneous reactions of fermentation, sulfate reduction, methanogenesis	CH_4	7 to 10	−23.2

Sources: Dean 1992; Lovely et al. 1994; Sawyer et al. 1994; Stumm and Morgan 1996.

microorganisms use molecular oxygen (O_2) as the TEA; facultative anaerobes may use O_2 or nitrate (NO_3^-). Strict anaerobes use inorganics such as sulfate or ferrous iron; for fermentative organisms, other organic molecules serve as the TEA. The amount of energy available for reactions is directly related to the TEA (Table 4-4), and microorganisms typically use the electron acceptors in order of decreasing energy availability, from O_2 to CO_2. Aerobic organisms, using O_2, can derive significantly more energy from respiration than can anaerobic organisms. This increased available energy has a pronounced influence on the rate and extent of contaminant degradation in an aerobic environment, as compared to an anaerobic environment (Smith 1997).

Nitrate is rare in most natural environments; however, it is a common anthropogenic groundwater contaminant. Nitrate is very soluble and mobile in groundwater and can be used as a TEA by many microorganisms. Iron is naturally abundant in many subsurface ecosystems, but the oxidized form Fe(III) has low solubility. Therefore, microorganisms must be in close proximity with Fe(III) to use it as an electron acceptor. The presence of the more soluble, reduced iron Fe(II) in groundwater is an indication that reduction of Fe(III) has occurred. Sulfate is common in naturally occurring forms (gypsum) and from anthropogenic sources. Like nitrate, sulfate is soluble and relatively mobile in groundwater. Finally, a closely cooperating consortium of microorganisms is involved in the relatively low-energy–yielding reduction of CO_2 and subsequent generation of methane ("methanogenesis"). Methane can also be generated by the fermentation of acetate, a slightly more energetically favorable reaction.

Halogenated organic compounds rank above Fe, in terms of electronegativity and free energy availability, and may also serve as TEAs (Table 4-4). Reductive dehalogenation has been recognized as a significant degradative pathway for many halogenated organics. Halogenated compounds are undoubtedly being removed by NR processes, either through halorespiration or cometabolism (Suflita et al. 1982; McCarty 1997).

Humic substances can also serve as TEAs for microorganisms (Lovely et al. 1996). Anaerobic microbial oxidation of VC and dichloroethylene (DCE), using humic acids as electron acceptors, was demonstrated by Bradley et al. (1998). This suggests that the pool of TEAs in natural environments may be very large, thus providing the requisite precursors for biodegradation reactions.

The amount of energy involved in an redox reaction is measured and reported in various units: Gibbs free energy (ΔG), electron activity ($p\varepsilon$), and redox potential (Eh). Available energy may be referenced to a complete redox reaction or to oxidation or reduction half-reactions. Common oxidation reactions used are the oxidation of H (Dolfing and Harrison 1992; Fathepure and Tiedje 1998), formaldehyde (Stumm and Morgan 1996), and glucose (Sims et al. 1991). The energy in reactions is typically expressed on the basis of moles of reactant oxidized or reduced (Sims et al. 1991; Dolfing and Harrison 1992; Fathepure and Tiedje 1998) or may be normalized to the number of electrons exchanged (Sawyer et al. 1994; Stumm and Morgan 1996).

Energies of reactions are derived from free energies of formation (which are referenced to a specific state of matter) and from unit activities (activity = activity coefficient × concentration). Gas is the standard state of many compounds of interest. For reactions occurring in water, some authors (e.g., Vogel et al. 1987; Dolfing and Harrison 1992) adjusted energy of reaction by using free energies of formation for an aqueous instead of gaseous state. However, this adjustment generally had little effect on the relative energies of reaction. A more important adjustment is to use the activities typical of aquatic systems for common ions: $\{H^+\} = 10^{-7}$, $\{Cl^-\} = 10^{-3}$, $\{Fe^{2+}\}$ and $\{Mn^{2+}\} = 10^{-6}$. Unit activity is used for the remaining reactants and products. Using unit activity is roughly equivalent to assuming that the reactants and products are all present at a concentration of 1 mol/L. Reported energies of reaction are therefore potentials; site-specific energies of reaction will depend on the actual activities of reactants and products.

Table 4-4 was constructed from free energies of formation compiled by Stumm and Morgan (1996) and Dean (1992) and primarily follows the conventions of Stumm and Morgan (1996). The range of energy released by reductive dehalogenation is based on chlorinated ethenes, while fermentation is based on formaldehyde. The reduction potential (Eh) is given per unit electron based on the reductive half-reaction only. The change in free energy (ΔG) is also given per unit electron but is based on the reductive half-reaction combined with the oxidation (mineralization) of formaldehyde.

Because hydrocarbon-degrading reactions using O_2 as the TEA are often orders of magnitude more rapid and energetically favorable than reactions using other electron acceptors, available O_2 may be the limiting factor for biodegradation in many environments. Borden and Bedient (1986) modeled the subsurface biodegradation of hydrocarbons, accounting for O_2 availability through transverse mixing, advection, and vertical exchange with the unsaturated zone, clearly illustrating how O_2 transport processes impact transformations. Their study suggests several factors that directly affect NR processes. First, biodegradation of a contaminant in an aquifer will sufficiently reduce available O_2 as to limit aerobic respiration. Second, a large microbial population is likely to develop adjacent to the contaminant source. Groundwater contaminant plumes appear much narrower than would be predicted by transport processes alone because transverse mixing with oxygenated aquifer water increases biodegradation at the margin of the plume. The adsorption of contaminant to soil particles may enhance the biodegradation process by reducing contaminant mobility and increasing exposure time to soil organisms. The movement of O_2 is not retarded, and O_2 will move into the more slowly moving hydrocarbon plume. Finally, vertical exchange with O_2 in the unsaturated zone may significantly enhance the rate of biodegradation.

Reductive dehalogenation, with halogenated organics serving as the TEA, is effective only in ecosystems with redox potentials near that seen under sulfate-reducing conditions (Table 4-4). In this redox environment, halogenated organics are energetically favorable reactants, and some microorganisms are able to capture that energy through halorespiration. Figure 4-2 shows the relationships of the various aerobic and anaerobic pathways of the dehalogenation process. Biodehalogenation was first recognized as an important environmental reaction in the degradation of pesticides in soil suspension systems (Castro and Belser 1968). Later, Suflita et al. (1982) demonstrated that haloaromatic compounds could also be dehalogenated in anaerobic lake sediment or sewage sludge cultures. In general, compounds with fewer halogens are more easily degraded by aerobic processes, while compounds with higher numbers of halogens are more easily degraded by anaerobic processes (Vogel et al. 1987).

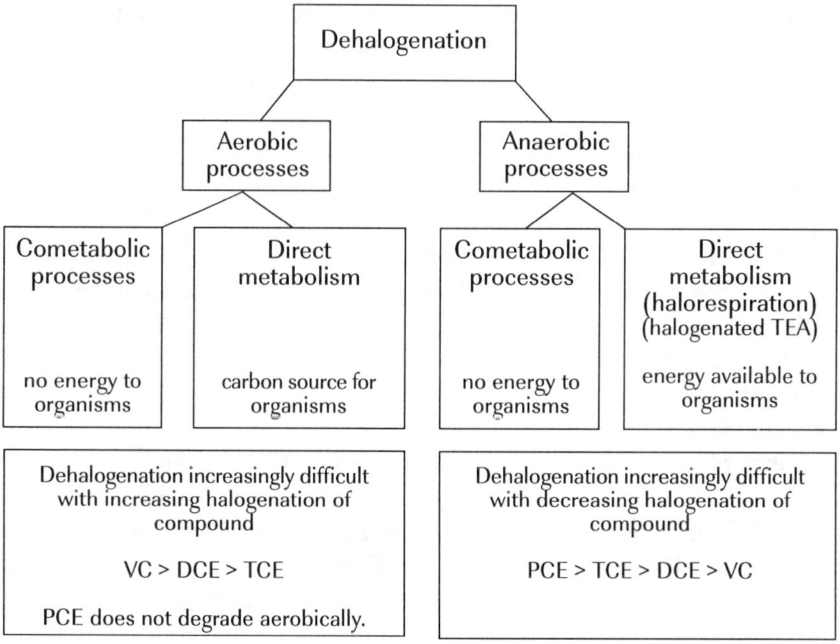

Figure 4-2 Dehalogenation processes
(VC = vinyl chloride, DCE = dichloroethylene, TCE = trichloroethylene, PCE = tetrachloroethylene)

Cometabolism is the fortuitous transformation of a contaminant; it occurs as a food source is degraded and results from the nonspecificity of enzymes such as methane monooxygenase (Alexander 1994). During cometabolism, the microbe does not obtain energy or C from the contaminant; therefore, an alternate food (energy) source must be available. Microorganisms may mineralize contaminants, as in the case of aerobic cometabolism of trichloroethylene (TCE; Wilson and Wilson 1985), or transform the contaminant into a simpler organic, as in the case of anaerobic reductive dehalogenation of DCE to form VC.

A very specialized group of microorganisms, the lignin-degrading white rot fungi, have the ability to break down several persistent environmental contaminants such as dichlorodiphenyltrichloroethane (DDT) or benzo(a)pyrene through the production of extracellular enzymes (Barr and Aust 1994). These relatively nonspecific enzymes attack phenolic compounds and cleave aromatic rings (Leatham et al. 1983). Other recalcitrant compounds, with complex C skeletons similar to the lignin polymer, are also degraded by the organisms (Bumpus et al. 1985).

Different microbial populations may interact to complete the biodegradation of contaminants. This commonly occurs under strict anaerobic conditions such as those associated with landfills or severely contaminated soil environments where available energy and available TEAs are often strictly limited. Reactions will go to completion only with the removal of limiting, potentially toxic reactants (short-chain fatty acids, H_2, etc.) and production of other reactants (CO_2, CH_4, etc.) by very specialized, interacting, microbial populations (Zinder 1993). Dolfing and Prins (1996) proposed the terms "waste chain" or "waste web" to describe some of these interactions in methanogenic ecosystems. These microorganisms depend upon the waste products of others as food and energy sources. If the waste products are not consumed, they accumulate and ultimately halt contaminant degradation.

Agricultural scientists have long recognized the impacts of soil microorganisms on the persistence and effectiveness of biodegradable herbicides. An initial field application of herbicides might be quite effective; however, subsequent applications become significantly less effective. In this case, the decrease in effectiveness of the herbicide is the result of enriched soil microbial populations capable of degrading the herbicide (Walker 1975).

Although plants have only recently been exploited for their remediation capabilities, their normal physiological activities are significant in the fate of environmental contaminants. The intentional use of higher plants for environmental cleanup is classified in the broad category of phytoremediation (Schnoor et al. 1995). The normal metabolic processes of higher plants move significant amounts of water from the soil, through the plant body, and into the atmosphere by translocation and transpiration. Environmental contaminants may also be carried in this transpiration stream. Enough water may be moved to alter the flow of a shallow plume in groundwater (Nobel 1991). Plants may be used directly for environmental cleanup or as enhancements to natural microbial processes.

Alteration of environmental contaminants by plant metabolic activity has been documented for several classes of organic compounds (Cunningham et al. 1996). Herbicides such as atrazine can be taken up, hydrolyzed, and dealkylated to less toxic metabolites by plants such as poplar trees. Burken and Schnoor (1997) were able to show an increase in atrazine metabolites along the transpiration stream, from the soil, through the roots, and into the leaves. Neither atrazine nor its metabolites were transpired from the leaves into the atmosphere. Burken and Schnoor (1997) reference other studies that show metabolites are bound with the lignin portion of the plant and are not bioavailable to rats that are fed the plants.

Like atrazine, TCE was found to be taken up and metabolized by plants. In plant cell cultures, TCE was converted to trichloroethanol and di- and trichloroacetic acid. However, in contrast to atrazine, TCE was also released to the atmosphere in the transpiration from leaves (Newman et al. 1997).

Animal–contaminant interactions. Animal–contaminant interactions are well known and are often to the detriment of the animal. However, positive interactions that contribute to remediation are also well documented. Darwin (1881) described the positive impacts of earthworms on the soil environment. Earthworms directly contribute to the transport of chemicals in soil. Material movement from the surface to the subsurface is enhanced through the formation of holes in soil. Movement from the subsurface to the surface is increased through the deposition of casts. Transformations may be enhanced, either directly through the earthworm's metabolism or indirectly through aeration of the soil, improving abiotic oxidation or aerobic biodegradation, or through mixing, thereby increasing contact between contaminants and soil microorganisms (Lee 1985).

Generally, uptake of organics by higher aquatic organisms is proportional to lipid solubility. The octanol–water partition coefficient (K_{OW}) is a useful predictor of the potential for uptake, with octanol serving as a model for typical lipids of animals. Once inside an organism, environmental contaminants may be stored, excreted, or metabolized. Storage may be of little broad environmental significance, although bioaccumulation may be important to the affected organism and organisms at higher levels of the food chain. Likewise, excretion is of minor environmental significance, unless transport to a new medium is involved. Metabolism, however, can be significant to both the organism and the environment. Metabolism often makes lipophilic compounds more water soluble, a process that enhances excretion from the organism (Walker et al. 1996), but the process may also make the material more available to other organisms in the environment.

Some aquatic filter feeders (e.g., clams and mussels) are surprisingly long-lived: several years for nearshore species and roughly 25 to 100 years for deep sea organisms associated with thermal vents (Turekian et al. 1975; Hart and Blusztajn 1998). This potentially long exposure time presents the possibility of partitioning and bioconcentrating low environmental levels of contaminants into the organism's lipid storage tissues. Some fish have become notorious for accumulation of organic and inorganic contaminants well above background levels. Birds, which usually feed a little higher in the food chain, have also accumulated materials at levels that have been harmful. Animal–contaminant interactions are probably less important in the ultimate environmental fate of contaminants than are microbial interactions, but economic or ecological impacts may make them topics for study (Crosby 1998).

Protozoans and biodegradation. Protozoan predation of bacteria in groundwater or soil has potential environmental significance in the remediation process. Kinner et al. (1998) found that, in a contaminated sandy aquifer, nanoflagellates could consume 12% to 74% of the unattached bacteria in a day. In the aquifer they studied, at least 30% of the total bacterial populations were unattached. This suggests that predation by the protozoans can have a significant impact on microbial biodegradation, simply by removing the bacteria from the process. Madsen et al. (1991) found that protozoans associated with a contaminant plume were 2 orders of magnitude more numerous than those found in an adjacent pristine environment. Organic contaminants in the saturated zone, where there was a high density of protozoans, were rapidly mineralized. Apparently, high protozoan (predator) numbers indicated a rapidly growing bacterial (prey) population that used the organic contaminant in the aquifer for growth.

DATA NEEDED TO PREDICT NATURAL REMEDIATION OF CONTAMINANTS

Contaminant characteristics: Physical state, mass, age

Assessment of NR requires knowledge of the physical and chemical characteristics of the contaminant. The physical and chemical properties of contaminants largely determine their environmental fate. Organic chain length, branching, and saturation strongly influence solubility and volatility, and therefore, environmental availability. Halogen additions, while favorably altering the utility of molecules, often complicate their environmental fate by decreasing degradability while sometimes increasing toxicity. Speciation of inorganics can change impacts from relatively innocuous to quite toxic, for example, Cr(III) versus Cr(VI). Metabolic fate is directly linked to molecular structure (Alexander 1981). The biodegradability of some synthetic detergents was dramatically improved with the relatively simple molecular change from branched to linear alkyl groups (Swisher 1963).

Contaminant concentration can be critical in the response of degrading organisms, with both maximum and minimum limits. In laboratory studies, Krumholtz et al. (1996) isolated an organism that degraded tetrachloroethylene (PCE) to DCE under anaerobic conditions at concentrations ranging from 25 to 100 μM (~ 4,150 to 16,600 mg/L); however, concentrations above 100 μM inhibited growth. In a field study, Hopkins et al. (1993) found toxicity with TCE at concentrations of only 7.6 μM (1,000 mg/L). Boethling and Alexander (1979) showed that below some threshold concentration, little or no biodegradation might occur. This threshold may result when not enough energy is available from the degradation of the contaminant for the microorganisms to reproduce in sufficient quantities to remove the contaminant.

The bioavailability of contaminants is not easily defined or measured but is critical in the degradation process. Much recent work has investigated the impact of sorption on bioavailability of organic contaminants in the subsurface (Ramaswami and Luthy 1997). It is hypothesized that in soil–water systems, only contaminants in the aqueous phase are bioavailable. Experimental data indicate that, as contaminants in soil age, the extent of their sorption increases, and therefore their bioavailability (and biodegradability) decreases (Hatzinger and Alexander 1995).

Monitoring environmental conditions

In retrospect, it is unfortunate that much of the environmental chemistry data from contaminated sites over the last 2 decades were collected only in response to regulatory requirements. Typically, little regard was given to the underlying environmental processes or the state of the ecosystems involved. Although the concentrations of environmental contaminants may be known with a high degree of confidence, little or no data may exist on the factors that influence remediation or confirm that NR is in progress, such as dissolved O_2, CO_2, methane, organic C, or major cation and anion concentrations. These additional data can be used to develop an understanding of the fate and transport of contaminants, the rates of change, and the significance of NR in the reduction of risk associated with the contaminants.

To assess NR, we need to understand the transport and fate of a contaminant in the environment. This requires information about both the characteristics of the contaminant and environmental conditions. The chemical composition of soil gas and groundwater (e.g., O_2, H_2, CO_2, H_2S, SO_4^{2-}, NO_3^-, NO_2^-, NH_3, pH, Fe^{2+}, organic acids, phenols, chlorinated metabolites, ethylene, ethane) provides direct evidence of the predominant conditions in the environment and reflects the presence and types of active microbial populations. Soil characteristics—organic C content, cation exchange capacity (CEC), and surface area—also help determine contaminant availability and movement, and influence the microbial populations.

Contaminant distribution in a single phase depends upon the characteristics and abundance of the phase. Hydrophobic compounds cannot move into soil organic material if there is little or no natural soil organic material available. Reactions need reactants and catalysts that are supplied by the environment. Different ecosystems have different microbial distributions and populations. Quantifying these environmental characteristics is critical to assessing NR.

Virtually all significant microbiological activity occurs in the liquid (water) phase. The water may be a thin layer intimately associated with solids, such as in a soil system, but liquid water is required for microbial metabolic processes (Stotzky

1997). Interfacial areas, temperature, and fluid velocity can all affect the rate and extent of reactions in water.

For both aerobic and anaerobic processes, threshold concentrations exist below which no degradation reactions appear to occur. Likewise, toxicity may produce limits above which there is little or no degradation activity. The available electron acceptors and the energy supply (electron donors) determine the dominant microbial populations. Available nutrients or ligands may limit growth and metabolism, and in many natural systems, C is the limiting nutrient. Microbial populations are dynamic and strongly interactive. Ecological succession at the micro level is perhaps even more important than at the macro level, but much more difficult to predict and trace.

CONCLUSIONS

Virtually all anthropogenic materials discharged into the environment are susceptible to some NR processes. An understanding of the potential effectiveness of NR processes and the collection of appropriate data enable assessment of the role of NR in the transport and ultimate disposition of contaminants in the environment.

Predictably and quantifiably, contaminants move and proportion themselves between the various compartments of an ecosystem: the atmosphere, surface water, and soil–groundwater systems. This movement alters the degradation, bioavailability, and reactivity of the contaminants. Partitioning of the contaminants can be predicted using physical and chemical parameters such as Henry's constants or octanol–water partition coefficients.

Contaminant transport is governed by clearly defined processes such as advection, dispersion, sedimentation, and diffusion. This movement is essential to the physical, chemical, or biological transformation processes that actually alter contaminant chemical structure and ultimately result in the degradation, reduced toxicity, or removal of the contaminants from key compartments of ecosystems. Abiotic processes, such as oxidation or reduction, hydrolysis, photolysis, precipitation or dissolution, or complexation, may be important for some contaminants. Abiotic conditions (pH, temperature, available TEAs, nutrients, etc.) may influence the rate and extent of biological activity, but the actions of microorganisms are the most important in the NR of environmental contaminants. Some processes are exclusively aerobic, while others rely on anaerobic conditions to complete reactions.

Phytoremediation, the use of higher plants for environmental cleanup, is still in the early stages of development but shows promise for several problems. Animal–contaminant interactions appear to have less impact on remediation but may pose important environmental concerns beyond remediation.

Chemical structure of contaminants affects both availability and degradability. Contaminant concentrations may have both maximum and minimum levels above or below which biological degradation and NR do not occur.

REFERENCES

Alexander M. 1981. Biodegradation of chemicals of environmental concern. *Science* 211:132–138.

Alexander M. 1994. Biodegradation and bioremediation. New York NY: Academic Pr.

Ashworth RA, Howe GB, Mullins ME, Rogers TN. 1988. Air–water partitioning coefficients of organics in dilute aqueous solution. *J Hazardous Materials* 18:25–36.

Barr DP, Aust SD. 1994. Mechanisms white rot fungi use to degrade pollutants. *Environ Sci Technol* 28:79A–87A.

Bedient PB, Rifai HS, Newell CJ. 1994. Ground water contamination: Transport and remediation. Englewood Cliffs NJ: PTR Prentice-Hall.

Boethling RS, Alexander M. 1979. Effect of concentration of organic chemicals on their biodegradation by natural microbial communities. *Appl Environ Microbiol* 37:1211–1216.

Borden RC, Bedient PB. 1986. Transport of dissolved hydrocarbon influenced by oxygen-limited biodegradation 1. Theoretical development. *Water Resources Res* 22:1973–1982.

Bradley PM, Chapelle FH, Lovley DR. 1998. Humic acids as electron acceptors for anaerobic microbial oxidation of vinyl chloride and dichloroethylene. *Appl Environ Microbiol* 64:3102–3105.

Brusseau ML, Rao PSC. 1989. Sorption nonideality during organic contaminant transport in porous media. *CRC Crit Rev Environ Cont* 19:33–99.

Bumpus JA, Tien M, Wright D, Aust SD. 1985. Oxidation of persistent environmental pollutants by a white rot fungus. *Science* 228:1434–1436.

Burken JC, Schnoor JL. 1997. Uptake and metabolism of atrazine by poplar trees. *Environ Sci Technol* 31:1399–1406.

Castro CE, Belser NO. 1968. Biodehalogenation: Reductive dehalogenation of the biocides ethylene dibromide, 1,2-dibromo-3-chloropropane, and 2,3-dibromobutane in soil. *Environ Sci Technol* 2:779–783.

Chen CS-H, Delfino JJ, Rao PCS. 1994. Partitioning of organic and inorganic components from motor oil into water. *Chemosphere* 28:1385–1400.

Chin Y-P, Aiken GR, Danielsen KM. 1997. Binding of pyrene to aquatic and commercial humic substances: The role of molecular weight and aromaticity. *Environ Sci Technol* 31:1630:1635.

Chiou CT, Peters LJ, Freed VH. 1979. A physical concept of soil–water equilibria for nonionic organic compounds. *Science* 206:831–832.

Cline PV, Delfino JJ, Rao PSC. 1991. Partitioning of aromatic constituents into water from gasoline and other complex solvent mixtures. *Environ Sci Technol* 25:914–920.

Crosby DG. 1998. Environmental toxicology and chemistry. New York NY: Oxford Univ Pr.

Cunningham SD, Anderson TA, Schwab AP, Hsu FC. 1996. Phytoremediation of soils contaminated with organic pollutants. *Adv Agronomy* 56:55–114.

Cussler EL. 1984. Diffusion: Mass transfer in fluid systems. Cambridge UK: Cambridge Univ Pr.

Darwin CR. 1881. The formation of vegetable mould through the action of worms, with observations on their habits. London UK: Murray.

Dean JA. 1992. Lange's handbook of chemistry. 14th ed. New York NY: McGraw-Hill.

Dolfing J, Harrison BK. 1992. Gibbs free energy of formation of halogenated aromatic compounds and their potential role as electron acceptors in anaerobic environments. *Environ Sci Technol* 26:2213–2218.

Dolfing J, Prins RA. 1996. Methanogenic "food chains." *ASM News* 62:117.

Fathepure BZ, Tiedje JM. 1998. Anaerobic bioremediation: Microbiology, principles, and applications. In: Adriano DC, Bollag J-M, Frankenburger WT, Sims R, editors. Bioremediation of soils. Madison WI: American Society of Agronomy.

Garnier J-M, Pham MK, Ciffroy P, Martin J-M. 1997. Kinetics of trace element complexation with suspended matter and with filterable ligands in freshwater. *Environ Sci Technol* 31:1597–1606.

Hatzinger PB, Alexander M. 1995. Effect of aging of chemicals in soil on their biodegradability and extractability. *Environ Sci Technol* 29:537–545.

Hart S. 1996. In situ bioremediation: Defining the limits. *Environ Sci Technol* 30:398A–401A.

Hart SR, Blusztajn J. 1998. Clams as recorders of ocean ridge volcanism and hydrothermal vent field activity. *Science* 280:883–886.

Hinchee RE, Reisinger HJ. 1987. A practical application of multiphase transport theory to ground water contamination problems. *Ground Water Monitoring Rev* 7:84–92.

Hopkins GD, Munakata J, Semprini L, McCarty PL. 1993. Trichloroethylene concentration effects on pilot field-scale in-situ groundwater bioremediation by phenol-oxidizing microorganisms. *Environ Sci Technol* 27:2542–2547.

Huang W, Young TM, Schlautman MA, Yu H, Weber Jr WJ. 1997. A distributed reactivity model for sorption of soils and sediments. 9. General isotherm nonlinearity and applicability of the dual reactive domain model. *Environ Sci Technol* 31:1703–1710.

Karickhoff SW. 1981. Semi-empirical estimation of sorption of hydrophobic pollutants on natural sediments and soils. *Chemosphere* 10:833–846.

Karickhoff SW, Brown DS, Scott TA. 1979. Sorption of hydrophobic pollutants on natural sediments. *Water Res* 13:241–248.

Kinner NE, Harvey RW, Blakeslee K, Novarino G, Meeder LD. 1998. Size-selective predation on groundwater bacteria by nanoflagellates in an organic-contaminated aquifer. *Appl Environ Microbiol* 64:618–625.

Krumholtz LR, Sharp R, Fishbain SS. 1996. A freshwater anaerobe coupling acetate oxidation to tetrachloroethylene dehalogenation. *Appl Environ Microbiol* 62:4108–4113.

Leatham GF, Crawford RL, Kirk TK. 1983. Degradation of phenolic compounds and ring cleavage of catechol by Phanerochaete chrysosporium. *Appl Environ Microbiol* 46:191–197.

Lee KE. 1985. Earthworms, their ecology and relationships with soils and land use. New York NY: Academic Pr.

Lee LS, Hagwall M, Delfino JJ, Rao PSC. 1992. Partitioning of polycyclic aromatic hydrocarbons from diesel fuel into water. *Environ Sci Technol* 26:2104–2110.

Lee LS, Rao PSC, Okuda I. 1992. Equilibrium partitioning of polycyclic aromatic hydrocarbons from coal tar. *Environ Sci Technol* 26:2110–2115.

Liu H, Amy G. 1993. Modeling partitioning and transport interactions between natural organic matter and polynuclear aromatic hydrocarbons in groundwater. *Environ Sci Technol* 27:1553–1562.

Lovely DR, Chapelle FH, Woodward JC. 1994. Use of dissolved H_2 concentrations to determine the distribution of microbially catalyzed redox reactions in anoxic groundwater. *Environ Sci Technol* 28:1205–1210.

Lovely DR, Coates JD, Blunt-Harris EL, Phillips EJP, Woodward JC. 1996. Humic substances as electron acceptors for microbial respiration. *Nature* 382:445–448.

Lyman WJ. 1995. Transport and transformation processes. In: Rand GM, editor. Fundamentals of aquatic toxicology, 2nd ed. Washington DC: Taylor & Francis. p 449–492.

Mackay D, Paterson S, Cheung B, Neely WB. 1985. Evaluating the environmental behavior of chemicals with a level III fugacity model. *Chemosphere* 14:335–374.

Madsen EL, Sinclair JL, Ghiorse WC. 1991. In situ biodegradation: Microbiological patterns in a contaminated aquifer. *Science* 252:830–833.

McCarty PL. 1997. Breathing with chlorinated solvents. *Science* 276:1521–1522

Means JC, Wood SG, Hassett JJ, Banwart WL. 1980. Sorption of polynuclear aromatic hydrocarbons by sediments and soils. *Environ Sci Technol* 14:1524–1528.

Newman LA, Strand SE, Choe N, Duffy J, Eduan G, Ruszaj M, Shurtleff BB, Wilmoth J, Heilman P, Gordon MP. 1997. Uptake and biotransformation of trichloroethylene of hybrid poplars. *Environ Sci Technol* 31:1062–1067.

Nobel PS. 1991. Physicochemical and environmental plant physiology. San Diego CA: Academic Pr.

Pignatello JJ, Xing B. 1996. Mechanisms of slow sorption of organic chemicals to natural particles. *Environ Sci Technol* 30:1–11.

Ramaswami A, Luthy RG. 1997. Measuring and modeling physicochemical limitations to bioavailability and biodegradation. In: Hurst, CJ, Knudsen GR, McInerney, MJ, Stetzenbach LD, Walter MV, editors. Manual of environmental microbiology. Washington DC: ASM Pr. p 721–729.

Roy SB, Dzomback DA. 1997. Chemical factors influencing colloid-facilitated transport of contaminants in porous media. *Environ Sci Technol* 31:656–664.

Sawyer CN, McCarty PL, Parkin GF. 1994. Chemistry for environmental engineering. New York NY: McGraw-Hill.

Schnoor JL, Licht LA, McCutcheon SC, Wolfe NL, Carreira LH. 1995. Phytoremediation of organic and nutrient contaminants. *Environ Sci Technol* 29:318A–323A.

Schwarzenbach RP, Gschwend PM, Imboden DM. 1993. Environmental organic chemistry. New York NY: J Wiley.

Scow KM, Hutson J. 1992. Effect of diffusion and sorption on the kinetics of biodegradation: Theoretical considerations. *Soil Sci Soc Am J* 56:119–127.

Sims JL, Suflita JM, Russell HH. 1991. Reductive dehalogenation of organic contaminants in soils and ground water. Washington DC: U.S. Environmental Protection Agency (USEPA). EPA/540/4-90/054.

Smith RL. 1997. Determining the terminal electron-accepting reaction in the saturated subsurface. In: Hurst CJ, Knudsen GR, McInerney MJ, Stetzenbach LD, Walter MV, editors. Manual of environmental microbiology. Washington DC: ASM Pr. p 577–585.

Stotzky G. 1997. Soil as an environment for microbial life. In: van Elsas JD, Trevors JT, Wellington EMH, editors. Modern soil microbiology. New York NY: Marcel Dekker. p1–20.

Stumm W, Morgan JJ. 1996. Aquatic chemistry. 3rd ed. New York NY: Wiley.

Suflita JM, Horowitz A, Shelton DR, Tiedje JM. 1982. Dehalogenation: A novel pathway for the anaerobic biodegradation of haloaromatic compounds. *Science* 218:1115–1117.

Swisher RD. 1963. The chemistry of surfactant biodegradation. *J Amer Oil Chem Soc* 40:648–656.

Turekian KK, Cochran JK, Kharkar DP, Cerrato RM, Viasnys JR, Sanders HL, Grassle JF, Allen JA. 1975. Slow growth rate of a deep-sea clam determined by 228Ra chronology. *Proc Nat Acad Sci USA* 72:2829–2832.

Vogel TM, Criddle CS, McCarty PL. 1987. Transformations of halogenated aliphatic compounds. *Environ Sci Technol* 21:722–736.

Walker CH. 1975. Environmental pollution by chemicals. 2nd ed. London UK: Hutchinson.

Walker CH, Hopkin SP, Sibly RM, Peakall DB. 1996. Principles of ecotoxicology. Bristol PA: Taylor & Francis.

Weber EJ. 1998. Abiotic transformation reactions. In: Schüürmann G, Markert B, editors. Ecotoxicology: Ecological fundamentals, chemical exposure, and biological effects. New York NY: Wiley.

Weber WJ, DiGiano FA. 1996. Process dynamics in environmental systems. New York NY: J Wiley.

Weber Jr WJ, Huang W. 1996. A distributed reactivity model for sorption of soils and sediments. 4. Intraparticle heterogeneity and phase-distribution relationships under nonequilibrium conditions. *Environ Sci Technol* 30:881–888.

Weber Jr WJ, McGinley PM, Katz LE. 1992. A distributed reactivity model for sorption of soils and sediments. 1. Conceptual basis and equilibrium assessments. *Environ Sci Technol* 26:1955–1962.

Whitman WB, Coleman DC, Wiebe WJ. 1998. Prokaryotes: The unseen majority. *Proc Nat Acad Sci USA* 95:6578–6583.

Wilson JT, Wilson BH. 1985. Biotransformation of trichloroethylene in soil. *Appl Environ Microbiol* 49:242–243.

Zinder SH 1993. Physiological ecology of methanogens. In: Ferry JG, editor. Methanogenesis: Ecology, physiology, biochemistry & genetics. New York NY: Chapman & Hall. p 128–206.

5

Modeling of Natural Remediation: Contaminant Fate and Transport

Brent M. Peyton, T. Prabhakar Clement, John P. Connolly

PREDICTIVE MODELS can play an important role in verifying the occurrence and significance of natural remediation (NR) and can significantly improve the design of monitoring and assessment procedures. Predictions resulting from the models can be used to address important questions that invariably arise during the assessment of NR. Such questions may include the following:

- Will environmentally important receptors be impacted by the contaminant?
- What are the expected average and maximum concentration levels, and what is the associated risk? How long will it take for the contaminant plume to degrade below regulated limits?

To answer these questions, we must adequately describe the key NR processes that are active at the site. To accomplish such description, robust mathematical models capable of predicting the transport and reaction of contaminants are required.

Typically, models used to evaluate NR modeling involve 2 distinct steps:

1) Modeling the general environmental system (e.g., sediment, surface water, and/or groundwater)
2) Modeling the transport and reaction of specific contaminants.

Natural Remediation of Environmental Contaminants. Michael Swindoll et al., editors.

Interactions between contaminants and environmental materials often control the rates of NR. Partitioning of fuel hydrocarbons to soil organic matter can limit the rate of hydrocarbon transport as well as the rate of bacterial degradation. The interactions may be sufficiently important that they modify the properties of the system itself. Interactions of this type can be important when, for example, rapid microbial growth during an in situ bioremediation effort might significantly reduce the permeability and porosity of the porous media (Taylor and Jaffe 1990; Clement et al. 1996b). However, for most NR modeling applications, the flow and reactive transport of contaminants may be considered as uncoupled processes.

For complex environmental systems such as those typically encountered at contaminated sites, computer models will never give answers that accurately reflect all aspects of the complex chemical and microbial interactions that may occur. For NR, rather than giving "answers," models should be used to predict future trends and to provide insight into the relative importance of various processes that occur at the site. In this way, an NR model can be used as a project mentor, rather than as a project director. Complex models such as these are subject to many conditions that may severely limit the accuracy or applicability of a particular model to a particular site. These limitations may or may not be known at the time the NR assessment begins. During an initial site characterization, for example, based on limited data, a model can predict the order of magnitude of the contaminant plume distribution and can give rough estimates for various rates of degradation and dispersion. This information can be very useful when a site characterization plan is designed or when initial sampling costs are estimated. Data obtained in the initial phase of the site characterization can be incorporated into the model to improve predictions regarding the rate of plume movement from the initial contaminating event to the current volumetric extent of the contamination. Time-phased sampling events may be used to calibrate the temporal aspects of the model and to predict the future extent of contaminant spreading, the rates of NR, and the eventual decline in concentration of the contaminant.

If NR is to be evaluated and accepted as a viable treatment option, the value of models to predict long-term risk cannot be overemphasized. In addition, a carefully assembled and validated model can help address stakeholder and regulator concerns over the rate and extent of contaminant movement. Models are important tools, and they provide valuable insight into the complex systems that bring about the NR of contaminants. The goal of this chapter is to provide a basis for understanding the issues that are important for modeling NR of contaminants in both surface water and groundwater.

SURFACEWATER BIOGEOCHEMICAL TRANSPORT MODELS

Surfacewater biogeochemical models describe the movement of water and sediments in relation to the transport and transformation of contaminants. These models can be valuable tools in evaluating the NR of contaminants in aquatic ecosystems. Typically, surfacewater biogeochemical transport models have 4 coupled components (Figure 5-1):

1) Hydrodynamics
2) Sediment transport
3) Physical and chemical partitioning processes
4) Contaminant fate through bioaccumulation and biodegradation.

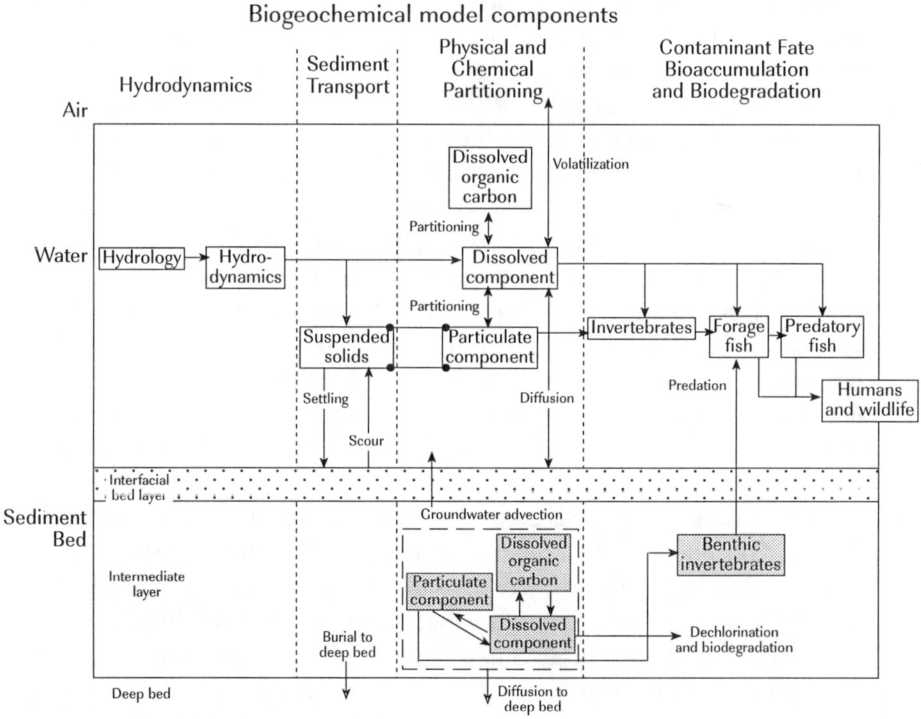

Figure 5-1 Processes that affect contaminant fate and transport in surface waters

The hydrodynamics component involves the movement of water and the friction or shear that this movement causes at the interface between the water and the sediment bed. A hydrodynamic model computes the velocity and depth of the water column, as well as the shear stress at the water–sediment bed interface, in response to upstream flows and flows entering from tributaries or the downstream boundary. The sediment transport component includes the movement of suspended and settled solids with the moving water and the settling and resuspension of solids that occurs at the water–sediment bed interface as a result of the shear caused by the moving water. A sediment transport model computes the concentration of solids in the water column and the rate at which sediment accumulates in the bed. The contaminant fate component includes the transport of contaminant dissolved in the water or sorbed to solids, the transfer between the dissolved and sorbed state, the transfer among chemical species of the contaminant, the transfer between the water and the atmosphere, and the degradation that occurs because of biotic or chemical reactions.

A contaminant fate model computes the concentrations of the contaminant in the water column and in the sediment. The models used to assess NR are systems of equations developed from the basic principles of conservation of mass, energy, and momentum; equations of state; and laboratory and field studies of individual phenomena. This system of equations is general and can be applied to various surfacewater systems. The application of the equations to a specific system involves the determination of appropriate values for each of the parameters in the equations. Site-specific data are the basis for assigning values, either directly or by the process of model calibration. Each of the 3 model components must be calibrated and validated using the available data. Good site-specific data are the key to the accurate prediction of NR; if these data are not available, the utility of model predictions may be limited. Nevertheless, in the absence of high quality data, modeling can still be instructive for identifying critical processes and future data collection needs.

A number of biochemical transport modeling approaches are in use. However, only a few of these approaches are appropriate for addressing NR. Because fundamental questions regarding NR often require the assessment of either the rate of contaminant decline or the time to achieve some endpoint, steady-state models are inappropriate. Because NR in surfacewater systems is invariably a function of cyclic or event-related phenomena such as temperature, light, flow, and solids loading, models that assume temporally constant rates of input or reaction are inadequate (Connolly 1997). A widely used framework that does not suffer these limitations is WASP (Ambrose et al. 1993). This framework uses a flexible compartment modeling approach that can represent a surface water in 1, 2, or 3 dimensions. The hydrodynamic and sediment transport components are separate from the fate component, allowing for convenient modification of the fate component to include reaction processes unique to the contaminant being modeled. The WASP modeling framework, or a variant of it, has been applied to evaluate NR in the lower Fox River (Velleux et al. 1995), Green Bay (Raghunathan et al. 1994), the James River (O'Connor et al. 1989), and the Hudson River (Thomann et al. 1991). The primary limitation of these WASP-like frameworks is that hydrodynamics and sediment transport are uncoupled. Rates of erosion and deposition, and input in the form of settling and resuspension velocities, are independent of input rates of flow and velocity. This limitation can be overcome by using a hydrodynamic–sediment transport model (e.g., TABS-2, Thomas and McAnally 1985; SEDZL, Ziegler and Nisbet 1994) to calculate a time series of erosion and deposition rates from a time series of flows that serve as inputs to the fate model.

Use of models to assess natural remediation in surfacewater ecosystems

Assessment of NR in surface waters often focuses on reducing contaminant concentration in surface sediments. Here, "surface sediments" refers to sediments from which contaminants are potentially available to biota. Natural remediation in surfacewater ecosystems is the cumulative result of reaction processes that destroy the contaminant, transfer processes that move the contaminant between the sediment and the water column and between the water column and the atmosphere, and sedimentation that buries and dilutes the contaminant (Figure 5-1). Data must exist on all of these processes if we are to have confidence in model predictions.

Often the primary mechanism of NR in surface waters is the burial of contaminated sediments by relatively clean sediments, and most of the solids loading that is responsible for the burial typically enters the system in short-term events that occur only a few times each year (Ager 1981). Accurate estimation of the relationship between flow and solids loading and simulation of sediment transport during the event periods is necessary for accurate prediction of burial rate and contaminant fate (Ziegler and Connolly 1995; Cardenas and Lick 1996). A practical example of this postulate is found in a model of the NR of Kepone in the James River estuary (O'Connor et al. 1983, 1989). The first version of this model assumed constant flow at the annual mean. This version significantly over-predicted the rate of decline of sediment Kepone concentrations. When the model was modified to account for flow variation and variable solids loading, the predicted rate of decline agreed with the observed rate.

Current state of the art

All of the physical, chemical, and biological processes that determine the fate of a contaminant in a surfacewater system have been the subject of extensive scientific investigation, which has allowed the development of sophisticated models. However, the information requirements of these models are formidable, and their computational requirements can be extreme. Consequently, aggregation of these models into a biogeochemical transport model remains beyond the current state of the art. Biogeochemical transport models have tended to use relatively simplistic descriptions of some or all of the components. The structure and application of the state-of-the art component models that make up a biochemical transport model are reviewed below.

Hydrodynamics

Hydrodynamics are described by 2- and 3-dimensional models that account for the major forces affecting water motion. These forces include horizontal pressure gradients associated with the slope of the water surface (due to channel slope, tides, or seiches), internal density gradients (due to salinity or temperature gradients), wind stresses at the water surface, bottom stresses at the water–sediment bed interface, internal friction or viscosity, and Coriolis acceleration (important only in coastal waters and oceans). The accuracy of the hydrodynamic calculation typically depends on the scale of the numerical grid, the resolution and accuracy of bathymetric data and boundary-forcing functions (stage height, salinity, wind speed and direction, tributary inflows), and the availability of sufficient current, temperature, salinity (if an estuary or coastal water), and water surface elevation data within the system to allow accurate estimation of bottom friction factors or equivalent bottom roughness heights.

The way in which a hydrodynamic model is calibrated depends on the available data. Typically, the bottom friction factor is adjusted to maximize the fit between computed and observed values of water surface elevation and current velocity. The approach is illustrated by 2 examples. Quantitative Environmental Analysis (1999) calibrated hydrodynamic models for each of 8 dammed reaches of the Upper Hudson River by fixing the dam stage height at the downstream limit of the model at the measured value and then adjusting the bottom friction factors until good agreement was achieved between the predicted and measured stage heights at an upstream location. The models were validated by simulating a flood that occurred in May 1983 and comparing computed and observed stage height measurements (Figure 5-2).

HydroQual (1998) calibrated a hydrodynamic model of Lavaca and Matagorda Bays on the Texas coast. A fine-scale numerical grid was employed, with 5,280 horizontal elements and 10 vertical layers to describe the approximately 80-km^2 bay system. A time series of water surface elevations at the connection to the Gulf of Mexico, wind velocities, and tributary inflows were used as forcing functions. The bottom roughness height was used as the calibration parameter. Calibration was assessed using a 1-month time series of water surface elevations at several locations within the bay system and current velocities measured at a single location in Lavaca Bay. A bottom roughness height of 0.6 mm yielded good results. The model predicted hourly water surface elevations at 3 locations with a mean error of 2% (Figure 5-3). Predicted current velocities also agreed well with observations (Figure 5-4).

Sediment transport

Sediment transport is simulated using a simplified distribution of particle sizes in a surface water. Typically, suspendable sediments are aggregated into 2 classes: one represents fine grain (cohesive) particles with diameters less than 62 μm, and the

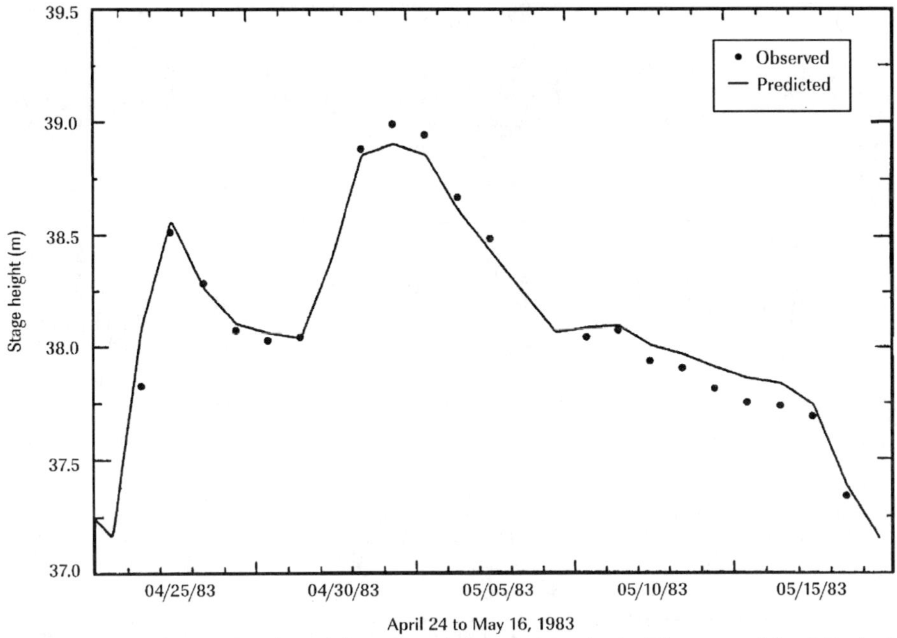

Figure 5-2 Observed and model-predicted stage heights in the Thompson Island Pool of the Upper Hudson River during a high flow event, 1983

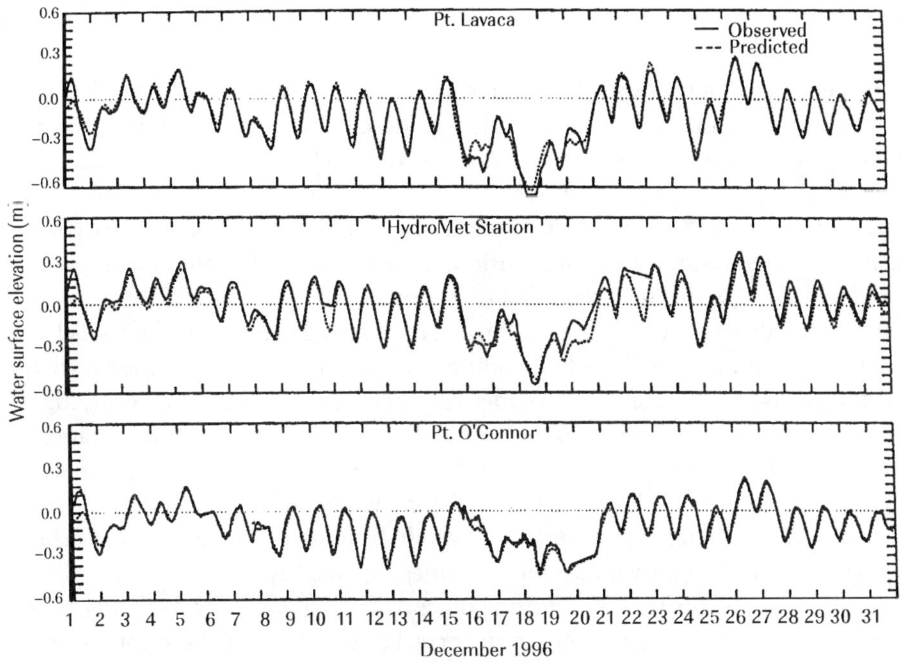

Figure 5-3 Observed and model-predicted water surface elevations at 3 locations in Lavaca and Matagorda Bays, Texas (HydroQual 1998 with permission from Aluminum Company of America)

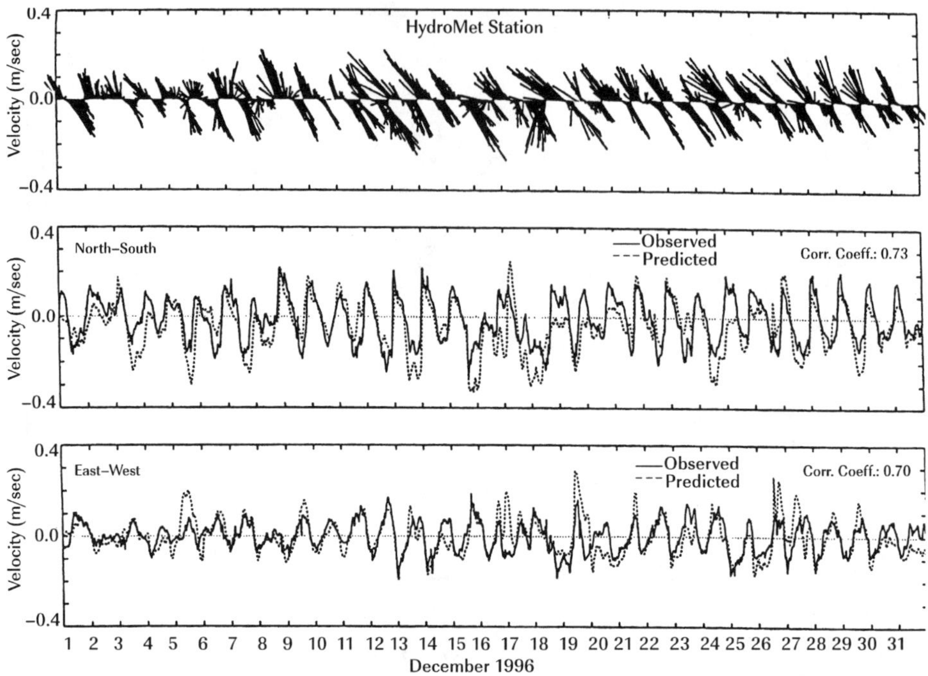

Figure 5-4 Observed and model-predicted water velocities and orthogonal component at a single location in Lavaca Bay, Texas (HydroQual 1998 with permission from Aluminum Company of America)

other represents fine sands with diameters between 62 and 250 μm (e.g., Ziegler and Nisbet 1994). Various empirical formulations describe the deposition and resuspension of these particle classes. The parameters in these formulations are site specific, particularly for resuspension, and require direct measurement (e.g., Tsai and Lick 1986; Jepsen et al. 1997). In addition to the obvious importance of accurate characterization of deposition and resuspension, the accuracy of the solids loading measurements (or the flow–solids loading correlation) and the accuracy of the particle size distribution of that loading are important determinants of model accuracy. In cases where the solids loading or the particle size distribution of that loading are poorly characterized, model calibration can result in incorrect estimates of the rates of resuspension and deposition. If solids loading is underestimated, calibration may result in too much solids resuspension to achieve the measured suspended solids levels. Further, it is likely that the burial rate would be underestimated, and consequently, the rate of NR would be underestimated. This difficulty occurred in a preliminary model for the Upper Hudson River (USEPA 1996). The solids loadings for 2 tributaries were significantly underestimated, resulting in an overestimate of resuspension and the incorrect calculation of net erosion rather than net burial (Schweiger et al. 1996).

The development of a sediment transport model begins by defining the characteristics of the sediment bed. A bed map is constructed in which the bed is divided into a minimum of 3 classifications:

1) Cohesive sediments
2) Noncohesive sediments
3) Hard bottom.

The noncohesive sediments may be further divided on the basis of median particle size. The erosion properties of the cohesive sediments are defined by measurement, and those of the noncohesive sediments are defined by specified values of an active layer depth and median particle diameter (Ziegler and Nisbet 1994). Tributary solids loading is defined by a data-based relationship between solids loading and tributary flow (e.g., Ferguson 1987; Walling and Webb 1988). Settling velocities of the cohesive sediment classes are defined by empirical correlations to particle size and concentration and water column turbulence (Ziegler and Nisbet 1995). The settling velocity of the noncohesive particles is a function of particle size (Cheng 1997). The model is calibrated by comparing its predictions to observed total suspended solids (TSS) data during flood conditions (e.g., Ziegler and Nisbet 1994) and by comparing predicted and observed rates of sedimentation (e.g., Ziegler and Nisbet 1995). Calibration parameters include the particle size composition of the solids loading and the median particle size and active layer depth of the noncohesive sediments.

The Upper Hudson River hydrodynamic models were used with a sediment transport model to predict erosion and deposition of sediment and associated polychlorinated biphenyls (PCBs) (Quantitative Environmental Analysis 1999; Ziegler et al. submitted). Suspended solids data from an April 1994 flood were used to calibrate the model. Comparisons of predicted and observed TSS at 4 locations covering a 55-km length of river are presented in Figure 5-5. The model closely approximated the observed data at all of the locations. It captured both the temporal variation and the general increase in TSS concentrations throughout the 55 km between the Thompson Island Dam and Waterford.

Contaminant fate

Contaminant fate models combine the water velocity and resuspension or deposition results of the previous 2 components with descriptions of the reaction and intermedia transfer processes that affect a contaminant. The transfer processes include sorption, exchange between the atmosphere and the dissolved phase in the water column, and exchange between the dissolved phase in the water column and the sediment bed pore water. The reaction processes include speciation, precipitation or dissolution, and biotic and abiotic degradation.

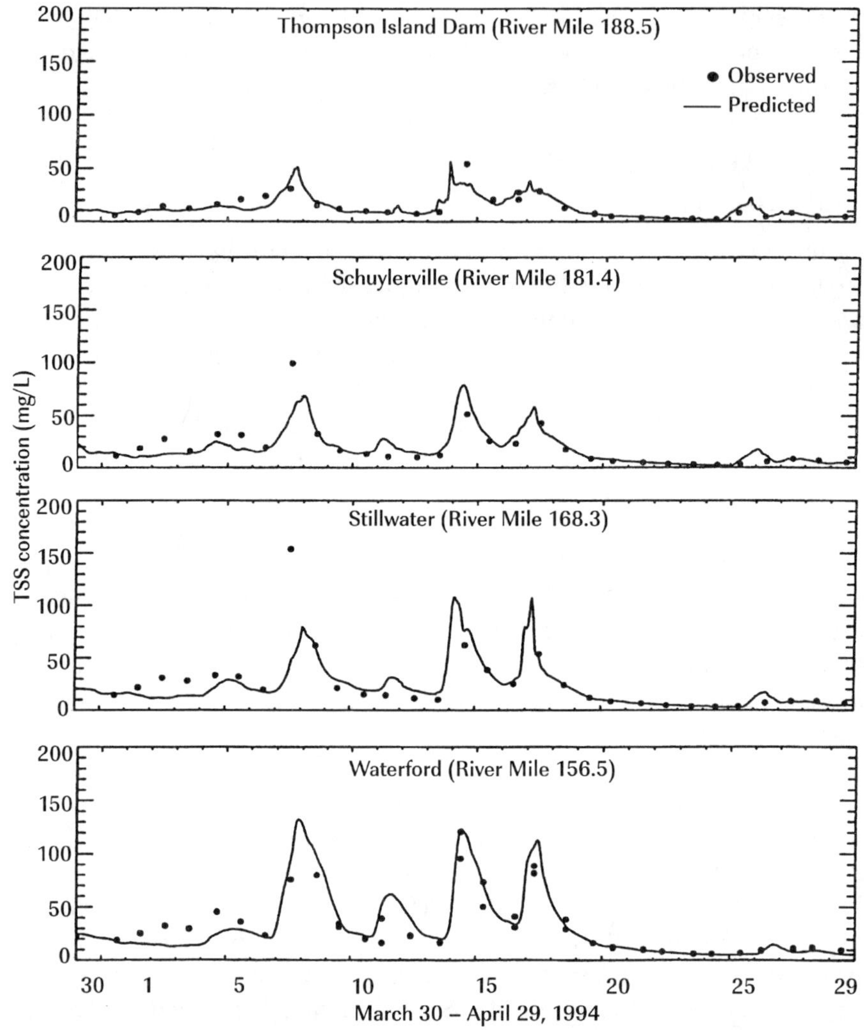

Figure 5-5 Observed and model-predicted suspended sediment concentrations at 4 locations in the Upper Hudson River, 1994 spring flood

Sorption is described as a reversible equilibrium process, most commonly by a partition coefficient. Laboratory experiments indicating relatively slow desorption rates (e.g., Pignatello and Xing 1996) or reductions in the bioavailability of sorbed contaminants with sediment aging (e.g., Loonen et al. 1997) suggest that the assumptions of equilibrium and reversibility may be inaccurate. The significance of any inaccuracies has not been rigorously evaluated in this area of ongoing research. Nonetheless, reversible equilibrium has been used with some success for more than 20 years.

Speciation is described using chemical equilibrium models such as MINTEQA2 (Allison and Perdue 1994), MINEQL+ (Schecher and McAvoy 1995), and WHAM (Tipping 1994). These models provide accurate estimates of speciation and precipitation or dissolution of inorganics, so long as the major chemical species in the water are well known and kinetic limitations do not influence speciation (i.e., thermodynamic equilibrium is assumed in these models). These models are computationally intensive, and their use in long-term NR simulations involves compromises between the frequency at which the speciation is updated and the accuracy of the solution. In addition, accurate application of these models to natural soils and sediments is quite difficult, since calibration of contaminant binding to heterogeneous sites is not straightforward.

Chemical and biochemical reactions that create or destroy a contaminant of concern are described in simple fashion. A second-order kinetic expression is used in which the reaction rate is proportional to the concentration of the contaminant and the concentration of another reactant such as the hydroxyl ion or bacteria. In the case of a biological reaction, the organism responsible for the reaction is not modeled. An organism concentration is input to the model, either as a constant or as a time function. This also may be the case for a chemical reactant, if a chemical equilibrium model is not incorporated within the modeling framework.

Because the influence that site characteristics have on many chemical and biochemical reactions is not well understood (Boethling and Alexander 1979; Lartiges and Garrigues 1995), it is common to use simple first-order reaction rates that are defined from laboratory experiments or from model calibration (e.g., Dilks et al. 1993; Tell and Parkerton 1997). The accuracy of such descriptions depends on how well the model describes all other processes: the greater the number of parameters that must be adjusted during calibration, the more uncertain the model predictions.

A volatilization mass transfer rate constant is used to compute air–water exchange. It is typically calculated from empirical formulations dependent on the molecular diffusivity of the contaminant, water velocity, water column depth, and wind velocity (e.g., Mackay and Yeun 1983; Rathbun 1990). The equations are fairly robust and yield accurate volatilization rates. The principal weakness of existing models is their inability to describe volatilization losses that occur at waterfalls. Such losses can be important for volatile chemicals (McLachlan et al. 1990).

Transfer at the water–sediment bed interface refers to the movement of contaminant between the water column and the sediment pore water. The models describe this process using a diffusion equation. A number of phenomena affect the transfer, including bioturbation, hydrodynamic pumping due to pressure gradients, and advection to or from groundwater. As a result, the transfer rate is site specific and varies temporally at a site (e.g., Riedel et al. 1988; Gill et al. 1999). The diffusion model is an obvious simplification. In cases where water–sediment bed transfer is an important mechanism, the accuracy of the model depends on the existence of data for calibration of this transfer rate constant.

We will again use the Hudson River model (Quantitative Environmental Analysis 1998; Connolly et al. in press) to illustrate model application. Daily values of river flow and water depth for the period 1977 to 1991 were obtained from the hydrodynamic model. Rates of resuspension and deposition were obtained from the sediment transport model. The sorption partition coefficient was determined from an analysis of dissolved and particulate PCB measurements taken by the USEPA as part of a field data collection program (USEPA 1997). A 20 °C organic C normalized partition coefficient (K_{OC}) of $10^{5.6}$ L kg^{-1} organic C was used in the model. The volatilization rate constant was calculated from two film theory using a Henry's law constant of 2×10^{-4} atm-m^3 mol^{-1}, a liquid film mass transfer coefficient calculated using the O'Connor-Dobbins reaeration equation, and a gas film mass transfer coefficient fixed at 100 m^{-1}. The vertical diffusion of PCBs between the pore waters of adjacent sediment segments was modeled using a diffusion coefficient of 1 $cm^{2\ -1}$. A time-dependent mass transfer coefficient was used to model the exchange of PCBs between the pore water and the water column. This coefficient was established from matched water column and sediment PCB data. It varied from about 3 cm/d in winter to 12 cm/d in early summer. PCBs entering from upstream were estimated from an extensive database of water column PCB and flow measurements extending from 1977 to 1998.

Figure 5-6 compares temporal profiles of calculated and observed average PCB_{3+} (PCBs with 3 or more Cl atoms) concentrations in surface (0 to 5 cm) cohesive and noncohesive sediments of the Thompson Island Pool, a 6-mile reach just downstream of the General Electric facilities that were the original sources of the PCBs. Cohesive sediment PCB_{3+} levels declined from 105 ppm in 1977 to 20 ppm in 1991 and 14 ppm in 1998, declines of 80% and 87%, respectively. The model closely reproduces this trend, predicting 18 ppm in 1991 and 10 ppm in 1998, declines of 83% and 90%, respectively. The concentrations computed by the model lie within the uncertainty bars shown on the plot, which indicate ± 2 standard errors of the mean. Thus, there is no statistically significant difference between the model and the data.

The noncohesive sediment PCB_{3+} levels declined from 40 ppm in 1977 to 12 ppm in 1991 and 7 ppm in 1998, declines of 70% and 83% respectively. This downward trend is slower than that of the cohesive sediments. The model reproduced the trend, computing concentrations of 11 ppm in 1991 and 8 ppm in 1998. Thus, the model accounts for the difference in the trend between the cohesive and noncohesive sediments, as well as the absolute concentration drops between 1977 and 1998.

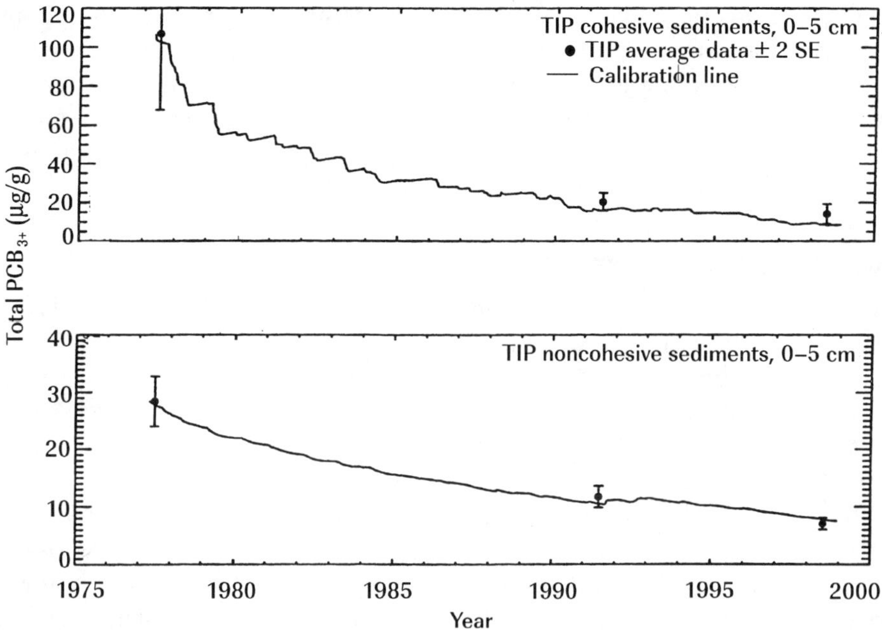

Figure 5-6 Averages of observed and model-predicted PCB concentrations in surface cohesive and noncohesive sediments in the Thompson Island Pool (TIP) of the Upper Hudson River

SUBSURFACE BIOCHEMICAL TRANSPORT MODELS

Subsurface biochemical transport models describe the movement and degradation of contaminants in aquifers. These models can be valuable tools in the evaluation of NR and in the prediction of risks associated with contaminants in aquifer ecosystems. To the casual observer, modeling subsurface NR may appear simple in comparison to modeling NR processes in surfacewater environments. However, while there are typically fewer processes to be accounted for in subsurface ecosystems (e.g., insignificant sediment transport), the underground systems are difficult to characterize with regard to distributions of geologic stratigraphy and its effects on water flow and contaminant partitioning. In the field, water flow and contaminant partitioning control contaminant transport and significantly affect biodegradation rates. Thus, significant effort is often required to accurately represent these complex and often difficult-to-characterize subsurface systems.

A typical subsurface NR-modeling task would involve 2 distinct modeling steps: groundwater flow modeling and reactive contaminant transport modeling. At some field sites, the transport and reaction of some system components might significantly modify the properties of porous media itself. For example, the introduction of oxygen into a zone of reduced Fe may cause rapid precipitation of oxidized Fe during the initiation of an active bioremediation effort. This precipitation event within the aquifer might grossly reduce the permeability and porosity of the porous media. Another case where interactions of this type can be important is when rapid microbial growth during an in situ bioremediation effort might significantly reduce the permeability and porosity of the porous media (Taylor and Jaffe 1990; Clement et al. 1996b, 1999). However, for most NR modeling applications, the groundwater flow and the reactive transport may be considered as uncoupled processes.

Modeling groundwater flow

The subsurface groundwater flow system is dynamically linked to the hydrological cycle through various natural or artificial recharge processes. As part of the hydrological cycle, groundwater is always in motion from regions of recharge to discharge points such as lakes, rivers, or oceans. The role of a groundwater flow model is to characterize the balance of aquifer withdrawal or recharge events so that changes in local groundwater flow rates and changes in water levels can be predicted.

In 1856, Henry Darcy performed experimental investigations on the flow of water through homogeneous, saturated porous media and developed what is known as the "Darcy formula" (Bear 1979). This formula is the basic law that governs the flow of water through porous media. When combined with a mass balance on water, the Darcy formula creates the basis of the groundwater flow model. It should be noted that Darcy's law is applicable for both saturated and unsaturated (vadose zone) groundwater flow systems. However, in this book, only saturated flow will be reviewed because many NR modeling applications involve prediction of contaminant fate in the saturated groundwater region.

Substituting the Darcy formula into the fluid mass balance equation and performing a water material balance around a control volume, the partial differential equation (Equation 5-1) that governs the flow of groundwater in multidimensional saturated porous media can be written as (Freeze and Cherry 1979; Domenico and Schwartz 1990):

$$S_s \frac{\partial h}{\partial t} = \frac{\partial}{\partial x}\left(K_x \frac{\partial h}{\partial x} \right) + \left(K_y \frac{\partial h}{\partial y} \right) + \frac{\partial}{\partial x}\left(K_z \frac{\partial h}{\partial z} \right) + q_s \qquad \text{(Equation 5-1)}$$

where

- h = hydraulic head [L]
- S_s = specific storage coefficient [L^{-1}]
- q_s = fluid sink or source term
- K_x, K_y, and K_z = principal components of the hydraulic conductivity [LT^{-1}]

It is assumed that the aquifer is represented by an orthogonal coordinate system whose axis directions are aligned along the principal direction of hydraulic conductivity. The distribution of groundwater head h (x, y, z, t) in a specified flow domain is obtained by solving Equation 5-1 with appropriate initial and boundary conditions. From the hydraulic head distribution, a groundwater transport velocity field can be estimated using this expression:

$$v_i = -\frac{K_{ii}}{\phi}\frac{\partial h}{\partial x_i} \qquad \text{(Equation 5-2)}$$

where

v = groundwater transport velocity $[LT^{-1}]$

ϕ = matrix porosity

Analytical models may be used to solve Equation 5-2 for specific initial and boundary conditions (Bear 1979). However, analytical models usually have limited applications because they are usually derived for a specific set of conditions. Several general purpose, finite-element and finite difference codes are available for solving Equation 5-1 under various initial and boundary conditions. Among available public domain groundwater flow computer codes, MODFLOW (McDonald and Harbaugh 1988) is the most popular and widely used. This is because 1) MODFLOW is a well-tested public domain code and 2) several commercial pre- and post-processing graphic-user interfaces (GUIs) are available for the MODFLOW code (e.g., GMS, VisualModflow, and Groundwater vistas) which facilitate integrating conceptual site models into MODFLOW's numerical modeling framework. Hence, MODFLOW is a useful and easy-to-use tool for solving saturated aquifer flow problems at NR sites.

Modeling reactive transport

Analytical models

Use of deterministic, single-species, advection–dispersion models for analyzing simple reactions in porous media has been well documented in the literature. The single-species models are relatively simple and, in some special cases, are amenable to analytical solutions. Ogata (1958) and Bear (1960) derived analytical solutions to contaminant transport equations for 1-dimensional problems. Van Genuchten and Alves (1982) and Toride et al. (1995) compiled various analytical solutions available for solving 1-dimensional solute transport equations. Beljin (1991) reviewed analytical solute transport models for 3-dimensional groundwater systems. However, in all of the above references, the fundamental partial differential equations represent the transport of either a nonreactive species or a single reactive species (Bear 1979).

To model practical natural attenuation problems, typically we need to solve coupled transport equations that involve multiple reactive species. For example, chlorinated solvent contaminants such as tetrachloroethylene (PCE) and trichloroethylene (TCE) are known to degrade and produce daughter products such as dichloroethylene (DCE) and vinyl chloride (VC). One of the most challenging aspects of managing a PCE- or TCE-contaminated site is the fact that the degradation daughter product VC is more toxic than the parent contaminant. Simple analytical models that predict the transport of single reactive species cannot be used to describe the formation of daughter products and to quantify their adverse effects. In the literature, only a few analytical solutions to multispecies reactive transport equations are found. Van Genuchten (1985) used the Laplace transform method to derive a 1-dimensional analytical solution to a 4-species transport problem that was coupled by first-order reactions. Lunn et al. (1996) used the Fourier transform method to derive an analytical solution to 3 species transport equations that were coupled by first-order reactions. Khandelwal and Rabideau (1999) presented an analytical solution to a sequentially decaying reaction chain (parent and 3 daughter products) influenced by linear nonequilibrium sorption; they also demonstrated the application of the new solution for evaluating permeable reactive barrier designs. However, almost all multispecies analytical models available in the literature can be used only for solving 1-dimensional problems, which greatly limits their use in realistic field-scale situations.

More recently, Sun et al. (1999) presented a novel transformation format to solve reactive transport equations that are coupled with sequential first-order reactions. The method is general enough to solve the reactive transport of any number of sequential reactive species in 1, 2, or 3 dimensions. Sun and Clement (1999) proposed a numbering scheme for systematically representing and solving multiple serial–parallel reactions, and they presented 2- and 3-dimensional validation results for the solution strategy under various initial and boundary conditions.

Clement (1999) presented a generalized solution methodology for deriving analytical solutions to multispecies transport equations coupled with a set of multiparent, serial, parallel, converging, diverging, and/or reversible first-order reactions. This method is also flexible for solving 1-, 2-, or 3-dimensional advection–dispersion equations that are coupled with any type of first-order reactions. They also presented derivations to demonstrate how the Sun et al. (1999) transformation format for sequential reaction chains, which was previously presented without any fundamental analysis, can be derived using linear transformation principles. However, a major limitation in these solutions is that the method cannot be used for solving the transport of species that have different retardation values or that exhibit nonequilibrium partitioning.

Analytical models are typically useful at the site-investigation stage for performing preliminary NR screening simulations to evaluate the impacts of contaminant plumes. Several such screening models are available in the literature, and the most popular ones used for NR design applications are the BIOSCREEN (Newell et al. 1996) and BIOCHLOR (Carol et al. 1999) tools. The BIOSCREEN decision support system is a Microsoft Excel spreadsheet–based screening tool for simulating the natural attenuation of dissolved hydrocarbons at fuel release sites. The BIOCHLOR natural attenuation model simulates chlorinated solvent (both ethene and ethane series) NR using an Excel spreadsheet–based interface similar to BIOSCREEN's. Both these models are public-domain tools, and the basic spreadsheets and the user manuals can be directly downloaded from various USEPA websites, e.g., http://www.epa.gov/ahaazvuc/bioscreen.html and www.epa.gov/ahaazvuc/biochlor.html.

Numerical models

Several public-domain numerical codes are also available for modeling field-scale transport in natural porous media (Konikow and Bredehoeft 1978; Voss 1984; Zheng 1990). However, none of these codes are capable of simulating coupled multispecies reactive transport, especially when the reactions are mediated by complex microbial systems.

Numerous researchers have numerically modeled multispecies, bioreactive transport in 1-dimensional soil columns (Molz et al. 1986; Zysset et al. 1994; Clement et al. 1996a, 1998). Chilakapati (1993) presents details of a 3-dimensional code for modeling reactive transport in a rectangular domain. De Blanc et al. (1996) modified the multiphase code UTCHEM to model multispecies bioreactive transport. Waddill and Widdowson (1998) present a 3-dimensional sequential electron acceptor model for modeling bioremediation of lighter-than-water nonaqueous phase liquid (LNAPL)-contaminated aquifers. Steefel and Yabusaki (1997) present a multidimensional code, OS3D, for modeling multicomponent reactive transport in groundwater systems. Steefel and MacQuarrie (1996) present an excellent review of different numerical approaches available for modeling reactive transport in porous media.

For in situ bioremediation, Rifai et al. (1987) were the first to present a multidimensional model for analyzing the natural fate and transport of hydrocarbon plumes in subsurface environments. This pioneering effort produced a practical computer code, BIOPLUME-II, for analyzing field-scale bioremediation scenarios. BIOPLUME-II uses a modified version of the U.S. Geological Survey (USGS) transport code MOC to simulate the 2-dimensional reactive transport of fuel-hydrocarbon and oxygen plumes. The code was recently updated to BIOPLUME-III, which allows for simulation of alternate TEAs such as nitrate, iron, and sulfate, and also simulates fuel hydrocarbon destruction (Rifai et al. 1998). Although these modifications are useful, the code still has limited applicability because, in BIOPLUME-III, depth-averaged 2-dimensional transport is assumed, bacterial growth is neglected, and only hydrocarbon kinetics are considered. Clement et al. (1997) recently developed a more comprehensive 3-dimensional reactive transport model, RT3D, which alleviates several of these practical limitations. Application of RT3D for several contaminant transport problems is discussed in Clement et al.

(1997). The code solves the coupled partial differential equations that describe the reactive flow and transport of multiple mobile and/or immobile species in 3-dimensional saturated groundwater systems. RT3D is a generalized multispecies version of the USEPA transport code MT3D (Zheng 1990).

The general macroscopic equations solved by the RT3D computer code are written as follows:

$$\frac{\partial C_k}{\partial t} = \frac{\partial}{\partial x_i}\left(D_{ij}\frac{\partial C_k}{\partial x_j}\right) - \frac{\partial}{\partial x_i}(v_i C_k) + \frac{q_s}{\phi}C_{s_k} + r_c - r_a + r_d \qquad \text{(Equation 5-3)}$$

$$\left(\frac{d\tilde{C}_{im}}{dt}\right) = \tilde{r}_c + r_a - r_d \qquad \text{(Equation 5-4)}$$

where

- k = 1,2,...m
- im = 1,2,...(n – m)
- n = total number of species
- m = total number of aqueous-phase (mobile) species (thus, n – m = total number of solid-phase or immobile species)
- C_k = aqueous-phase concentration of the kth species [ML^{-3}]
- $\tilde{C}_{im}$ = solid-phase concentration of the im^{th} species [MM^{-1}]
- D_{ij} = hydrodynamic dispersion coefficient [L^2T^{-1}]
- v = pore velocity [LT^{-1}]
- q_s = volumetric flux of water per unit volume of aquifer representing sources and sinks [T^{-1}]
- C_s = concentration of source–sink [ML^{-3}]
- r_c = reaction rate that describes the mass of the species removed or produced per unit volume per unit time [ML^3T^{-1}]
- $\tilde{r}_c$ = reaction rate at the solid phase [$MM^{-1}T^{-1}$]
- r_a = attachment (or adsorption) rates that describe the kinetic exchange of the transported species between aqueous and solid phases [$ML^{-3}T^{-1}$]
- r_d = detachment (or desorption) rates that describe the kinetic exchange of the transported species between aqueous and solid phases [$ML^{-3}T^{-1}$]

Numerical solution procedure

The RT3D code was developed to solve the multispecies reactive transport Equations 5-3 and 5-4. The code uses a reaction operator-split (OS) numerical strategy to solve any number of coupled transport equations (of the form of Equations 5-3 and 5-4). Previously, Walter et al. (1994) successfully used a similar OS approach to solve multicomponent transport with geochemical reactions. Clement et al. (1996a) used the OS strategy to solve a biologically reactive flow problem in a radial system. Valocchi and Malmstead (1992) and Kaluarachchi and Morshed (1995) have noted that splitting the reaction terms using the standard OS strategy may have numerical limitations. They recommended an improved, alternating OS strategy that may yield more accurate numerical results. However, Barry et al. (1995) states that the improvement provided by the alternating OS may not be applicable for multicomponent nonlinear problems. They also demonstrated the efficiency of the standard OS approach by solving a 2-species reactive transport problem. In this book, we use a standard OS strategy, similar to the one used by Zheng (1990), to develop a general numerical solution scheme for solving the coupled partial/ordinary differential Equations 5-3 and 5-4.

The USEPA code MT3D uses a similar OS approach to solve a mathematical model of the form of Equation 5-3, which describes single-species transport with first-order reaction (Zheng 1990). RT3D uses all original advection–dispersion and source–sink mixing routines available in MT3D for solving the advection–dispersion problem. These routines are invoked by RT3D multiple times to compute the transport of multiple mobile species. However, the original MT3D-reaction solver (an explicit solver) was replaced by a new reaction routine that has an improved implicit reaction solver. Appropriate code modifications were also implemented to input multiple initial and boundary conditions and the multispecies reaction information.

The logical steps involved in the numerical solution procedure are illustrated in Figure 5-7. As shown in Figure 5-7, the use of the OS strategy helps solve the complex coupled reactive transport system in a modular fashion. The solution algorithm initially solves the advection, dispersion, and source–sink mixing steps for all mobile components for a transport time step Δt. The length of the transport step is restricted by the constraints posed from the advection, dispersion, and source–sink mixing solvers (Zheng 1990). After solving the transport equations, the coupled reaction equations are solved implicitly by using multiple reaction-time steps. Computation of the required reaction-time step sizes to precisely integrate the differential equations is automated within the differential equation solver. Use of this modular OS approach for solving the reactive transport problem facilitates representation of different contaminant transport systems through a set of preprogrammed reaction packages. Further, other user-defined reaction kinetics may also be easily incorporated into the simulator using a user-defined reaction package (Clement 1997). Currently, the RT3D computer code supports 7 reaction packages for modeling different types of reactions (Clement 1997; Clement and Jones 1997). Details of several of these reaction packages, which could be used directly for NR modeling purposes, are described below.

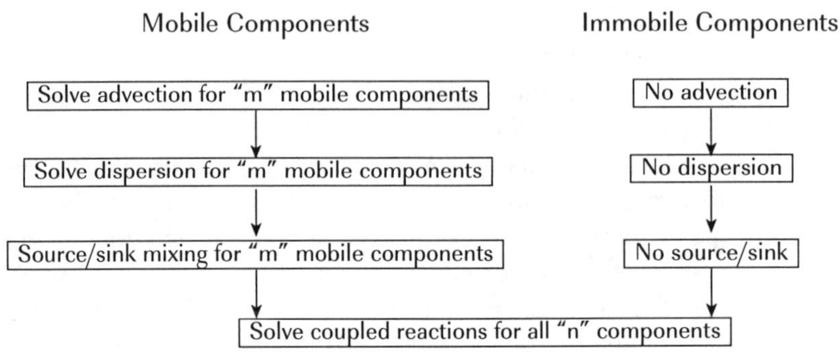

Figure 5-7 Numerical solution procedure

Numerical modeling of natural remediation of fuel hydrocarbon plumes

Several million underground storage tanks (USTs) are used throughout the U.S. to store petroleum and other hazardous chemicals. USEPA estimates that tens of thousands of these tanks and their piping have leaked or are currently leaking, discharging hazardous material into soil and associated groundwater environments. A vast majority of the leaking UST sites have discharged or are discharging some form of petroleum products initially as nonaqueous phase liquids (NAPLs). When groundwater comes in contact with these NAPLs, it typically becomes contaminated with benzene, toluene, ethylbenzene, and xylene (BTEX) compounds because these petroleum species are the most soluble in water. Therefore, dissolved BTEX compounds are typically the most common groundwater contaminants associated with spilled or leaked fuels such as gasoline, diesel, and jet fuel. As an additional concern, methyl tertiary butyl ether (MTBE), a volatile gasoline oxygenate added to reduce air pollution, has been detected at many sites. MTBE is very soluble, slowly biodegraded, and potentially carcinogenic; thus, for more recently fuel-contaminated sites, MTBE as well as BTEX compounds may be regulated contaminants of concern (Landmeyer et al. 1998).

The need to remediate fuel-contaminated sites has led to the development of a new remediation technology, now known as "bioremediation." In the early years, researchers were primarily interested in developing methods to stimulate active bioremediation under aerobic or denitrification conditions. However, recent field studies have identified that natural attenuation processes (including biodegradation) may provide a low-cost alternative for managing these sites. Rice et al. (1995) studied more than 1,000 fuel hydrocarbon plumes and found that most subsurface formations possess the inherent potential to naturally biodegrade fuel hydrocarbons. Wiedemeier et al. (1995) developed detailed protocols for implementing the NR technology at fuel-contaminated sites.

Because factors that control fuel hydrocarbon decay depend on multiple coupled geochemical, biological, and transport processes, it is difficult to estimate the overall effectiveness of NR at field sites. Therefore, the natural attenuation potential of petroleum sites must be analyzed on a site-specific basis, and multispecies reactive solute transport models can significantly aid in such analysis. In the literature, 3 approaches are described for modeling the fuel hydrocarbon degradation processes that are coupled to multiple electron acceptors. The first approach uses an instantaneous reaction model (Rifai et al. 1987, 1998). In this approach, the degradation rate of contaminants (such as BTEX compounds) and geochemical indicators (such as oxygen, nitrate, manganese, iron, or sulfate) is assumed to occur at very high rates and hence are modeled as instantaneous reactions. The second approach is the first-order kinetic model (Clement et al. 1997; Lu et al. 1999); in this approach, the reaction rates are assumed to be first-order with respect to hydrocarbon concentration. A rate-limiting term and an inhibition term, respectively, are used to simulate the presence and absence of different electron acceptors. The third approach is a detailed Monod-kinetic model in which the growth and transport of microbial populations are tracked to describe various reaction mechanisms (Essaid et al. 1995; Clement et al. 1996a, 1997; Waddill and Widdowson 1998). This microbial-growth kinetic-dependency concept is more complex to simulate numerically. It also requires extensive field data for calibration and hence is difficult to apply at field-scale sites.

A general reactive transport solver, such as the RT3D code (Clement 1997), can be used to build and solve all 3 types of models discussed above. However, based on available literature, it is unclear which of the 3 modeling approaches is the most appropriate for modeling field-scale natural attenuation systems. While the more complex models seem to better reflect the underlying biological processes, they tend to have too many model parameters that are difficult to estimate. For example, it is not cost effective to measure the growth and transport parameters of different microbial species under field conditions. Moreover, without extreme care, experimental kinetic values measured in laboratory reactors have little use for field-scale applications. On the other hand, simpler models, such as the instantaneous reaction ones, tend to oversimplify the underlying biogeochemical reaction conditions and thus pose the risk of incorrectly modeling critical processes. For example, it may not be realistic to model methanogenesis, which is a slow and complex fermentation process mediated by multiple microbial populations, as a single instantaneous reaction. More field-scale modeling studies are required to develop guidelines for selecting appropriate models based on the scale and complexity of the problem and the contaminant type. In addition, management questions to be addressed by the model vary significantly from site to site. Presently, it appears that for understanding and managing well-documented contaminants such as fuel hydrocarbons, which are found to degrade ubiquitously (Rice et al. 1995), complex modeling efforts may not be required.

In the development of hydrocarbon natural attenuation models, another important difficulty arises from the complexities associated with Fe reduction reactions. Unlike other mobile electron donors, iron in the form of Fe(III) is always present in the immobile oxide form attached to the soil matrix. This introduces some unique difficulties. To model Fe reduction, the first question to answer is this: Will it be possible to quantify the bioavailable Fe(III), which would provide a reasonable estimate for the amount of Fe-based electron acceptor available for microbial degradation of the hydrocarbon? The second question to address is related to the geochemical stability of the reduced iron Fe(II) that is produced. It is well known that Fe(II) is a reactive species that can interact with other geochemical species to form several Fe(II)-containing mineral precipitates (Lovley 1991). Therefore, the concentration levels of Fe(II) observed in the field may not reflect the true Fe-reduction potential of the aquifer. In summary, our current understanding of field-scale processes related to the Fe system is limited, and thus more focused research efforts are needed to further quantify the Fe reactions in relation to NR. Similar arguments can also be made for other solid phase electron-acceptor species such as Mn.

Numerical modeling of natural remediation of chlorinated solvents plumes

Waste sites where the primary concern is chlorinated solvent plumes are common throughout North America. Most of these chlorinated solvent plumes originated from waste disposal pits where industrial solvents such as PCE and TCE were indiscriminately disposed. Recent demonstrations of natural degradation of petroleum hydrocarbon, in virtually all groundwater systems, have raised the prospects that chlorinated solvents might also be amenable to NR. Wiedemeier et al. (1997) presented a technical protocol that documents the conditions under which NR of chlorinated solvent may be feasible. The degree of NR will obviously depend on the biodegradation potential of the aquifer. Based on previous laboratory and field-scale results, Wiedemeier et al. (1997) report that representative first-order biodegradation rates for chlorinated solvents in the presence of aquifer materials may range from 0.00068 to 0.54 d^{-1} for PCE, 0.0001 to 0.021 d^{-1} for TCE, 0.00016 to 0.026 d^{-1} for DCE, and 0.0003 to 0.012 d^{-1} for VC. Obviously, there is a large range in the degradation kinetics, and site-specific information should be obtained to select appropriate values for these first-order rate coefficients.

Numerous field and laboratory studies have demonstrated that microorganisms can degrade chlorinated solvents (Bower et al. 1981; Freeman and Gossett 1989; Semprini et al. 1995; Klier et al. 1998). The most important process for the natural degradation of the more highly chlorinated solvents is the anaerobic reductive dehalogenation process (Wiedemeier et al. 1997). During this process, the chlorinated solvent is used as an electron acceptor, and Cl atoms are sequentially removed and replaced with H atoms. Under favorable environmental conditions, other biochemical processes, in addition to anaerobic decay, may also degrade the chlorinated organics. McCarty and Semprini (1994) indicate that dichloro- and monochloroethenes have a good potential for degradation through both direct or cometabolic aerobic pathways. Based on radiolabeled microcosm studies, Klier et al. (1998) have shown evidence for aerobic dichloroethene degradation. Based on this information, we formulated a conceptual model for describing all the biochemical reaction steps involved in dechlorination of various chlorinated solvent chemicals. In this model, the degradation reactions are assumed to be mediated by both aerobic and anaerobic dechlorination processes, as shown in Figure 5-8.

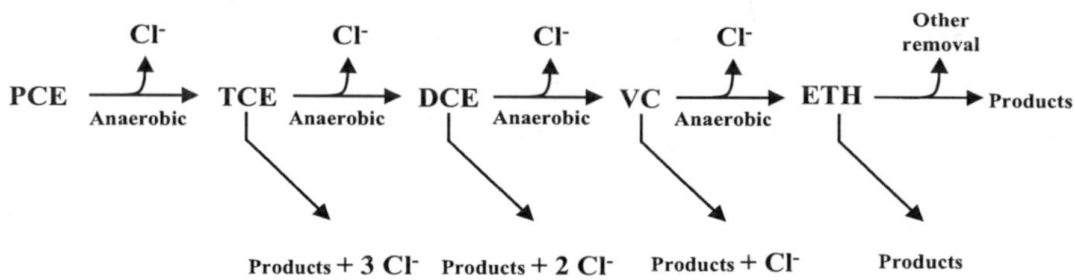

Figure 5-8 Conceptual biochemical model for biodegradation of chlorinated ethene (ETH) compounds

Assuming first-order biodegradation kinetics for every reaction step, the transport and transformation of PCE, TCE, DCE, VC, ethene (ETH), and chloride (Cl) can be simulated by solving the following set of partial differential equations (Equations 5-5 through 5-10):

$$R_P \frac{\partial[PCE]}{\partial t} = \frac{\partial}{\partial x_i}\left(D_{ij}\frac{\partial[PCE]}{\partial x_j}\right) \frac{\partial(v_i[PCE])}{\partial x_i} + \frac{q_s}{\phi}[PCE]_s - K_P[PCE] \quad \text{(Equation 5-5)}$$

$$R_T \frac{\partial[TCE]}{\partial t} = \frac{\partial}{\partial x_i}\left(D_{ij}\frac{\partial[TCE]}{\partial x_j}\right) \frac{\partial(v_i[TCE])}{\partial x_i} + \frac{q_s}{\phi}[TCE]_s + Y_{T/P} K_P[PCE] - K_{T1}[TCE] - K_{T2}[TCE] \quad \text{(Equation 5-6)}$$

$$R_D\frac{\partial[DCE]}{\partial t}=\frac{\partial}{\partial x_i}\left(D_{ij}\frac{\partial[DCE]}{\partial x_j}\right)\frac{\partial(v_i[DCE])}{\partial x_i}+\frac{q_s}{\phi}[DCE]_s$$
$$+Y_{D/T}K_T[TCE]-K_{D1}[DCE]-K_{D2}[DCE]$$ (Equation 5-7)

$$R_V\frac{\partial[VC]}{\partial t}=\frac{\partial}{\partial x_i}\left(D_{ij}\frac{\partial[VC]}{\partial x_j}\right)\frac{\partial(v_i[VC])}{\partial x_i}+\frac{q_s}{\phi}[VC]_s$$
$$+Y_{V/D}K_{D1}[DCE]-K_{V1}[VC]-K_{V2}[VC]$$ (Equation 5-8)

$$R_E\frac{\partial[ETH]}{\partial t}=\frac{\partial}{\partial x_i}\left(D_{ij}\frac{\partial[ETH]}{\partial x_j}\right)\frac{\partial(v_i[ETH])}{\partial x_i}+\frac{q_s}{\phi}[ETH]_s$$
$$+Y_{E/V}K_{V1}[VC]-K_{E1}[ETH]-K_{E2}[ETH]$$ (Equation 5-9)

$$R_C\frac{\partial[Cl]}{\partial t}=\frac{\partial}{\partial x_i}\left(D_{ij}\frac{\partial[Cl]}{\partial x_j}\right)\frac{\partial(v_i[Cl])}{\partial x_i}+\frac{q_s}{\phi}[Cl]_s$$
$$+Y1_{C/P}K_{P1}[PCE]+Y1_{C/T}K_{T1}[TCE]+Y1_{C/D}K_{D1}[DCE]$$
$$+Y1_{C/V}K_{V1}[VC]+Y2_{C/T}K_{T2}[TCE]+Y2_{C/D}K_{D2}[DCE]$$
$$+Y2_{C/V}K_{V2}[VC]$$ (Equation 5-10)

where

[PCE], [TCE], [DCE], [VC], [ETH], [Cl] = contaminant concentrations of various species [mg/L]

K_P, K_{T1}, K_{D1}, and K_{V1}, K_{E1} = first-order anaerobic degradation rates [d^{-1}]

K_{T2}, K_{D2}, and K_{V2}, K_{E2} = first-order aerobic degradation rates [d^{-1}]

R_P, R_T, R_D, R_V, R_E, R_C = retardation factors

$Y_{T/P}$, $Y_{D/T}$, $Y_{V/D}$, $Y_{E/V}$ = chlorinated compound yields under anaerobic reductive dechlorination conditions: 0.79, 0.74, 0.64, and 0.45, respectively

$Y1_{C/P}, Y1_{C/T}, Y1_{C/D}, Y1_{C/V}$ = yield values for chloride under anaerobic conditions: 0.21, 0.27, 0.37, and 0.57, respectively

$Y2_{C/T}, Y2_{C/D}, Y2_{C/V}$ = yield values for chloride under aerobic conditions: 0.81, 0.74, and 0.57, respectively

The yield values are estimated from the reaction stoichiometry and molecular weights. The anaerobic degradation of 1 mole of PCE would yield 1 mole of TCE; therefore, $Y_{T/P}$ equals the molecular weight of TCE / molecular weight of PCE (131.4 / 165.8 = 0.79). Note the reaction models presented assume that the biological degradation reactions only occur in the aqueous phase, which is a conservative assumption.

To apply this model to field-scale problems, we need to first assign values for the first-order rate constants, which are spatially variable parameters. Clearly, in the above model description, it is meaningless to use a (spatially) constant value for any of the rate parameters because several of the processes described in the model are mutually exclusive. For example, the anaerobic reductive dechlorination reaction cannot occur in highly aerobic regions; therefore, the rates associated with reductive dechlorination processes need to turn off in highly aerobic regions. Hence, all the rate parameters in the model must be specified as spatially variable parameters to turn appropriate reactions on or off in different regions. In a typical field model, the spatial extent of different regions can be delineated by using a conceptual site model for reaction zones (Clement et al. 2000). Figure 5-9 shows the conceptual model used to delineate reaction zones at a chlorinated solvent release site in Dover Air Force Base (DAFB), Delaware. The reaction-zone model was developed based on the groundwater geochemistry data and on the spatially changing anaerobic and aerobic degradation patterns observed at the site (Klier et al. 1999). The spatial distribution of the primary contaminants (PCE and TCE) and the degradation by-products (DCE, VC, and Cl) observed at the site are shown in Figures 5-10 and 5-11. The spatial patterns of other geochemical indicators such as oxygen and methane are shown in Figure 5-12. The field data presented in Figure 5-12 indicate the presence of anaerobic regions in the vicinity of the source area. The presence of the anaerobic zone is likely due to the oxygen demand posed by the fuel hydrocarbon and other biodegradable organic compounds that were co-disposed at or near the source areas. In addition, natural C in the sediment material might also be serving as a potential C source.

In a numerical modeling study for the DAFB site, the reaction-zone model described in Figure 5-9 was used as a guide to vary the reaction parameters from zone to zone (Clement et al. 2000). As shown in the figure, the site was assumed to have 4 distinct reaction zones: 2 anaerobic zones of different characteristics, a transition zone, and an aerobic zone. In the anaerobic zones, anaerobic reactions are

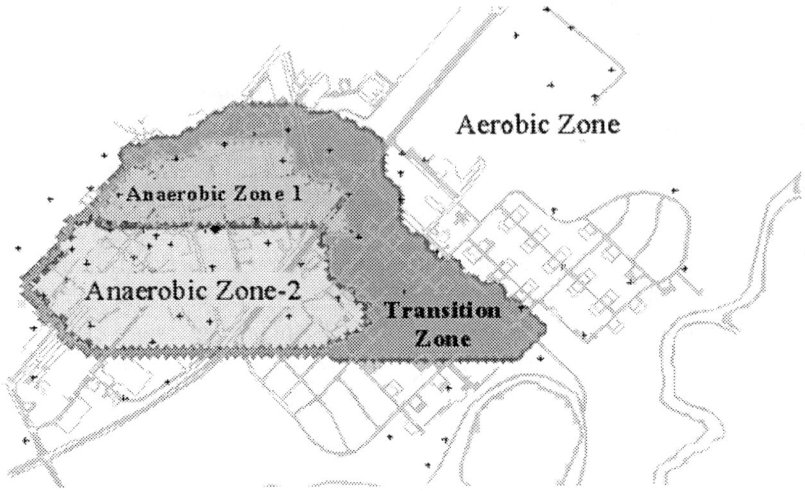

Figure 5-9 Conceptual model for delineating biological reaction zones at Dover Air Force Base, Delaware (reprinted from Clement et al. 2000 with permission from Elsevier Science)

assumed to be the predominant degradation reactions, and in the oxidized zone, aerobic reactions are assumed to be the predominant reactions. Also, within a given reaction zone, the parameters were assumed to remain both spatially and temporally constant over the entire simulation time. The values of reaction parameters estimated for different regions are summarized in Table 5-1.

It is important to note that the approach employed above for delineating the spatial extent of the reaction zones has several limitations. First, the approach is a subjective procedure that is based on available site-specific data. By imposing the field data directly into the model, the method tightly constrained the problem to a specific field condition and hence the model may not be applicable at other sites where such intensity of data is not available. Moreover, it is hypothesized that the reaction zones must have been formed by the co-disposed hydrocarbons (or other electron donors) rather instantaneously and that the zones remain stable during both calibration and forecasting time periods. Because almost all field projects are restricted by the availability of time and resources, the amount of data collected during typical NR evaluation projects will usually be insufficient to validate the stable reaction-zone hypothesis. Although this hypothesis seems to be a "pragmatic" engineering approximation for modeling chlorinated solvent natural attenuation problems, its validity should be carefully evaluated before the model is used for forecasting.

More research is clearly needed to develop procedures to automatically predict the formation of transient reaction zones based on geochemical considerations.

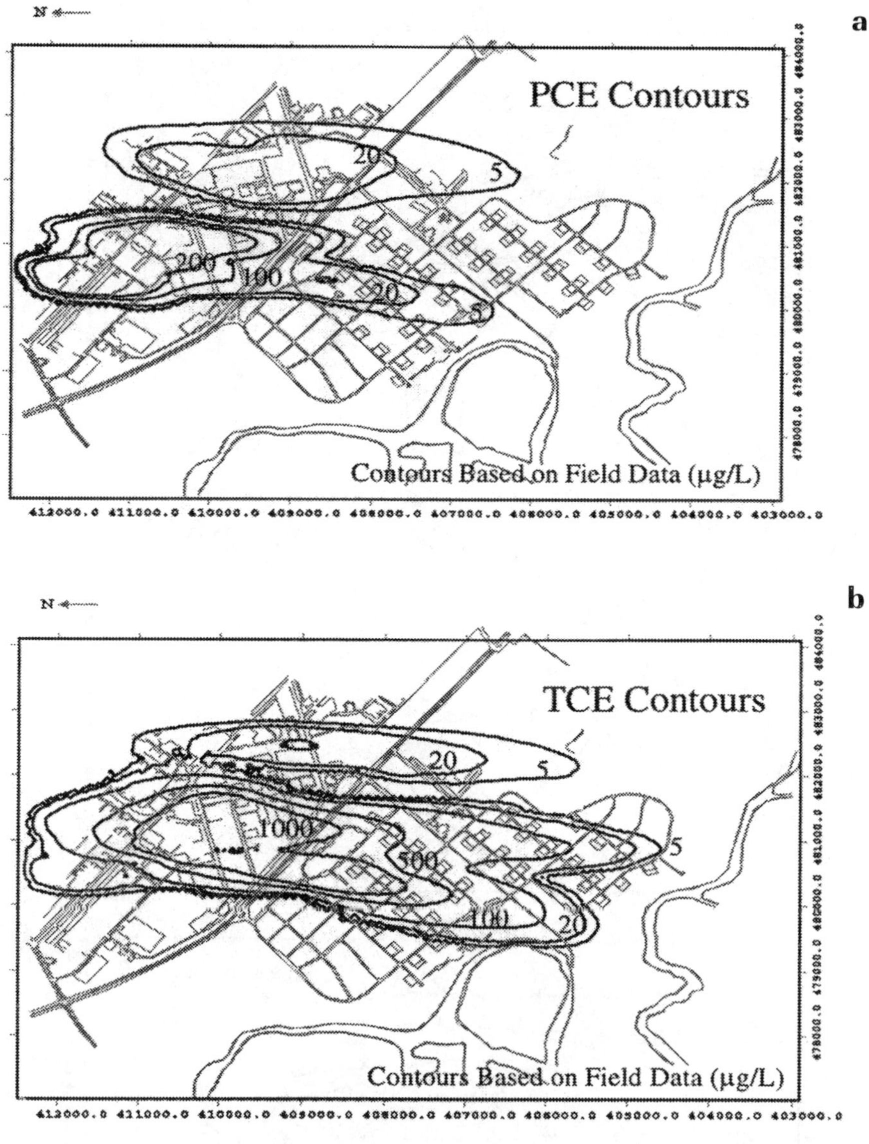

Figure 5-10 Field-measured dissolved PCE (a) and TCE (b) concentrations at Dover Air Force Base, Delaware (reprinted from Clement et al. 2000 with permission from Elsevier Science)

a

b

c

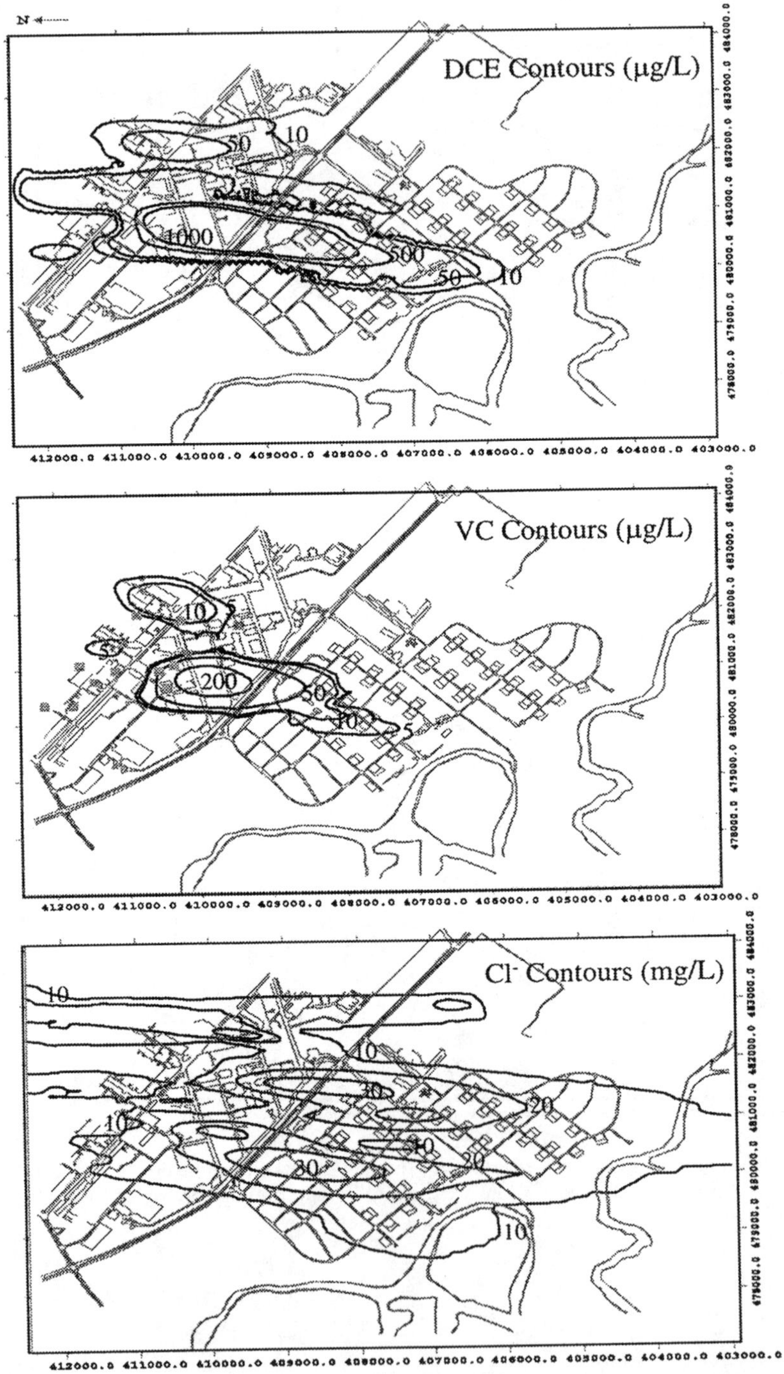

Figure 5-11 Field-measured dissolved DCE (a), VC (b), and Cl^- (c) concentrations at Dover Air Force Base, Delaware (reprinted from Clement et al. 2000 with permission from Elsevier Science)

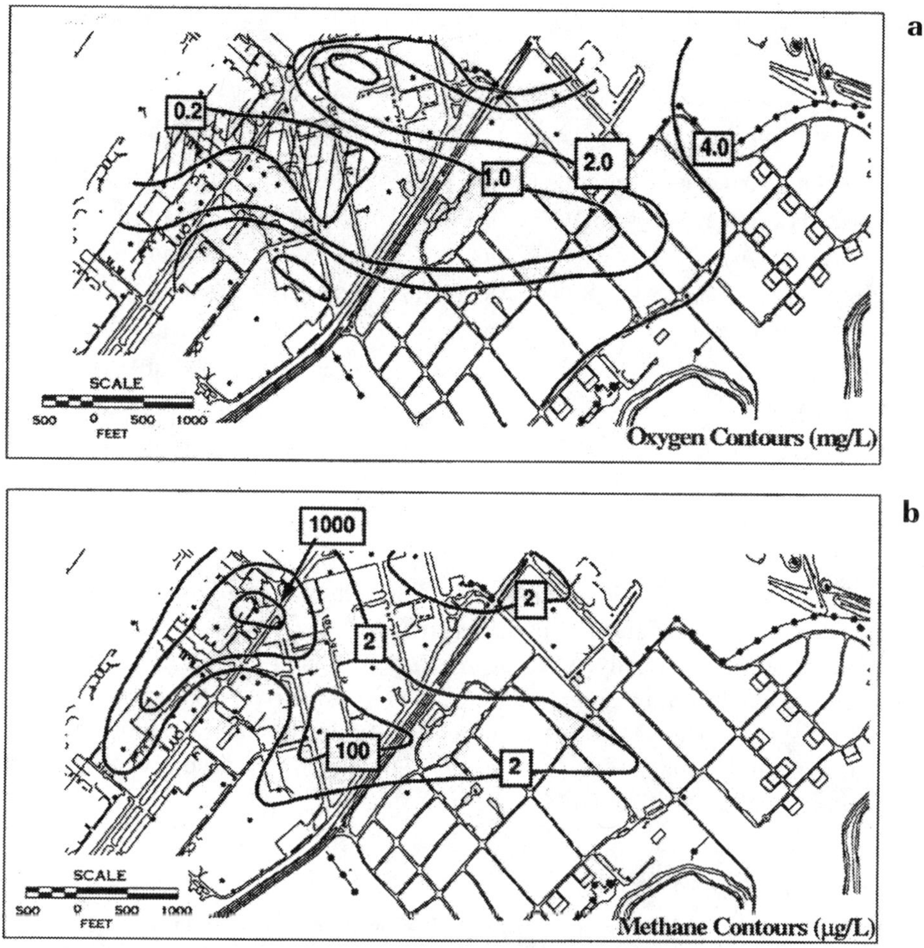

Figure 5-12 Field-measured dissolved oxygen (a) and methane (b) concentrations at Dover Air Force Base, Delaware (reprinted from Clement et al. 2000 with permission from Elsevier Science)

McNab and Narasimhan (1994) proposed a partial redox disequilibrium approach to model sequential chemical transformation in natural groundwater systems. They solved a hypothetical test problem to demonstrate the use of their model. However, direct application of such research models to field-scale bioremediation problems may be somewhat problematic because there is very little information available about the complex chemical interactions that may take place between inorganic chemicals as a result of the progress of the contaminant degradation reactions. In addition, such complex models will add a long list of additional unknown model parameters, which will be difficult to quantify under field conditions. Therefore, the feasibility of such models should be carefully considered before complexity is added

Table 5-1 Estimated degradation rate constants for Dover Air Force Base, Delaware, plumes (d^{-1})

First-order rate constant	Associated contaminant	Anaerobic zone: Zone 1	Anaerobic zone: Zone 2	Transition zone	Aerobic zone
K_P (anaerobic)	Tetrachloroethylene (PCE)	3.2×10^{-4}	4.0×10^{-4}	1.0×10^{-4}	0.0
K_{T1} (anaerobic)	Trichloroethylene (TCE)	9.0×10^{-4}	4.5×10^{-4}	1.125×10^{-4}	0.0
K_{T2} (aerobic)	TCE	0.0	0.0	0.4×10^{-5}	1.0×10^{-5}
K_{D1} (anaerobic)	Dichloroethylene (DCE)	8.45×10^{-4}	6.5×10^{-4}	1.625×10^{-4}	0.0
K_{D2} (aerobic)	DCE	0.0	0.0	1.6×10^{-3}	4.0×10^{-3}
K_{V1} (anaerobic)	Vinyl chloride (VC)	8.0×10^{-3}	4.0×10^{-3}	1.0×10^{-3}	0.0
K_{V2} (aerobic)	VC	0.0	0.0	0.8×10^{-3}	2.0×10^{-3}
K_{E1} (anaerobic)	Ethene (ETH)	2.4×10^{-2}	1.2×10^{-2}	0.3×10^{-2}	0.0
K_{E2} (aerobic)	ETH	0.0	0.0	0.4×10^{-2}	1.0×10^{-2}

to the biogeochemical reaction framework at the field scale. In addition to the limited field studies available for modeling and predicting complex organic or inorganic interactions, relatively recent studies on contaminant partitioning indicate that equilibrium assumptions about soil–contaminant partitioning may not be valid in some cases and that, particularly for complex sites, nonequilibrium partitioning models may need to be considered to improve the accuracy of long-term predictions of contaminant fate and transport.

Modeling nonequilibrium partitioning

Nonequilibrium contaminant partitioning to soils and sediments has the potential to significantly affect predictions of contaminant fate and transport in both surface and subsurface NR models. Many models assume equilibrium partitioning (EqP) between contaminants and their local surroundings. However, assumptions of equilibrium and reversibility may be inaccurate (Loonen et al. 1997) and may lead to gross overestimation of contaminant transport and release rates. In addition, nonequilibrium partitioning may affect contaminant bioavailability, which in turn may control the rate of biodegradation by microorganisms. Understanding the factors that affect partitioning and bioavailability of soil and groundwater contaminants is critical to accurately estimating contaminant persistence and NR.

Many chemical, physical, and microbiological properties of contaminated subsurface environments are reported to influence nonequilibrium partitioning of contaminants. Important soil properties that have been implicated include specific surface area, organic matter content and type, soil matrix porosity, and pore size distribution (Norris and Matthews 1994; Alexander 1995). Luthy et al. (1997) provided an excellent study on the role of soil and sediments (geosorbents) in the sequestration of hydrophobic organic contaminants (HOCs). Farrell and Reinhard (1994) classified geosorbents into 3 scales based on pore size distribution. As shown in Figure 5-13, macropores (pore diameter > 500Å) occupy the space between soil and sediment grains and will only fill with water when submerged. Advection and dispersion dominate mass transfer in macropores. Second, mesopores (20Å < diameter < 500Å) comprise the majority of pore space within soil and sediment grains. At environmentally relevant pressures and humidity, these pores fill with water because of capillary condensation (Werth and Reinhard 1997a), even when the macropores are in the vadose or unsaturated zone. Finally, micropores (diameter < 20Å) are the pores inside fine particles and are suspected to control slow desorption of HOC (Steinberg et al. 1987).

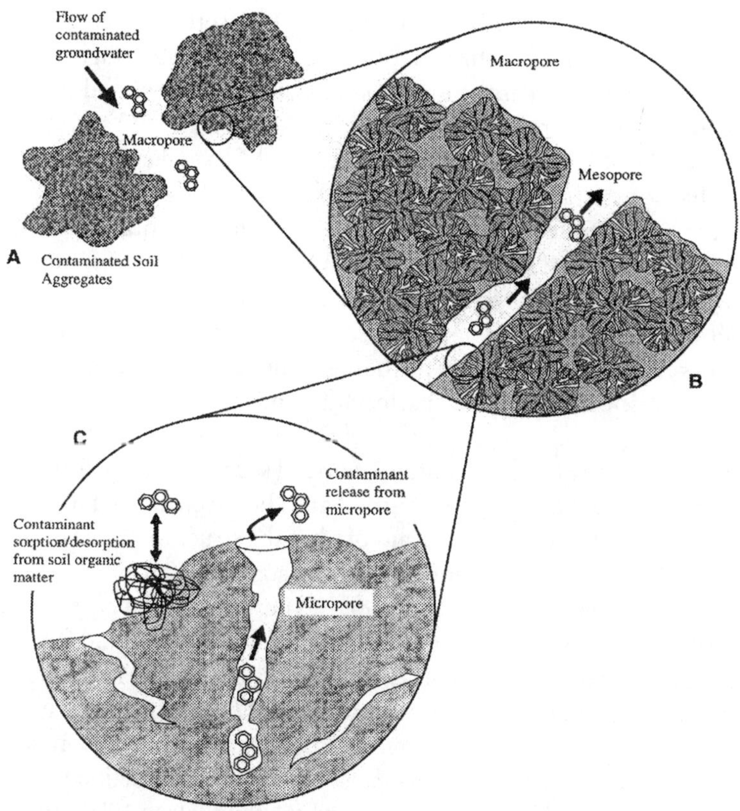

Figure 5-13 Conceptual schematic of processes that may control nonequilibrium partitioning of hydrophobic organic contaminants in soil matrices (A = Groundwater flow and contaminants in macropores [diameter > 500 Å] between contaminated soil aggregates. B = Contaminants leaving the mesopore [500 Å< diameter > 20 Å] to enter the bulk groundwater flow within the mesopores. C = Contaminant sorption or desorption with soil organic matter and contaminant release from a micropore [diameter < 20 Å] to a mesopore.)

The organic fraction of soils and sediments is responsible for the sorption of many compounds, particularly those that are hydrophobic (Chung and Alexander 1998; Cornelissen et al. 1998). Many polynuclear aromatic hydrocarbons (PAHs) and other nonpolar pollutants are sorbed primarily by the native organic matter found in soils and sediments (Fu and Alexander 1994; Laor et al. 1998).

Contaminant properties that can affect nonequilibrium partitioning are the hydrophobicity (typically expressed as the octanol–water partitioning coefficient [K_{OW}]) and the molecular diameter. Hydrophobic organic contaminants typically exhibit lower desorption rates in soil and aquifer solids than do their nonhydrophobic counterparts (Steinberg et al. 1987; Ball and Roberts 1992). Reduced bioavailability has also been attributed to the partitioning of contaminants onto soils over time (Alexander 1994; Cornelissen et al. 1998). Moreover, an additional and much less defined process, termed "contaminant sequestration," can also further reduce the bioavailability (Luthy et al. 1997; Nam and Alexander 1998). Sequestration is loosely defined as the process by which some, usually small, fraction of contaminant is entrapped within soil particles and aggregates (Pignatello and Xing 1996). The concept of contaminant sequestration has been used to explain the slow or incomplete biodegradation of pesticides (Alexander 1995) and PAHs in soils (Weissenfels et al. 1992; Fu et al. 1994; Hatzinger and Alexander 1995). It has been postulated that sequestration involves slow partitioning of hydrophobic contaminants into organic matrices or the slow diffusion of contaminants into remote portions of subsurface particles (Alexander 1994, 1995).

Through a process known as "aging," contaminants that are in contact with soils for long periods of time may have a higher sequestered fraction. Much empirical data describes the existence of aging phenomena, but the mechanisms of aging are still unknown (Steinberg et al. 1987; Pavlostathis and Jaglal 1991; Carroll et al. 1994; Kan et al. 1994). Aging may result in physical transport-related sequestration and/or strong chemical binding between contaminant and sorbents. While little evidence points to the true causes of aging, the term "aging" has been defined to exclude reactions that mineralize the contaminant or alter the structure of the original contaminant (Hatzinger and Alexander 1995; Burgos et al. 1996).

In micropores, steric-restricted diffusion can occur when the pore diameter is of the same order of magnitude as the diffusing molecule (Limbach and Wei 1990). Diffusion coefficients of contaminants in micropores can be much smaller than those measured in the bulk liquid (Satterfield et al. 1973; Smith 1986; Limbach and Wei 1990). The phenomenon of restricted diffusion can be explained in part by the fact that the constrained space of a pore causes the molecular friction coefficient to exceed the value that would be observed in the bulk fluid (Deen 1987), meaning that the molecular Brownian motion is significantly hindered by the pore walls. The theory of restricted diffusion of rigid spherical solutes in cylindrical pores has been examined and experimentally verified (Limbach and Wei 1990).

More empirically, many investigators have reported that release rates of organic contaminants from soils exhibit 2 distinct time phases: an initial period of rapid contaminant release (on the order of minutes) followed by a slow rate of change on the order of weeks to years (Wu and Gschwend 1986; Farrell and Reinhard 1994; Young and Weber 1995; Pignatello 1996). The initial fast rate of release has been more thoroughly studied; however, the later phase of slow contaminant release is often of more significance to the assessment of NR (which often takes years).

Sorption and desorption to solid surfaces also play a major role in controlling the fate of organic pollutants in an aquatic environment. Wu and Gschwend (1988) developed and experimentally verified a model of HOC sorption kinetics based on molecular diffusion and phase partitioning. Wu and Gschwend (1986) suggested the rate of contaminant solution–solid exchange should be described as a radial diffusive penetration of organic pollutants into solid porous natural particles. In this model, sorbate molecules are retarded by local EqP between mobile and immobile phases during diffusion through the pore fluids held in the interstices of natural soil aggregates.

Numerous other transport models have been proposed to represent the release of contaminants in soils (Wu and Gschwend 1988; Ball and Roberts 1992; Connaughton and Stedinger 1993; Farrell and Reinhard 1994; Pedit and Miller 1995; Arocha et al. 1997; Bai et al. 1997). These models provide a conceptual framework to describe controlling mechanisms in contaminant desorption and serve to connect theoretical understanding with practical observations. The contaminant release models can be classified into homogeneous and heterogeneous models, depending on assumptions regarding the nature of the soil structure. A homogeneous model assumes a uniform soil matrix, an assumption that is oversimplified and may lead to inaccuracies (Farrell and Reinhard 1994). Although some researchers have been successful in applying a homogeneous model to the uptake and desorption of organic contaminants (Wu and Gschwend 1988; Ball and Roberts 1992), the data in these experiments were measured over limited concentration ranges and short time periods and are thus of minimal value to NR modeling.

In contrast to homogeneous models, heterogeneous models are often based on quantifying the effects of soil pore size distribution, nonequilibrium desorption, and diffusion mechanisms on the rate of contaminant release from solids. Studies have shown that heterogeneous models can often fit the experimental data better in a wide contaminant concentration range for a much longer period of time. Werth and Reinhard (1997b) improved the homogeneous pore-diffusion model of Ferrell and Reinhard (1994) into a more complex triple porosity model. In the former work, the homogeneous model was accurate only for the first 10 minutes of desorption experiments; this short time scale is not useful for predictions applicable to NR. In the latter work, the triple porosity model accounted for steric hindrance and contaminant sorption in a radial diffusion model to describe hindered contaminant transport in micropores. The diffusion model successfully fit the slow desorption over at least 3 orders of magnitude in time and up to 3 orders of magnitude change in flux. Their work helped to clarify the potentially dominant mechanisms of slow HOC desorption in micro- and meso-pores, where both steric hindrance and sorptive forces may control mass transfer.

SUMMARY

Natural remediation is the reduction in contaminant concentration that naturally occurs as a result of contaminant diffusion and dispersion and as a result of natural degradation through biotic and chemical reactions in the environment. This chapter describes important aspects of the use and understanding of numerical models in the evaluation of NR scenarios. For this chapter, NR models were divided into 2 broad classes: surfacewater models and groundwater models. Each environment, and thus each model type, has its own unique characteristics that control the rate of contaminant transport, dilution, and degradation.

Surfacewater models have 4 components that must be integrated to produce an accurate representation of contaminant fate and transport:

1) Surfacewater hydrodynamics
2) Sediment transport
3) Contaminant sorption and desorption processes
4) Biota contaminant transformation and accumulation.

The hydrodynamics of surfacewater systems describe processes that affect the movement of water and subsequently the movement of soluble and suspended contaminants from one location to another. In these surfacewater systems, sediment transport is often a critical component of the model, since many contaminants are strongly sorbed to sediment particles. The deposition of sediment may significantly slow the spreading of a contaminant. In contrast, resuspension events such as floods or annual spring runoff may significantly increase, and may dominate, the contaminant transport properties in surface systems. Contaminant sorption and desorption processes and transformation rates often are very site and contaminant specific, and they may not follow idealized assumptions of equilibrium. Sorption and desorption processes control the distribution of a contaminant between the solid and aqueous phases of the system. Transformation rates describe the chemical or biological reactions that affect contaminant concentrations. Without laboratory testing, the sorption, desorption, and reaction parameters are often difficult to quantify. Therefore these processes, though very complex and the subject of significant research, are often modeled as simple equilibrium and first-order processes, respectively.

In contrast to surfacewater modeling, subsurface or groundwater modeling is slightly less complex because of the significantly lower sediment transport load in aquifer systems and the typically lower level of macrobiotic diversity in aquifer systems. Natural remediation modeling of groundwater systems requires knowledge of the subsurface hydrodynamics of groundwater flow in response to natural hydrostatic gradients. Groundwater flow patterns can also be influenced by wells used for irrigation or drinking water. Like surfacewater systems, groundwater contaminant fate and transport is affected by contaminant sorption and desorption and by chemical and biotic transformations. Significant research has improved our understanding of the chemical and biological processes that occur naturally in groundwater. These processes can decrease contaminant concentrations and may eventually limit further spreading of contaminant plumes.

Overall, modeling the NR of contaminants in surface and subsurfacewater systems has the potential to improve our estimates of the degradation rates and the extent of spreading of contaminants released into our environment. In providing these estimates, the real risks associated with NR of environmental contaminants may be better balanced against the costs of other remediation options.

REFERENCES

Ager DV. 1981. The nature of the stratigraphical record. New York NY: Wiley.

Alexander M. 1994. Biodegradation and bioremediation. San Diego CA: Academic Pr.

Alexander M. 1995. How toxic are toxic chemicals in soil? *Environ Sci Technol* 29:2713–2717.

Allison JA, Perdue EM. 1994. Modeling metal-humic interactions with MINTEQA2. In: Senesi N, Miano TM, editors. Humic substances in the global environment and implications on human health. Amsterdam, the Netherlands: Elsevier. p 927–942.

Ambrose Jr RB, Wool TA, Martin JL. 1993. The water quality analysis simulation program, WASP5, Part A: Model documentation. Athens GA: Center for Exposure Assessment Modeling (CEAM), U.S. Environmental Protection Agency (USEPA). p 209.

Arocha MA, Jackman AP, McCoy BJ. 1997. Numerical analysis of sorption and diffusion in soil micropores, macropores, and organic matter. *Computer Chem Eng* 21(5):489–499.

Bai M, Roegiers JC. 1997. Triple-porosity analysis of solute transport. *J Contam Hydrol* 28(3):247–266.

Barry DA, Miller CT, Culligan PJ, Bajracharya K. 1995. Split operator methods for reactive chemical transport in groundwater. MODSIM95 conference proceedings.

Ball W P, Roberts PV. 1992. Comment on "Long-term Sorption of halogenated organic chemicals by aquifer material. 1. Equilibrium." *Environ Sci Technol* 26:1223–1236.

Bear J. 1960. The transition zone between fresh and salt waters in coastal aquifers [PhD thesis]. Berkley CA: Univ Calif.

Bear J. 1979. Hydraulics of groundwater. New York NY: McGraw-Hill.

Beljin MS. 1991. Review of three-dimensional analytical models for solute transport in ground-water system. Indianapolis IN: International Ground Water Modeling Center, Holcomb Research Institute, Butler Univ.

Boethling RS, Alexander M. 1979. Effect of concentration of organic chemicals on their degradation by natural microbial communities. *Appl Environ Microbiol* 37:1211–1216.

Bower EJ, Rittman BE, McCarty PL. 1981. Anaerobic degradation of halogenated and 2-carbon organic compounds. *Environ Sci Technol* 15(5):596–599.

Burgos WD, Novak JT, Berry DF. 1996. Reverse sorption and irreversible binding of naphthalene and alpha naphthol to soil: Elucidation of processes. *Environ Sci Technol* 30(4):1205–1211.

Cardenas M, Lick W. 1996. Modeling the transport of sediments and hydrophobic contaminants in the lower Saginaw River. *J Great Lakes Res* 22:669–682.

Carol EA, Newell CJ, Gonzales JR, Haas P, Clement TP, Sun Y. 1999. BIOCHLOR: Natural attenuation decision support system v1.0; user's manual. Air Force Center for Environmental Excellence Report.

Carroll KM, Harkness MR, Bracs AA, Balcarcel RR. 1994. Application of permanent/polymer diffusional model to the desorption of polychlorinated biphenyls from Hudson River sediments. *Environ Sci Technol* 28(2):253–258.

Cheng N-S. 1997. Simplified settling velocity formula for sediment particle. *J Hydr Eng* 123:149–152.

Chilakapati A. 1993. Numerical simulation of reactive flow and transport through the subsurface [PhD thesis]. Houston TX: Rice Univ.

Chung N, Alexander M. 1998. Differences in sequestration and bioavailability of organic compounds aged in dissimilar soils. *Environ Sci Technol* 32:855–869.

Clement TP. 1997. RT3D: A modular computer code for simulating reactive multi-species transport in 3-dimensional groundwater systems. Draft report. Richland WA: Pacific Northwest National Laboratory. PNNL-SA-28967. http://bioprocess.pnl.gov/rt3d.htm.

Clement TP. 1999. A generalized analytical method for solving multi-species transport equations coupled with a first-order reaction network. *Water Resources Res*. In press.

Clement TP, Hooker BS, Skeen RS. 1996a. Numerical modeling of biologically reactive transport near a nutrient injection well. *ASCE J Environ Eng Div* 122(9):833–839.

Clement TP, Hooker BS, Skeen RS. 1996b. Macroscopic models for predicting changes in saturated porous media properties caused by microbial growth. *Ground Water* 34(5):934–942.

Clement TP, Johnson CD, Sun Y, Klecka GM, Bartlett C. 2000. Natural attenuation of chlorinated solvent compounds: Model development and field-scale application. *J Contam Hydrol* 42:113–140.

Clement TP, Jones NL. 1997. RT3D tutorials for GMS users. Richland WA: Pacific Northwest National Laboratory, PNNL-11805.

Clement TP, Peyton BM, Ginn TR, Skeen RS. 1999. Modeling bacterial transport and accumulation processes in saturated porous media: A review. In: Lewins J, Becker M, editors. Advances in nuclear science and technology. New York NY: Kluwer Academic/Plenum. p 59–78.

Clement TP, Peyton BM, Skeen RS, Jennings DA, Petersen JN. 1997. Microbial growth and transport in porous media under denitrification conditions: Experiment and simulations. *J Contam Hydrol* 24:269–285.

Clement TP, Sun Y, Hooker BS, Petersen JN. 1998. Modeling multi-species reactive transport in groundwater aquifers. *Groundwater Monitor Remed* 18(2):79–92.

Connaughton DF, Stedinger JR. 1993. Description of time-varying desorption kinetics: Release of naphthalene from contaminated soils. *Environ Sci Technol* 27:2397–2403.

Connolly JP. 1997. Modeling the environmental fate and transport of metals. Society of Environmental Toxicology and Chemistry (SETAC) Pellston Workshop on Reassessment of Metals Criteria for Aquatic Life Protection; 1996 Feb 10–14. Pensacola FL: SETAC Pr. 114.

Connolly JP, Zahakos HA, Benaman J, Ziegler CK, Rhea JR, Russell K. A model of PCB fate in the Upper Hudson River. *Environ Sci Technol.* In press.

Cornelissen G, Paul CM, Noort V. 1998. Mechanism of slow desorption of organic compounds from sediments: A study using model sorbents. *Environ Sci Technol* 32:3124–3131.

de Blanc PC, McKinney DC, Speitel GE. 1996. A 3-D flow and biodegradation model. Proceedings of the ASCE Conference on Nonaqueous Phase Liquid (NAPLs) in the Subsurface Environment: Assessment and Remediation. Washington DC: ASCE. p 478–489.

Deen WM. 1987. Hindered transport of large molecules in liquid-filled pores. *AIChE J* 33(9):1409–1425.

Dilks DW, Helfand JS, Bierman Jr VJ, Burkhard L. 1993. Field application of a steady-state mass balance model for hydrophobic organic chemicals in an estuarine system. *Wat Sci Tech* 28:263–271.

Domenico PA, Schwartz FW. 1990. Physical and chemical hydrogeology. New York NY: Wiley.

Essaid HI, Bekins BA, Godsy EM, Warren E, Baedecker MJ, Cozzarelli IM. 1995. Simulation of aerobic and anaerobic biodegradation processes at a crude oil spill site. *Water Resources Res* 31(12):3309–3327.

Farrell J, Reinhard M. 1994. Desorption of halogenated organics from model solids, sediments, and soil under unsaturated conditions. 2 Kinetics: *Environ Sci Technol* 28:63–72.

Ferguson RI. 1987. Accuracy and precision of methods for estimating river loads. *Earth Surf Proc Land* 12:95–104.

Freeman DL, Gossett JM. 1989. Biological reductive dehalogenation of tetrachloroethylene and trichloroethylene to etylene under methanogenic conditions. *Appl Environ Microbiol* 55(4):1009–1014.

Freeze RA, Cherry JA, 1979. Groundwater. Inglewood Cliffs NJ: Prentice Hall.

Fu MH, Alexander M. 1994. Desorption and biodegradation of sorbed styrene in soil and aquifer solids. *Environ Toxicol Chem* 13:749–753.

Gill GA, Bloom NS, Cappellino S, Driscoll CT, Dobbs C, McShea L, Mason R, Rudd JWM. 1999. Sediment-water fluxes of mercury in Lavaca Bay, Texas. Submitted to *Environ Sci Technol.*

Hatzinger PB, Alexander M. 1995. Effect of aging of chemicals in soil on their biodegradability and extractibility. *Environ Sci Technol* 29(2):537–545.

HydroQual Inc. 1998. Development and application of a modeling framework to evaluate hurricane impacts on surficial mercury concentrations in Lavaca Bay: Final report to Aluminum Company of America. Mahwah NJ: HydroQual.

Jepsen R, Roberts J, Lick W. 1997. Effects of bulk density on sediment erosion rates. *Water Air Soil Poll* 99:21–31.

Kaluarachchi J, Morshed J. 1995. Critical assessment of the operator-splitting technique in solving the advection-dispersion-reaction equation: 1. First-order reaction. *Adv Water Resources* 18:89–100.

Kan AT, Fu G, Tomson MB. 1994. Adsorption/desorption hysteresis in organic pollution and soil sediment interactions. *Environ Sci Technol* 28:859–867.

Khandelwal A, Rabideau AJ. 1999. Transport of sequentially decaying reaction products influenced by linear nonequilibrium sorption. *Water Resources Res* 35(6):1939–1945.

Klier NJ, Klecka GM, Lutz EJ, Ellis DE, Chapelle FH, Witt ME. 1999. The groundwater geochemistry of Area 6, Dover Air Force Base, Dover, Delaware. National Technical Information Service. In press.

Klier NJ, West RJ, Donberg PA. 1998. Aerobic biodegradation of dichloroethylene in subsurface soils. *Chemosphere* In press.

Konikow LF, Bredehoeft JD. 1978. Computer model of two dimensional solute transport and dispersion in groundwater. Reston VA: U.S. Geological Survey (USGS). TWRI USGS Report I-19.15/5.

Landmeyer JE, Chapelle FH, Bradley PM, Pankow JF, Church CD, Tratnyek PG. 1998. Fate of MTBE relative to benzene in a gasoline-contaminated aquifer (1993–98). *Ground Water Monitor Remed* 18(4)93–102.

Laor Y, Farmer WJ, Aochi Y, Strom PF. 1998. Phenanthrene binding and sorption to dissolved and to mineral-associated humic acid. *Water Res* 32:1923–1931.

Lartiges SB, Garrigues PP. 1995. Degradation kinetics of organophosphorus and organonitrogen pesticides in different waters under various environmental conditions. *Environ Sci Technol* 29:1246–1254.

Limbach KW, Wei J. 1990. Restricted diffusion through granular materials. *AIChE J* 36(2):242 –248.

Loonen H, Muir DCG, Parsons JR, Govers HAJ. 1997. Bioaccumulation of polychlorinated dibenzo-*p*-dioxins in sediments by oligochaetes: Influence of exposure pathway and contact time. *Environ Toxicol Chem* 16:1518–1525.

Lovley DR. 1991. Dissimilatory Fe(III) and Mn(IV) reduction. *Microbiol Rev* 55(2):259–287.

Lu G, Clement TP, Zheng C, Wiedemeier TH. 1999. Natural attenuation of BTEX compounds: Model development and field-scale application. *Ground Water* 37(5):707–717.

Lunn M, Lunn R J, Mackay R. 1996. Determining analytic solution of multiple species contaminant transport, with sorption and decay. *J Hydrology* 180:195–210.

Luthy RG, Aiken GR, Brusseau ML. 1997. Sequestration of hydrophobic organic contaminants by geosorbents (critical review). *Environ Sci Technol* 31:3341.

Mackay D, Yeun ATK. 1983. Mass transfer coefficient correlations for volatilization of organic solutes from water. *Environ Sci Technol* 17:211–217.

McCarty PL, Semprini L. 1994. Groundwater treatment for chlorinated solvents. In: Handbook of bioremediation. Boca Raton FL: Lewis.

McDonald MD, Harbaugh AW. 1998. A modular three-dimensional finite-difference flow model. Techniques in water resources investigations of the U.S. Geological Survey, Book 6. Reston VA: U.S. Geological Survey (USGS). 586 p.

McLachlan M, Mackay D, Jones PH. 1990. A conceptual model of organic chemical volatilization at waterfalls. *Environ Sci Technol* 24:252–257.

McNab WW, Narasimhan TN. 1994. Modeling reactive transport of organic compounds in groundwater using a partial redox disequilibrium approach. *Water Resources Res* 30(9):2619–2635.

Molz FJ, Widdowson MA, Benefield LD. 1986. Simulation of microbial growth dynamics coupled to nutrient oxygen transport in porous media. *Water Resources Res* 22:1207–1216.

Nam K, Alexander M. 1998. Role of nanoporosity and hydrophobicity in sequestration and bioavailability: Tests with model solids. *Environ Sci Technol* 32:71–74.

Newell C J, Gonzales J, McLeod RK. 1996. BIOSCREEN natural attenuation decision support system. Ada OK: U. S. Environmental Protection Agency (USEPA), Center for Subsurface Modeling Support. EPA/600/R-96/087.

Norris RD, Matthews JE. 1994. Handbook of bioremediation. Boca Raton FL: Lewis.

O'Connor DJ, Connolly JP, Garland EJ. 1989. Mathematical models-fate, transport, and food chain. In: Levin SA, Harwell MA, Kelly JR, Kimball KD, editors. Ecotoxicology: problems and approaches. New York NY: Springer-Verlag. p 221 243.

O'Connor DJ, Mueller JA, Farley KJ. 1983. Distribution of kepone in the James River estuary. *J Environ Eng* 109:396–413.

Ogata A. 1958. Dispersion in porous media [PhD thesis]. Chicago IL: Northwestern Univ.

Pavlostathis SG, Jagal K. 1991. Desorptive behavior of trichloroethylene in contaminated soil. *Environ Sci Technol* 25(2):274–279.

Pedit JA, Miller C T. 1995. Heterogeneous sorption processes in subsurface systems: 2. Diffusion modeling approaches. *Environ Sci Technol* 29:1766–1772.

Pignatello JJ, Xing B. 1996. Mechanisms of slow sorption of organic chemicals to natural particles. *Environ Sci Technol* 30:1–11.

Quantitative Environmental Analysis, LLC. 1998. Thompson Island Pool sediment PCB sources: Final report to General Electric Company. Montvale NJ: Quantitative Environmental Analysis.

Quantitative Environmental Analysis, LLC. 1999. PCBs in the Upper Hudson River: Report to General Electric Company. Montvale NJ: Quantitative Environmental Analysis.

Raghunathan RJ, DePinto JV, Bierman Jr VJ, Rogers PJ, Young TC, Maritn SC, Zhang X. 1994. Development of a toxic chemical dynamics model (GBTOX) for the Green Bay Mass Balance Study: Part two-model diagnosis and interpretation. *J Great Lakes Res.*

Rathbun RE. 1990. Prediction of stream volatilization coefficients. *J Environ Eng* 116:615–631.

Rice DE, Grose RD, Michaelsen JC, Dooher BP, Macqueen DH, Cullen SJ, Kastenberg WE, Everett LG, Marino MS. 1995. California leaking underground fuel tank (LUFT) historical case analyses. California State Water Resources Publication, UCRL-AR-122207.

Riedel GF, Sanders JG, Osman RW. 1988. The effect of biological and physical disturbances on the transport of arsenic from contaminated estuarine sediments. *Est Coastal Shelf Sci* 23:693–706.

Rifai SH, Bedient PB, Borden RC, Haasbeek JF. 1987. BIOPLUME II: Computer model of two-dimensional contaminant transport under the influence of oxygen limited biodegradation in ground water. National Center for Ground Water Research. Houston TX: Rice Univ.

Rifai SH, Newell CJ, Gonzales JR, Dendrou S, Kennedy L, Wilson J. 1998. BIOPLUME-III: Natural attenuation decision support system. Version 1.0 user manual. Prepared for the U.S. Air Force Center for Environmental Excellence, Brooks Air Force Base, San Antonio, Texas.

Satterfield CN, Colton CK, Pitcher WH. 1973. Restricted diffusion in liquids within fine pores *AIChE J* 19:628–635.

Schecher WD, McAvoy DC. 1995. MINEQL$^+$: A chemical equilibrium program for personal computers. Version 3.01. Hallowell ME: Environmental Research Software.

Schweiger MB, Haggard JG, Connolly JP. 1996. Comments of General Electric Company on Phase 2 report. Review copy: Further site characterization and analysis. Volume 2B, Preliminary model calibration report. Albany NY: Hudson River PCBs Reassessment RI/FS, General Electric Company.

Semprini L, Kitanidis P, Kampbell D, Wilson J. 1995. Anaerobic transformation of chlorinated aliphatic hydrocarbons in a sand aquifer based on spatial chemical distribution. *Water Resources Res* 31:1051–1062.

Smith DM. 1986. Restricted diffusion through pores with periodic constrictions. *AIChE J* 32(6):1039–1042

Steefel CI, MacQuarrie KTB. 1996. Approaches to modeling of reactive transport in porous media. In: Lichtner PC, Steefel CI, Oelkers EH, editors. Volume 34, Reviews in mineralogy. p 84–129.

Steefel CI, Yabusaki SB. 1997. OS3D/GIMRT, Software for multicomponent-multidimensional reactive transport. User manual and programmers guide. Richland WA: Pacific Northwest National Laboratory. PNL-11166.

Steinberg SM, Pignatello JJ, Sawhney BL. 1987. *Persist Environ Sci Technol* 21:1201–1208

Sun Y, Clement TP. 1999. A generalized decomposition method for solving coupled multi-species reactive transport problems. *Transport in Porous Media J* 37(3):327–346.

Sun Y, Petersen JN, Clement TP, Skeen RS. 1999. Development of analytical solutions for multi-species transport with serial and parallel reactions. *Water Resources Res* 35:(1):185–190.

Taylor SW, Jaffe PR. 1990. Biofilm growth and the related changes in the physical properties of a porous medium. 1. Experimental investigations. *Water Resources Res* 26:2153–2159.

Tell JG, Parkerton TF. 1997. A comparison of multi-medial models used in regulatory decision-making. *SAR QSAR Env Res* 6:29–45.

Thomann RV, Mueller JA, Winfield RP, Huang C-R. 1991. Model of fate and accumulation of PCB homologues in Hudson Estuary. *ASCE J Environ Engr* 117:161–178.

Thomas WA, McAnally WH. 1985. Open channel flow and sedimentation, TABS-2. Vicksburg MS: U.S.A.E. Waterways Experiment Station. Instructional report HL-85-1.

Tipping E. 1994. WHAM: A chemical equilibrium model and computer code for waters, sediments and soils incorporating a discrete site/electrostatic model of ion-binding by humic substances. *Computers Geosci* 6:973–1023.

Toride N, Leij FJ, van Genuchten MTh. 1995. The CXTFIT code for estimating transport parameters from laboratory and field tracer experiments. U.S. Salinity Laboratory Report no. 137.

Tsai CH, Lick W. 1986. A portable device for measuring sediment resuspension. *J Great Lakes Res* 12:314–321.

[USEPA] U.S. Environmental Protection Agency. 1996. Phase 2 report. Review copy: Further site characterization and analysis volume. 2B, Preliminary model calibration report Hudson River PCBs Reassessment RI/FS. New York NY: Region II USEPA.

[USEPA] U.S. Environmental Protection Agency. 1997. Phase 2 report. Review copy: Further site characterization and analysis. Volume 2C, Data evaluation and interpretation report Hudson River PCBs Reassessment RI/FS. New York NY: Region II USEPA.

Valocchi AJ, Malmstead M. 1992. Accuracy of operator-splitting for advection-dispersion-reaction problems. *Water Resources Res* 28:1471–1476.

van Genuchten MTh, Alves W J. 1982. Analytical solutions of the one-dimensional convective-dispersive solute transport equation. Riverside CA: Agricultural Research Service, U.S. Department of Agriculture (USDA).

Velleux M, Endicott D, Steuer J, Jaeger S, Patterson D. 1995. Long-term simulation of PCB export from the Fox River to Green Bay. *J. Great Lakes Res* 21:359–372.

Voss CI. 1984. Two-dimensional density dependent flow and transport of either dissolved solute or thermal energy. Reston VA: U.S. Geological Survey (USGS) report.

Waddill DW, Widdowson MA. 1998. Three-dimensional model for subsurface transport and biodegradation. *ASCE J Environ Engr* 124(4):336–344.

Walling DE, Webb BW. 1988. The reliability of rating curve estimates of suspended sediment yield: Some further considerations. In: Sediment budgets. Wallingford, Oxfordshire UK: International Association of Hydrological Sciences (IAHS). IAHS Pub. No. 174. p 337–350.

Walter AL, Frind EO, Blowes DW, Ptacek CJ, Molson JW. 1994. Modeling of multicomponent reactive transport in groundwater. 1. Model development and evaluation. *Water Resources Res* 30:3137–3148.

Weissenfels WD, Klewer KJ, Langhoff J. 1992. Adsorption of polycyclic aromatic hydrocarbons (PAHs) by soil particles: Influence on biodegradability and biotoxicity. *Appl Microbiol Biotechnol* 36:689–696.

Werth CJ, Reinhard M. 1997a. Effects of temperature on trichloroethylene desorption from silica gel and natural sediments. 1. Isotherm. *Environ Sci Technol* 31:689–692.

Werth CJ, Reinhard M. 1997b. Effects of temperature on trichloroethylene desorption from silica gel and natural sediments. 2. Kinetics. *Environ Sci Technol* 31:697–703.

Wiedemeier TH, Swanson MA, Moutoux DE, Gordon EK, Wilson JT, Wilson BH, Kampbel DH, Hansen J, Haas P. 1997. Technical protocol for evaluating natural attenuation of chlorinated solvents in groundwater. San Antonio TX: Air Force Center for Environmental Excellence, Technology Transfer Division, Brooks AFB.

Wiedemeier TH, Wilson JT, Kampbell DH, Miller RN, Hansen J. 1995. Technical protocol for implementing intrinsic remediation with long-term monitoring for natural attenuation of fuel contamination dissolved in groundwater. Volumes 1 and 2. San Antonio TX: Air Force Center for Environmental Excellence, Technology Transfer Division, Brooks AFB.

Wu S, Gschwend PM. 1986. Sorption kinetics of hydrophobic organic compounds to natural sediments and soils. *Environ Sci Technol* 20:717–725.

Wu S, Gschwend PM. 1988. Numerical modeling of sorption kinetics of organic compounds to soil and sediment particles. *Environ Sci Technol* 24:1373–1383.

Young TM, Weber W J. 1995. A distributed reactivity model for sorption by soils and sediments 3. Effects of diagenetic processes on sorption energetics. *Environ Sci Technol* 29:92–97.

Zheng C. MT3D. 1990. A modular three-dimensional transport model for simulation of advection, dispersion and chemical reactions of contaminants in groundwater systems, Washington DC: U.S. Environmental Protection Agency (USEPA) report.

Ziegler CK, Connolly JP. 1995. The impact of sediment transport processes on the fate of hydrophobic organic chemicals in surface water systems. Water Environment Federation Specialty Conference; 1995 May 14–17; Cincinnati OH.

Ziegler CK, Israelsson PH, Connolly JP. Modeling of flood and long-term sediment transport dynamics in Thompson Island Pool, Upper Hudson River. Submitted to *ASCE J Hyd Engr*.

Ziegler CK, Nisbet B. 1994. Fine-grained sediment transport in Pawtuxet River, Rhode Island. *ASCE J Hyd Engr* 120:561–576.

Ziegler CK, Nisbet B. 1995. Long-term simulation of fine-grained sediment transport in large reservoir. *ASCE J Hydr Engr* 121:773–781.

Zysset A, Stauffer F, Dracos T. 1994. Modeling of reactive groundwater transport governed by biodegradation. *Water Resources Res* 30:2423–2434.

6

The Use of Ecotoxicology and Population Models in Natural Remediation

David Glaser and John P. Connolly

ENVIRONMENTAL CONTAMINATION represents a threat to the health and well-being of ecosystems. Risk arises insofar as exposure to contaminants affects the sustainability of natural populations, which includes their long-term probability of survival, their ability to resist chemical and other stresses, and their resilience or the ability to recover once affected by stressors. The effects of contaminants can potentially ramify through ecosystems, putting multispecies communities and ecosystems at risk.

One solution to the problem of ecological risks posed by environmental contaminants is natural remediation (NR), which can occur as a result of changes in contaminant concentration, bioavailability, toxicity, and mobility or as a result of a biological response to contamination, such as the development of resistance. To make informed management decisions, a risk manager should be able to quantify both the effects of contaminants on ecological communities and how these effects are likely to change as a result of NR or engineered remediation activities. Ecotoxicology and population models can provide that information. These models link the exposure estimates produced by the fate and transport models described in Peyton et al. (2000) to actual effects on individuals and populations. In contrast, approaches based solely on estimating the toxicity to individuals leave population-level effects unexplored, often resulting in conservative risk assessments.

Population and ecosystem models can be used effectively to assess sustainability only in conjunction with ecotoxicology models that quantify chemical exposure and

Natural Remediation of Environmental Contaminants. Michael Swindoll et al., editors.
 ISBN 1-880611-33-3.

toxicity to individuals. The focus of this chapter is the use of models to understand these 3 processes — toxicity to individuals, exposure in the field, and population dynamics — and their roles in predicting the rate and dynamics of NR.

Models of toxicity to individuals date back at least to the 1960s and the growth of interest in the effects of environmental chemicals on people and ecological communities. Examples include statistical models used to extrapolate from the results of laboratory toxicity experiments to the wide range of chemicals in the environment and the large number of species to be protected. The growth of exposure modeling began later, in the late 1970s, as it became clear that, following the banning of some of the most toxic and ubiquitous chemicals, the exposure sources for many chemicals would change from point-source discharges to sediments and soils, and therefore the exposure of natural communities and people to these chemicals would become more difficult to quantify. Population models have the longest history, dating back at least to Malthus (1978), with the extensive development of theory and practice in the 20th century following Lotka (1925). However, the application of population models to the assessment of chemical impacts has begun to grow only in the last decade.

The effective use of these tools rests upon a foundation of experimental and field observation, supplemented with an understanding of biological and chemical mechanisms, and integrated using mathematics (primarily statistics), differential equations, and matrix algebra. Toxicology models require an experimental database of measured toxicity, along with statistics used to extrapolate to other chemicals and species. Models used to extrapolate the results of toxicity tests to other chemicals (e.g., quantitative structure-activity relationship [QSAR] models) also require an understanding of chemical structure. Exposure models require an understanding of the processes underlying food web transfer as well as natural history information: diet, food web structure, movement, and growth. They generally involve algebraic expressions or differential equations that calculate a dose or body burden from exposure concentrations. Population models require quantification of the key factors underlying population dynamics and thus rely on field measurements of reproduction and survival. Effective use of population models also relies on the quantification of exposure as well as the availability of toxicological relationships between exposure and vital rates (survival and reproduction). Thus, the effective use of population models rests upon toxicity and exposure models.

Models vary in complexity and sophistication. Simple models of toxicity extrapolation involve multiplying by a safety factor or summing the concentrations of similarly acting chemicals. These models can be used with very limited expertise, but their appropriate use relies upon an understanding of the underlying toxicity data. Simple exposure models are in common use, involving, for example, an algebraic formula for calculating dose based upon feeding rate, movement patterns, and chemical levels in prey. Simple models of population dynamics require more mathematical sophistication. Perhaps the most common and most useful type of population model involves linear matrices, the use of which requires an understanding of their interpretation, of the methods for estimating parameter values, and of the dynamics of the population of interest. Performing the calculations is not a limitation; the computer software required for solving matrices or performing numerical simulations is now readily available.

At the other extreme, complex, sophisticated models can reduce uncertainty but require a great deal of expertise. Application of QSARs can require a working knowledge of chemistry. Application of complex food web bioaccumulation models requires knowledge of bioenergetics and toxicokinetics. Application of matrix models with nonlinear terms and migrating subpopulations requires considerable knowledge concerning the dynamics of the species of interest. Because complex models require more effort to develop and analyze, their value in ecological risk management is subject to greater scrutiny. This means that, as the complexity of models increases, the need for uncertainty analysis increases. A meaningful uncertainty analysis requires considerably more data, as well as a greater understanding of mechanism, than does application of a model to develop a best estimate of NR.

This chapter provides an overview of ecotoxicology models and population models, including examples of their usefulness in assessing NR. Ecotoxicology models are discussed first; these describe the relationship between contaminant levels in the environment, for example, in surface sediments and water, and toxicological effects on individual organisms. There are 2 components of ecotoxicology models: exposure and toxicity. Models that predict the impacts of exposure to toxic contaminants on population and community sustainability are discussed second. The types of models discussed here are summarized in Table 6-1.

ECOTOXICOLOGY MODELS

Estimating exposure for organisms in the field invariably involves some form of modeling, ranging in sophistication from a simple steady-state calculation to a time-variable food web computer simulation. Exposure models relate concentrations of contaminants in exposure media (usually soils, sediments, and water) to levels in biota. To assess toxicity, these models can compute concentration or the time integral of concentration in the body of the species of interest, dose to the species of interest, or concentration in the prey of the species of interest. The choice of model depends on the nature of the question at hand, the available information, the size of the project, and the complexity of the exposure sources and food web.

Models of toxicological effects use measured or estimated exposure to compute toxicity. In most studies, toxicity is assumed to occur whenever exposure exceeds a critical value, whenever the critical value is exceeded for a specified time period, or whenever the time integral of concentration exceeds a critical value; however, this last is rare in environmental studies. Toxicological models and their use in ecological risk analysis were reviewed by Suter (1993) and McCarty and Mackay (1993).

Table 6-1 Exposure and population models

Type of model	Strengths	Limitations	Example references
Exposure models			
Steady-state	Simple to develop; program not required	Time- and age-dependent dynamics not considered	Great Lakes Water Quality Initiative Criteria to Protect Wildlife (USEPA 1995)
Dynamic	Mechanistic; time- and age-dependent dynamics considered	Experience, computer program, and values for several parameters required	DDE in bald eagles of the Southern California Bight (see text)
Population models			
Phenomenological	Simple, often with analytical solutions; minimum number of parameter values to estimate	Parameter values must be estimated from observations of population size and growth under relevant conditions	Kolar et al. 1997
Demographic	Survival and reproductive rates can be estimated based upon independent information; mathematics well developed; commercial programs exist	Metapopulation dynamics and multispecies dynamics not considered	Munns et al. 1997
Individual-based (IBM)	Parameters and equations have clear and specific biological meaning; stochastic dynamics for small population size considered	Data requirements: relationships between individual behaviors, vital rates, and contaminant levels	Glaser et al. 1996; Hallam et al. 1996
Metapopulation	More realistic prediction of impacts for many populations	Data requirements: subpopulation sizes and arrangement, migration patterns	
Multispecies	Permits evaluation of indirect effects	Information to structure the model properly is usually limiting	Bartell et al. 1992

Historically, relationships between toxicity and exposure have been expressed in several ways: concentration in food (e.g., in units of mg/kg), concentration in water (mg/L), dose (mg/kg-d), body burden (mg/kg), or concentration in a particular tissue (mg/kg). Dose, concentration in food, and concentration in water provide at best indirect estimates of exposure. The uptake and loss rates of the contaminant, as well as the physiological response of the organism to contaminant in the body, are subsumed in relationships between these metrics and toxicity. Recent efforts have stressed the use of body burden as the best measure of exposure (McCarty and Mackay 1993). This is because of improvements in our ability to estimate body burden using exposure models, as well as improvements in our understanding of the relationships between body burden and toxicity.

Toxicology models

Several classes of toxicological models are available. Statistical models are used to interpret the results of toxicity tests, for example, to estimate no-effect levels or to compute concentration–response relationships for use in population models (Newman and Dixon 1996). QSARs predict the toxicity of chemicals for which no data exist (e.g., McCarty et al. 1991; Ankley et al. 1997; Russom et al. 1997; Bradbury et al. 1998; Ward 1998). Statistical relationships are used to compute toxicity for species for which no data exist (e.g., Watanabe et al. 1992; Calabrese and Baldwin 1994).

In addition, there are several models of the toxicity of mixtures. Narcotic effects are computed by comparing a critical body burden with the sum of the molar concentrations of all chemicals that cause Type I narcosis (McCarty et al. 1992). The impact of compounds with dioxin-like effects is often computed by summing the products of the concentrations of individual chemicals and their toxic equivalency factors (TEFs) to arrive at the toxicity of a mixture relative to 2,3,6,7 tetrachloro-dibenzo-*p*-dioxin (TCDD) (e.g., Ahlborg et al. 1994; Newsted et al. 1995). Models have been developed to quantify the overall toxicity of multiple metals in sediments (DiToro et al. 1990; Campbell and Tessier 1996).

In the following section, models that predict the rate and dynamics of changes in exposure and therefore the time course of NR are evaluated. The time component of toxicological models derives mainly from their exposure component. Once exposure is estimated, toxicological analyses are static, in the sense that the response of the organism is essentially instantaneous compared with the time course of NR. Therefore, the rest of this discussion concentrates on exposure models. The reader is directed to recent reviews of the state of the art of toxicological models (McCarty and Mackay 1993; Suter 1993; Newman and Jagoe 1996).

Exposure models

Both steady-state and dynamic exposure models are used in risk assessment. Steady-state models are considerably easier to use but do not provide information on the rate of response of contaminant levels in organisms to changes in their exposure or to changes in contaminant levels with age. The use of dynamic models requires familiarity with the principles of bioenergetics and toxicokinetics but can provide considerably more information and greater accuracy.

Steady-state exposure models

Generic bioaccumulation factors (BAFs) are the simplest method of estimating exposure. Their strength lies in the number and quality of studies and measurements upon which they are based. Alternatively, site-specific measured BAFs can be used to assess NR. The obvious advantage of site specificity must be balanced against the often great uncertainty associated with limited sample sizes. In either case, BAFs are appropriate if the food web has one dominant source of carbon, that source is tied exclusively to one exposure source (e.g., contaminated surface sediments), the food web is unchanging, and long-term changes in contaminant levels in adults are the focus of an analysis. With no site- or species-specific information, a steady-state model is equivalent to a generic BAF.

The use of a steady-state model of bioaccumulation is illustrated by the procedure used to derive the Great Lakes Initiative Water Quality Criteria to Protect Wildlife (USEPA 1995). Equation 6-1 is a steady-state calculation of the chemical dose received by a bird or mammal (TD, mg/kg-d). In this procedure, the concentration of the chemical in the water (C_W, pg/L) is multiplied by a BAF (L/kg) to compute the level of contaminant in the prey of the species of interest. The model incorporates species-specific feeding rate (F_i, kg/kg-d), drinking rate (W, L/kg-d), and dietary composition (each F_i represents 1 prey type) in computing the total dose. Through the use of an additional area use factor (AUF), species-specific patterns of movement can be incorporated. For example, if the species spends only one-half of the year in the contaminated area, then the average dose can simply be divided by 2.

$$TD = \left[W + \sum \left(F_i \times BAF_i\right)\right] \times C_W \times AUF \times 10^9 \qquad \text{(Equation 6-1)}$$

Early life stages are often the most sensitive to contaminants. Contaminant levels can change dramatically as juveniles depurate contaminant received from the mother and grow rapidly, thus limiting the accuracy of steady-state exposure models. An additional limitation of this approach to assessing NR arises from the fact that short-term changes in exposure concentrations are not considered. For example, if the focus of the analysis is the response to chemical releases that occur during dredging or the response to a chemical spill, then a steady-state analysis is inappropriate. Finally, the calibration of a steady-state model will suffer if the exposure concentrations change during the calibration period on a time scale that is commensurate with the response time of the organisms. If the organisms are in the process of responding to changes in exposure concentrations, then by definition, they are not in steady state with respect to their exposure sources.

Steady-state exposure models do not require special software or mathematical expertise. These simple algebraic equations can be solved in a spreadsheet. The more important requirements for their use include an understanding of how to structure the equation, how to determine the best values for parameters, and how to assess uncertainty. These requirements involve knowledge of natural history (feeding behavior, home range, and migration patterns), understanding of the basic principles of toxicokinetics, and the availability and interpretation of laboratory and field measurements of BAFs for the base of the food web (e.g., BAF_i in Equation 6-1).

Hundreds of published and unpublished studies conducted over the past 30 years include field and laboratory measurements of contaminant levels in biota and in their exposure sources. However, many studies do not include all of the information needed to properly estimate BAFs, so their interpretation is a critical issue, in particular the following:

1) Natural variability results in uncertainty in the mean values, and sample sizes usually are relatively small.
2) Contaminant measurements in the exposure sources (e.g., sediments and water in an aquatic system) must be matched spatially and temporally with biota measurements. For example, water column contaminant concentrations in rivers can vary dramatically day to day and seasonally, and sediment contaminant levels can vary orders of magnitude over small distances. BAFs and biota–sediment accumulation factors (BSAFs) that are based upon inappropriately matched water, sediment, and biota measurements are subject to considerable uncertainty.
3) Bioaccumulation factors for organic compounds should be estimated using carbon-based contaminant levels in sediments, lipid-based levels in biota, and dissolved concentrations in water. Field and laboratory estimates of BAFs (properly measured in units of mg contaminant/kg lipid divided by mg dissolved contaminant/L) and BSAFs (properly measured in units of mg contaminant/kg lipid divided by mg contaminant/kg organic carbon) often do not include matched lipid measurements in biota, organic carbon measurements in sediments, or estimates of the dissolved fraction in the water column. These are critical parameters for estimating trophic transfer, especially for hydrophobic contaminants.

Thus, the most important limiting factors associated with the use of steady-state exposure models are data requirements.

Dynamic exposure models

Dynamic exposure models have been in use for more than 20 years (Norstrom et al. 1976). The most common applications have been to the accumulation of hydrophobic compounds in fish (Thomann and Connolly 1984; Connolly 1991; Gobas 1993). However, models have also been applied to the study of metals in aquatic ecosystems (Cd, Cu, and Pb, HydroQual 1990; Hg, Norstrom et al. 1976; Cd, Thomann et al. 1997). In addition, the bioenergetic components of these models have a long history in the study of fish population dynamics (Hanson et al. 1997). The models are well-accepted tools for use in developing water quality criteria (WQC; e.g., the Great Lakes Water Quality Initiative), in natural resource damage assessment (NRDA; e.g., HydroQual 1997), and in assessing remedial alternatives in contaminated bodies of water (Connolly et al. 1992; HydroQual 1995; Limno-Tech et al. 1996; Glaser et al. 1997).

Dynamic exposure models can also be an important tool for predicting the food web transfer of nonbioaccumulative chemicals. For example, early life stages are often the most sensitive to chemicals in the environment. Levels of nonbioaccumulative chemicals may not be in steady state with respect to the environment during early life stages because of rapid growth, and dynamic exposure models can be used to compute them. In addition, WQC are written in terms of the frequency with which contaminant levels exceed a critical threshold. There is some degree of temporal averaging of exposure levels in food webs, even for nonbioaccumulative chemicals. A dynamic exposure model can be used to predict the frequency with which body burdens exceed a specified level.

An example of the use of dynamic exposure models

The use of dynamic exposure models to assess NR is illustrated here with a model of dichlorodiphenyldichloroethylene (*p,p′*-DDE) in bald eagles in the Southern California Bight (HydroQual 1997; Glaser and Connolly submitted).

Large quantities of the pesticide DDT were released into the coastal waters of the Southern California Bight before these chemicals were banned in the 1970s. The dominant source was the Whites Point outfall of the Los Angeles County Sanitation District wastewater treatment plant in Los Angeles. The DDT was deposited on the sediment bed, resulting in a plume of contaminated sediments several km long and approximately 2 km wide off the coast of Los Angeles, on the Palos Verdes Shelf. Concentrations of *p,p′*-DDE, the most abundant metabolite of DDT, in surface sediments and in intertidal mussels still exhibit a large peak on the Palos Verdes Shelf 30 years later, indicating that these sediments comprise the dominant source of this chemical to the waters of the Southern California Bight (Connolly and Glaser 1999).

This study involved the development of a computer simulation model of the transfer of *p,p′*-DDE to bald eagles (*Haliaeetus leucocephalus)* living on Santa Catalina Island, approximately 30 km from the Palos Verdes Shelf. The study was designed to aid in exploring the sources of these contaminants to the eagles. This work was performed as part of the Southern California Bight Natural Resource Damage Assessment.

An existing time-variable mechanistic bioaccumulation model that had been developed for use in aquatic ecosystems (Thomann and Connolly 1984; Connolly and Tonelli 1985; Connolly 1991; Connolly et al. 1992) was enhanced to model the bioaccumulation of *p,p′*-DDE in birds. The bioaccumulation model was used with the observed dietary composition of the eagle and the field-measured contaminant levels in its prey to compute contaminant levels in the eagles' eggs. Comparison between the computed and measured levels in the eggs provided a test of the realism of the dietary composition that formed the basis for the calculation.

Bald eagles were observed in the Southern California Bight in the 1940s. By 1960, none were left. A reintroduction program began in the early 1980s on Santa Catalina Island, and as of the mid 1990s, there still was no natural reproduction by the reintroduced birds and contaminant levels were high (Garcelon 1994): *p,p′*-DDE levels averaged 36 mg/g wet in the eggs (SD = 15, n = 10). The *p,p′*-DDE is likely the stressor responsible for the lack of natural reproduction. The concentration–response relationship developed from field data for bald eagles by Wiemeyer et al. (1993) indicates that *p,p′*-DDE levels greater than approximately 3 mg/g wet in eggs result in reduced reproduction, and levels greater than approximately 20 mg/g result in reduction of 90% or more in productivity.

The food web pathways by which *p,p′*-DDE may be transferred from the sediments of the Palos Verdes Shelf to the bald eagle are illustrated in Figure 6-1. Bald eagles consume a mixture of birds, fish, and sea lion (*Zulophus californianus*) carcasses; sea lions consume fish.

The fish and invertebrates accumulate contaminants from a combination of surface sediments and the water column. The link from the sediments of the Palos Verdes Shelf to benthic fish living on the shelf was demonstrated in a modeling analysis of the white croaker (*Genyonemus lineatus*) and Dover sole (*Microstomus pacificus*) (HydroQual 1997). The observed peak in contaminant levels in mussels on the Shelf indicates that these sediments are likely to be the dominant source of *p,p′*-DDE to the waters of the Shelf and therefore to water column–based food webs on the Shelf (e.g., of kelp bass [*Paralabrax clathratus*]). The link between the sediments of the Palos Verdes Shelf and the sediments and waters elsewhere in the Bight was not quantified here; this is the reason for the dashed lines in Figure 6-1. Measured *p,p′*-DDE levels measured in the surface sediments of Southern California exhibit a clear peak on the Shelf (HydroQual 1997; Connolly and Glaser 1999), providing evidence that these sediments are the one identifiable source of *p,p′*-DDE to the

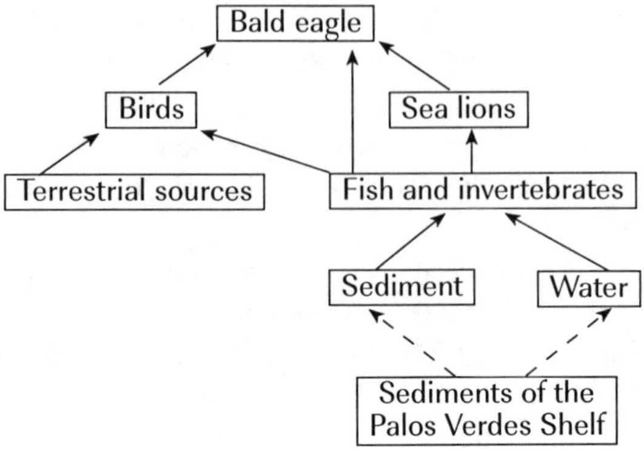

Figure 6-1 Simplified food web of the Southern California Bight

Bight. Nonetheless, the link from the sediments of the Shelf is represented by a dashed line, indicating uncertainty in this portion of the pathway.

In the model, the bald eagle is represented as having 2 body compartments between which a contaminant moves: blood (an aqueous compartment) and lipid (Clark et al. 1987). The blood is the compartment into which contaminant enters from the gut and out of which contaminant is excreted. Contaminant is transferred between blood and body fat. The basic equation of the model is as follows:

$$\frac{dv}{dt} = \alpha C v_p - (k+G)v \qquad \text{(Equation 6-2)}$$

where

n = contaminant concentration in the body (mg/g wet)
a = contaminant assimilation efficiency
C = food consumption rate (g wet/g wet-d)
n_p = contaminant concentration in the prey (mg/g wet)
k = contaminant elimination rate (1/d)
G = growth rate (1/d)

Chemical uptake and loss rates are controlled by the respiration rate, the growth rate and the body composition of the birds. Contaminant dose (in units of mg/g wet-d) is computed by multiplying concentration in prey (n_p, mg/g wet) by the rate of food consumption (C, g wet/g wet-d). The rate of food consumption is calculated from the rate of energy usage by the animal. Energy usage is estimated from the rates of production and metabolism of body tissue by the animal. Growth rate defines the net production of body tissue (G, g wet/g wet-d). In birds, growth occurs during the nestling stage. Egg production by adults acts similarly to growth and is therefore included in the estimate of energy needs.

Chemical elimination mechanisms (k in Equation 6-2) in eagles differ from those in fish. Fish lose chemicals across the gill surface and to some degree across the gut wall (Gobas et al. 1989). In contrast, metabolic modification of *p,p′*-DDE and subsequent renal excretion of the metabolic products is probably the dominant loss mechanism in birds (Walker 1987). Values for the elimination rates were determined using published toxicokinetic and bioaccumulation data (HydroQual 1997; Glaser and Connolly submitted).

Growth of the egg and contaminant transfer to it act as a loss mechanism in the same way that body growth acts to dilute the body burden. In addition, contaminant transfer to eggs represents the primary exposure route for newly hatched birds. Lipid-based contaminant levels in bird eggs are lower than those in the mother (Norstrom et al. 1986). It has been suggested that this difference exists because a portion of dietary lipid is shunted directly to egg production during the few days each year when eggs are being manufactured (Clark et al. 1987). Thus, the egg is considered to contain a mixture of dietary and maternal lipid and therefore a similar mixture of dietary and maternal contaminant. Contaminant concentration in the egg on a lipid basis (mg/g lipid) is equal to a weighted average of concentrations in dietary and body lipid. The annual average loss rate from egg production is calculated as follows:

$$LOSS = \frac{v_e W_e N_e N_{clutches}}{v_{wb} W_{wb}} \qquad \text{(Equation 6-3)}$$

where

$LOSS$ = loss rate (y^{-1})
n_e = concentration of the contaminant in the eggs (mg/g wet)
W_e = weight of each egg (g wet/egg)
N_e = number of eggs produced in each clutch (eggs/clutch)
$N_{clutches}$ = number of clutches per female produced each year (clutches/female-y)
n_{wb} = concentration of the contaminant in the female (mg/g wet)
W_{wb} = weight of female (g wet/female)

Values for all bioenergetic and toxicokinetic parameters were determined from site-specific measurements and published data.

Food habits of the bald eagle were studied by Garcelon and others (Garcelon, Romsos, Golightly 1994; Garcelon, Tomassi et al. 1994) on Santa Catalina Island between December 1991 and July 1993. Prey species were identified from field observations of foraging eagles and at nesting sites. The dietary composition was based upon the total number of observations from all techniques. Fish represented the greatest proportion of the eagle diet numerically, although birds and mammals were seasonally important (Table 6-2). More than 28 species of fish were identified during the study. Contaminant concentrations have been measured in 9 of these species in the vicinity of Santa Catalina Island. California sea lion carrion was the predominant mammal, although feral goats were also taken. The birds were divided into 4 groups, based on the dietary composition data of Garcelon and others (Garcelon, Romsos, Golightly 1994; Garcelon, Tomassi et al. 1994): western gulls, other gulls, other waterbirds, and land birds.

Table 6-2 Dietary composition, contaminant levels, and body composition for prey of the bald eagle

Prey	Portion of diet on an energy basis[b]	Contaminant levels (ppm wet weight)[a]	
		p,p′-DDE	PCBs
Fish and invertebrates	0.79	0.11	0.07
Sea lions	0.058	26.	5.2
Other mammals	0.0073	0.0	0.0
Western gulls	0.013	5.4	1.3
Other gulls	0.013	5.4	1.3
Water birds	0.085	1.7	0.60
Land birds: resident	0.0078	0.0	0.0
Land birds: migratory	0.0053	3.1	0.30

[a] Arithmetic means used.
[b] Portion of the diet on an energy basis = proportion on a numerical basis × energy content, standardized to a total of 1.0.

The bald eagles of Santa Catalina Island forage over a 3- to 5-km distance along the coast (Garcelon, Romsos, Golightly 1994; Garcelon, Tomassi et al. 1994). In the model, they are considered to be resident and to feed on fish, birds, and sea lion carrion found near Santa Catalina Island. Bald eagle eggs were collected from Santa Catalina Island between 1989 and 1993 (Costa et al. 1994). Contaminant levels for fish, birds, and sea lions for Santa Catalina Island were calculated using data collected since 1985.

The computed *p,p′*-DDE levels in the eggs of the bald eagles are consistent with measured egg levels: The computed concentration (37 ppm) lies within 5% of the measured level (36 ppm, uncertainty range based on 2 standard errors of the mean [SEM] = 27 to 45).

There is some uncertainty concerning the average contaminant level in sea lion carcasses to which the bald eagles are exposed. The data upon which the average was based included one very high contaminant level (1570 m/g lipid *p,p′*-DDE). The average concentration in sea lion prey of the bald eagle including this value was 26 mg/g (n = 26); this value was used in the model. Excluding this value, the average was 8.3 mg/g (n = 25). This high value was measured in a sea lion carcass upon which bald eagles were actually observed feeding (Garcelon, Romsos, Golightly

1994; Garcelon, Tomassi et al. 1994). It is conceivable that such a high concentration could result from the starvation of a young animal. Therefore, it is realistic to include this value in the estimate of overall exposure. The importance of this value to the computed egg concentrations was assessed by testing alternative assumptions concerning the frequency of such highly contaminated carcasses in the diet. If, for example, such highly contaminated carcasses are present in 1 out of 50 carcasses fed upon by the eagles (instead of 1 out of 26, as in the average used in the model), then the computed bald eagle egg level would be approximately 30 mg/g wet, which lies within 2 SEM of the data (36 ± 9.5). Therefore, this would still be consistent with the data. If this value is excluded from the average, then the computed egg level is 22 mg/g wet, which lies outside of the range encompassed by 2 SEM, and therefore it is inconsistent with the measured egg levels. In conclusion, it is likely that bald eagles are occasionally exposed to sea lion carcasses with very high contaminant levels, although it is uncertain how frequently they consume them.

One of the unexpected results of this analysis is that, although sea lion carcasses were observed to comprise about 6% of the diet of the bald eagle on an energy basis (3% on a mass basis), *p,p′*-DDE associated with the sea lions comprised 60% of the dose to the eagles sampled in the early 1990s. A more surprising conclusion relates to the sources of contaminant within the Bight. The eagles are nonmigratory and forage within 5 km of the nest (Garcelon, Tomassi et al. 1994), and the vast majority of the diet is composed of fish, most of which are nearshore species and therefore probably accumulate much of their contaminant around Santa Catalina Island. Nonetheless, much of the dose of *p,p′*-DDE to these eagles actually originated over a wider area of the Bight, including the Palos Verdes Shelf and Santa Monica Bay because of the extensive movements of the sea lion (Connolly and Glaser submitted).

Limitations in the bald eagle data hinder our ability to estimate the course of natural recovery. It might be expected that the rate of natural recovery could be estimated based upon *p,p′*-DDE measurements in the waters and fish in the vicinity of Santa Catalina Island, where the bald eagle feeds. However, the bioaccumulation model has indicated that most of the dose of *p,p′*-DDE to these bald eagles originated in the sea lions, which can roam over hundreds of kilometers. Thus, the rate of natural recovery in the sea lions and therefore in these bald eagles is controlled to some extent by changes in *p,p′*-DDE levels in the sediments and waters in other areas of the Southern California Bight, including the more contaminated regions over and to the north of the Palos Verdes Shelf.

Simulation models such as the one described here can be used to project the course of natural recovery and the impacts of remediation activities, given projected contaminant concentrations in the prey. In addition, a bioaccumulation model, linked with a dose–response relationship and a population dynamics model, can provide quantitative estimates of the impacts of contaminants on populations

(Swartzmann 1994; Glaser et al. 1996) and the potential for the natural recovery of impacted populations.

Additional applications of dynamic exposure models

Dynamic models have been applied most frequently to organic compounds in aquatic ecosystems. The 2 most widely applied computer models of bioaccumulation for aquatic systems were developed by Frank Gobas (e.g., Gobas 1993; Gobas et al. 1995) and John Connolly (e.g., Thomann and Connolly 1984; Connolly 1991; Connolly et al. 1992).

Similar models have been used to represent the accumulation of organic compounds in birds and mammals. Clark et al. (1987) developed a model of dynamics of several organic compounds in the herring gull. A similar model for nestling tree swallows was developed by Nichols et al. (1995). The model described in the section titled "An example of dynamic exposure models" was also applied to PCBs and to peregrine falcons and double-crested cormorants (Glaser and Connolly 1994; HydroQual 1997; Glaser and Connolly in preparation). The key uncertainties in models of adult bird toxicokinetics are the excretion rates. These are slow for hydrophobic compounds (on the order of several months to a few years), and the mechanisms are incompletely understood, although it appears that metabolic modification is involved (Walker 1987).

Models of bioaccumulation in mammals include loss due to production of young and to lactation. Lactation can be the dominant loss mechanism for females, resulting in reduced concentrations in the adult female and increased concentrations in nursing young (McLachlan 1994; HydroQual 1997; Connolly and Glaser in preparation). Therefore, lactation can obviate the need to estimate elimination rates in adult females.

POPULATION DYNAMIC MODELS

It is notoriously difficult to decide on the acceptable level of a contaminant in the environment, and the available information is usually used in a conservative fashion. That is, analyses are usually performed in such a way as to minimize the probability of adverse impacts, as opposed to estimating the most likely degree of impact. A common methodology involves comparing the estimated field exposure with a critical exposure level determined from laboratory and field toxicity information (the "hazard quotient"). Because any change in the survival or reproduction rates of individuals of a species may have important consequences for population dynamics, in the absence of any quantification of those consequences, decision-makers often use the conservative criterion of no-observed-adverse effect.

Population models provide an additional level of sophistication for interpreting exposure and toxicity information and therefore provide a more realistic basis for making environmental management decisions. Such models estimate the likely impacts of changes in survival and reproduction on population dynamics. Population models do not provide definitive answers to management questions; value judgements will always be required, and the information provided by these analyses is still likely to be used conservatively. However, such models improve our understanding of the impacts of contaminants on our environment and make the basis for decision-making more realistic. Uncertainty leads to conservative decisions. More realistic information should lead to less uncertainty in predicted impacts and, therefore, to less conservative decisions.

Population models have a relatively short history in their application to environmental contamination problems, for both historical and technical reasons. The development of realistic population models and their field validation is hindered by uncertainties in both the appropriate model structure and the best parameter values. In addition, the disciplines of ecotoxicology and ecology are separate, with distinct histories. Population modeling arose in response to problems associated with wildlife management, harvesting, insect pest control, forest dynamics, and aquatic ecology. The primary goal was to characterize the key factors determining population dynamics. The field of ecotoxicology was developed to explore the impacts of one type of stressor: chemical contamination. Ecotoxicological studies have focused primarily on impacts on individual organisms, because these must be determined before population effects can be predicted and because they can often be determined in the laboratory.

The mathematics of population dynamics has advanced beyond our ability to apply models to the real world. As a consequence, we have a relatively good understanding of the realm of possible behaviors of populations and communities, but useful applications to real populations are less common. Recently, there have been several applications concerning contaminant impacts on field populations. In general, these applications have not addressed NR questions directly. However, once a realistic population model is structured and parameterized, and once the time course of exposure levels and the toxicological relationships between parameter values and contaminant levels have been estimated, virtually any population model can be used to assess the impacts of natural and anthropogenic remediation.

Recently, Barnthouse (1998) reviewed population models in the context of assessing the ecological risks associated with pesticides. That review offers additional discussion and additional examples of the use of population models in ecological risk assessment (ERA).

Types of population models

The choice of population model depends on the factors regulating the population of interest, on the nature of the question being asked, on the amount of available information, and to some degree, on personal preference. We describe the major types of population models that have been applied to the analysis of contaminant impacts and their potential application for evaluating NR.

Phenomenological models

The simplest models consist of parameters that describe population-level phenomena (Caswell 1996). A key example in the history of ecology is the logistic equation (Equation 6-4). This equation describes the rate of change with time of the number of individuals in the population.

$$\frac{dN}{dt} = rN\left(\frac{K-N}{K}\right) \quad \text{(Equation 6-4)}$$

where

- N = number of individuals in the population
- r = rate of exponential increase in the number of individuals at low population density
- $\left(\frac{K-N}{K}\right)$ = density-dependent population limitation, that is, for a decrease in the population growth rate as the number of individuals increases
- K = carrying capacity reached by population at steady state

The primary advantage of these simple models is that few parameter values are needed. Analytical results can often be derived. This simplicity is also the primary disadvantage of these models. Each parameter lumps together many biological processes, processes that vary in time and space. This usually makes it impossible to provide one realistic value for "r" and for "K" for a field population. An example of the use of the logistic equation in an ERA is a model of the recovery of the mayfly population in Lake Erie developed by Kolar et al. (1997) described below.

Demographic models

In demographic models, the rate of increase of the population at low densities (the parameter "r" in the logistic equation) is split into age- or stage-specific reproduction and survival rates. There are more parameters in these models than in phenomenological models, but the parameters have more biological meaning. Matrix models are one type of demographic model and have proven to be especially useful in modeling natural populations (Equation 6-5; Leslie 1945; Caswell 1989; Tuljapurkar and Caswell 1997).

$$\mathbf{n}(t+1) = \mathbf{A}\mathbf{n}(t) \quad \text{(Equation 6-5)}$$

where

$\mathbf{n}$	=	number of individuals in each age class or life stage of a population
$\mathbf{A}$	=	matrix of age- or stage-specific reproduction and survival rates
t	=	time
$\mathbf{n}(t)$	=	population size at time t
$\mathbf{n}(t+1)$	=	population size at next time step

Matrices with linear terms can be solved analytically for the long-term population growth rate and other population-level variables.

These models can be modified to include nonlinear terms, which makes them more realistic but which adds parameters along with the potential additional uncertainty associated with estimating their values. In addition, they can require numerical simulation. Munns et al. (1997) used a stage-classified model to assess the impacts of contaminants on the population dynamics of the mummichog (described below).

Individual-based models

Individual-based models (IBMs; DeAngelis and Gross 1992) are simulation models in which the survival and reproduction of each individual in a population is followed through time. In contrast to other types of population models, a great deal of biological detail is usually included, for example, feeding and mating behavior. The parameters of IBMs have the most tightly defined biological meaning. These models are useful for exploring the dynamics of very small populations for which random events can have significant impacts. They are also potentially useful when a considerable amount is known about the impacts of contaminants on specific behaviors, and between those behaviors and vita rates (survival and reproduction). IBMs are used in a stochastic fashion, which means that not only the best estimates of parameter values are specified but their distributions as well, and many simulations are performed with alternative parameter sets. The results are presented as central tendencies with uncertainty bounds. Hallam et al. (1996) applied an individual-based population model to explore the consequences of Hg contamination on a wood stork population, and Glaser et al. (1996) modeled the dynamics of a bald eagle population exposed to multiple organic chemicals (described below).

Metapopulation models

Metapopulation models represent the dynamics of subpopulations rather than individuals. The important processes in metapopulation models are the extinction of subpopulations and the recolonization of habitat patches (Levins 1969; Gilpin and Hanski 1991). These models are appropriate when the population of interest is composed of subpopulations, that is, groups of individuals that interbreed freely but are partially isolated from other such groups. Their use is determined by both the structure of the population of interest and the spatial and temporal scales of interest. For example, Lande (1988) employed a metapopulation model to assess the effects of habitat loss and fragmentation on population dynamics of the northern spotted owl. Spromberg et al. (1998) explored the impacts of contaminants on a model metapopulation and found that, because of migration between subpopulations, contamination of one or a few subpopulations can affect the dynamics of the metapopulation as a whole. This effect is mediated not through transfer of contaminant but through effects on the movement of migrants. The contaminated subpopulations act as sinks for migrants and produce few if any emigrants, reducing the availability of migrants to other uncontaminated subpopulations. Spromberg et al. called this effect "action at a distance"; it is important to consider in predicting NR in metapopulations.

Ecosystem models

Any of the population models described above can be modified to include interactions between populations, such as competition and predation, as well as other components of ecosystems, for example, N concentration and human impacts. For example, the Lotka-Volterra system of equations describes a simple predator–prey relationship between 2 populations (Lotka 1925; Equation 6-6). In this equation, species Y preys on species X. This equation results in continuous population cycles of prey and predator:

$$\frac{dX}{dt} = rX - pXY$$
$$\frac{dY}{dt} = qXY - dY \qquad \text{(Equation 6-6)}$$

where

- X = prey population
- Y = predator population
- r = exponential growth rate of the prey population (X) in the absence of predation
- p = negative impact of predation on X
- q = positive impact of predation on Y
- d = death rate of Y in the absence of any prey

Ecosystem models can include multiple populations as well as nonliving resources. For example, ecosystem models of aquatic systems may include many linked equations describing the growth of phytoplankton populations, sources and sinks of their limiting nutrients (N, P, Si), consumption by zooplankton, fish population dynamics, and dissolved oxygen. Obviously, the number of parameters that must be estimated increases with each additional variable.

Multispecies models can aid in management decisions when interspecific interactions exert important controls on population dynamics, or when contaminants have indirect effects on the population of interest. Multispecies models usually include relatively simple representations of individual population dynamics, partly for tractability and partly to focus on the dynamics of population interactions. Qualitative multispecies models developed by Levins (1975) can be used to explore the range of such indirect effects of contaminants. The chief advantage of qualitative multispecies models is that analytical solutions can be achieved for relatively complex communities. A great deal of insight can be achieved with limited effort. The 2 chief disadvantages are these: 1) without quantification, results are often ambiguous, and 2) the models are linear. All biological interactions are nonlinear. For example, the rate at which a predator consumes prey saturates at sufficiently high prey levels. If prey levels tend to remain relatively low, then a linear approximation for the consumption rate may be reasonable. If the prey population reaches the saturation level on a frequent basis, then the linear approximation may reduce the accuracy of model predictions significantly.

Nonlinear systems of differential equations do not have this restriction, but they usually require numerical simulation. In addition, the number of parameters can be so great that uncertainties in each of them limit the usefulness of model results. A multispecies model of an aquatic ecosystem was developed by Bartell et al. (1992) and used to study ecological risks associated with contaminants (described below).

Stochastic models

Deterministic models use single values for parameters and produce a best estimate of population size and growth rate. Stochastic models use distributions of values for parameters and produce distributions of predicted population-level variables. IBMs are inherently stochastic. The other model types described above can be either deterministic or stochastic. The most important advantage of stochastic models is that they permit a more realistic treatment of certain endpoints, for example, the probability of population decline or the probability of population extinction within a given time period. Their chief disadvantage is that distributions must be estimated for each parameter, and model results can be sensitive to distribution shape as well as to correlations between variables.

Use of models in natural remediation

In principle, virtually any population model can be used to assess the impacts of NR on risks to biota. The key issues associated with the application of a model in assessing NR include these:

- Choosing appropriate endpoints, developing a toxicological relationship
- Developing an understanding of the factors regulating the dynamics of the populations of interest
- Characterizing the spatial domain of the population
- Assessing the importance of indirect impacts in communities.

Endpoints

The purpose of applying population models to risk assessment is to make measurement endpoints better reflect assessment endpoints (USEPA 1992). Deterministic endpoints such as population size and growth rate are the simplest population-level measurement endpoints to use in assessing risk. However, because of the variability inherent in natural population dynamics and our limited knowledge of the factors governing population dynamics, there is usually a great deal of uncertainty associated with model predictions. For this reason, stochastic measurement endpoints provide information that is more useful in making management decisions. Virtually any simulation model can be used to assess the sustainability of the population, which in a model analysis can be represented by the risk or probability of extinction of the population within a given time period. Specific measurement endpoints include these:

- Probability of population decline
- Probability of population size declining below a critical value
- Tendency of the population to resist decline in the face of specific types of environmental stresses
- Probability of recovery of a population following, for example, a decline due to a catastrophic stress.

The interpretation of model estimates of any of these endpoints is subjective. Determining an acceptable probability of extinction, or the reduction in probability of extinction that is sufficient to warrant a specific remediation action, is a decision that must be made in the social and economic context of the problem at hand. However, endpoints such as these represent the best information that models can provide for making remediation decisions.

Toxicological relationship

The requirements for toxicological information for population models are much more rigorous than for ERAs that use hazard quotients (HQs). A concentration–response relationship is required for population models, as opposed to a point estimate such as a no-observed-adverse-effect level (NOAEL) that suffices for ERAs based upon HQs. In addition, to be relevant to population dynamics, the results of toxicological studies must include a relationship with either survival, growth, or reproduction. While certain specific enzyme assays may indicate that an organism has been exposed to a toxicant, the relationship between assay results and impacts on reproduction and survival must be determined if the results are to be useful. For example, the induction of P450IA1-associated catalytic activity in the H4IIE rat hepatoma cells indicates that an organism has been exposed to a compound with dioxin-like activity. A relationship between enzyme induction in the laboratory and productivity of double-crested cormorants in the field provides the information needed for a population model (Tillitt et al. 1992).

The life table response experiment (LTRE) is a protocol for exploring the impacts of chemicals on vital rates (Caswell 1996). In an LTRE, a population is exposed to a set of treatments, and the survival and reproduction of each age class or stage is measured. These measured vital rates are then used in a population model to compute the long-term population growth rate or other measures of population health. LTREs can be conducted in the laboratory or can be based upon field observations. They can be thought of as a population-level bioassay (Caswell 1996), which is labor intensive but which provides information that is directly relevant to the assessment of population-level impacts.

Factors regulating population dynamics

A useful population model includes all of the processes that significantly affect natural populations. The response of a population to changes in the concentration of a contaminant may be sensitive to other processes, including chemical and nonchemical factors. For example, a small change in the reproductive rate because of chemical contamination may make relatively little difference to the probability of extinction under normal conditions. However, such a change may have a significant effect on the ability of a population to recover following another relatively rare but more extreme stress.

The primary factors that regulate a given population must be understood to some degree in order to structure the model, that is, to formulate the equations: Should density-dependent adult mortality be included? What controls the probability that eggs or newborns will reach adulthood? Are the trends in the population of interest strongly tied to trends in predators, parasites, prey, or competitors? This requirement differentiates population models from other models used in the assessment of NR. In contrast, the physical and chemical processes that dominate chemical fate and transport are well understood. They may be difficult to quantify, but the same model structure can be used across a wide range of sites.

Spatial domain

At the outset of a population modeling study, a decision must be made as to whether the population of interest should be treated as part of a larger metapopulation. This decision is typically driven by both technical information regarding the structure of the population and judgments regarding the degree of conservatism required in the management of risk. A probability of extinction, generally considered unacceptable when only one site is considered, may actually be considered acceptable in situations where it is known that the organisms at the site constitute a subpopulation of a larger metapopulation. Occasional local extinction may not affect the dynamics of the metapopulation significantly, if the extinct subpopulation may be recreated by immigration from other subpopulations. A requirement for persistence of every subpopulation will provide results that are more conservative than a metapopulation analysis.

Indirect impacts of contaminants in ecological communities

Contaminants can have unexpected impacts on populations because of the interactions that occur in ecological communities (Levins 1975). For example, a decrease in adult survivorship in one species may lead to an increase, a decrease, or no impact in other species. This complexity is due to the specific relationships among species (competitive, predatory, symbiotic) as well as to the overall structure of the food web (Puccia and Levins 1985). This example illustrates how our ability to develop models of population dynamics is often ahead of our ability to apply them. We know these effects are possible, although it is usually very difficult to assess their applicability to specific sites. The decision on the importance of impact ramification through communities determines the type of model that is most appropriate.

For example, indirect effects would probably be a relatively unimportant consideration in a study of a bioaccumulative compound that is present at low levels in the environment and to which a relatively rare top predator is especially sensitive, such as might be the case for DDE in some locations. Indirect effects might be important if the predator controlled the abundance of one of its prey species; in its absence, the structure of the prey community might change significantly.

Examples of population models used to assess contaminant impacts

In this section, we present examples of population models that have been used to assess the impacts of chemical stressors on field populations. Only studies that involved comparisons of model results with field population data are included, because such comparisons are necessary to develop the confidence needed to use the models to assess NR and the impacts of remediation actions.

Phenomenological models

Kolar et al. (1997) developed a model describing the population dynamics of mayflies (*Hexagenia*) in the western basin of Lake Erie. The purpose of the analysis was to predict the recovery time of the population, that is, the course of NR. The model was relatively simple, based upon the logistic equation, with added terms for immigration as well as losses from sediment toxicity, competition with chironomids, and fish predation. The impacts of occasional low-oxygen events were included in the model as mass mortality. Sensitivity analyses were used to explore the impacts of uncertainty in parameter values. Model computations indicated that the time to recovery of the mayfly population lay between 48 and 81 years. An attempt was made to verify the model by performing a simulation for the period from 1970 to 1990. Computed mayfly population densities ranged from 3 to 23 nymphs/m^2, compared with estimated densities of between approximately 4 and 21 nymphs/m^2. The overlap of the computed and observed densities was used to indicate that the model provided a realistic picture of mayfly population dynamics.

The results of the calibration or verification simulation suggest that the parameter values used in that simulation provide a realistic description of the population dynamics, and therefore that the predicted range of recovery times is reasonable. The model is realistic in that it includes multiple stressors, in particular, competition, predation, occasional low-oxygen events, and sediment toxicity.

This analysis suffered from the limitation that very little information was available to estimate site- and species-specific parameter values that have significant effects on model results. No quantitative dose–response relationships were available, so the effect of sediment toxicity was relatively uncertain. Similarly, the impact of competition appears to have been based on qualitative appraisals of studies conducted using other insect species. The uncertainty range in computed times to recovery was based on the range in growth rates and in the uncertainty in the year in which a low-oxygen event would occur. Inclusion of uncertainty in other parameters would certainly widen the range of computed recovery times. A key question is whether alternative sets of parameter values could be found that would result in a similar model–data comparison but different predictions of time to recovery.

Matrix models

Levin et al. (1996) developed a population model for estuarine polychaetes to quantify the effects of fuel oil, sewage, and blue-green algae on survival, reproduction, and age at maturity. They used the LTRE approach, measuring the impacts on these parameters in the laboratory and then computing the effects on the long-term population growth rate. Field populations were observed to grow at much slower rates than those in the laboratory, probably due to different physical conditions as well as the absence in the laboratory of parasites, predators, and competitors. Therefore, the model has limited utility in predicting the actual rate of NR in the field. However, this study provides insight into the relative effects of these stressors on vital rates and the sensitivity of population growth rate to these effects. Thus, the model can be used to assess the relative changes in long-term population growth rate that might occur as contaminant levels decline during NR. This is important and useful information for risk managers.

Munns et al. (1997) used a stage-classified model of the population dynamics of the mummichog (*Fundulus heteroclitus*) to assess the impacts of dioxin and PCBs on population growth rate. Parameterization of the model was based upon published studies of survival and reproduction of the mummichog. The concentration–response relationship for dioxin was developed from a laboratory investigation of survival and reproduction in this species; the relationship for PCBs was based upon laboratory measurements of survival and reproduction in fish collected from New Bedford Harbor, Massachusetts, a site contaminated with PCBs. The results of the model analysis consisted of a relationship between body burden and population growth rate for each contaminant. The relative impacts of these contaminants on population growth rate were estimated, and these were used to conclude that the potential for long-term population impacts in the field existed.

Model results for PCBs were related to population conditions in New Bedford Harbor. In the most contaminated site, the model predicted significant impacts on population dynamics. The observation of an apparently healthy population of mummichogs indicated that either parameterization of the model was incorrect, compensatory mechanisms that ameliorate toxic effects were active, or immigration helped to maintain the population. In summary, the conclusions regarding the relative impacts of PCBs and dioxin on population growth rate form part of a full assessment of NR and apparently would result in an overestimate of population impacts in the field. Results could be made more realistic, and therefore potentially less conservative, with additional field data on population processes such as immigration and on population size.

Barnthouse et al. (1990; see also Barnthouse 1993) developed matrix models of striped bass and menhaden population dynamics and exposed the model populations to a hypothetical toxicant. The model indicated that the 2 species differed in their responses to toxicants because of differences in their life histories. This model could also be used to compare the responses of these fish populations to changes in the levels of toxicants, (i.e., to NR) for example, by computing the reduction in the probability of extinction following a given reduction in the average contaminant level. The key uncertainty in this analysis was the extrapolation from the available toxicity data. In addition, the model results are valid for long-term simulations under conditions similar to those historically observed because recruitment rates are strongly influenced by random environmental events, and the model was calibrated by adjusting recruitment rates to match the observed population distribution.

Individual-based models

Hallam et al. (1996) used an IBM of a population of wood storks in the Florida Everglades to assess the population consequences of Hg contamination. The model included 3 major components:

1) The toxicokinetic component estimated Hg body burden in the birds from levels in their fish prey.

2) A series of 6 concentration–response relationships related Hg body burden to flight, foraging, and searching behaviors; to egg production; to hatching success; and to hatchling survival.
3) Finally, the population model integrated foraging behavior and computed body burdens and the assumed toxicological relationships to predict population size and food consumption rate.

Unfortunately, the data needed to constrain the functions and their parameter values are very limited. The toxicokinetic model was not tested or calibrated because of the lack of Hg data. The 6 concentration–response relationships were developed based upon an extensive review of existing toxicological studies for Hg in birds; however, no measured concentration–response relationships could be gleaned from that literature, so the relationships used in the model were considered only as examples. Finally, the model simulations were performed for only 120 days because information on migration and other factors was not available for this population. Therefore, the long-term impacts of Hg contamination could not be assessed. These limitations resulted in a study that, while it provides a step in the right direction, cannot yet be used to assess quantitatively the process of NR.

However, this model demonstrates the potential for developing models that can predict the impacts of chemicals on population dynamics. The mathematics involved are simple, the important processes affecting population dynamics can be listed, and sets of reasonable mathematical functions for these processes can be developed. This means that the world of possible behaviors can be explored mathematically, providing insights into the ecotoxicology of Hg in wood stork populations.

Glaser et al. (1996) developed a framework consisting of linked exposure, toxicology, and IBMs and applied this framework to the resident bald eagle population of the lower Columbia River. The bioaccumulation model was used to compute average levels of PCBs and DDE in eagle eggs as well as the within-population variation in contaminant levels. Concentration–response relationships measured in other field populations of birds were used to compute the effects of the chemicals on fledgling production. Finally, the IBM was used to compute population size and the probability of population decline and extinction. Monte Carlo simulations were performed, incorporating uncertainty in survival rates.

Model computations were compared with 20 years of field observations of population size, and the model was calibrated using immigration rate as the key calibration parameter. Model results indicated that immigration must have accounted for most of the recent population growth that has been observed, that DDE is probably the main cause of the observed poor reproductive rate, and that the population should now be self-sustaining in the absence of immigration. The calibrated model was then used to illustrate the projection of population dynamics into the future, subject to projected temporal trends in DDE, PCB, and dioxin and furan levels in fish.

The calibration of the exposure and population models to field data and the availability of a concentration–response relationship make this study unusual. The immigration rate was the unconstrained calibration parameter; the results would be strengthened by direct measurement of the immigration rate. In addition, an analysis of the uncertainty associated with the concentration–response relationship and the reproduction and survival rates used in the model would provide an indication in the uncertainty of the overall conclusions. Another strength of the analysis is the direct linkage to DDE and PCB exposure levels in the prey provided by the bioaccumulation model, which was calibrated to field-measured contaminant levels. A weakness is that the bioaccumulation of dioxins could not be modeled because of a lack of data. The PCB and DDE bioaccumulation model could be expanded to link levels in fish to levels in sediment and water, as was done in the Hudson River model described by Peyton et al. (2000) and Quantitative Environmental Analysis (QEA 1999). Combined with a physical or chemical fate model, such a linked set of models would provide a calibrated tool for directly assessing the effects of NR and of potential remediation activities on population-level processes in the bald eagle.

Ecosystem models

A review of ecosystem-level models was provided by Suter (1993). This discussion concentrates on one family of models that has been used to evaluate the impacts of chemicals in field enclosures and shows promise of being a useful tool for the assessment of NR.

Bartell et al. (1992) developed the Comprehensive Aquatic System Model (CASM) for estimating risks associated with contaminants. The state variables of the model included nutrients, phytoplankton, zooplankton, planktivorous fish, and piscivorous fish. The results of toxicity bioassays were incorporated by using the model itself to simulate the results of the bioassays. In this way, the changes in individual parameters that were necessary to produce the adverse effects observed during each toxicity study were estimated. The model was then used to assess the impacts of phenol on a series of experimental ponds at the Oak Ridge National Laboratory in Tennessee. The predictive capabilities of the model were limited, in part because of the relatively great variability in the pond data and in part because of a lack of information concerning the appropriate model structure and parameterization for the ponds.

The value of the model lies in the insight it can provide into the dynamics of aquatic ecosystems, especially the interactions between ecological processes and toxicological impacts. For example, piscivore growth rate was found to be sensitive to different parameters at different times of the year, suggesting the ways in which the observed impacts of a short-term exposure to a toxicant may depend on season. In addition, the model can be used to focus research strategies for improving predictive capabilities. For example, sensitivity analysis indicated that additional information concerning the structure of the food web in the ponds and the actual exposure concentration was particularly helpful in improving model performance.

The Littoral Ecosystem Risk Assessment Model (LERAM) was developed by Hanratty and Stay (1994) by modification of the CASM. It was used to estimate the effects of chlorpyrifos on littoral enclosures studied by Brazner et al. (1989). The effects of chlorpyrifos on individual populations were determined by using LERAM to model the results of toxicity studies, as was done with the phenol model. Model parameters were calibrated using data from control enclosures, and then the calibrated model was used to compute population biomasses in the presence of chlorpyrifos. Ecological risk was estimated by observing changes in the computed total biomass and production of each taxon. Calculations were generally within a factor of 2 for biomass. Model estimates of production were not significantly different from observed values, although the data were quite variable.

Hanratty and Liber (1996) continued the development of LERAM, using it to predict the persistence and ecological effects of diflubenzuron on littoral enclosures. Laboratory toxicity studies were analyzed using the model itself to compute the changes in model parameter values necessary to produce the changes in biomass observed in the toxicity studies. In this study, model simulations were performed before the enclosure experiments were performed, to determine the appropriate range of diflubenzuron concentrations for use in the field enclosure experiments. LERAM correctly predicted that changes in the community would begin at 0.8 mg/L, and that concentrations greater than 35 mg/L would yield negligible new information on chemical effects. Within this range of concentrations, some effects on individual taxa were not predicted correctly, and the model over-predicted the importance of indirect effects.

So far, the model has been applied to small littoral enclosures with some success. Application to open water bodies will no doubt add an additional level of uncertainty to calculations. As experience with the model increases and as more toxiciological data become available, the ability of LERAM to predict ecosystem effects may make it a useful tool for projecting NR.

UNCERTAINTY

The use of models to help solve environmental problems is often resisted because of the uncertainty associated with their conclusions. Uncertainty is always difficult and often impossible to quantify in exposure and population models. It is hard enough to develop best estimates of model parameters; the data are generally insufficient to compute the distributions of key parameter values.

However, even if uncertainty cannot be quantified, models can be important tools for assessing NR. Any information that suggests a central tendency for parameter values results in a central tendency for model predictions. That is, if the distributions of values for key parameters exhibit peaks, even broad ones, then the model results based upon the most likely parameter values (the peaks) provides an estimate of risk that is more likely than any other estimate. Thus, the first question in an uncertainty analysis is semiquantitative or even qualitative: Is there information that indicates that the chosen model structure and parameter values are more likely than other possibilities? If the answer is yes, then the risk manager is better off with the best estimate provided by the model than without it. That is, even in the absence of a quantitative uncertainty analysis, a model can provide a basis for risk management decision-making. The next question is quantitative and concerns the degree to which a model can reduce the uncertainty in predictions and therefore provide a basis for less conservative risk management decisions.

Exposure models

The sources of uncertainty include the structure of the model, that is, the equations, and the site-specific data. The key mechanisms underlying exposure models are relatively well understood. This means that model structure and, to some degree, parameterization are often translatable from one site to the next. For example, the chemical assimilation efficiencies are likely to be similar across predatory fish species; generic weight-based relationships that described respiration rates permit parameterization for a large range of species.

The key uncertainties in exposure models are the site-specific data requirements: food web structure, dietary composition, depth of bioavailable sediment, spatial and temporal domains over which the food web averages exposure, and average exposure levels. Because of the extensive experience now available for hydrophobic compounds such as PCBs and DDE in fish, uncertainties in model parameters (bioenergetics, contaminant assimilation efficiencies) are secondary to uncertainties associated with site-specific data.

For the assessment of NR, the most important issue associated with food web structure and dietary composition is the ultimate source of contaminant, for example, surface sediments versus water column in aquatic systems and the bioavailable depth in the sediment bed. In addition, uncertainty in exposure levels arises from the fact that water column levels can be very variable temporally and sediment levels can be very variable spatially. Another important source of uncertainty lies in the possibility that changes to the ecological community might alter the food web. For example, the invasion of lakes by zebra mussels may affect the availability of organic chemicals to ducks (Mazak et al. 1997). Finally, changes in the abundance of prey can affect the physiological condition of a predator, for example, lipid content, which in turn can change the degree of bioaccumulation.

Among the parameters of aquatic bioaccumulation models, the greatest uncertainty is associated with the bioaccumulation factors describing the relationship between the water column and primary producers (e.g., Swackhamer and Skoglund 1993; Glaser and Connolly 1995) and sediments and benthic invertebrates (e.g., Tracey and Hanson 1996). Models of the accumulation of chemicals by algae and benthic and pelagic invertebrates have been developed (e.g., Gabric et al. 1990; Landrum et al. 1991; Boese et al. 1997; Wang et al. 1997; Endicott et al. 1998). To the extent that these models permit the use of site- and species-specific data, they may help reduce this uncertainty and may improve estimates of the impacts of large, short-term changes in exposure concentrations on contaminant levels throughout the food web.

Estimating the degree of uncertainty in models of exposure increases in difficulty with the complexity of the model. The uncertainty associated with simple steady-state models can be calculated directly from the data used to develop the model. For example, the uncertainty associated with a BAF can be estimated using the SEM of the data that were used to compute the factor. At the other extreme, estimating the uncertainty associated with a calibrated, time-variable, food web bioaccumulation model is an active area of research. Traditional Monte Carlo procedures are common methods but suffer from data limitations: Results can be very sensitive to the shapes of the distributions of parameter values and the correlations between them. In addition, calibration itself reduces uncertainty considerably, because the acceptable range of parameter values becomes restricted to those that produce a reasonable calibration. In fact, the uncertainty in a calibrated model arises only insofar as there are variables to which the calibration is insensitive but to which projections are sensitive.

Uncertainty in a calibrated model can be estimated in 3 ways. First, a "brute force" method can be used. This is a modified Monte Carlo analysis in which the distributions of parameter values are estimated based on a priori information, model simulations are performed for the calibration period, and the resulting simulations are culled for those that provide what is determined to be a reasonable calibration. The acceptable sets of parameter values are then used in projecting future contaminant levels, and the distribution of projected levels provides an estimate of model uncertainty. This method is very labor and computationally intensive. It has been applied in a Bayesian framework, using a likelihood function to compare the calibration simulations with data, by Dilks et al. (1992).

Second, an alternative method was developed by DiToro and Parkerton (1993) as part of the Green Bay Mass Balance Study. This is considerably more involved mathematically but requires fewer simulations. It has been applied to the Green Bay fate and transport model (Bierman et al. 1992).

Finally, a less labor-intensive procedure for estimating the uncertainty associated with a calibrated model can be developed based upon an understanding of the important questions under consideration and the key processes controlling bioaccumulation. For example, in studies of contamination in aquatic systems, the most important issue is often the relative contributions of sediment– and water column–associated contaminant to the food web, because this determines the response to NR and to anthropogenic remediation. Therefore, the uncertainty associated with model projections can be bounded by developing 2 calibrations. The first calibration includes the best estimates of each parameter in the model. The alternative calibration includes different parameter values chosen so as to result in contributions of sediment and water column contaminant that are as different as possible from the best calibration. These alternative parameter values must still be consistent with the available experimental measurements and must result in an acceptable relationship between model results and field data. Model projections performed using both the best and the alternative calibrations provide an indication of model uncertainty relevant to the issues at hand. In addition to requiring an understanding of the key processes underlying bioaccumulation, such an analysis is based to some degree upon professional judgment of reasonable bounds for parameter values.

Population models

Population models are inherently more uncertain than exposure models because the structure, or equations, of the model are population or site specific. In addition, field measurements of population size, productivity, and survival are often much more limited than are measurements of contaminant levels. A few measurements of chemical levels can go a long way toward calibrating an exposure model, while measured population trends are required to calibrate a population model. Overall, population models can now provide insight into mechanisms underlying populations dynamics and can be used to explore the range of behaviors likely to occur in natural populations. To a limited degree, projections of the absolute effects of NR on population dynamics can be performed in some cases. However, the lack of field data for key processes and limited long-term population trend data usually result in relatively great uncertainty in model projections. They are most useful for predicting relative changes in population size and growth rate due to changes in exposure, and in this way, they can provide risk managers with essential information. Because population models are becoming more common in the field of ecotoxicology, the usefulness of the information gleaned from population analyses is likely to increase rapidly in the near future.

CONCLUSIONS

In assessing NR, the issue is not whether or not to use a model but what types of models will be applied. Anyone involved in risk management for environmental contaminants, including the assessment of NR, uses models, either explicitly or implicitly. The simplest risk assessment may rely upon an HQ computed using a laboratory-measured NOAEL. If the species used in the toxicity study differs from the species of interest in the field, then an extrapolation model is in effect being used, which can be stated as follows: Responses of taxonomically related species to chemical exposure are similar. If contaminant levels in the prey form the basis for comparison between the results of the laboratory study and the field exposure, then an exposure model has been applied implicitly: Toxicokinetics in the lab and field species are assumed to be sufficiently alike that similar prey levels result in similar concentrations at the site of action within the body. Risk assessments involving multiple chemicals often result in at least one HQ greater than 1, in which case the estimation of the population impacts becomes qualitative: How much greater than 1 is significant to the population? Such an interpretation implies a relationship between the HQ and population sustainability, which is in effect an unquantified population model.

The explicit use of mathematical models clarifies the assumptions behind risk management decisions. This is their primary value. Models are often criticized for their uncertainty, and therein lies another value: By focusing attention on the uncertainties associated with predicting the results of NR or potential remediation activities, models provide a context for the comprehensive evaluation of costs, benefits, and risks. Mechanistic models are superior to purely statistical models because they inform the user what processes are most uncertain and therefore what further data collection would be of use for improving management decisions.

REFERENCES

Ahlborg UG, Becking GC, Birnbaum LS, Brouwer A, Derks HJGM, Feeley M, Golor G, Hanberg A, Larsen JC, Liem AKD, Safe SH, Schlatter C, Waern F, Younes M, Yrjanheikki E. 1994. Toxic equivalency factors for dioxin-like PCBs: Report on a WHO-ECEH and IPCS consultation, December 1993. *Chemosphere* 28:10–49.

Ankley GT, Erickson R, Sheedy B, Kosian PA, Mattson V, Cox JS. 1997. Evaluation of models for predicting the phototoxic potency of polycyclic aromatic hydrocarbons. *Aqua Toxicol* 37:37–50.

Barnthouse LW. 1993. Population-level effects. In: Suter GW, editor. Ecological risk assessment. Boca Raton FL: Lewis. p 247–274.

Barnthouse LW. 1998. Modeling ecological risks of pesticides: A review of available approaches. In: Schuurmann G, Markert B, editors. Ecotoxicology. New York NY: J Wiley. p 769–798.

Barnthouse LW, Suter G, Rosen A. 1990. Risks of toxic contaminants to exploited fish populations: Influence of life history, data uncertainty and exploitation intensity. *Environ Toxicol Chem* 9:297–311.

Bartell SM, Gardner RH, O'Neil RV. 1992. Ecological risk estimation. Boca Raton FL: Lewis.

Beamish FWH. 1964. Respiration of fishes with special emphasis on standard oxygen consumption. *Can Zool* 42:177–188.

Bierman VJ, DePinto JV, Young TC, Rodgers PW, Martin SC, Raghunathan R, Hinz SC. 1992. Development and validation of an integrated exposure model for toxic chemicals in Green Bay, Lake Michigan. Final report to the USEPA Large Lakes and River Research Branch, Grosse Ile MI. Cooperative Agreement CR-814885.

Boese BL, Lee H, Echols S. 1997. Evaluation of a first-order model for the prediction of the bioaccumulation of PCBs and DDT from sediment into the marine deposit-feeding clam Macoma nasuta. *Environ Toxicol Chem* 16:1545–1553.

Bradbury SP, Mekenyan OG, Ankley GT. 1998. The role of ligand flexibility in predicting biological activity: Structure-activity relationships for aryl hydrocarbon, estrogen, and androgen receptor binding affinity. *Environ Toxicol Chem* 17:15–25.

Brazner J, Heinis L, Jensen D. 1989. A littoral enclosure for replicated field experiments. *Environ Toxicol Chem* 8:1209–1216.

Calabrese EJ, Baldwin L. 1994. A toxicological basis to derive a generic interspecies uncertainty factor. *Environ Health Perspect* 102:14–17.

Campbell PGC, Tessier A. 1996. Ecotoxicology and metals in the aquatic environment: geochemical aspects. In: Newman MC, Jagoe CH, editors. Ecotoxicology, a hierarchical treatment. Boca Raton FL: CRC Pr.

Caswell H. 1989. Matrix population models: construction, analysis, and interpretation. Sunderland MA: Sinauer Assoc.

Caswell H. 1996. Demography meets ecotoxicology: untangling the population level effects of toxic substances. In: Newman MC, Jagoe CH, editors. Ecotoxicology: A hierarchical treatment. Boca Raton FL: CRC Pr.

Clark TP, Norstrom RJ, Fox GA, Woon HT. 1987. Dynamics of organochlorine compounds in herring gulls (*Larus argentatus*): II. A two-compartment model and data for ten compounds. *Environ Toxicol Chem* 6:547–559.

Connolly JP. 1991. Application of a food chain model to polychlorinated biphenyl contamination of the lobster and winter flounder food chains in New Bedford Harbor. *Environ Sci Technol* 25:760–770.

Connolly JP, Glaser D. 1999. DDE and PCB bioaccumulation in female sea lions of the California Channel Islands. Submitted to *Continental Shelf Research*.

Connolly JP, Parkerton TF, Quadrini JD, Taylor ST, Thuman AJ. 1992. Development and application of a model of PCBs in the Green Bay, Lake Michigan walleye and brown trout and their food webs. Report for Large Lakes Research Station, USEPA, Grosse Ile MI. Cooperative Agreement CR-815396.

Connolly JP, Tonelli R. 1985. A model of kepone in the striped bass food chain of the James River estuary. *Est Coastal Shelf Sci* 20: 349–366.

Costa H, Wade T, Bailey A. 1994. Analytical chemistry data report for the Southern California Natural Resource Damage Assessment. U.S. Department of Justice.

DeAngelis DL, Gross LJ. 1992. Individual-based models and approaches in ecology. New York NY: Chapman & Hall.

Dilks DW, Canale RP, Meier PG. 1992. Development of Bayesian Monte-Carlo techniques for water quality model uncertainty. *Ecolog Model* 62:149–162.

DiToro DM, Mahony JD, Hansen DJ, Scott KJ, Hicks MB, Mayr SM, Redmond MS. 1990. Toxicity of cadmium in sediments: The role of acid volatile sulfide. *Environ Toxicol Chem* 9:1487–1502.

DiToro DM, Parkerton TF. 1993. Uncertainty analysis methodology for Green Bay models. Final report to the USEPA Large Lakes Research Station, Grosse Ile MI. Cooperative Agreement CR-815397-01-0.

Endicott D, Kreis RG, Mackelburg L, Kandt D. 1998. Modeling PCB bioaccumulation by the zebra mussel (*Dreissena polymorpha*) in Saginaw Bay, Lake Huron. *Great Lakes Res* 24:411–426.

Erickson R, McKim J. 1990. A simple flow-limited model for exchange of organic chemicals at fish gills. *Environ Toxicol Chem* 9:159–165.

Evans D. 1984. Temperature independence of the annual cycle of standard metabolism in the pumpkinseed. *Trans Am Fish Soc* 113:494–512.

Gabric AJ, Connell DW, Bell P. 1990. A kinetic model for bioconcentration of lipophilic compounds by oligochaetes. *Water Resources* 24:1225–1231.

Garcelon DK. 1994. Effects of organochlorine contaminants on bald eagle reproduction at Santa Catalina Island. Expert report, Southern California Bight Natural Resource Damage Assessment, U.S. Department of Justice.

Garcelon DK, Romsos JS, Golightly P. 1994a. Food habits of bald eagles on Santa Catalina Island, January–July 1993. Expert report, Southern California Bight Natural Resource Damage Assessment, U.S. Department of Justice.

Garcelon DK, Tomassi S, Kristan D, Delaney D. 1994b. Food habits of the bald eagle on Santa Catalina Island. Expert report, Southern California Bight Natural Resource Damage Assessment, U.S. Department of Justice.

Gilpin M, Hanski I. 1991. Metapopulation dynamics empirical and theoretical investigations. San Diego CA: Academic Pr.

Glaser D, Connolly JP. 1994. Fish and wildlife toxicokinetic modeling and risk analysis. Presentation at Annual Meeting of the Society for Risk Analysis; December 1994.

Glaser D, Connolly JP. 1995. Uncertainty in bioaccumulation modeling. In: Toxic substances in water environments: Assessment and control. Water Environment Federation Specialty Conference; 1995. p 8-49 – 8-60.

Glaser D, Connolly JP, Bergoffen A. 1996. Population impacts of contaminants on the resident bald eagles of the Lower Columbia River. Presentation at Society for Environmental Toxicology and Chemistry (SETAC) Annual Meeting; 1996 Nov 17–21; Washington DC. Pensacola FL: SETAC.

Glaser D, Connolly JP, Haggard J. 1997. Development of a quantitative bioaccumulation model for PCBs in fish from the Upper Hudson River. Presentation at Annual Meeting of the Society for Environmental Toxicology and Chemistry (SETAC); 1997 Nov 16–20; San Francisco CA. Pensacola FL: SETAC.

Gobas FAPC. 1993. A model for predicting the bioaccumulation of hydrophobic organic chemicals in aquatic food-webs: Application to Lake Ontario. *Ecolog Model* 69:1–17.

Gobas FAPC, McCorquodale JR, Haffner GD. 1993. Intestinal absorption and biomagnification of organochlorines. *Environ Toxicol Chem* 12:567–576.

Gobas FAPC, Muir DCG, Mackay D. 1989. Dynamics of dietary bioaccumulation and faecal elimination of hydrophobic organic chemicals in fish. *Chemosphere* 17:943–962.

Gobas FAPC, Z'Graggen MN, Zhang X. 1995. Time response of the Lake Ontario ecosystem to virtual elimination of PCBs. *Environ Sci Technol* 29:2038–2046.

Gress F. 1994. Reproductive performance, eggshell thinning, and organochlorines in brown pelicans and double-crested cormorants breeding in the Southern California Bight. Expert report, Southern California Bight Natural Resource Damage Assessment, U.S. Department of Justice.

Hallam T, Trawick T, Wolff W. 1996. Modeling effects of chemicals on a population: Application to a wading bird nesting colony. *Ecolog Model* 92:155–178.

Hanratty M, Liber K. 1996. Evaluation of model predictions of the persistence and ecological effects of diflubenzuron in a littoral ecosystem. *Ecolog Model* 90:79–95.

Hanratty M, Stay F. 1994. Field evaluation of the littoral ecosystem risk assessment model's predictions of the effects of chlorpyrifos. *Appl Ecol* 31:439–453.

Hanson PC, Johnson TB, Schindler DE, Kitchell JF. 1997. Fish bioenergetics 3.0. Madison WI: Univ Wisconsin-Madison Center for Limnology.

Hayton WL, Barron MG. 1990. Rate-limiting barriers to xenobiotic uptake by the gill. *Environ Toxicol Chem* 9:151–157.

HydroQual Inc. 1990. Analysis of PCBs and metals in the flounder and lobster and their food chains in New Bedford Harbor, Massachusetts. Report prepared for Battelle Ocean Sciences, Duxbury, MA. Report # BATL0054.

HydroQual Inc. 1995. Addendum to Green Bay final report: Food chain model projections. Report for Large Lakes Research Station, USEPA, Grosse Ile MI.

HydroQual Inc. 1997. Southern California bight damage assessment food web/pathways study. Report for the National Oceanic and Atmospheric Administration (NOAA).

Kolar C, Hudson P, Savino J. 1997. Conditions for the return and simulation of the recovery of burrowing mayflies in Western Lake Erie. *Ecolog Appl* 7:665–676.

Lande R. 1988. Demographic models of the northern spotted owl (*Strix occidentalis caurina*). *Oecologia (Berlin)* 75:601–607.

Landrum PF, Eadie B, Faust W. 1991. Toxicokinetics and toxicity of a mixture of sediment-associated polycyclic aromatic hydrocarbons to the amphipod *Diporeia* sp. *Environ Toxicol Chem* 10:35–46.

Leslie PH. 1945. On the use of matrices in certain population mathematics. *Biometrika* 33:183–212.

Levin L, Caswell H, Bridges T, DiBacco C, Cabrera D, Plaia G. 1996. Demographic responses of estuarine polychaetes to pollutants: Life table response experiments. *Ecolog Appl* 6:1295–1313.

Levins R. 1969. Some demographic and genetic consequences of environmental heterogeneity for biological control. *Bull Entomol Soc America* 15:237–240.

Levins R. 1975. Evolution in communities near equilibrium. In: Cody ML, Diamond JM, editors. Ecology and evolution of communities. Cambridge MA: Belknap Pr.

Limno-Tech Inc., Menzie Cura & Associates, The Cadmus Group Inc. 1996. Phase 2 report: Further site characterization and analysis. Volume 2, Preliminary model calibration report, Hudson River PCBs reassessment RI/FS. Prepared for the USEPA, Region II.

Lotka AJ. 1925. Elements of mathematical biology. New York NY: Dover.

Malthus RT. 1978. An essay on the principle of population as it affects the future improvement of society. London UK: Johnson.

Mazak EJ, MacIsaac HJ, Servos MR, Hesslein R. 1997. Influence of feeding habits on organochlorine contaminant accumulation in waterfowl on the Great Lakes. *Ecolog Appl* 7:1133–1143.

McCarty L, Mackay D. 1993. Enhancing ecotoxicological modeling and assessment. *Environ Sci Technol* 27:1719–1728.

McCarty L, Mackay D, Smith A, Ozburn G, Dixon D. 1991. Interpreting aquatic toxicity QSARs: The significance of toxicant body residues at the pharmacologic endpoint. *Sci Total Environ* 109/110:515–525.

McCarty L, Mackay D, Smith A, Ozburn G, Dixon D. 1992. Residue-based interpretation of toxicity and bioconcentration QSARs from aquatic bioassays: Neutral narcotic organics. *Environ Toxicol Chem* 11:917–930.

McLachlan M. 1994. Model of the fate of hydrophobic contaminants in cows. *Environ Sci Technol* 28:2407–2414.

Munns W, Black D, Gleason T, Salomon K, Bengtson D, Guthjahr-Gobell R. 1997. Evaluation of the effects of dioxin and PCB on *Fundulus heteroclitus* populations using a modeling approach. *Environ Toxicol Chem* 16:1074–1081.

Newman MC, Dixon PM. 1996. Ecologically meaningful estimates of lethal effect in individuals. In: Newman MC, Jagoe CH. Ecotoxicology: A hierarchical treatment. Boca Raton FL: Lewis. p 225–254.

Newman MC, Jagoe CH. 1996. Ecotoxicology: A hierarchical treatment. Boca Raton FL: Lewis.

Newsted JL, Giesy JP, Ankley GT, Tillitt DE, Crawford RA, Gooch JW, Jones PD, Denison MS. 1995. Development of toxic equivalency factors for PCB congeners and the assessment of TCDD and PCB mixtures in rainbow trout. *Environ Toxicol Chem* 14:861–871.

Nichols JW, Larsen CP, McDonald ME, Niemi GJ, Ankley GT. 1995. Bioenergetics-based model for accumulation of polychlorinated biphenyls by nestling tree swallows/ *Tachycineta bicolor*. *Environ Sci Technol* 29:604–612.

Norstrom RJ, Clark TP, Jeffrey DA, Won HT, Gilman AP. 1986. Dynamics of organochlorine compounds in herring gulls (*Larus argentatus*: I. Distribution and clearance of (^{14}C)DDE in free-living herring gulls (*Larus argentatus*). *Environ Toxicol Chem* 5:41–48.

Norstrom RJ, McKinnon AE, DeFreita ASW. 1976. A bioenergetics-based model for pollutant accumulation by fish. Simulation of PCB and methylmercury residue levels in Ottawa River yellow perch (*Perca flavescens*). *J Fish Res Bd Can* 13:248–267.

Opperhuizen A, Schrap SM. 1988. Uptake efficiencies of two polychlorobiphenyls in fish after dietary exposure to five different concentrations. *Chemosphere* 17:253–262.

Peyton B, Connolly JP, Clement P. 2000. Modeling of natural remediation. In: Swindoll CM, Stahl RG, editors. Chapter X, Natural remediation of environmental contaminants: Its role in ecological risk assessment and risk management. New York NY: Wiley.

Puccia CJ, Levins R. 1985. Qualitative modeling of complex systems. Cambridge MA: Harvard Univ Pr.

[QEA] Quantitative Environmental Analysis. 1999. PCBs in the Upper Hudson River. Volume 2, A model of PCB fate, transport and bioaccumulation. Prepared for the General Electric Company.

Rice JA. 1981. Derivation and application of a bioenergetics model for largemouth bass (*Micropterus salmoides*) [PhD thesis]. Madison WI: Univ Wisconsin.

Russom C, Bradbury S, Broderius S, Hammermeister D, Drummond R. 1997. Predicting modes of toxic action from chemical structure: Acute toxicity in the fathead minnow (*Pimephales promelas*). *Environ Toxicol Chem* 16:948–967.

Spromberg JA, John BM, Landis WG. 1998. Metapopulation dynamics: Indirect effects and multiple distinct outcomes in ecological risk assessment. *Environ Toxicol Chem* 17:1640–1649.

Suter GW. 1993. Ecological risk assessment. Boca Raton FL: Lewis.

Swackhamer DL, Skoglund RS. 1993. Bioaccumulation of PCBs by algae: Kinetics versus equilibrium. *Environ Toxicol Chem* 12:831–838.

Swartzmann G. 1994. Simulating the long-term effects of DDE on Channel Islands bald eagle and peregrine falcon populations. Expert report, Southern California Bight Natural Resource Damage Assessment, U.S. Department of Justice.

Thomann RV, Connolly JP. 1984. Model of PCB in the Lake Michigan lake trout food chain. *Environ Sci Technol* 18:65–71.

Thomann RV, Shkreli F, Harrison S. 1997. A pharmacokinetic model of cadmium in rainbow trout. *Environ Toxicol Chem* 16:2268–2774.

Tillitt DE, Ankley GT, Giesy JP, Ludwig JP, Kurita-Matsuba H, Weseloh DV, Ross PS, Bishop CA, Sileo L, Stromborg KL, Larson J, Kubiak TJ. 1992. Polychlorinated biphenyl residues and egg mortality in double-crested cormorants from the Great Lakes. *Environ Toxicol Chem* 11:1281–1288.

Tracey GA, Hansen DJ. 1996. Use of biota-sediment accumulation factors to assess similarity of nonionic organic chemical exposure to benthically-coupled organisms of differing trophic mode. *Arch Environ Contam Toxicol* 30:467–475.

Tuljapurkar S, Caswell H. 1997. Structured-population models in marine/terrestrial/and freshwater systems. New York NY: Chapman & Hall.

[USEPA] U.S. Environmental Protection Agency. 1992. Framework for ecological risk assessment. Washington DC: USEPA.

[USEPA] U.S. Environmental Protection Agency. 1995. Great Lakes water quality initiative criteria documents for the protection of wildlife. Washington DC: USEPA.

Walker CH. 1987. Kinetic models for predicting bioaccumulation of pollutants in ecosystems. *Environ Pollut* 44:227–240.

Wang XL, Ma YJ, Yu WJ, Geyer HJ. 1997. Two-compartment thermodynamic model for bioconcentration of hydrophobic organic chemicals by alga: Quantitative relationship between bioconcentrations factor and surface area of marine algae or octanol/water partition coefficient. *Chemosphere* 35:1781–1797.

Ward CH. 1998. The role of ligand flexibility in predicting biological activity: Structure-activity relationships for aryl hydrocarbon, estrogen, and androgen receptor binding affinity. *Environ Toxicol Chem* 17:15–25.

Watanabe K, Bois F, Zeise L. 1992. Interspecies extrapolation: A reexamination of acute toxicity data. *Risk Anal* 12:301–310.

Wiemeyer SN, Bunck CM, Stafford CJ. 1993. Environmental contaminants in bald eagle eggs 1980–84 and further interpretations of relationships to productivity and shell thickness. *Arch Environ Contam Toxicol* 24:213–227.

7

Natural Remediation of Polynuclear Aromatic Hydrocarbons and Other Petroleum Hydrocarbons

Robert A. Pastorok, Christina Noftsker, Timothy J. Iannuzzi, David F. Ludwig, Robert C. Barrick, Michael V. Ruby, D. Frederick Bodishbaugh

NATURAL REMEDIATION (NR) is a potentially viable option for management of sites with elevated concentrations of polynuclear aromatic hydrocarbons (PAHs) and other petroleum hydrocarbons (PHs) that originate from creosoting operations, gas manufacturing, and spills of crude oil or refined products (Neff 1987; Baker et al. 1990; Booth et al. 1991). Natural remediation can work as effectively as active cleanup and is less damaging than invasive techniques (e.g., sediment removal), especially in shoreline or wetland habitats sensitive to physical disturbances (Johnson and Pastorok 1985; Baker et al. 1990; Foster et al. 1990; Booth et al. 1991). Preslo et al. (1989) indicated that NR is effective for soils containing gasolines, fuel oils (No. 2, diesel, kerosene), coal tar residues, and chlorinated solvents, all of which are amenable to dispersive processes and biodegradation.

"Natural remediation" is the overall process by which chemical concentrations and/or bioavailability in environmental media are decreased, with a corresponding reduction in toxicity or risk, without human intervention. "Enhanced natural remediation" involves human intervention to improve natural processing of chemicals and increase the recovery rate of biological communities. At a minimum, NR of PAHs and other PHs involves physical–chemical processing and biodegradation, which break down organic contaminants and generally reduce the toxicity in

Natural Remediation of Environmental Contaminants. Michael Swindoll et al., editors.
 ISBN 1-880611-33-3.

habitats such as surface waters, sediments, and soils. Physical–chemical processing includes dispersal, sorption, migration, volatilization, and chemical degradation. In cases where macrobiota are affected, NR also involves recovery of the higher biological community.

"Biological recovery" is defined as "the re-establishment of a healthy biological community in which the plants and animals characteristic of that community are present and functioning normally" (Baker et al. 1990). Because natural populations normally exhibit fluctuations in abundance, age structure, sex ratios, and other parameters, biological communities are not static in either their structure (e.g., relative abundance of species) or demographics of species populations (Sutherland 1974; Sousa 1984). Therefore, it is not possible to precisely define a "recovered" community in terms of species composition, community structure, and stable age structure of populations.

Natural remediation consists of 3 phases, each of which is defined by the dominant process that improves the physical–chemical habitat or reestablishes biological community structure (Figure 7-1). The 3 dominant phases are these:

1) Physical–chemical processing
2) Microbial biodegradation
3) Biological community recovery.

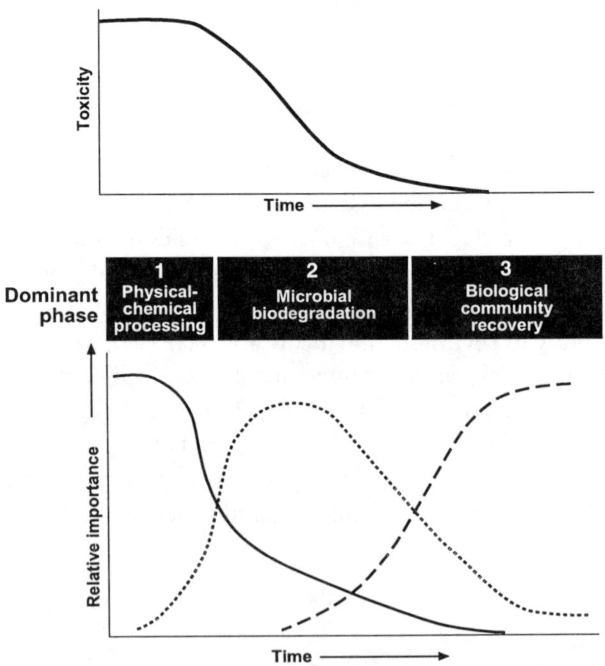

Figure 7-1 Phases of natural remediation for an ecological system impacted by a biodegradable contaminant

Physical or chemical processing and biodegradation together lead to decreases in contaminant concentrations and/or bioavailability, and are therefore responsible for "physical–chemical habitat recovery." For example, immediately following an acute release of PAHs or other PHs to the environment, physical dispersal (including phase changes such as volatilization) and abiotic degradation (e.g., photodegradation) are often the most important processes controlling the concentrations and fates of the chemicals. After concentrations attenuate to levels not toxic to tolerant microbial species, the biodegradation process becomes more important and may dominate the NR process (Phase 2 in Figure 7-1). Eventually, when bioavailability or concentrations decrease below toxicity thresholds for macrobiota, higher organisms (e.g., vascular plants, macroinvertebrates, fish, amphibians, reptiles, birds, and mammals) recolonize, and the process of biological community recovery dominates (Phase 3 in Figure 7-1).

The sequence shown in Figure 7-1 does not occur in all cases; one or another phase may dominate during the NR process, depending on the rate at which contaminant toxicity decreases and on the habitat- and site-specific conditions. For example, after an oil spill in an exposed rocky intertidal habitat, natural wave action might quickly disperse almost all of the oil. Hence, this process would be much more important than biodegradation during the recovery sequence. In cases where a PAH or other PH release does not significantly affect the biological community, physical–chemical processing and/or biodegradation would dominate during the recovery sequence.

Rates of biodegradation of PAHs and other PHs have been measured extensively under laboratory and field conditions (e.g., Guerin and Boyd 1992; Efroymson and Alexander 1994; Bosma et al. 1997). The results of these studies show wide variations in microbial activity and the rate of toxicity reduction, depending on the presence of acclimated populations of hydrocarbon-degrading species, concentrations of nutrients and dissolved oxygen, temperature, and other environmental factors (Bartha and Atlas 1987; Prince 1993; Braddock et al. 1996; Venosa 1996). Natural remediation rates are also influenced by the presence of higher organisms. For example, the presence of vascular plants or the reworking of soils by earthworms may enhance natural biodegradation of PHs in soil (Gilbert 1996; Rathbone et al. 1997; Schnoor 1997; Wetzel et al. 1997). Thus, the potential for NR varies greatly among sites and is partly dependent on physical and chemical conditions of the ecological habitat, as well as the abundance of certain microbes and other species that influence biodegradation rates.

In cases where the biological community is initially altered by the presence of PH contamination, the overall rate of NR depends not only on physical and microbial processes but also on recolonization of impacted species and reestablishment of community structure. Despite the importance of biological community recovery when NR is being considered at petroleum-contaminated sites, little attention has been given to rate-limiting steps in community assembly. Moreover, the interaction among microbial community development, biodegradation rates, and establishment of macrobiota community structure has not been addressed sufficiently by either theoretical modeling or empirical studies.

OBJECTIVES

The focus of this chapter is on NR of PHs in soils and sediments. Because of their importance in environmental toxicology and data availability, the emphasis of this review is on PAHs. The objectives of this chapter are as follows:

- Review the environmental sources, chemistry, fate, and toxicity of PHs.
- Identify key processes and rate-limiting steps in NR, including important ecological processes.
- Develop a general model of natural physical–chemical processing and biological recovery.

We discuss key issues, including the types of PHs and products released into the environment, the role of biodegradation relative to other fate processes in soil and sediment systems, and the importance of constraints on bioavailability, food web transfer, and toxicity of PHs. These issues are followed by a discussion of the interaction between habitat and biological recovery. Then, based on the transition from dominance by physical–chemical processes to biodegradation to biological recovery, we present a model of NR.

PETROLEUM PRODUCTS AND CHEMICALS

Incomplete combustion of fossil fuels and wood, pyrolysis processes, diagenesis of sedimentary organic material to form oil or coals, manufacture of carbon black and coke, and conversion of coal or oil shale to synthetic fuels all produce a homologous series of PAHs. Dozens of investigators have conducted comprehensive studies of these processes and source materials, primarily in the 1970s and 1980s. Some PAHs also form as the result of microbial processes during the early diagenesis of modern sediments and occur naturally in wood and plants and in lipid material in bacteria (e.g., Stoll and Schener 1937; Blumer 1965; Stonecipher and Turner 1970; Albrecht and Ourisson 1971; Haas and Hites 1976; Prahl and Carpenter 1979; Wakeham et al. 1979; Garrigues et al. 1986). More recently, attention has shifted from characterization studies to PAH metabolism, bioavailability, toxicological effects, and computerized pattern recognition for source allocation (e.g., Cerniglia and Heitkamp 1989; Burns et al. 1997; Tang et al. 1998).

PAH structure

The fused- or condensed-ring structure of PAHs resembles graphite (Herbes 1976), which is simply a natural progression of the diagenesis of organic material under temperature and pressure over time. PAHs that are formed during combustion are dominated by 4-, 5-, and 6-ring aromatic compounds without alkyl side chains. A small amount of coronene, a 7-ring aromatic compound, is also typically found but rarely quantified in chemical analyses. In contrast, 2- and 3-ring aromatic compounds, with and without alkyl side chains, dominate the PAHs in fossil fuels. Relatively low concentrations of 4-ring PAHs (and even lower concentrations of 5- and 6-ring PAHs) may also be present in fossil fuels.

Sources of PAHs

The presence of some PAHs in fossil fuels has been known for decades, although detection and isolation techniques were crude (e.g., Donath 1908; Pictet et al. 1916; Egloff 1947; Sakabe and Sassa 1952). Reliable detection of PAHs in environmental media and more extensive analyses in fossil fuels and other source material have been possible only since the 1970s and 1980s, when the majority of PAH studies were conducted (e.g., Giger and Blumer 1974; Bentz 1976; Hites et al. 1977; Giger and Schaffner 1978). Major sources of PAHs in sediments, soils, surface water, and groundwater include spills of petroleum or direct discharge of petroleum-containing wastes in industrial effluents or domestic wastewaters, disposal of wood processing and manufactured gas plant wastes, natural oil seeps, and anthropogenic or natural combustion processes. Lightning, industrial high-temperature pyrolysis (such as coking processes and the manufacture of carbon black), automobile fuel combustion and exhaust (especially diesel), and wood burning are major examples of these latter processes (Figure 7-2) (Blumer and Sass 1972; Blumer and Youngblood 1975; Neff 1979; Josephson 1984).

Composition of petroleum products

In oils, the concentration and composition of PAHs vary depending on the geographic location and history of the source material (U.S. Coast Guard [USCG] 1977; Radke 1987; Peters and Moldowan 1993). Petroleum contains thousands of different chemicals. Most of these chemicals are hydrocarbons, of which PAHs are a small subset of aromatic compounds (Figure 7-3) (Austin 1984). The percentage of PAHs in petroleum varies considerably among crude oils but is low (e.g., less than 3% in several crude oils [Neff 1978]). At least 90% of the aromatic hydrocarbons in crude oil are 1-ring (benzene) through 3-ring (phenanthrene) compounds. Additional 1-ring aromatic compounds are produced in the refining process (Austin 1984), and some heavy distillates in the refining of crude oils, such as bunker fuel oils, have larger amounts of 4-ring compounds and higher homologues. In coals, PAHs are

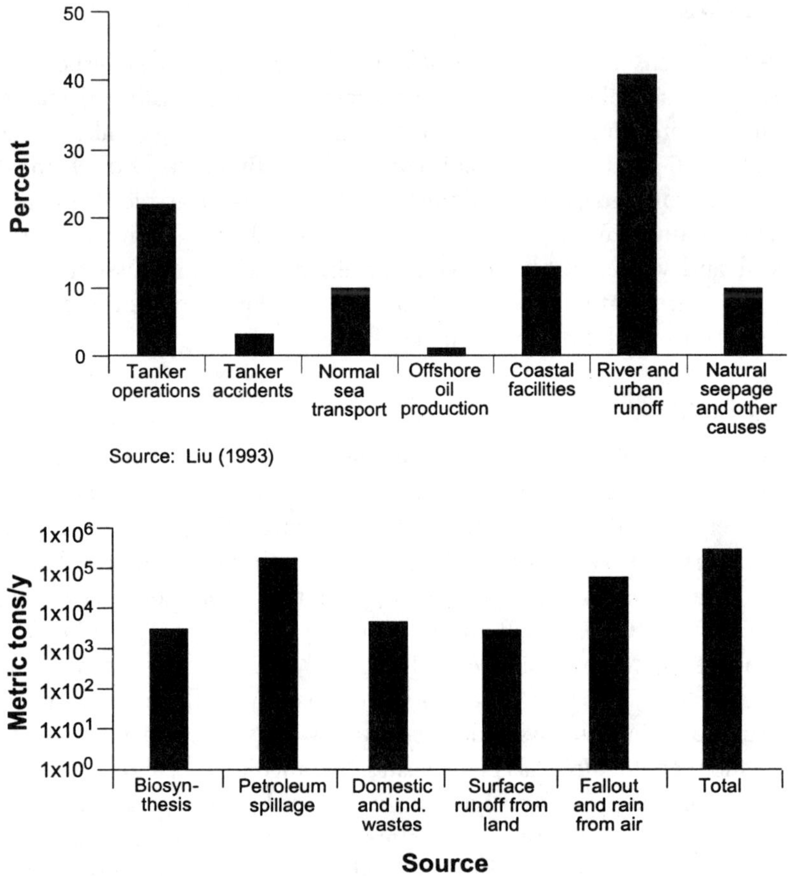

Source: Neff (1979)

Figure 7-2 Sources of petroleum hydrocarbons in the environment (Liu 1993 used with permission from the author)

present in the highest concentrations in immature soft brown coal (lignite) and subbituminous coal. Only moderate levels of selected PAHs occur in bituminous coals, and few if any PAHs are detected in hard coal such as anthracite (Greiner et al. 1976; Allan et al. 1977; Barrick et al. 1984).

PAHs in most fossil fuels are found in much lower concentrations than are aliphatic hydrocarbons. In contrast, creosote, which is prepared from either coal or wood ash, may consist of 80% to 90% PAH by weight. Creosote can have a PAH composition similar to that of oils, but is distinguished by its low concentration of aliphatic hydrocarbons. Various isomers of methyl- and ethyl-substituted PAHs are the principal alkylated compounds in both fossil fuels and creosote. These compounds, particularly those with side chains containing 1, 2, or 3 carbons, are often present in concentrations similar to those of the "parent" non-alkylated compounds.

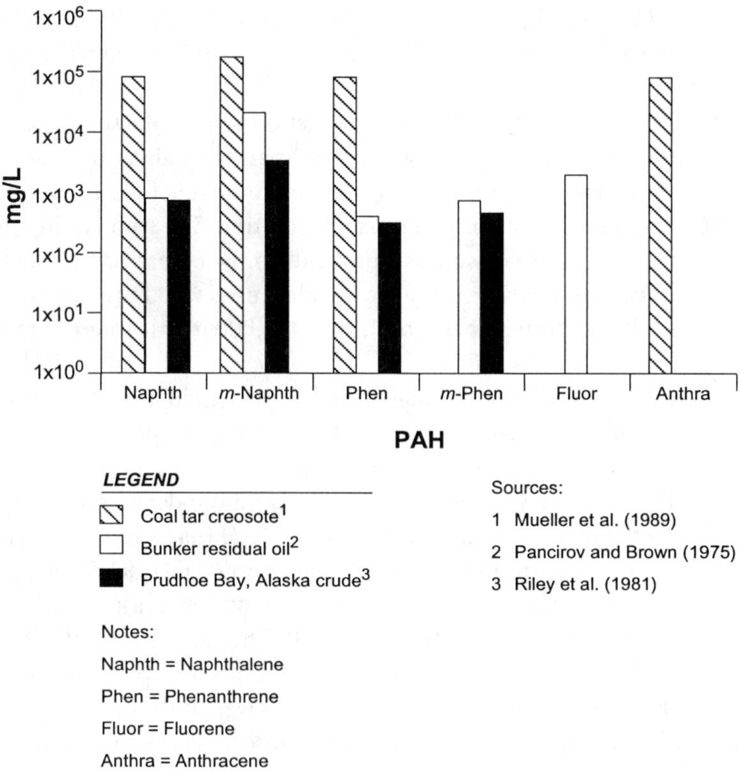

Figure 7-3 Predominant PAHs in coal tar creosote, Bunker C residual oil, and Prudhoe Bay, Alaska, crude

Use of petroleum products containing PAHs

The low-molecular-weight and high-molecular-weight PAHs differ considerably in their use. Naphthalene, a 2-ring PAH, is the simplest of the PAHs and is a natural component of fossil fuels and creosote. It also forms during the burning of tobacco. Naphthalene for human use is produced either from petroleum or as a distillation product from coal tar (Agency for Toxic Substances and Disease Registry [ATSDR] 1993). Naphthalene is important in industry as an intermediate in the production of phthalate anhydride, which is used as a chemical intermediate in the plastics industry. Other products made from naphthalene are moth repellents, leather tanning agents, and the insecticide Carbaryl.

Gasoline is also an important product that contains aromatic compounds. Gasoline is a highly refined product of crude oils consisting of 46% by volume isoalkanes and 29% by volume aromatics (USEPA 1985). Single-ring aromatic compounds (including benzene) make up the majority of aromatic hydrocarbons in gasoline, followed by naphthalene, which makes up less than 1%. The concentration of benzo[a]pyrene in regular and premium gas is 0.212 and 0.480 mg/L, respectively (Sullivan 1974).

Creosote, which is a by-product of the production of manufactured gas and coke from bituminous coal, consists largely of PAHs and contains a low concentration of aliphatic hydrocarbons. Industrial creosote production falls into 2 categories: distillate (100% creosote) and creosote in coal tar solution. The wood industry is one of the primary consumers of creosote as a pesticide for wood preservation. On a per-volume basis, coal tar creosote is the most widely used wood preservative in the U.S., with 414 to 550 creosoting operations using 454,000 metric tons of creosote annually (Mattraw and Franks 1986).

Carcinogenic PAHs consisting of higher molecular weight 4- and 5-ring compounds are unwanted components of crude oil and creosote. These compounds include benzo[*a*]pyrene, benz[*a*]anthracene, methylchrysene, 7,12-dimethylbenz[*a*]-anthracene, and indeno(1,2,3-*c,d*)pyrene. Of these compounds, however, methylchrysene is the only well-known carcinogenic PAH that is present in crude oil at greater than trace concentrations, such as in Qatar crude oil (194.5 mg/kg) (Neff et al. 1994). In comparison, the average concentration of benzo[a]pyrene in 5 other crude oils was approximately 1.5 mg/L (Neff 1979). The carcinogenic PAHs are typically unwanted by-products with no commercial value, except as a major component of creosote. It is possible, however, that lower concentration phenolics in creosote may be the most toxic component of creosote. These compounds are characteristic components of soot and other combustion or pyrolytic particles, but they are considerably less bioavailable than in creosote or oils (Barrick 1982; Prahl and Carpenter 1983; Socha and Carpenter 1987; McGroddy and Farrington 1994).

Because both crude oils and coal tar distillates such as creosote are easily spilled, they are most important when NR is being considered. The chemistry, fate, and toxicity of these PAH mixtures as they apply to remediation activities are discussed in more detail in the following section.

ENVIRONMENTAL CHEMISTRY, FATE, AND TOXICITY

The viability of NR as a risk management option for petroleum-contaminated sites depends on the susceptibility of PAHs and other PHs to physical, chemical, and biological transformation to nontoxic products. The potential for biological uptake and food web transfer of petroleum products or their components also influences the risk to biological communities. Key processes and rate-limiting steps in NR of PHs are illustrated in Figure 7-4 and are discussed below.

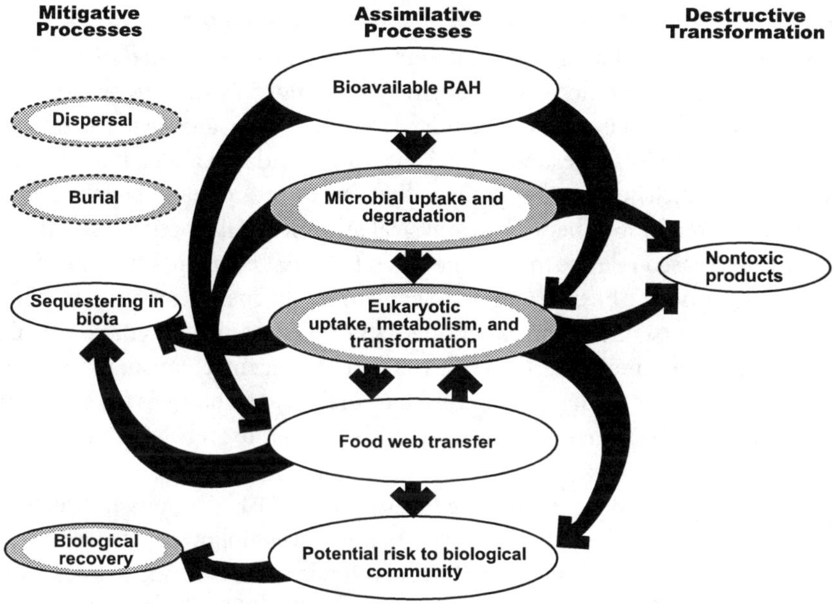

Note: Mitigative processes within dotted lines affect the fate of petroleum hydrocarbons in conjunction with many of the assimilative processes. For graphic simplicity, links from these mitigative processes to assimilative processes are not shown. Shading indicates a primary rate-limiting step in natural remediation.

Figure 7-4 Key processes and rate-limiting steps in natural remediation of sites contaminated with PAHs

Physical, chemical, and microbial processes

Physical–chemical processes

The fused aromatic ring structures of PAHs have relatively low water solubility, reflecting their nonpolar hydrophobic nature. The solubility of PAH compounds decreases as the number of aromatic rings, degree of alkylation, or molecular weight increases. In comparison, 2-ring naphthalene has a solubility of 30,000 μg/L, whereas 5-ring PAHs have solubilities in the 0.5 to 5.0 μg/L range. Other physical–chemical processes that affect PAHs in aquatic and terrestrial environments include dilution, sorption, diffusion, volatilization, and chemical degradation.

The movement of dissolved PHs in water along the flow path is termed "advection." "Dispersion" occurs primarily as PHs spread along and away from the flow path during advection (e.g., in groundwater) or as they mix with clean material (e.g., when a contaminated stream enters clean receiving water). "Diffusion" is the movement of contaminants away from areas of higher concentration to areas of lower concentration (e.g., in soil). The result of advection, dispersion, and diffusion is a net dilution of the contaminant concentration in an aqueous environment.

Weathering and sequestration of PAHs in soils and sediments

Various studies conducted with PAH-contaminated soils have indicated that, as PAHs persist in the environment, they become increasingly resistant to desorption (Linz and Nakles 1996; Kelsey and Alexander 1997) and less available to microorganisms and invertebrates (Alexander 1995; Hatzinger and Alexander 1995; Kelsey and Alexander 1997). These studies indicate that, over time, a fraction of the PAHs becomes sequestered in soils because of physical and chemical reactions, and their availability is decreased relative to the time when they first contacted the soil. Several mechanisms have been proposed to explain this sequestering of hydrophobic organic compounds, including partitioning into soil organic matter and entrainment in inorganic micropores (Luthy et al. 1997). The exact mechanism remains to be elucidated because of the lack of microscopic analytical techniques for observing the sequestration process; therefore, mechanistic studies must rely on inferences drawn from macroscopic observations (Luthy et al. 1997).

Sorption of hydrophobic organic chemicals, such as PAHs, to soils and sediments often entails an initially rapid and reversible process followed by a period of slow sorption over weeks to years. This slow sorption leads to a chemical fraction that resists desorption (Ball and Roberts 1991; Alexander 1995). The result of this slow sorption is a substantial reduction in the rate of desorption, which correlates with a net reduction in the bioavailability of the contaminant (Kelsey and Alexander 1997).

Although the exact mechanisms of sequestration are not known, it is clear that time, soil properties, the chemistry of individual PAHs, and environmental factors (temperature, soil moisture content) control the extent of PAH sequestration (White et al. 1997). Increasing amounts of organic matter and decreased soil particle size (i.e., greater clay content) appear to lead to an increase in the sequestered PAH fraction. Sequestration reactions for PAHs in sediments appear to be controlled by the same basic set of parameters as for soils.

Microbial processes and biodegradation

In addition to physical and chemical processes, microbial degradation is another key factor in natural attenuation of PAHs. Biodegradation is considered the principal process responsible for the removal of the nonvolatile fraction of oil from the environment after a marine oil spill (Prince 1993). The unaugmented microbial degradation of environmental contaminants is commonly referred to as "intrinsic bioremediation." The conditions necessary for intrinsic bioremediation include the presence of appropriate microorganisms; accessibility of the contaminant; appropriate conditions for microbial growth such as pH, temperature, carbon source, electron acceptors (e.g., oxygen, nitrate, sulfate, iron); and nutrient availability. Oxidation–reduction (redox) reactions mediated by aerobic and anaerobic bacteria and fungi are responsible for degradation of PAHs. The actual mechanism of degradation is determined by redox potential and the availability of competent microbes in the environment, while other factors such as nutrient availability drive the rates of removal (Kawatsuka and Igarashi 1975; Nicholson et al. 1992).

In general, the aliphatic compounds in petroleum are simpler to degrade than PAHs, which display greater resistance to degradation with increasing molecular weight and alkyl substitution on the ring structure (McDonald et al. 1997). Although not as rapidly as aliphatic hydrocarbon degradation, PAH biodegradation does occur. Bacteria found in soils with a history of exposure to PAHs are very capable of degrading 2- and 3-ring PAHs such as naphthalene and phenanthrene (Dean-Raymond and Bartha 1975; Herbes 1981; Carmichael and Pfaender 1997). Although post-spill field data show very slow degradation rates for 3-ring PAHs, degradation has been demonstrated in the laboratory for 4-ring compounds such as fluoranthene, pyrene, and chrysene (Bouchez et al. 1995; Carmichael and Pfaender 1997; Table 7-1).

Though bacteria found in contaminated sediments and soils show PAH degradation capacity (Bartha and Atlas 1987; Croft et al. 1995), an acclimation period is often necessary before degradation occurs. In pristine soils, this ecological adaptation period is often more pronounced for refractory compounds such as PAHs than it is for aliphatic hydrocarbons (Coates, Anderson, Woodward et al. 1996; Carmichael and Pfaender 1997). The duration may vary from days to months, depending on the type and concentration of contaminant, background microbial activity, and whether environmental conditions are conducive to microbial growth. If a second exposure occurs soon after the first one, acclimation may not be observed, and degradation rates will frequently be 1 to 2 orders of magnitude greater when compared with previously unexposed organisms (Wilson et al. 1985). For example, no adaptation phase was seen for PAH degradation by bacteria from previously exposed soils, whereas microorganisms from pristine soils showed no ability to degrade the PAH contaminants in the same time frame (approximately 8 weeks) (Wilson et al. 1985; Carmichael and Pfaender 1997).

Aerobes. After the microorganisms are acclimated, aerobic or anaerobic degradation will occur depending on the predominant redox conditions. Two mechanisms of aerobic PAH degradation are known. First, the PAH may serve as a growth substrate. Second, some microorganisms that are unable to gain energy from the metabolism of PAHs are capable of partially metabolizing the compound if an alternative growth substrate is available, in a process known as "co-oxidation" or "cometabolism." An example of cometabolism of PAHs is seen for low solubility 4- and 5-ring PAHs such as chrysene and benzo[a]pyrene. These compounds do not appear to support bacterial growth; however, they may be cometabolically degraded under aerobic conditions (Gibson et al. 1975).

In addition to the bacteria, fungi and other eukaryotic organisms are also capable of aromatic oxidation, forming a *trans*-diol during oxidation of aromatic

Table 7-1 Bacteria and fungi capable of degrading PAHs[a]

Organism	Polynuclear aromatic hydrocarbon (PAH)	References
Bacteria		
Pseudomonas	Naphthalene Phenanthrene Anthracene	Guerin and Boyd 1992; Crocker et al. 1995; Dagher et al. 1997
Mycobacteria	Phenanthrene Fluorene Fluoranthene Pyrene	Heitkamp and Cerniglia 1988; Boldrin et al. 1993
Sphingomononas paucimobilis	Benzo[*a*]pyrene Chrysene Pyrene Dibenzo[*a*]pyrene	Cerniglia 1992; Ye et al. 1996
Flavobacteria	Phenanthrene Anthracene	Cerniglia 1984
Alcaligenes	Phenanthrene	Kiyohara et al. 1982
Aeromonas	Naphthalene Phenanthrene	Kiyohara et al. 1976
Vibrio	Phenanthrene	Kiyohara and Nagao 1978
Beijernickia	Naphthalene Phenanthrene Anthracene Benz[*a*]anthracene Benzo[*a*]pyrene	Gibson et al. 1975; Mahaffey et al. 1988
Bacillus	Naphthalene	Cerniglia et al. 1984
Nocardia	Phenanthrene Anthracene	Treccani et al. 1954
Corynebacterium	Naphthalene	Dua and Meera 1981
Micrococcus	Phenanthrene	Ghosh and Mishra 1983
Sulfate reducers	Naphthalene Phenanthrene	Coates et al. 1996a
Fungi		
Phanerocheate chrysosporium	Phenanthrene Anthracene Benzo[*a*]pyrene	Meulenberg et al. 1997
Trichosporon penicillatum	Phenanthrene	MacGillivray and Shiaris 1993

[a] Modified and reprinted with permission from Cerniglia CE, Heitkamp MA. 1989. Microbial degradation of polycyclic aromatic hydrocarbons (PAH) in the aquatic environment, p 48. In: Varanasi U, editor. Metabolism of polycyclic aromatic hydrocarbons in the aquatic environment. ©CRC Press, Boca Raton, Florida.

hydrocarbons (Cerniglia 1984). One disadvantage of eukaryotic transformation is that *trans*-diols of various PAHs are carcinogenic, whereas *cis*-diols formed during bacterial degradation are not (Atlas and Bartha 1993; Figure 7-5).

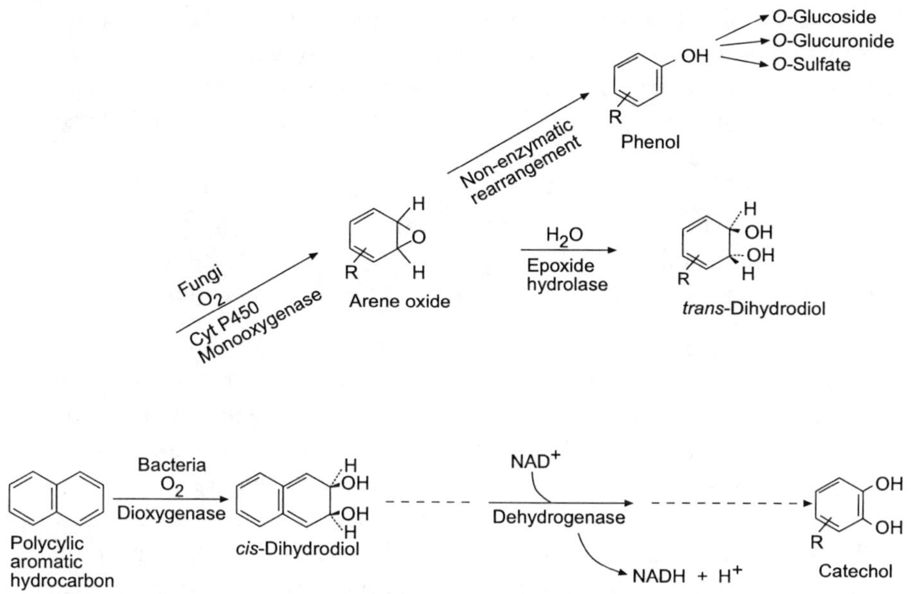

Figure 7-5 Typical monooxygenase vs. dioxygenase mediated oxidation of PAHs by eukaryotic and prokaryotic microorganisms

Anaerobes. Anaerobic bacteria are capable of transforming and, in some cases, completely mineralizing PAHs. The biochemical degradation pathway for anaerobic microbial degradation has not yet been elucidated (Coates, Anderson, Lovley 1996; Coates, Anderson, Woodward et al. 1996). Nitrate-reducing bacteria are capable of degrading naphthol and, more slowly, naphthalene (Mihelcic and Luthy 1988). Sulfate-reducing bacteria have been found to mineralize [^{14}C]naphthalene and [^{14}C]phenanthrene to $^{14}CO_2$ (Table 7-1). Interestingly, in one study of sites with varying degrees of exposure, only sediments from the site that had long-term exposure to high concentrations of PAHs had sulfate-reducing bacterial populations large enough to degrade PAHs (Coates, Anderson, Lovley 1996; Coates, Anderson, Woodward et al. 1996). The lag time necessary for these sulfate-reducing bacteria was 90 days before degradation began, implying the time frame necessary for enzyme induction under anaerobic conditions may be sufficient to meet natural recovery goals.

Fungi. Very few accounts of fungal mineralization of PAHs are found in the literature (Bezalel et al. 1996). The predominant degradation pathway yields an oxidized PAH intermediate (Figure 7-5). The ability to oxidize PAHs appears to be fairly widespread among fungi. Researchers isolated 13 yeasts from heavily contaminated marine coastal sediments, of which more than 50% transformed phenanthrene

(MacGillivray et al. 1993). Although fungi cannot usually mineralize PAHs completely, a symbiotic relationship with bacteria can lead to complete degradation of some compounds (Meulenberg et al. 1997). Soil bacteria degraded some of these oxidized PAHs more readily than did the parent PAH compound. Some intermediates of phenanthrene, however, have been shown to strongly inhibit bacterial respiration and were not degraded. Researchers cited the 1,000- to 100,000-fold increase in water solubility of the oxidized intermediates as the primary reason for increased biodegradability (Meulenberg et al. 1997). This increased solubility may also explain the increased toxicity associated with certain metabolites.

Rate-limiting factors. If bacteria have been acclimated to PAHs, rate-limiting factors include availability of nutrients and electron acceptors, bioavailability and degradability of the contaminants, soil moisture content, and predation. For the PAHs that are biodegradable (primarily the smaller, more water-soluble compounds), the most important environmental factor limiting biodegradation is the slow rate of mass transfer from the solid phase, nonaqueous-phase liquids (NAPLs), or soil (Guerin and Boyd 1992; Efroymson and Alexander 1994; Bosma et al. 1997). Recent bioremediation studies reveal that PAH degradation in soils is seldom limited by intrinsic bacterial activities. The mass-transfer rates of PAHs in soils are slower than the microbial degradation rates, resulting in a mass-transfer equilibrium that controls degradation rates (Guerin and Boyd 1992; Efroymson and Alexander 1994; Shuttleworth and Cerniglia 1995; Bosma et al. 1996).

The recalcitrance of sorbed PAHs is highlighted by reports of high concentrations of PAHs, even low-molecular-weight compounds such as naphthalene and phenanthrene, that were sequestered in soils contaminated 60 years earlier (Carmichael and Pfaender 1997). Adding a fresh mixture of PAHs to these soils resulted in degradation, demonstrating that microorganisms capable of degrading these compounds were present, but that sequestered PAHs in the weathered mixture were not bioavailable. These studies suggest the fate of aged in situ compounds should not be predicted from measurements of degradation rates of freshly added compounds in the laboratory (Carmichael and Pfaender 1997).

Rates of intrinsic hydrocarbon removal in marine sediments range from 0.5 µg/g-d for sediments polluted by the Amoco Cadiz spill (Prince 1993) to 1.9 to 10 µg/g-d for Prince William Sound sediments contaminated with Exxon Valdez oil (Braddock et al. 1996). A summary of published studies of hydrocarbon biodegradation rates in marine systems under nutrient-enriched conditions reported a range of 5 to 2,500 µg/g-d (Bartha and Atlas 1987).

If sufficient PAHs and other PHs are bioavailable, O, N, and P are the most likely nutrients to limit bacterial growth. Enhanced NR through the addition of fertilizers or oxygen can overcome nutrient limitations of degradation rates. Biodegradation of oil released from the Exxon Valdez increased 2- to 3-fold with the addition of fertilizers (Prince 1993). The rate of increase of biodegradation in

Prince William Sound, Alaska, with fertilizers was not as high as other researchers have found (Bartha and Atlas 1987) because rates of intrinsic biodegradation of the spilled oil were relatively high anyway. In marine systems, the indigenous bacteria may maintain relatively high degradation rates as a result of nutrients delivered with every tidal cycle. In a study of alkane and PAH degradation in a marine environment, background N levels (1 to 2 mg/L N) in the interstitial pore spaces were sufficient to support high rates of degradation (Venosa et al. 1996). Increases of only 50% over background PAH degradation rates and a 2-fold increase in alkene degradation rates occurred under enhanced conditions with the addition of nitrate and phosphate to a marine oil spill. This disparity between increases in rates of PAH and aliphatic hydrocarbon degradation was also noted by Swannell et al. (1997). Aerobic degradation of aromatic hydrocarbons in buried oil was not enhanced by the addition of inorganic N and phosphate (Swannell et al. 1997), whereas rates of aliphatic hydrocarbon degradation were significantly enhanced.

Predation on hydrocarbon-degrading microorganisms by protozoa may sometimes limit overall rates of biodegradation. At one petroleum-contaminated site, researchers found more than 19,000 protozoa per gram of soil, rivaling the protozoa numbers found in activated sewage sludge facilities (Madsen et al. 1991). Protozoa have also been found in elevated numbers in the subsurface sediments where jet fuel contamination has occurred (Sinclair 1991). Despite limiting the numbers of hydrocarbon degraders, the presence of protozoa may also allow for community succession by providing an abundant food source for nematodes and other organisms at higher trophic levels.

Contaminant removal by bacteria may affect recovery of higher organisms in several, sometimes opposing ways. For example, microbial biodegradation not only reduces toxicity, which would foster overall ecosystem recovery, but it may also create anoxic environments, which could constrain development of a community of higher organisms. Depletion of oxygen by aerobic bacteria and generation of hydrogen sulfide in subsurface soil or sediment by sulfate-reducing bacteria may create an anoxic zone. Such a situation could limit the rate of further contaminant removal by bacteria and hinder recovery rates for aerobic multicellular organisms during community succession.

Comparison of soil and sediment

The key factors that control NR of PHs are similar in soil and sediment systems, although the relative importance of some factors varies between these systems (Table 7-2). For example, because of the direct contact of soil with the atmosphere, volatilization is more important in soil than in sediment. In contrast, the influence of currents in aquatic ecosystems enhances dispersion of nonvolatile compounds in sediment compared with soil. Biodegradation is probably more important for soil systems than for sediments because of the slow rates of microbial activity in the anaerobic environment of most sediments. Changes in bioavailability due to

Table 7-2 Key factors that influence natural remediation rates

Influencing factor	Soil	Sediment
Polynuclear aromatic hydrocarbon (PAH) concentration and type	X	X
Volatilization	X	—
Other dispersive processes	—	X
Sorption	X	X
Bioavailability	X	X
Biodegradation	X	—
Sediment deposition	—	X
Bioturbation	X	X

weathering and bioturbation are probably equally important in sediment and soil. PAH concentration and type determine the toxicity of the contaminants and their susceptibility to volatilization, phototransformation, biodegradation, and bioaccumulation. The influence of PAH concentration and type is largely independent of the particular soil or sediment matrix.

Physical factors are known to be important in determining the fate of PHs in marine sediments. Swannell et al. (1996) followed an experimental oil spill in a marine environment and found physical processes dominated as PHs affected beach sediments. Physical removal of oil, particularly from sedimentary pores and the beach surface, played a significant role in reducing shore concentrations and toxicity. Biodegradation occurred primarily after physical removal processes were largely complete.

The predominance of physical processes in sediments influenced by tides was demonstrated in a U.S. Environmental Protection Agency (USEPA) study of nutrient additions to an experimental spill site on a sandy beach in Delaware Bay (Venosa et al. 1996). The closer to the water the beached oil was, the less beneficial nutrient additions were. The microbial populations did not become well established in the tidal area and were, therefore, unable to use nutrients either from the seawater or from fertilizers. Higher up the beach, in the dune area where sediments were not constantly saturated and reworked by tides, biodegradation accounted for significantly more contaminant removal than it did in the more seaward stretches, with and without fertilizer additions. The carrying capacity for hydrocarbon-degrading bacteria in the subtidal marine environment was strongly influenced by the physical removal of bacteria, caused by scouring of the biomass, when breaking waves tumble sand grains.

The importance of sediment deposition is illustrated in Figure 7-6, which shows long-term chemical and biological changes that may occur within the sediment because of NR. For example, immediately after an oil spill, a sediment core profile may show substantial contamination of surface sediment only. After several years, natural sedimentation covers the original upper layer of the core with clean sediments, and the constituents that have not attenuated through natural processes are displaced downward in the core. Also, sediment mixing by bioturbation and

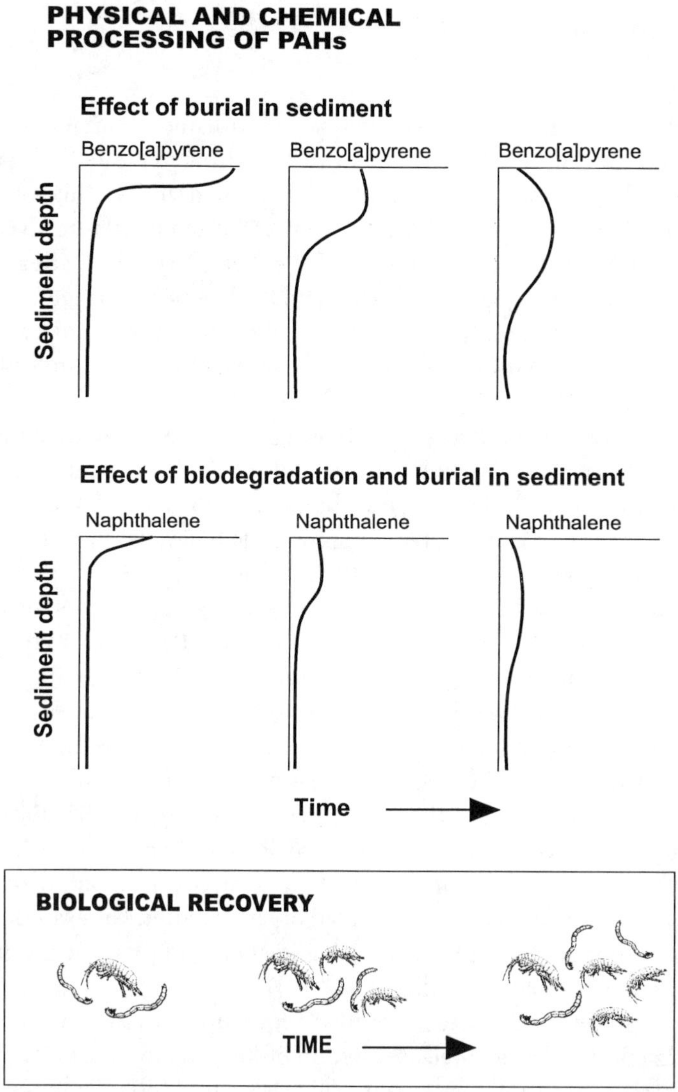

Figure 7-6 Long-term chemical and biological changes within a sediment core profile due to natural remediation

currents distributes the remaining PHs throughout a broader portion of the core and thereby also contributes to lowering the peak concentration. These processes continue over time until the hydrocarbons related to the spill are completely covered and below the mixing depth or are attenuated through other NR mechanisms. At that time, the pattern of remaining hydrocarbon concentrations related to the spill stabilizes in the sediment core profile.

Biological uptake, fate, and effects

Bioavailability

Recent research has demonstrated that hydrophobic organic contaminants, such as PAHs, can become sequestered in soils over time, resulting in limited bioavailability to most ecological receptors (Alexander 1995). "Bioavailability" in this context refers to the fractional uptake of a particular PAH from soil or sediment into an ecological receptor. Bioavailability varies as a function of the chemistry of individual PAHs, soil matrix parameters (e.g., total organic C), and the length of soil aging time. Reduction of PAH bioavailability with the soil-aging process has been observed experimentally in both microorganisms and soil invertebrates (Alexander 1995; Kelsey and Alexander 1997; White et al. 1997).

The reduction of PAH availability with soil aging has also been demonstrated with extraction tests designed to mimic uptake into biological organisms. In extraction tests with both C-18 resins and *n*-butanol, both of which mimic the lipophilic biological compartment, reductions in PAH bioavailability have been observed with soil aging (Lane and Loehr 1992; Kelsey et al. 1997). In addition, the *n*-butanol extractions have been observed to correlate with the fraction of PAHs available to earthworms (Kelsey et al. 1997). The residual PAH fraction remaining after the initial extraction is highly resistant to further leaching.

The bioavailability of PAHs from sediments depends on the chemistry of individual PAHs, the amount and type of organic C present, the sediment particle size distribution, the length of time the PAH has been in contact with the sediment (aging time), and lipid content of the organism (Landrum et al. 1985, 1987, 1995; Varanasi et al. 1985; McCarthy 1989; Meador, Stein et al. 1995). For PAHs, the predominant environmental factor affecting bioavailability from sediments is adsorption to particles and particle-associated organic matter. For example, free PAH in sediment pore water is the bioavailable fraction for uptake by the amphipod *Rhepoxynius abronius* (Meador, Casillas et al. 1995).

The PAH sequestration process described above produces a fast-release, highly bioavailable fraction and a slow-release fraction of limited bioavailability. The reduction of PAH concentrations generally follows such a 2-phase rate curve during bioremediation, with the initial phase producing a significant reduction within a short time. This phase is followed by a discrete second phase that occurs at a slower rate and is gradually reduced to yield a residual concentration of PAHs that is not amenable to biodegradation (the sequestered fraction) (Alexander 1995).

In summary, the available studies indicate that PAH bioavailability from soil and sediments is reduced through the sequestration process and that this reduction occurs in a manner consistent with the postulated sequestration mechanism.

Food web transfer and fate

The fate of PHs in biological systems depends on rates of uptake and elimination by organisms, the capacity for metabolic transformation and detoxification, and the potential for bioaccumulation, food web transfer, and biomagnification (Landrum et al. 1992; Meador, Stein et al. 1995).

Uptake, metabolism, and elimination. Differential uptake and elimination of PAHs and other PHs depend on chemical properties such as solubility and log K_{OW}, bioavailability, and species-specific physiological processes. In sediment and soil environments, direct uptake occurs largely by passive diffusion; chemical bioavailability and the amount of membrane surface area ultimately control uptake rates. Ingestion of contaminated prey, detritus, and soil or sediment particles is also an important uptake route and may dominate over uptake through water (or pore water) for extremely hydrophobic PAHs (log $K_{OW} > 5.5$) (Meador, Casillas et al. 1995). Conflicting reports about the relative importance of water- and dietary-uptake routes (e.g., Corner et al. 1976; Roesijadi et al. 1978) may be related to confounding factors in study designs. Though uptake from water is relatively efficient, contaminant body burden is frequently overestimated by equilibrium partitioning (EqP) theory (Driscoll et al. 1997). Relatively high concentrations of PAHs in prey species and high ingestion rates enhance the importance of the dietary route, despite low assimilation efficiencies for PAHs. For example, dietary uptake by fishes is generally an inefficient process (estimates of 1% to 32% in rainbow trout) (Niimi and Palazzo 1986; Niimi and Dookhran 1989). Assimilation efficiencies for ingestion of sediment-sorbed benzo[a]pyrene were 0% to 26% in *Lumbriculus variegatus* and 11% to 57% in *Diporeia* spp. (Kukkonen and Landrum 1995).

The biological fate of PAHs varies among phyla, depending primarily on ring structure and the existence and chemical-specific activity of the cytochrome P450 enzyme system. PAHs such as benzo[a]pyrene, phenanthrene, and naphthalene are rapidly metabolized to mutagenic, tumorigenic, genotoxic, or carcinogenic agents (Jerina et al. 1984; James 1989; Varanasi et al. 1989; Stegeman and Lech 1991). Metabolism occurs at different rates in various species, depending on many of the same factors that affect uptake rates (i.e., physiology, environmental conditions, life stage, etc.). Vertebrates tend to metabolize PAHs at higher rates than do invertebrates. Within the invertebrate groups, there are clear differences in metabolic capability. Studies have shown that the first stage of PAH metabolism occurs very slowly or not at all in the more primitive invertebrates, such as protozoa, cnidaria, and mollusks, and rapidly in phylogenetically higher invertebrates, such as arthropods, echinoderms, and annelids (James 1989).

During metabolism, the parent PAH compounds are transformed to one or more compounds that are more polar and readily excreted. PAHs are metabolized by organisms in 2 stages called "Phase I reactions" and "Phase II reactions." Phase I reactions transform the PAH to a more polar intermediate compound through oxidative, hydrolytic, or reductive processes (Livingstone 1985; Buhler and Williams 1989). Phase II enzymes use a variety of endogenous molecules (glutathione, sugars, amino acids, sulfate, etc.) and in association with the Phase I polar intermediary, form a highly water-soluble, excretable compound. PAHs require initial metabolism by the mixed-function oxidase system before they can be acted on by the Phase I and Phase II enzymes (Livingstone 1985). Metabolism of PAHs occurs primarily in the livers of vertebrates, the hepatopancreas of crustacea and arthropods, and the digestive glands of other invertebrates such as mollusks and annelids (Buhler and Williams 1989). Many different metabolic pathways exist for each PAH, and each pathway may produce numerous metabolites and intermediaries (Livingstone 1985; Buhler and Williams 1989).

Elimination of PAHs occurs through passive diffusion or active excretion of metabolites. Variables that control elimination rates include physiological factors such as enzyme induction, age-related metabolic rates, and reproductive state, as well as environmental factors such as temperature and salinity. Meador, Stein et al. (1995) compiled estimates of half-lives of PAHs in marine organisms, which ranged from 0.8 days for fluoranthene and pyrene in shrimp to 77 days for benzofluoranthene in oysters. For various organisms and compounds, most half-lives are on the order of 5 to 20 days. For areas that were once chronically contaminated but have since been remediated, populations may still be found with persistent PAHs in tissue, especially metabolites (Meador, Stein et al. 1995).

Landrum et al. (1992) compared several toxicokinetic models for use in aquatic hazard assessment, including simple compartment models, pharmacokinetic models, and bioenergetics models. Some of the models described by Landrum et al. (1992) address the toxicokinetics of contaminants associated with sediment, including uptake from pore water, direct contact with sediment, and sediment ingestion. Meador, Stein et al. (1995) modeled the relative contribution of sediment ingestion and ventilation routes to PAH uptake for a filter-feeding bivalve, showing that the results were sensitive to assimilation efficiency for the 2 routes. Kinetic models are most useful because they do not require equilibrium assumptions. Such models hold promise for use in evaluations of NR scenarios when an understanding of the dynamics of PAH concentrations in organisms relative to environmental concentrations is desired.

Accumulation and food web transfer. The behavior of PAHs in the marine food web has been extensively reviewed (Meador, Stein et al. 1995). Though marine organisms at all trophic levels are capable of accumulating PAH tissue residues from their diet, biomagnification (an increase in tissue concentrations at higher trophic levels because of food web transfer) generally does not occur. This absence of biomagnification applies mostly to food webs that include vertebrates, apparently because fish and other vertebrate species are able to rapidly metabolize assimilated PAHs. Many studies have demonstrated that fish do not accumulate higher tissue

PAH concentrations than their prey (Neff 1979; Eisler 1987; USEPA 1991; Suedel et al. 1994). In controlled studies of linear aquatic food chains, PAH tissue residues decrease at higher trophic levels of the food chain (Broman et al. 1990). Biomagnification in invertebrate-to-invertebrate trophic transfers is not well described but has been reported (Dillon 1981; Biddinger and Gloss 1984). The potential for biomagnification in invertebrate consumer species exists because invertebrates are less able to metabolize and eliminate PAHs than are vertebrates (Eisler 1987; Meador, Stein et al. 1995). This ability appears to be highly variable among invertebrate species. As a group, the mollusks, especially bivalves, are generally less capable of PAH metabolism than are crustaceans or polychaetes.

Accumulation and possible magnification of PAH metabolites by vertebrate consumer species have not been well described. The process is potentially important, especially for carcinogenic PAHs, because some metabolites possess much higher potency than the parent compound. Reactive PAH metabolites that covalently bind to DNA can be persistent in fish (Sikka et al. 1991; Stein et al. 1993; Meador, Stein et al. 1995). Dietary uptake of carcinogenic PAH metabolites by fish is reportedly significant but may be less than that of the parent compounds (McElroy and Sisson 1989; Gardner et al. 1991). Most controlled studies of PAH trophic transfer do not report the fate of metabolites, however.

Bioconcentration factors (BCFs) for PAHs may be less than expected based on relationships to K_{OW}, especially for more hydrophobic compounds (log $K_{OW} > 5.5$) (Meador, Stein et al. 1995). A stronger relationship between the BCF calculated from the free PAH fraction and K_{OW} has been observed. The biota–sediment accumulation factor (BSAF) for any given PAH derived from lipid-normalized tissue concentration and organic C–normalized sediment concentration theoretically should be approximately 1.7 (Lee et al. 1993). However, BSAFs are often less than expected because of reduced bioavailability, PAH metabolism, or disequilibrium conditions.

Toxicity, carcinogenicity, and mutagenicity of PAHs

Scientific literature contains considerable information on the toxic effects observed in organisms that inhabit systems contaminated with high concentrations of PAHs. However, studies that present unequivocal concentration- or dose-response data are limited. It is difficult to relate PAH concentrations (or doses) to effects, except through the use of laboratory toxicity bioassays on field-collected soil and sediment (e.g., Pastorok et al. 1994). This issue is further complicated because some organisms metabolize PAHs into compounds that are even more toxic than the parent compound (Jerina et al. 1984; James 1989; Varanasi et al. 1989; Steseman and Lech 1991). In addition, PAHs are reportedly one of the many groups of chemicals that may be endocrine modulators in wildlife (Peakall et al. 1981; Fries and Lee 1984; Hontela et al. 1995). In general, low-molecular-weight PAHs (LPAHs) containing 2 or 3 rings appear to be significantly more toxic to soil or sediment organisms than high-molecular-weight PAHs (HPAHs) with 4 to 7 rings. As a group, HPAHs tend to exhibit more carcinogenic or mutagenic responses in higher organisms than do LPAHs.

Toxicity values for sediment and soil organisms vary considerably for most PAHs. For example, MacDonald (1994) reported that acenaphthene caused toxic effects in aquatic invertebrates in concentrations ranging from 12.5 to 39,557 μg/kg in sediment. The variation in toxicity values is most likely due to a myriad of factors, including site-specific variables such as the organic C content of sediment, simultaneous exposure to other potentially toxic chemicals, differential photo-induced toxicity, inherent variability in species sensitivities, and differences in study design.

Results of field studies from several investigators have consistently suggested a relationship between elevated concentrations of PAHs in sediments and the development of abnormal tissue lesions in fish (Malins et al. 1984, 1988; Baumann and Harshbarger 1985; Hargis et al. 1989; Baumann et al. 1991). The reported toxicity of PAHs to vertebrates tends to fall into 3 primary categories: carcinogenicity, genotoxicity, and reproductive effects. Parent PAH compounds do not appear to be directly carcinogenic. However, the polar metabolites of some PAHs, which are hydrophilic and extremely reactive, may undergo a variety of spontaneous or enzyme-mediated reactions in the cells (Neff 1985). The reactions of metabolites often include covalent binding to proteins and DNA, which leads to the formation of DNA adducts and cellular damage, including mutagenesis, teratogenesis, and carcinogenesis.

Some PAHs may readily absorb energy from the ultraviolet (UV) portion of sunlight, changing the parent compound into a more toxic, water-soluble form (Bowling et al. 1983; Newsted and Giesy 1987). Photoactivated PAHs have been shown to cause direct damage to tissues and to indirectly affect organisms by promoting the formation of oxygen-free radicals (Newsted and Giesy 1987). Oxidative stress has been suggested as one of the most important mechanisms of PAH phototoxicity in aquatic organisms. Turbidity, pH, hardness, and alkalinity have been identified as important environmental variables in assessing UV exposure by benthic organisms that are in contact with sediment-bound PAHs (Jacher and Burton 1995). Although little work has been done to investigate photoactivation in soils, the mechanisms and magnitude of toxicity should be similar to those for aquatic organisms.

INTERACTION OF HABITAT AND BIOLOGICAL RECOVERY

Following Booth et al. (1991), 3 kinds of recovery related to NR processes are discussed in this section:

1) Habitat recovery: the return of physical and chemical habitat characteristics to nontoxic conditions within the limits of natural variability.

2) Biological recovery: "the re-establishment of a healthy biological community in which the plants and animals characteristic of that community are present and functioning normally" (Baker et al. 1990).
3) Recovery of resource services: the return of services provided by public resources to baseline levels within the limits of expected variability.

Endpoints and recovery times can be very different for habitats, biological assemblages, and resource services (Booth et al. 1991). Habitat recovery depends on the reduction in concentrations of PHs through physical–chemical processing and biodegradation. In cases of severe contamination (e.g., coating of a shoreline with crude oil after a spill), biological recovery usually must be preceded by substantial recovery of the physical–chemical habitat. Otherwise, biological recovery may begin during the initial stages of habitat recovery.

Recovery of resource services may or may not depend on habitat recovery (Booth et al. 1991). For example, recovery of a fishery damaged by an oil spill depends on habitat and biological recovery. In this instance, resource service recovery would be expected to take longer than habitat recovery (note that habitat recovery in this case is defined by the return of hydrocarbon concentrations to levels below the toxic threshold for the primary fishery). In contrast, human use of a beach for swimming and sunbathing could begin as soon as oil is no longer visible on the sand. Thus, the recovery period for services related to swimming and sunbathing could be considerably shorter than the period of time required for habitat recovery (Booth et al. 1991).

From the standpoint of natural resource damage assessment (NRDA), recovery is related to biological recovery, but because it is based on resource services, it does not necessarily depend on the full recovery of the injured resources (43 Code of Federal Regulations [CFR] Part 11). Because recovery of resource services is dependent not only on site-specific ecological conditions but also on temporally varying economic conditions, recovery times for resource services are not addressed further. Booth et al. (1991) provides more information on approaches to evaluating recovery of resource services.

Importance of ecological processes in control of NR rates

Our view of NR encompasses the interaction of the 3 phases of recovery:

1) Physical–chemical processing,
2) Microbial biodegradation
3) Biological community recovery (Figure 7-1).

Previous researchers have focused on relationships between PH concentrations and biodegradation or between concentrations and biological community recovery. The emphasis here is the potential role of the macrobiological community in regulating the rate of NR after habitat recovery has occurred. Macrobiota include plants such as algae and vascular plant species; various invertebrate species such as annelid worms, crustaceans, and mollusks; fish; and wildlife such as reptiles, amphibians, birds, and mammals.

Natural remediation relies not only on the reestablishment of populations of single species but also on interactions among species and on the physical and chemical changes in the environment resulting from their nutritional and biochemical activities. Thus, the macrobiological community and its successional development affect the physical–chemical processing of chemicals, which in turn enhances recovery of the macrobiota. Plant ecologists and soil scientists have long recognized the ability of plants and soil invertebrates to modify the physical–chemical properties of the substrate such that conditions are improved for certain other species. Similar mechanisms have been studied in the aquatic environment (e.g., substrate conditioning by polychaete worms and other macroinvertebrates in soft sediments [Rhodes et al. 1977; Aller 1982]). Moreover, certain species act as foundation species or "ecosystem engineers" (Lawton and Jones 1995) by creating physical habitat for a diverse assemblage of other species. For example, the presence of a mussel bed in an exposed rocky intertidal habitat provides protected physical habitat for many species in the interstices between the individual mussels (Suchanek 1986).

Aside from micro- and macrobiological interactions and their relationships to the physical–chemical environment, the complexity of macrobiological communities dictates that colonization and population growth of normally rare or moderately abundant species will likely determine recovery times. The study of recovery processes and assembly rules for biological communities occupies a prominent place in modern ecology (e.g., Diamond 1975; Drake 1991; Law and Morton 1996) but has generally not been applied to ecological risk assessment (ERA) and NRDA.

The following ecological principles illustrate the role of biological community development in determining mechanisms of NR and rates of recovery:

- Colonization of species and the mechanics of community assembly (e.g., invasion sequence) ultimately determine the rate of community development (Drake 1991).
- The recolonization rate of a species depends on its primary form of reproduction (e.g., vegetative growth vs. propagule dispersal), the timing of its reproduction relative to the impact, its fecundity, the area of impact, and the existence of favorable environmental conditions for establishment of colonizers.
- Substrate conditioning by certain species (e.g., microbiota, earthworms) is necessary for establishment of other biota (e.g., some plants), which in turn provide habitat for other species.
- Establishment of mutualistic or symbiotic interactions between species (e.g., mycorrhizal fungi and N-deficient plants) that are necessary for

normal functioning of a community requires time for development of multiple interacting populations.

- Biological communities may not return to a pre-impact state. Natural ecological systems are in a state of disequilibrium, and changes are the rule, not the exception (Sousa 1979; Paine and Levin 1981; Baker et al. 1990). After an impact, the biological community is not expected to return to exactly the same state as the pre-impact community (Drake 1991; Law and Morton 1993). The new community that establishes itself during NR may have more or less value to humans than the pre-impact community. Establishment of a functioning biological system with some desired level of productivity and biodiversity should be the goal of enhanced NR.

Model of habitat and biological recovery

We propose a general model of NR that consists of 2 broad components: physical–chemical recovery and macrobiological recovery (Figure 7-7). For the purposes of illustrating the general model, the development of a biological community is defined by idealized curves showing increases in the abundance of key species or increases in the number of species (Figure 7-7). For simplicity, a severely impacted community (Curve 3 in Figure 7-7) is assumed to reach a quasi-equilibrium state after some period of recovery defined by recolonization of key species and successional changes.

The rate of development of a recovering macrobiological community and the time to achieve quasi-equilibrium depend on the initial conditions after introduction of the chemical, mainly the magnitude of acute ecological effects and the duration of toxic conditions. Initially, acute toxicity may kill many individuals of key species, thus disrupting the trophic links important for energy flow. When the abundance of key species decreases because of the toxicity of the released chemical, dramatic shifts can occur in the relative abundances of other species through trophic cascades (Carpenter and Kitchell 1993). For example, reductions in sea otter abundance due either to toxic effects or to behavioral avoidance of an area after an oil spill could result in large increases in populations of sea urchins and essentially "clearcutting" of kelp beds by these efficient grazers. In the case of severe acute effects, a complete sequence of community development must occur during recovery, consisting of the following phases:

- Recolonization by tolerant, opportunistic species
- Rapid growth of populations of colonizers
- Conditioning of the substrate by early colonizers
- Invasion by less tolerant forms as physical–chemical and microbial processes reduce concentrations of chemicals

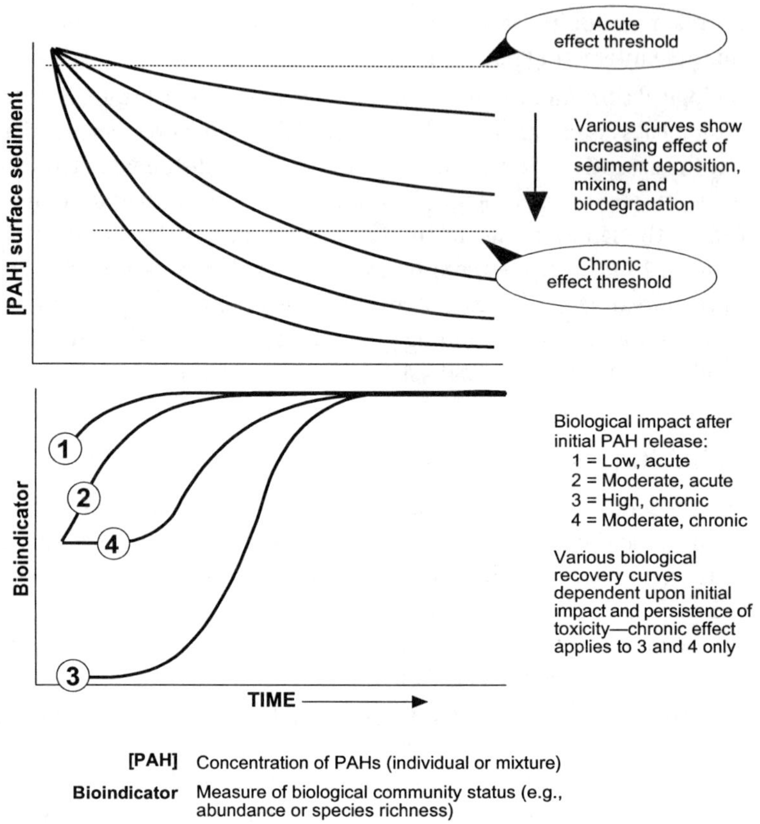

Figure 7-7 Graphical model of reductions in PAH concentration and biological recovery after PAH release

- Assembly of community structure and development of temporal and spatial patterns of species abundance through ecological processes, such as competition, predation, and mutualism.

Assuming that chemical concentrations are not limiting biological recovery in a late phase of NR and that the community is essentially starting from bare substrate, we hypothesize that the community development curve follows a sigmoidal shape. To illustrate community development, we use species richness as a key indicator because of its ecological importance, societal value, and sensitivity to PH contamination. Evidence for the sigmoidal shape of the community development curve comes from species colonization experiments (e.g., Figure 7-8) and from monitoring studies after oil spills (Figure 7-9). In reality, the "plateau" of the community development curve is never perfectly flat because stochastic demographic processes

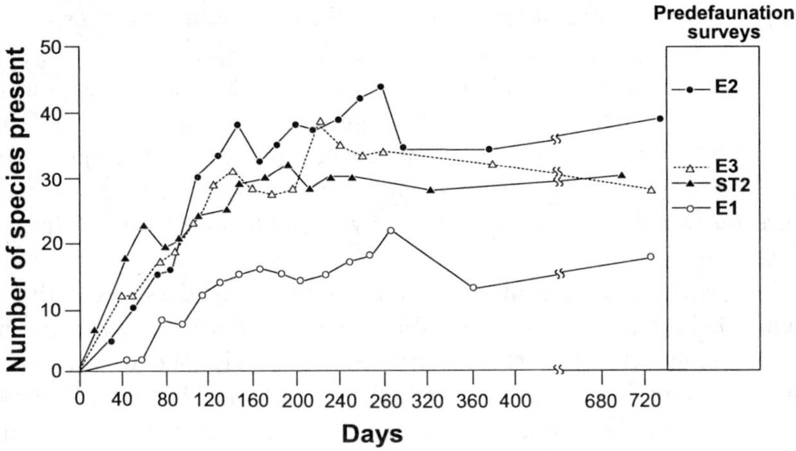

Note: Following complete extermination of the arthropod fauna, there was relatively little damage to the vegetation. At first, the number of species on each islet rose rapidly, then soon leveled off to species densities fairly close to those before defaunation (plotted on the right vertical axis).

Figure 7-8 Colonization curves (number of arthropod species through time) of 4 tiny mangrove islets in the Florida Keys (reprinted from Simberloff and Wilson 1970 with permission. © Ecological Society of America)

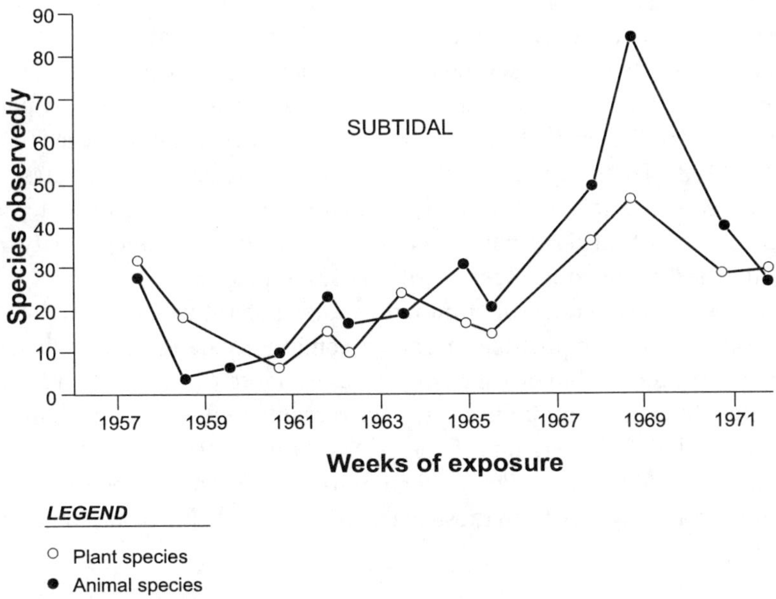

Figure 7-9 Long-term variation in species richness after the *Tampico Maru* spill (North 1973)

and natural environmental disturbances cause fluctuations in populations and communities (e.g., Sousa 1979, 1984; Paine and Levin 1981).

The early development of the community is limited by colonization, which can be a slow process (lower part of Curve 3 in Figure 7-7). After species invade, population growth and community development may be rapid (e.g., steep slope of Curve 3, Figure 7-7). Finally, as biological interactions control the development of community structure, the rate of recovery and community assembly slows (asymptotic portion of curve).

Differences in the magnitude of the initial impact affect the shape of the community development curve by truncating the lower portions of the curve in cases of low to moderate impact (i.e., when the substrate is not bare because tolerant species survived the contamination). Curves 1 and 2 in Figure 7-7 show the relative difference in biological recovery for acute effects only, where the initial impact is less for Curve 1. When the initial impact is small or moderate, the remaining organisms include key species that constitute the essential elements of biological community structure. In this case, the initial rate of recovery is faster and the total recovery time is shorter (Curve 1).

Chronic effects delay recovery, but when chemical concentrations fall below critical thresholds of toxicity to key species, the community may develop along a recovery trajectory whose general shape is similar to that after acute effects. The recovery curve for a community subjected to chronic effects is shifted to the right (later in time) of that for the acute-effects case. For example, in cases of moderate acute toxic impacts, Curve 4 (compared to Curve 2 in Figure 7-7) shows the effect of initially persistent toxicity. The curves in Figure 7-7 are idealized; an impacted community may follow one of several recovery trajectories depending on stochastic processes (Drake 1991). Natural disturbances such as storms and fires may perturb communities as well, such that the idealized recovery curve is not achieved (e.g., Sousa 1979, 1984; Paine and Levin 1981). Moreover, even when recovery has occurred, species richness may normally fluctuate around some steady-state value between environmental perturbations because of time lags and stochasticity in colonization and extinction of species within habitat patches.

Considering the tolerance distribution of species sensitivities within the biological community (or possible alternative community states) leads to uncertainty bands around the biological recovery curve (Figure 7-10). The recovery curves representing different communities may have very different uncertainty bands, depending on chemical tolerances of species. At the extremes of the uncertainty distribution (e.g., 95th percentile), predictions of relative rates of recovery among community types may vary from those obtained by considering the average-tolerance case.

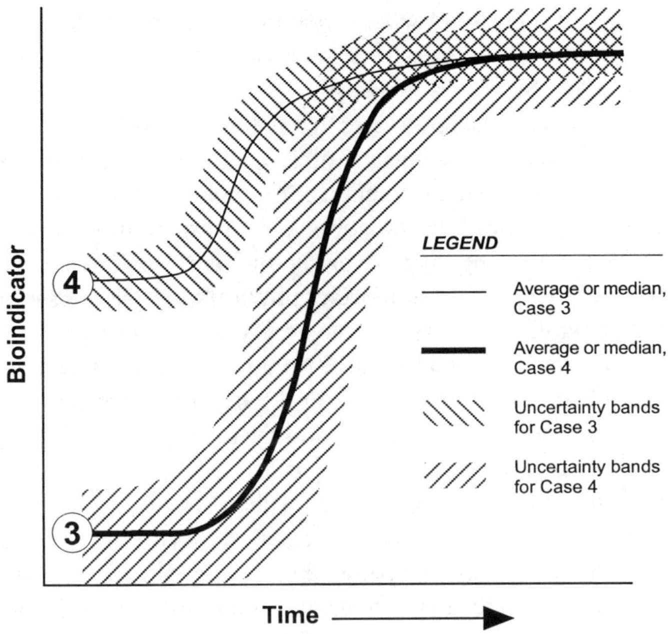

Note: Biological impact after initial PAH release: Case 3 = High; Case 4 = Moderate

Figure 7-10 Uncertainty in biological recovery because of variability of PAH tolerance among species in a community

CONCLUSIONS

The rate of NR of PAHs and other PHs in soil or sediment is controlled by physical–chemical processes that disperse and degrade hydrocarbons to nontoxic levels, by microbial transformation and biodegradation, and by biological recovery. Sequestration of PAHs and other PHs in soils and sediments helps the process of NR by removing a fraction of the hydrocarbons as an ecological exposure source, allowing the ecological system to recover more rapidly. Ecological processes such as microbial degradation, colonization rates of affected species, and evolutionary adaptation (e.g., chemical tolerance of indigenous species) influence the rate of contaminant degradation and the subsequent recovery of biological communities. Biological recovery includes colonization by species and successional changes in species composition and diversity over time. In high-energy habitats (e.g., exposed rocky coasts) or environments with a high potential for biodegradation of contaminants (e.g., calm, warm waters), the rate of biological recolonization and community development may limit NR after initial physical–chemical recovery. Long-term monitoring data would allow for broader conclusions to be made about biological recovery and the impact of varying rates of physical, chemical, and microbial processes on recovery time.

The basic physical–chemical processes that determine the fate of PAHs and other PHs in soil and sediment systems are similar, but certain processes play a larger role in one or the other environment. For example, NR rates for PAHs in soils may be determined largely by biodegradation. In contrast, dilution of contaminated-sediment PAHs by mixing (hydrodynamic and bioturbation) and ultimately burial by natural sedimentation are important physical processes for sediments.

Information on natural recovery of biological communities after PH contamination is inadequate for many habitats. Despite progress in our understanding of the fate and effects of PH in the environment, little attention has been given to the interface between the microbial community and the macrobiological community (interface between Phase 2 and Phase 3 in Figure 7-1). Moreover, the role of the macrobiota in regulating the latter phase of NR (Phase 3 in Figure 7-1) is poorly understood relative to the current knowledge of physical–chemical mechanisms that modify concentrations and forms of hydrocarbons. Especially after a severe impact, recovery processes related to the macrobiological community may regulate the overall rate of NR. In some instances (e.g., high degree of contamination, inaccessible environment), active remedial action will be insignificant relative to NR, regardless of how great the effort to actively clean up the environment. In these cases, an understanding of NR is essential to assess the potential for long-term impact of the contamination and to minimize effects on valued biological resources to the extent possible.

ACKNOWLEDGMENTS

Portions of this manuscript were based on a poster presentation at the 1997 annual meeting of the Society of Environmental Toxicology and Chemistry (SETAC). Authors of the poster were Robert Pastorok, Lucinda Jacobs, Mike Ruby, Shawn Severn, Scott Becker, and Bob Barrick of Exponent. Marie BenKinney, Lucinda Jacobs, and Shawn Severn of Exponent provided helpful comments on drafts of this paper.

REFERENCES

Albrecht P, Ourisson G. 1971. Biogenic substances in sediments and fossils. *Angew Chem Int Ed* 10:209–224.

Alexander M. 1995. How toxic are toxic chemicals in soil? *Environ Sci Technol* 29(11):2713–2717.

Allan JB, Bjoroey M, Douglas AG. 1977. Variation in the content and distribution of high molecular weight hydrocarbons in a series of coal macerals of different ranks. *Proc 7th*

Intl Meeting Adv Org Geochem; 1975. Madrid, Spain: Empresa Nac Adaro Invest Mineras SA.

Aller RC. 1982. The effects of macrobenthos on chemical properties of marine sediment and overlying waters. In: McCall PL, Tevesz MJS, editors. Animal-sediment relations. New York NY: Plenum Pr. p.53–102.

Atlas RM, Bartha R. 1993. Microbial ecology: Fundamentals and applications. Redwood City CA: Benjamin Cummings.

[ATSDR] Agency for Toxic Substances and Disease Registry. 1993. Toxicological profile for naphthalene. Atlanta GA: U.S. Department of Health and Human Services.

Austin GT. 1984. Petroleum processing. In: Austin GT, editor. Chapter 37: Shreve's chemical process industries. 5th ed. New York NY: McGraw Hill.

Baker JM, Clark RB, Kingston PF, Jenkins RH. 1990. Natural recovery of cold water marine environments after an oil spill. Presented at the Thirteenth Annual Arctic and Marine Oilspill Program Technical Seminar; 1990 June.

Ball WP, Roberts PV. 1991. Long-term sorption of halogenated organic chemicals by aquifer material. 2. Intraparticle diffusion. *Environ Sci Technol* 25:1237–1249.

Barrick RC. 1982. Flux of aliphatic and polycyclic aromatic hydrocarbons to central Puget Sound from Seattle (West Point) primary sewage effluent. *Environ Sci Technol* 16:682–692.

Barrick R, Furlong E, Carpenter R. 1984. Hydrocarbon and azaarene markers of coal transport to aquatic sediments. *Environ Sci Technol* 18:846–854.

Bartha R, Atlas RM. 1987. Transport and transformation of petroleum: Biological processes. In: Boesch DM, Rabalais NN, editors. The long-term effects of offshore oil and gas development: An assessment and a research strategy. New York NY: Elsevier Applied Science. p 287–341.

Baumann PC, Harshbarger JC. 1985. Frequencies of liver neoplasia in a feral fish population and associated carcinogens. *Mar Environ Res* 17:324–327.

Baumann PC, Mac MJ, Smith SB, Harshbarger JC. 1991. Tumor frequencies in walleye (*Stizostedion vitreum*) and brown bullhead (*Ictalurus nebulosus*) and sediment contaminants in tributaries of the Laurentian Great Lakes. *Can J Fish Aquat Sci* 48:1804–1810.

Bentz A. 1976. Oil spill identification. *Anal Chem* 48:454–472.

Bezalel L, Hadar Y, Cerniglia CE. 1996. Mineralization of polycyclic aromatic hydrocarbons by the white rot fungus *Pleurotus ostreatur*. *Appl Environ Microbiol* 62:292–295.

Biddinger GR, Gloss SP. 1984. The importance of trophic transfer in the bioaccumulation of chemical contaminants in aquatic ecosystems. *Residue Reviews* 91:104–145.

Blumer M. 1965. Organic pigments: Their long-term fate. *Science* 149:722–726.

Blumer M, Sass J. 1972. Oil pollution: Persistence and degradation of spilled fuel oil. *Science* 176:1120–1122.

Blumer M, Youngblood WW. 1975. Polycyclic aromatic hydrocarbons in soils and recent sediments. *Science* 188:53–55.

Boldrin B, Tiehm A, Fritzsche C. 1993. Degradation of phenanthrene, fluorene, fluoranthene, and pyrene by a *Mycobacterium* sp. *Appl Environ Microbiol* 59:1927–1930.

Booth PN, Becker DS, Pastorok RA, Sampson JR, Graham WJ. 1991. Evaluation of restoration alternatives for natural resources injured by oil spills. Health and Environmental Affairs Department. Washington DC: American Petroleum Institute. API Publication No. 304.

Bosma TNP, Ballemans E, Marlies W, Hoekstra NK, te Welscher RAG, Smeenk JGMM, Schraa G, Zehnder AJB. 1996. Biotransformation of organics in soil columns and an infiltration area. *Ground Water* 34:49–56.

Bosma TNP, Middeldorp PJM, Schraa G, Zehnder AJB. 1997. Mass transfer limitation of biotransformation: Quantifying bioavailability. *Environ Sci Technol* 31:248–252.

Bouchez M, Blanchet D, Vandecasteele J-P. 1995. Degradation of polycyclic aromatic hydrocarbons by pure strains and by defined strain associations: Inhibition phenomena and cometabolism. *Appl Microbiol Biotechnol* 43:156–164.

Bowling, JW, Leversee GJ, Landrum PF, and Giesy JP. 1983. Acute mortality of anthracene-contaminated fish exposed to sunlight. *Aquat Toxicol* 3:79–90.

Braddock JF, Lindstrom JE, Yeager TR, Rasley BT, Brown EJ. 1996. Patterns of microbial activity in oiled and unoiled sediments in Prince William Sound. In: Rice SD, Spies RB, Wolfe DA, Wright BA, editors. Proceedings from the Exxon Valdez Oil Spill Symposium. Anchorage AK: American Fisheries Society. p 95–108.

Broman DC, Naf C, Lundbergh I, Zebuhr Y. 1990. An in situ study on the distribution, biotransformation and flux of polycyclic aromatic hydrocarbons (PAHs) in an aquatic food chain (seston-*Mytilus edulis* L. - *Somateria mollissima* L.) from the Baltic: An ecotoxicological perspective. *Environ Toxicol Chem* 11:331–345.

Buhler DR, Williams DE. 1989. Enzymes involved in metabolism of PAH by fishes and other aquatic animals: Oxidative enzymes (or Phase I enzymes). In: Varanasi U, editor. Metabolism of polycyclic aromatic hydrocarbons in the aquatic environment. Boca Raton FL: CRC Pr. p 151–184.

Burns WA, Mankiewicz PJ, Bence AE, Page DS, Parker KR. 1997. A principal-component and least-squares method for allocating polycyclic aromatic hydrocarbons in sediment to multiple sources. *Environ Toxicol Chem* 16:1119–1131.

Carmichael LM, Pfaender FK. 1997. Polynuclear aromatic hydrocarbon metabolism in soils: Relationship to soil characteristics and preexposure. *Environ Toxicol Chem* 16(4):666–675.

Carpenter SR, Kitchell JF, editors. 1993. The trophic cascade in lakes. Cambridge UK: Cambridge Univ Pr.

Cerniglia CE. 1984. Microbial transformation of aromatic hydrocarbons. In: Atlas RM, editor. Petroleum microbiology. New York NY: Macmillan. p 99–128.

Cerniglia CE. 1992. Biodegradation of PAHs. *Biodegradation* 3:351–368.

Cerniglia CE, Freeman JP, Evans FE. 1984. Evidence for an arene oxide-NIH shift pathway in the transformation of naphthalene to 1-naphthol by *Bacillus cereus*. *Arch Microbiol* 138:283–286.

Cerniglia CE, Heitkamp MA. 1989. Microbial degradation of polycyclic aromatic hydrocarbons (PAH) in the aquatic environment. In: Varanasi U, editor. Metabolism of polycyclic aromatic hydrocarbons in the aquatic environment. Boca Raton FL: CRC Pr. p. 42–68.

Coates JD, Anderson RT, Lovley DR. 1996. Oxidation of polycyclic aromatic hydrocarbons under sulfate-reducing conditions. *Appl Environ Microbiol* 62:1099–1101.

Coates JD, Anderson RT, Woodward JC, Phillips EJP, Lovley DR. 1996. Anaerobic hydrocarbon degradation in petroleum-contaminated harbor sediments under sulfate-reducing and artificially imposed iron-reducing conditions. *Environ Sci Technol* 30:2784–2789.

Corner EDS, Harris RP, Kilvington CC, O'Hara SCM. 1976. Petroleum compounds in the marine food web: Short-term experiments on the fate of naphthalene in *Calanus*. *J Mar Biol Assoc UK* 56:121–133.

Crocker FH, Guerin WF, Boyd SA. 1995. Bioavailability of naphthalene sorbed to cationic surfactant-modified smectite clay. *Environ Sci Technol* 29:2953–2958.

Croft B, Swannell RPJ, Grant AL, Lee K. 1995. Effect of bioremediation agents on oil biodegradation in medium-fine sand. In: Hinchee RE, Kittel JA, Reisinger HJ, editors. Applied bioremediation of petroleum hydrocarbons. Columbus OH: Battelle Pr. IR-469. p 423–434.

Dagher F, Deziel E, Lirette P, Paquette G, Bisaillon J-G, Villemur R. 1997. Comparative study of five polycyclic aromatic hydrocarbon degrading bacterial strains isolated from contaminated soils. *Can J Microbiol* 43:368–377.

Dean-Raymond D, Bartha R. 1975. Biodegradation of some polynuclear aromatic petroleum components by marine bacteria. *Dev Ind Microbiol* 16:97–110.

Diamond JM. 1975. Assembly of species communities. In: Cody ML, Diamond JM, editors. Ecology and evolution of communities. Cambridge MA: Belknap Pr. p 342–444.

Dillon TM. 1981. Effects of dimethylnaphthalene-contaminated food source and daily fluctuating temperatures on the survival potential of the estuarine shrimp *Palaemonetes pugio*. *Mar Biol* 63:227–233.

Donath E. 1908. *Chem Zig* 32:1271. [as cited in Sakabe and Sassa 1952]

Drake JA. 1991. Community-assembly mechanics and the structure of an experimental species ensemble. *Amer Natur* 137:1–26.

Driscoll SK, Harkey GA, Landrum PF. 1997. Accumulation and toxicokinetics of fluoranthene in sediment bioassays with freshwater amphipods. *Environ Toxicol Chem* 16:742–753.

Dua RD, Meera S. 1981. Purification and characterization of naphthalene oxygenase from *Corynebacterium renale*. *Eur J Biochem* 120:461–465.

Efroymson RA, Alexander M. 1994. Biodegradation in soil of hydrophobic pollutants in nonaqueous-phase liquids (NAPLs). *Environ Toxicol Chem* 13:405–411.

Egloff G. 1947. Physical constraints of hydrocarbon IV. New York NY: Reinhold.

Eisler R. 1987. Polycyclic aromatic hydrocarbon hazards to fish, wildlife, and invertebrates: a synoptic review. Washington DC: U.S. Fish & Wildlife Service. Biological Report No. 85(1.17). 72 p.

Foster MS, Tarpley JA, Dearn SL. 1990. To clean or not to clean: The rationale, methods, and consequences of removing oil from temperate shores. *Northwest Environ J* 6:105–120.

Fries CR, Lee RF. 1984. Pollutant effects on the mixed function oxygenase (MFO) and reproductive systems of the marine polychaete *Nereis virens*. *Mar Biol* 79:187–193.

Gardner GR, Yevich PP, Harshbarger JC, Malcolm AR. 1991. Carcinogenicity of Black Rock Harbor sediment to the eastern oyster and trophic transfer of Black Rock Harbor carcinogens from the blue mussel to the winter flounder. *Environ Health Perspect* 90:53–66.

Garrigues P, Saptorahardjo A, Gonzales C, Wehrung P, Albrecht P, Saliot A, Ewald M. 1986. Biogeochemical aromatic markers in the sediments from Mahakam Delta (Indonesia). *Org Geochem* 10:959–964.

Ghosh DK, Mishra AK. 1983. Oxidation of phenanthrene by a strain of *Micrococcus*: Evidence of protocatechuate pathway. *Curr Microbiol* 9:219–224.

Gibson DT, Mahadevan V, Jerina DM, Yagi H, Yeh HJC. 1975. Oxidation of the carcinogens benzo[a]pyrene and benz[a]anthracene to dihydrodiols by a bacterium. *Science* 189:295–297.

Giger W, Blumer M. 1974. Polycyclic aromatic hydrocarbons in the environment: Isolation and characterization by chromatography, visible ultraviolet, and mass spectrometry. *Anal Chem* 46:1663–1671.

Giger W, Schaffner C. 1978. Determination of polycyclic aromatic hydrocarbons in the environment by glass capillary gas chromatography. *Anal Chem* 50:243–249.

Gilbert F, Stora G, Bertrand JC. 1996. In situ bioturbation and hydrocarbon fate in an experimental contaminated Mediterranean coastal ecosystem. *Chemosphere* 33:1449–1458.

Greiner AC, Spyckerelle C, Albrecht P. 1976. Aromatic hydrocarbons from geological sources. I. New naturally occurring phenanthrene and chrysene derivatives. *Tetrahedron* 32(2):257–260.

Guerin WF, Boyd SA. 1992. Differential bioavailability of soil-sorbed naphthalene to two bacterial species. *Appl Environ Microbiol* 54:1142–1152.

Haas A, Hites R. 1976. On the origin of polycyclic aromatic hydrocarbons in recent sediments: Biosynthesis by anaerobic bacteria. *Geochim Cosmochim Acta* 40:1141–1143.

Hargis Jr W J, Zwerner DE, Thoney DA, Kelly KL, Warinner JE III. 1989. Neoplasms in mummichogs from the Elizabeth River, Virginia. *Aquat Animal Health* 1:165–172.

Hatzinger PB, Alexander M. 1995. Effect of aging of chemicals in soil on their biodegradability and extractability. *Environ Sci Technol* 29:537–545.

Heitkamp MA, Cerniglia CE. 1988. Polycyclic aromatic hydrocarbon degradation by a *Mycobacterium* sp. in microcosms containing sediment and water from a pristine ecosystem. *Appl Environ Microbiol* 54:1968–1973.

Herbes SE. 1976. Partitioning of polycyclic aromatic hydrocarbons between dissolved and particulate phases in natural waters. *Water Res* 11:493–496.

Herbes SE. 1981. Rates of microbial transformation of polycyclic aromatic hydrocarbons in water and sediments in the vicinity of a coal coking wastewater discharge. *Appl Environ Microbiol* 41:20–28.

Hites RA, LaFlamme RE, Farrington JW. 1977. Polycyclic aromatic hydrocarbons in marine/aquatic sediments: Their ubiquity. *Adv Chem Ser* 185:289–311.

Hontela A, Dumont P, Duclos D, Fortin R. 1995. Endocrine and metabolic dysfunction in yellow perch, *Perca flavescens*, exposed to organic contaminants and heavy metals in the St. Lawrence River. *Environ Toxicol Chem* 14:725–731.

Jacher KA, Burton GA. 1995. In situ toxicity assessment on nonpoint source runoff using the freshwater amphipod *Hyallela azteca*. In: Proceedings of the Society of Environmental Toxicology and Chemistry; 1993; Houston TX. Pensacola FL: SETAC.

James MO. 1989. Biotransformation and disposition of PAH in aquatic invertebrates. In: Varanasi U, editor. Chapter 3, Metabolism of polycyclic aromatic hydrocarbons in the aquatic environment. Boca Raton FL: CRC.

Jerina, DM, Yagi H, Thakker DR., Sayer JM, van Bladeren PJ, Lehr RE, Whalen DL, Levin W, Chang RL, Wood AW, Conney AH. 1984. Identification of the ultimate carcinogenic metabolites of the polycyclic aromatic hydrocarbons: Bay-Region (R,S)-diol-(S,R)-epoxides. In: Caldwell J, Paulson GD, editors. Foreign compound metabolism. London UK: Taylor and Francis. p 257–266.

Johnson TL, Pastorok RA. 1985. Oil spill cleanup: Options for minimizing adverse ecological impacts. Health and Environmental Sciences Department. Washington DC: American Petroleum Institute. API Publication No. 4435.

Josephson J. 1984. Polycyclic aromatic hydrocarbons. *Environ Sci Technol* 18:93A–95A.

Kawatsuka S, Igarashi M. 1975. Degradation of PCP in soils. *Soil Sci and Plant Nutri* 21(4):405–414.

Kelsey JW, Alexander M. 1997. Declining bioavailability and appropriate estimation of risk of persistent compounds. *Environ Toxicol Chem* 16:582–585.

Kelsey JW, Kottler BD, Alexander M. 1997. Selective chemical extractants to predict bioavailability of soil-aged organic chemicals. *Environ Sci Technol* 31:214–217.

Kiyohara H, Nagao K. 1978. The catabolism of phenanthrene and naphthalene by bacteria. *J Gen Microbiol* 105:69–75.

Kiyohara H, Nagao K, Kuono K, Yano K. 1982. Phenanthrene degrading phenotype of *Alcaligenes faecalis* AFK2. *Appl Environ Microbiol* 43:458–461.

Kiyohara H, Nagao K, Nomi R. 1976. Degradation of phenanthrene through *o*-phthalate by an *Aeromonas* sp. *Agric Biol Chem* 40:1075.

Kukkonen J, Landrum PF. 1995. Measuring assimilation efficiencies for sediment-bound PAH and PCB congeners by benthic organisms. *Aquat Toxicol* 32:75–92.

Landrum PF, Harkey GA, Kukkonen J. 1995. Evaluation of organic contaminant exposure in aquatic organisms: The significance of bioconcentration and bioaccumulation. In: Newman M, Jagoe C, editors. Ecotoxicology: A hierarchical treatment. Chelsea MI: Lewis. p 85–131.

Landrum PF, Lee H II, Lydy MJ. 1992. Toxicokinetics in aquatic systems: Model comparisons and use in hazard assessment. *Environ Toxicol Chem* 11:1709–1725.

Landrum PF, Nihart SR, Eadie BJ, Herche LR. 1987. Reduction in bioavailability of organic contaminants to the amphipod *Pontoporeia hoyi* by organic matter of sediment interstitial waters. *Environ Toxicol Chem* 6:11–20.

Landrum PF, Reinhold MD, Nihart SR, Eadie BJ. 1985. Predicting the bioavailability of organic xenobiotics to *Pontoporeia hoyi* in the presence of humic and fulvic materials and natural dissolved organic matter. *Environ Toxicol Chem* 4:459–467.

Lane WF, Loehr RC. 1992. Estimating the equilibrium aqueous concentrations of polynuclear aromatic hydrocarbons in complex mixtures. *Environ Sci Technol* 26:983–990.

Law R, Morton RD. 1993. Alternative permanent states of ecological communities. *Ecology* 74(5):1347–1361.

Law R, Morton RD. 1996. Permanence and the assembly of ecological communities. *Ecology* 77(3):762–775.

Lawton JH, Jones CG. 1995. Linking species and ecosystems: Organisms as ecosystem engineers. In: Jones CG, Lawton JH, editors. Linking species and ecosystems. New York NY: Chapman & Hall. p 141–150.

Lee II H, Boese BL, Pelletier J, Winsor M, Specht DT, Randall RC. 1993. Guidance manual: Bedded sediment bioaccumulation tests. Narragansett RI: U.S. Environmental Protection Agency (USEPA), ERL. EPA 600/R-93/183.

Linz DG, Nakles DV. 1996. Environmentally acceptable endpoints in soil: Risk-based approach to contaminated site management based on availability of chemicals in soil. American Academy of Environmental Engineers. ISBN 1 883767 14 8.

Livingstone DR. 1985. Responses of the detoxification/toxification enzyme systems of molluscs to organic pollutants and xenobiotics. *Mar Pollut Bull* 16:158–170.

Lui PI-F. 1993. Introduction to energy and the environment. New York NY: Van Nostrand Rheinhold. p 202.

Luthy RG, Aiken GR, Brusseau ML, Cunningham SD, Gschwend PM, Pignatello JJ, Reinhard M, Traina SJ, Weber WJ, Westall JC. 1997. Sequestration of hydrophobic organic contaminants by geosorbents. *Environ Sci Technol* 31(12):3341–3347.

MacDonald DD. 1994. Approach to the assessment of sediment quality in Florida coastal waters. Volume 1, Development and evaluation of the sediment quality assessment guidelines. Tallahassee FL: Florida Department of Environmental Protection.

MacGillivray AR, Shiaris MP. 1993. Biotransformation of polycyclic aromatic hydrocarbons by yeasts isolated from coastal sediments. *Appl Environ Microbiol* 59:1613–1618.

Madsen EL, Sinclair JL, Ghiorse WC. 1991. In situ biodegradation: Microbiological patterns in a contaminated aquifer. *Science* 252:830–833.

Mahaffey WR, Gibson DT, Cerniglia CE. 1988. Bacterial oxidation of chemical carcinogens: Formation of polycyclic aromatic acids from benz[a]anthracene. *Appl Environ Microbiol* 54:2415–2423.

Malins DC, McCain BB, Brown DW, Chan S-L, Myers MS, Landahl JT, Prohaska PG, Friedman AJ, Rhodes LD, Burrows DG, Gronlund WD, Hodgins HO. 1984. Chemical pollutants in sediments and diseases of bottom-dwelling fish in Puget Sound, Washington. *Environ Sci Technol* 18:705–713.

Malins DC, McCain BB, Landahl JT, Myers MS, Krahn MM, Brown DW, Chan S-L, Roubal WT. 1988. Neoplastic and other diseases in fish in relation to toxic chemicals: An overview. *Aquat Toxicol* 11:43–67.

Mattraw HCJ, Franks BJ. 1986. USGS water supply. In: Mattraw AC, Franks BJ, editors. Washington DC: U.S. Government Printing Office. Paper No. 2285. p 1–8.

McCarthy JF. 1989. Bioavailability and toxicity of metals and hydrophobic organic contaminants. In: Suffet IH, MacCarthy P, editors. Aquatic humic substances: Influence on fate and treatment of pollutants. Advances in Chemistry Series 219. Washington DC: American Chemical Society (ACS). p 263–277.

McDonald TJ, Aldrett S, Bonner JS, Autenrieth RL, Morris PJ. 1997. Differential biodegradation of polynuclear aromatic hydrocarbons. In: Alleman BC, Leeson A, editors. In situ and on-site bioremediation. Columbus OH: Battelle Pr. p 213–224.

McElroy AE, Sisson JD. 1989. Trophic transfer of benzo[a]pyrene metabolites between benthic marine organisms. *Mar Environ Res* 28:265–269.

McGroddy SE, Farrington JW. 1994. In situ and desorption partition coefficients measured for PAHs and PCBs in Boston Harbor sediments. In: Proceedings of 207th American Chemical Society (ACS) national meeting; 1994 Mar 13–18; San Diego CA. Preprinted extended abstract 34(1):123–126.

Meador JP, Casillas E, Sloan CA, Varanasi U. 1995. Comparative bioaccumulation of polycyclic aromatic hydrocarbons from sediment by two infaunal organisms. *Mar Ecol Prog Ser* 123:107–124.

Meador JP, Stein JE, Reichert WL, Varanasi U. 1995. Bioaccumulation of polycyclic aromatic hydrocarbons by marine organisms. *Rev Environ Contamin Toxicol* 143:79–163.

Meulenberg R, Rijnaars HHM, Doddema HJ, Field JA. 1997. Partially oxidized polycyclic aromatic hydrocarbons show an increased bioavailability and biodegradability. *FEMS Microbiology Letters* 152:45–49.

Mihelcic JR, Luthy RG. 1988. Degradation of polycyclic aromatic hydrocarbon compounds under various redox conditions in soil-water systems. *Appl Environ Microbiol* 54:1182–1187.

Mueller JG, Chapman PJ, Pritchard PH. 1989. Creosote-contaminated sites. *Environ Sci Technol* 23:1197–1201.

Neff JM. 1979. Polycyclic aromatic hydrocarbons in the aquatic environment: Sources, fates and biological effects. Barking, Essex, UK: Applied Science. 266 p.

Neff JM. 1985. Polycyclic aromatic hydrocarbons. In: Rand GM, Petrocelli SR, editors. Fundamentals of aquatic toxicology. London UK: Taylor and Francis. p 416–454.

Neff JM. 1987. Biological effects of oil in the marine environment. *Chem Eng Prog* November:27–330.

Neff JM, Langseth DE, Graham EM, Sauer Jr. TC, Gnewuch SC, Little AD. 1994. Transport and fate of non-BTEX petroleum chemicals in soils and groundwater. Washington DC: American Petroleum Institute. API Publication No. 4593.

Newsted JL, Giesy JP. 1987. Quantitative structure activity model of photodynamic toxicity of polycyclic aromatic hydrocarbons to *Daphnia magna*. *Environ Toxicol Chem* 6:1829–1834.

Nicholson DK, Woods SL, Istok JD, Peek DC. 1992. Reductive dechlorination of chlorophenols by a PCP-acclimated methanogenic consortium. *Appl Environ Microbiol* 58(7):2280–2286.

Niimi AJ, Dookhran GP. 1989. Dietary absorption efficiencies and elimination rates of polycyclic aromatic hydrocarbons (PAHs) in rainbow trout *(Salmo gairdneri)*. *Environ Toxicol Chem* 8:719–722.

Niimi AJ, Palazzo V. 1986. Biological half-lives of eight polycyclic aromatic hydrocarbons (PAHs) in rainbow trout (*Salmo gairdneri*). *Water Res* 20:503–507.

North WJ. 1973. Position paper on effects of acute oil spills. In: Background papers, workshop on inputs, fates, and effects of petroleum in the marine environment. Washington DC: National Academy of Sciences (NAS). p 745–765.

Paine RT, Levin SA. 1981. Intertidal landscapes: Disturbance and the dynamics of pattern. *Ecol Monographs* 51:145–178.

Pancirov RJ, Brown RA. 1975. Analytical methods for PAH in crude oils, heating oils, and marine tissues. Washington DC: American Petroleum Institute. p 103–113.

Pastorok RA, Peek DC, Sampson JR, Jacobson MA. 1994. Ecological risk assessment for river sediments contaminated by creosote. *Environ Toxicol Chem* 13(12)1929–1941.

Peakall DB, Tremblay T, Kinter WB, Miller DS. 1981. Endocrine dysfunction in seabirds caused by ingested oil. *Environ Res* 24:6–14.

Peters KE, Moldowan JM. 1993. The biomarker guide: Interpreting molecular fossils in petroleum and ancient sediments. Englewood NJ: Prentice Hall.

Pictet A, Ramaeyer L, Kalser O. 1916. Compt, rend. 165:58 [as cited in Sakabe and Sassa 1952]

Prahl FG, Carpenter R. 1979. The role of zooplankton fecal pellets in the sedimentation of polycyclic aromatic hydrocarbons in Dabob Bay, Washington. *Geochim Cosmochim Acta* 43:1959–1972.

Prahl FG, Carpenter R. 1983. PAH-phase associations in Washington coastal sediment. *Geochim Cosmochim Acta* 47:1013–1023.

Preslo L, Miller M, McLearn M, Suyama W, Kostecki P, Fleischer E. 1989. Available remedial technologies for petroleum contaminated soils. In: Bonazountas M, Hetrick DM, Kostecki PT, Calabrese EJ, editors. SESOIL in environmental fate and risk modeling. Amherst MA: Amherst Scientific.

Prince RC. 1993. Petroleum spill bioremediation in marine environments. *Crit Rev Microbiol* 19:217–242.

Radke M. 1987. Organic geochemistry of aromatic hydrocarbons. In: Brooks J, Whelte D, editors. Advances in petroleum geochemistry 2:141–205.

Rathbone K, Banks MK, Kulakow P, Schwab AP. 1997. Sorghum variety impact on biodegradation of diesel fuel contaminated soil. Abstracts from the 1997 Conference on Hazardous Waste Research. Manhattan KS: Great Plains/Rocky Mountain Hazardous Substance Research Center.

Rhodes DC, Aller RC, Goldhaber MB. 1977. The influence of colonizing benthos on physical properties and chemical diagenesis of the estuarine seafloor. In: Coull BC, editor. Ecology of marine benthos. Columbia SC: Univ South Carolina Pr. p 113–138.

Riley RG, Thomas BL, Anderson JW, Bean RM. 1981. Changes in the volatile hydrocarbon content of Prudhoe Bay crude treated under different simulated weathering conditions. *Mar Environ Res* 4:109–119.

Roesijadi G, Anderson JW, Blaylock JW. 1978. Uptake of hydrocarbons from marine sediments contaminated with Prudhoe Bay crude oil: Influence of feeding type of test species and availability of polycyclic aromatic hydrocarbons. *J Fish Res Board Can* 35:608–614.

Sakabe T, Sassa R. 1952. An aromatic hydrocarbon isolated from coal bitumen. *Bull Chem Soc Jpn* 25:353–355.

Schnoor JL. 1997. Design of phytoremediation at contaminated sites: Abstracts from the 1997 Conference on Hazardous Waste Research. Manhattan KS: Great Plains/Rocky Mountain Hazardous Substance Research Center.

Shuttleworth KL, Cerniglia CE. 1995. Environmental aspects of PAH biodegradation. *Appl Biochem Biotechnol* 54:291–302.

Sikka HC, Steward AR, Kandaswami C, Rutkowski JP, Zaleski J, Earley K, Gupta RC. 1991. Metabolism of benzo[a]pyrene and persistence of DNA adducts in the brown bullhead (*Ictalurus nebulosus*). *Comp Biochem Physiol* 100C:25–28.

Simberloff DS, Wilson ED. 1970. Experimental zoogeography of islands. A two-year record of colonization. *Ecology* 51:934–937.

Sinclair JL. 1991. Eukaryotic microorganisms in subsurface environments. In: Fliermans CB, Hazen TC, editors. Proceedings of the First International Symposium on Microbiology of the Deep Subsurface. Aiken SC: WSRC Information Services. p 3-39–3-51.

Socha SB, Carpenter R. 1987. Factors affecting pore water hydrocarbon concentrations in Puget Sound sediment. *Geochim Cosmochim Acta* 51:1273–1284.

Sousa WP. 1979. Disturbance in marine intertidal boulder fields: The nonequilibrium maintenance of species diversity. *Ecology* 60:1225–1239

Sousa WP. 1984. The role of disturbance in natural communities. *Ann Rev Ecol System* 15:353–391.

Stegeman JJ, Lech JJ. 1991. Cytochrome P450 monooxygenase systems in aquatic species: Carcinogen metabolism and biomarkers for carcinogen and pollutant exposure. *Env Health Perspectives* 90:101–109.

Stein JE, Reichert WL, French B, Varanasi U. 1993. ^{32}P-postlabeling analysis of DNA adduct formation and persistence in English sole exposed to benzo[a]pyrene and 7H-dibenzo[c,g]carbazole. *Chem Biol Interact* 88:55–69.

Stoll, M, Schener W. 1937. Concrete of oak moss. C.R 17me *Congr Chim Ind Paris* 17:206–212.

Stonecipher WD, Turner RW. 1970. Rosin and rosin derivatives. *Encycl Polm Sci Technol* 12:139–161.

Suchanek TH. 1986. Mussels and their role in structuring rocky shore communities. In: Moore PG, Seed R, editors. The ecology of rocky coasts. New York NY: Columbia Univ Pr. p 70–96.

Suedel BC, Boraczek JA, Peddicord RK, Clifford PA, Dillon TM. 1994. Trophic transfer and biomagnification potential of contaminants in aquatic ecosystems. *Rev Environ Contam Toxicol* 136:21–89.

Sullivan JB. 1974. Marine pollution by carcinogenic hydrocarbons. In: Marine pollution monitoring (petroleum). Washington DC: Natl Bur Stand. p 261–263.

Sutherland JP. 1974. Multiple stable points in natural populations. *Am Nat* 108:859–873.

Swannell RPJ, Lee K, McDonaugh M. 1996. Field evaluations of marine oil spill bioremediation. *Microbiol Rev* 60(2):342–365.

Swannell RPJ, Mitchell DJ, Jones DM, Willis A, Lee K, Lepo JE. 1997. Field evaluation of bioremediation to treat crude oil on a mudflat. In: Alleman BC, Leeson A, editors. In situ and on site bioremediation. Columbus OH: Battelle Pr. p 401–406.

Tang J, Carroquino MJ, Robertson BK, Alexander M. 1998. Combined effect of sequestration and bioremediation in reducing bioavailability of polycyclic aromatic hydrocarbons in soil. *Environ Sci Technol* 32:3586–3590.

Treccani V, Walker N, Wiltshire GH. 1954. The metabolism of naphthalene by soil bacteria. *J Gen Microbiol* 11:341–348.

[USCG] U.S. Coast Guard. 1977. Oil spill identification system. National Technical Information System. Groton CT: USCG. Final Report NTIS ADA 003803 (CG-D-41-75).

[USEPA] U.S. Environmental Protection Agency. 1985. Information review. Unleaded gasoline. TSCA Interagency Testing Committee. Washington DC: USEPA. IR-469.

[USEPA] U.S. Environmental Protection Agency. 1991. Assessment and control of bioconcentratable contaminants in surface waters. Washington DC: Office of Water.

Varanasi U, Reichert WL, Stein JE, Brown DW, Sanborn HR. 1985. Bioavailability and biotransformation of aromatic hydrocarbons in benthic organisms exposed to sediment from an urban estuary. *Environ Sci Technol* 19:836–841.

Varanasi U, Stein JE, Nishimoto M. 1989. Biotransformation and disposition of polycyclic aromatic hydrocarbons (PAH) in fish. In: Varanasi U, editor. Metabolism of polycyclic aromatic hydrocarbons in the aquatic environment. Boca Raton FL: CRC Pr. p 94–149.

Venosa AD, Suidan MT, Wrenn BA, Strohmeier KL, Haines JR, Eberhart BL, King D, Holder E. 1996. Bioremediation of an experimental oil spill on the shoreline of Delaware Bay. *Environ Sci Technol* 30:1764–1775.

Wakeham SG, Schaffner C, Giger W. 1979. Polycyclic aromatic hydrocarbons in recent lake sediments – II. Compounds derived from biogenic precursors during early diagenesis. *Geochim Comochim Acta* 44:415–429.

Wetzel SC, Banks MK, Schwab AP. 1997. Rhizosphere effects on the degradation of pyrene and anthracene in soil. Chapter 18. In: Phytoremediation of soil and water contaminants. American Chemical Society (ACS) Symposium Series 664.

White JC, Kelsey JW, Hatzinger PB, Alexander M. 1997. Factors affecting sequestration and bioavailability of phenanthrene in soils. *Environ Toxicol Chem* 16(10):2040–2045.

Wilson JT, McNabb JF, Cochran JW, Wang TH, Tomson MB, Bedient PB. 1985. Influence of microbial adaptation on the fate of organic pollutants in groundwater from symposium of biodegradation of chemicals in the subsurface environment. *Environ Toxicol Chem* 4:721–726.

Ye D, Siddiqi AM, Maccubbin AE, Kumar S, Sikka HC. 1996. Degradation of PAHs by *Sphingomononas paucimobilis*. *Environ Sci Technol* 30:136–142.

8

Natural Remediation of Chlorinated Organic Compounds

Michael D. Lee and John W. Davis

THE UBIQUITY OF CHLORINATED HYDROCARBONS in the environment continues to be the basis for public and regulatory concern because of their recalcitrance and potential for bioaccumulation. Although a number of chlorinated compounds are produced naturally in soil and marine environments (Gribble 1994, 1996), the vast majority is anthropogenic. Currently, more than 15,000 different chlorinated organic compounds are in commercial use (Hileman 1993). Of these, large volumes of chlorinated aliphatic and aromatic compounds are produced each year for a variety of industrial and commercial uses, including solvents, lubricants, intermediates in the chemical industry, plasticizers, crop protection, pharmaceuticals, and medical equipment. The U.S. International Trade Commission (1995) reported that more than 32,443,000 kg of chlorinated organic chemicals were produced in the United States in 1994. A subset of this group, the chlorinated solvents, is used in the manufacture of polychlorinated vinyl chloride (PVC), degreasing agents, and chlorine intermediates. As a result of the large-scale industrial production and use of chlorinated compounds, these chemicals have a potential for introduction into the environment. The Toxic Release Inventory (U.S. Environmental Protection Agency [USEPA] 1996a, 1996b) reports releases of more than 192,000,000 kg of chlorinated organic compounds in the U.S. for the years 1988 and 1994 to 1996.

Chlorinated organics are a heterogeneous chemical family with a wide range of physical and chemical properties. As a result, the various chlorinated organic compounds differ in their toxicity, persistence, or bioaccumulation potential. Some

Natural Remediation of Environmental Contaminants. Michael Swindoll et al., editors.
 ISBN 1-880611-33-3.

chlorinated organics are associated with specific risks to health and the environment, as a result of their persistence and lipophilicity, or as a result of their genotoxic or carcinogenic effects. Some chlorinated organic compounds, such as polychlorinated biphenyls (PCBs), dioxins, furans, and dichlorodiphenyltrichloroethane (DDT), tend to accumulate because of their very low rate of biodegradation, are very toxic in nature, and are bioaccumulative.

The introduction of a Cl substituent into an organic compound generally leads to increased toxicity (Freitag et al. 1994; Henschler 1994). For example, chloromethanes have a central anesthetic effect, and some chlorinated ethanes, chlorinated ethenes, and propenes are reported to be carcinogenic (Henschler 1994). Some chlorinated ethenes are genotoxic, with the genotoxic effects usually the result of a metabolically reactive intermediate. Introduction of a Cl substituent to aromatic compounds has been reported to increase not only acute toxicity but also subchronic and chronic toxicity. Many chlorinated aromatic hydrocarbons, including hexachlorobenzene, PCBs, and dioxins, have been classified as nongenotoxic carcinogens. Of the 50 compounds on the German list of carcinogenic substances, one-third are chlorinated organic compounds (MAK-Werte-List 1974).

Many chlorinated organic compounds resist both chemical and biological attack, which contributes to the environmental persistence of these materials. Some compounds, such as PCBs and tetrachloroethene (PCE), are produced specifically for their persistence and stability. Several types of degradation processes, including biodegradation, hydrolysis, photolysis, and photooxidation, can impact or reduce the level of a chemical within environmental compartments. Howard et al. (1991) classified the relative persistence of chemicals based upon their environmental half-life ($t_{1/2}$). Compounds with $t_{1/2}$ of 1 to 7 days are considered to degrade rapidly; those with $t_{1/2}$ of 1 to 4 weeks degrade moderately fast; and those with $t_{1/2}$ of 4 weeks to 6 months degrade slowly. Compounds with $t_{1/2}$ of 6 months to 1 year are considered resistant to degradation. In general, the shorter the $t_{1/2}$ of a chlorinated organic compound, the less likely the compound will accumulate in the environment and adversely impact environmental or human receptors. Many of the chlorinated aliphatic and aromatic compounds have $t_{1/2}$ greater than 4 weeks. However, half-life is not an inherent property of a compound but ultimately depends upon the interaction between the chemical and the physical, chemical, and biological parameters of the environment.

Because of their lipophilic nature, many chlorinated organic compounds have been identified as being bioaccumulative chemicals. "Bioaccumulation" in the broadest sense is defined as "the extent to which an organism accumulates a compound from its surrounding environment by all processes" (Isenesse 1991). Bioaccumulation has been shown to be inversely correlated to the aqueous solubility of these compounds. Bioaccumulation of chlorinated organics occurs primarily as a physical or chemical process. Partitioning of the compound into the lipids is the most important mechanism, hence the uptake and sequestration in body lipids reflects the hydrophobic characteristics of these compounds. The octanol–water partition coefficient (K_{OW}), which is based on a chemical's hydrophobicity or tendency to partition into lipids, is often used to estimate bioconcentration. Bioconcentration potential can be categorized according to the following scheme: low with a bioconcentration factor (BCF) of < 100 and a log K_{OW} of < 3.0; moderate with a BCF of 100 to 3,000 and a log K_{OW} of 3.0 to 5.0; high with a BCF of > 3,000 and a log K_{OW} of > 5.0. Based on this scheme, many of the chlorinated aromatic compounds would be classified as having a moderate to high bioconcentration potential and therefore are of potential ecological concern.

The adsorption of a compound in a particular environmental compartment is a function of the organic content of the soil or sediment as well as the intrinsic properties (e.g., hydrophobicity) of the organic compound. The adsorption coefficient K_{OC} describes the relationship between the mass of material adsorbed per unit C. Chemical and physical properties important to the ultimate fate and transport of a number of chlorinated organic compounds are identified in Table 8-1.

USEPA's definition of "natural attenuation processes" includes "...a variety of physical, chemical, or biological processes that, under favorable conditions, act without human intervention to reduce the mass, toxicity, mobility, volume, or concentration of contaminants in soil or groundwater. These in-situ processes include biodegradation; dispersion; dilution; sorption; volatilization; radioactive decay; and chemical or biological stabilization, transformation, or destruction of contaminants" (USEPA 1997). Thus, it is important to understand the role that these physical, chemical, and biological processes play in natural attenuation of chlorinated organic compounds.

TRANSPORT AND ENVIRONMENTAL PARTITIONING OF CHLORINATED ORGANIC COMPOUNDS

The transport and partitioning of a chemical can effectively reduce the overall concentration of a contaminant, thereby reducing or increasing the potential for exposure. For example, dilution and dispersion can reduce the overall concentration of a chemical in aqueous environments. Dispersion occurs as contaminated water moves at different rates through layers of an aquifer or surface water, resulting in mechanical mixing and molecular diffusion (Freeze and Cherry 1979). Dilution results from the mixing of contaminated water with uncontaminated water from recharge or transport.

The hydrophobicity of the chlorinated aryl or alkyl compounds plays an important role in determining the environmental distribution of these compounds. Soils and sediments often act as sinks for many of the highly chlorinated hydrocarbons. Numerous studies have reported chlorinated aromatics in sediments in

Table 8-1 Physical and chemical properties of selected chlorinated organics

Chlorinated organic	Physical–chemical property							
	Melting point (°C)	Water solubility (mg/L)	Vapor pressure (mm Hg)	Henry's law constant (atm-m^3·mol)	Density (g/cc)	Log K_{ow}	Log K_{oc}	Maximum contaminant level (mg/L)
Halogenated volatile organics								
Liquid solvents								
Carbon tetrachloride (CT)	−23	8 E+02	9.13 E+01	2 E−02	1.5947	2.83	2.64	5
Chlorobenzene	−45	4.9 E+02	8.8 E+00	3.46 E−03	1.106	2.84	2.2	100
Chloroform (CF)	−64	8.22 E+03	1.6 E+02	3.75 E−03	1.485	1.97	1.64	nd
cis-1,2-dichloroethylene	−81	3.5 E+03	2 E+02	7.5 E−03	1.284	1.86	1.5	70
1,1-dichloroethylene	−97.4	5.5 E+03	1.82 E+02	5.7 E−03	1.175	1.79	1.48	nd
1,2-dichloroethane	−37.4	8.69 E+03	6.37 E+01	1.1 E−03	1.253	1.48	1.15	5
1,1-dichloroethylene	−122.5	4 E+02	5 E+02	1.54 E−01	1.214	2.13	1.81	7
1,2-dichloropropane	−90	2.7 E+03	3.95 E+01	3.6 E−03	1.158	2.02	1.71	5
Ethylene dibromide	9.97	3.4 E+03	1.1 E+01	3.18 E−04	2.172	1.76	1.45	0.05
Methylene chloride	−97	1.32 E+04	3.5 E+02	2.57 E−03	1.325	1.25	0.94	5
1,1,2,2-tetrachloroethane	−43	2.9 E+03	4.9 E+00	5 E−04	1.600	2.39	2.34	nd
Tetrachloroethylene (PCE)	−22.7	1.5 E+02	1.4 E+01	2.27 E−02	1.625	3.14	2.82	5
trans-1,2-dichloroethylene	−50	6.3 E+03	2.65 E+02	6.6 E−03	1.257	2.09	1.77	70
1,1,1-trichloroethane	−32	9.5 E+02	1 E+02	2.76 E−03	1.325	2.49	2.18	200
1,1,2-trichloroethane	−36	4.5 E+03	1.88 E+01	1.17 E−03	1.4436	2.17	1.75	nd
Trichloroethylene (TCE)	−87	1 E+03	5.87 E+01	8.92 E−03	1.462	2.42	2.10	5
Gases								
Chloroethane (b.p. 12.5 °C)[a]	−138.3	5.7 E+03	1 E+03	1.1 E−02	0.9414	1.43	1.17	nd
Vinyl chloride (VC) (b.p. −13.9 °C)	−157	1.1 E+03	2.3 E+03	6.95 E−01	0.9121	0.60	0.91	2

Table 8-1 (continued)

Chlorinated organic	Physical–chemical property: Melting point (°C)	Water solubility (mg/L)	Vapor pressure (mm Hg)	Henry' s law constant (atm-m^3·mol)	Density (g/cc)	Log K_{ow}	Log K_{oc}	Maximum contaminant level (mg/L)
Halogenated semivolatile organics								
Pesticides								
Chlordane	106	5.6 E–02	1 E–05	2.2 E–01	1.6	5.48	4.58	2
DDD	112	1.60 E–01	1 E–06	7.96 E–01	1.385	5.56	5.38	nd
DDE	88.4	4.0 E–02	6.40 E–06	1.9 E–01	nd	5.69	5.41	nd
DDT	108	3.1 E–03	1.5 E–07	2.8 E–01	0.985	6.36	5.48	nd
Dieldrin	176.5	1.86 E–01	1.78 E–07	9.7 E–01	1.75	5.34	3.23	nd
Chlorinated benzenes								
1,2-dichlorobenzene	–17	1 E+02	9.6 E–01	1.88 E–01	1.306	3.38	3.06	600
1,4-dichlorobenzene	53	8 E+01	6 E–01	1.58 E–01	1.2475	3.39	3.07	750
Chlorinated phenols								
Pentachlorophenol	190	1.4 E+03	6.1 E–04	2.8 E–01	1.978	5.12	4.80	nd
2,3,4,6-tetrachlorophenol	69.5	1.00 E+03	nd	nd	1.839	4.1	2.0	nd

b.p. = boiling point
nd = No data

freshwater and marine environments (Oliver and Nicol 1982; Durham and Oliver 1983; De Leer 1985; Golden et al. 1993; Palm and Lammi 1995). Sorption of the chlorinated organics can often be predicted based on the K_{OW}, which increases with the degree of Cl atoms present. For example, Scharp et al. (1994) reported the sorption of chlorobenzenes in sediments was linear with increasing Cl substitution. Conversely, the lower chlorinated congeners of chlorobenzene (mono- and di-) have lower K_{OW}, are more water soluble, and thus their residence time in the water column would be longer.

The chlorinated alkanes and alkenes have been identified as among the most prevalent groundwater contaminants. The physical or chemical properties of these compounds contribute to their relative mobility and persistence in the subsurface. In pure form, most C_1–C_3 chlorinated aliphatics liquids have densities greater than water (Table 8-1). Chemicals with high density and low viscosity have a higher potential mobility in the subsurface (Feenstra 1994). Also, many of these chlorinated solvents have relatively low sorptive capacity and are soluble in water at concentrations higher than the U.S. maximum contaminant levels (MCLs) drinking water standards (see Table 8-1).

The environmental fate of chlorinated organic compounds is impacted primarily by the environment where the chemical is released (i.e., point of entry into the environment) and by the physical and chemical properties of the chemical. The inherent physical and chemical properties influence a chemical's reactivity and dictate how these compounds move among the different environmental compartments of air, water, soil, sediments, and biota. For example, organic compounds denser than water will sink to the bottom of an aquifer or other body of water, compounds with a high water solubility will dissolve into the water, and compounds with a high Henry's law constant will preferentially partition into air rather than water.

Multimedia fate models are often employed to predict the fate and lifetime of compounds on a local, regional, and global level. Many of these models rely on the physical and chemical properties of a chemical to predict environmental partitioning and reactive tendencies. Mackay, Shiu, Ma (1992) identified 5 key physical–chemical properties that differentiate how chemicals behave in the environment:

1) Solubility in water
2) Vapor pressure
3) Octanol–water partition coefficient
4) Disassociation constants in water
5) Susceptibility to transformation (e.g., degradation reactions).

A listing of some of the properties that affect the transport of chlorinated organics is given in Table 8-2. Using the Level 1 fugacity model of Mackay et al. (1996), one can calculate the primary compartments where a chemical will partition, the relative concentration, and the tendency to bioaccumulate.

Table 8-2 Threshold values for key parameters, indicating material has high potential to leach[a,b]

Parameter	Value or range
Water solubility	> 30 mg/L
Soil adsorption (K_d)	< 5 (usually < 1)
K_{oc}	< (300 to 500)
Volatility (Henry's law constant)	< 10^{-2} atm-m^{-3} mol
Speciation	Negatively charged, fully or partially at ambient pH
Hydrolysis half-life	> 25 weeks
Photolysis half-life	> 1 week
Field dissipation half-life	> 3 weeks

[a] Source: USEPA 1986.

[b] Note: No single threshold will indicate leachability.

ABIOTIC PROCESSES

A number of abiotic processes impact the natural remediation (NR) of chlorinated organic compounds, including oxidation–reduction (redox) reactions, sorption, volatilization, precipitation, and hydrolysis. Typically, abiotic processes result in a change in a chemical's physical or chemical properties, which can reduce potential environmental risks by altering contaminant mobility, bioavailability, or toxicity. The specific interactions that occur among chemicals, soil, and water dictate which abiotic processes occur and are more important to the fate of the chemical.

Hydrolysis has been identified as an important abiotic mechanism responsible for the transformation of a range of chlorinated organic compounds. For example, Vogel and McCarty (1987) reported that 1,1,1-trichloroethane (TCA) was transformed to 1,1-dichloroethene and acetic acid through hydrolytic substitution reaction. Jeffers and coworkers (1989) reported hydrolysis rate constants for several chlorinated methanes, ethanes, ethenes, and propanes. Half-lives ranged from a low of 0.1 year for pentachloroethane (PCA) to a high of 2.1×10^{10} years for *cis*-dichloroethylene (DCE). They found that ethenes were generally unreactive, while the ethanes' reactivity increased with increasing chlorination if each C compound contained at least one Cl.

Redox reactions play an important role in determining the fate and transport of chemical contaminants. Knowing the redox potential of the aqueous environment allows for accurate predictions as to what reactions and chemical species will likely occur. Redox potential is important from an environmental context because the properties of a chemical will vary at different redox states. For instance, Fe in a reduced state can exist as the soluble ferrous (Fe^{2+}) form; under oxidizing conditions, Fe may exist as the insoluble ferric (Fe^{3+}) form. Both abiotic- and biotic-mediated transformations of ferrous iron to ferric iron result in the formation of an Fe precipitate. Precipitation and complexation that can occur through redox reactions can alter the environmental mobility and bioavailability of both organic and inorganic contaminants.

Oxidation reactions are important to the environmental chemistry of many contaminants (Manahan 1994) and can alter a chemical's structure, resulting in changes in properties, including solubility and mobility. For example, Ulrich and Stone (1989) reported the oxidation of chlorinated phenols by manganese oxides under environmentally relevant conditions. The resulting dechlorinated products, quinones and diphenoquinones, readily reacted into humus-like materials. Manganese oxides are found in many soils and may be responsible for the immobilization and detoxification of chlorinated organics in soils and sediments.

Although oxidative processes can play a role in the NR of chlorinated organic compounds, introduction of a Cl atom into an organic compound renders it more oxidized and less susceptible to oxidative attack. In contrast, studies conducted over the last 10 years have shown the reduction reactions of chlorinated organics are one of the most significant abiotic mechanisms resulting in the transformation of these compounds in the environment. Vogel et al. (1987) reviewed the abiotic transformation of halogenated aliphatic compounds and reported that the relative rate of reduction followed specific patterns related to their standard reductive potential. The more highly chlorinated aliphatic compounds, such as PCE or hexachloroethane (HCA), have higher standard potentials, and more energy is released upon reduction of these compounds. Thus, thermodynamic considerations dictate that more highly chlorinated compounds are more likely to undergo reductive transformations rather than oxidative reactions under environmentally relevant conditions.

The abiotic reductive transformation of chlorinated organic compounds may be particularly important in aquatic sediments. Peijnenburg and coworkers (1992) reported on the transformation of more than 45 halogenated aromatic hydrocarbons in sterile, gamma-irradiated, anoxic sediment systems. The disappearance of these compounds from the anoxic sediments was described as a pseudo–first-order process, with $t_{1/2}$ ranging from 1 to 78 days. These results demonstrate that the abiotic transformation process must be considered when we assess the fate and lifetime of chlorinated organic compounds in aquatic sediments where anoxic conditions predominate.

Compounds such as minerals (Kriegman-King and Reinhard 1994) and metals have been shown to react with chlorinated organics, resulting in less toxic end products. For example, TCE reacts with zero valent Fe as described by the 2 half-reactions (Focht et al. 1996):

$$Fe^0 \rightarrow Fe^{2+} + 2e^-$$

$$C_2HCl_3 + 3H^+ + 6e^- \rightarrow C_2H_4 + 3Cl^-$$

The reaction of TCE with Fe to produce ethene requires no additional input of energy and thus may have significant environmental consequences for the reduction of contaminant mass.

Sorption as an NR process includes both adsorptive and absorptive mechanisms. "Adsorption" refers to the adherence of the chemical to the surface of a solid (such as sand), while "absorption" refers to the diffusion of the chemical into the solid's matrix. The mechanisms responsible for sorption are complex, being influenced by chemical and physical factors of both the chemical and the solid (Alexander 1995). Sorption occurs as the result of partitioning of a chemical between the water and the solid and the formation of bonds between the chemical and the solid. For example, chemicals with low water solubility, such as PCE, PCBs, and dioxins, will readily partition from water and sorb to a solid such as soil. The organic content, electric charge, and surface area of the solid control the sorption of chemicals.

The sorption–desorption process often results in sequestering organic compounds within the soil or sediment particles, which results in an overall decrease in bioavailability of the sorbed compound. Alexander (1995) proposed 3 lines of evidence that sequestration plays an important role in reducing the bioavailability of toxic chemicals in soils. First, abiotic processes render the chemical less available to microorganisms, which are more able to assimilate organic chemicals than are plants and animals. The lack of bioavailability of these compounds can be observed by an overall decrease in the rates and extent of biodegradation of sorbed organic compounds in aged soils. For example, Knabel and coworkers (1996) reported the decrease in the mineralization of a variety of surfactants if they were first preadsorbed to different soil components (e.g., clays, humics).

Second, as organic compounds persist in soil, they become more resistant to extraction by solvents. This fact is particularly noticeable for some chlorinated organics that are highly reactive with organic substances in soil or sediments. For example, Sarkar and coworkers (1988) demonstrated the enzymatic coupling of 2,4-dichlorophenol to fulvic acid by several oxidoreductases. Beyerle-Pfnur and Lay (1990) reported that chloroanilines can react with a variety of soils and sediments and remain stably bound; subsequent extraction with solvent resulted in recovery of less than 10% of the chloroanilines. These strong interactions were likely the outcome of covalent bonding between the amino groups and the oxygen-containing group of the humic fraction, resulting in an overall decrease in the bioavailability of the chloroanilines.

Third, there is an observed decrease in toxicity to test organisms after the toxic material is added to soil or sediment particles. Recently it has been demonstrated that bioavailability, and hence toxicity, of a compound is not directly related to the total concentration of a material in soil or sediment systems. For example, Roberston and Alexander (1998) observed a decrease in the toxicity of Dieldrin and DDT to several species as the resident time in soil increased. In contrast to the decline in bioavailability, there was very little change in the total mass of these chlorinated pesticides. Taken together, these 3 lines of evidence demonstrate that, as chlorinated organic chemicals become sequestered in inaccessible microsites or covalently bound to the organic fraction of the soil or sediment particles, their overall bioavailability and toxicity diminish.

Toxicity reduction by sorption of nonionic organic chemicals onto sediments has long been recognized, and sediment quality criteria (SQC) for the compounds have been proposed using the equilibrium partitioning (EqP) method (Toro et al. 1991). EqP establishes that the level of bioavailability of a chemical is related to the interstitial water concentration and not to the total sediment–chemicals concentration. Thus, organic chemicals that persist in soils and sediments can exhibit a wide range of toxicities across different sediments.

"Precipitation" refers to the formation of a solid phase (mineral) from a dissolved phase of the chemical and occurs as the result of the chemical exceeding its aqueous solubility. Precipitation is a commonly occurring process and controls the solubility of many elements, including Al, Fe, Mn, Mg, and Ca. Chemical reactions, either abiotically or biologically mediated, can result in changes in chemical structure, which in turn may alter chemical solubility. Changes in the geochemical equilibrium, which can result from redox reactions, complexation, or sorption, can result in the precipitation of a chemical because of a change in the chemical's structure and solubility.

BIODEGRADATION PATHWAYS FOR CHLORINATED COMPOUNDS

Chlorinated compounds are naturally degraded by bacteria through a number of pathways, including metabolic oxidation, co-oxidation, cometabolic reduction, and metabolic reduction reactions (Fantroussi et al. 1988; Mohn and Tiedje 1992; Fathepure and Tiedje 1998; Fetzner 1998; Lee et al. 1998; Reinhard et al. 1998; Wiedemeier et al. 1998). Electron acceptors that are involved in the degradation of chlorinated compounds include oxygen, nitrate, iron, sulfate, and carbon dioxide. Chlorinated compounds can also be biodegraded under fermentative conditions. The most energy is obtained by bacteria when oxygen is used as an electron acceptor. Once oxygen is depleted by aerobic biodegradation processes, microbes use nitrate, the next most energetically favorable electron acceptor, followed by iron, sulfate, and carbon dioxide.

The electron acceptors under which a number of chlorinated compounds are degraded and their degradation products are shown in Table 8-3. However, this list is not exhaustive, as other chlorinated compounds are biodegradable, and degradation of these chlorinated compounds may occur under other electron-accepting conditions, which may yield different degradation products. Highly chlorinated com-

Table 8-3 Electron acceptors for chlorinated compound biodegradation and their degradation products

Compound	Electron acceptor[a]	Products	Reference
Chlorinated ethenes			
Tetrachloroethene (PCE)	SO_4	TCE, cDCE	Bagley and Gossett 1989
	CO_2	TCE, cDCE, VC, ethene, ethane, carbon dioxide	Vogel and McCarty 1985; de Bruin et al. 1992
Trichloroethene (TCE)	O_2	not identified	Leahy et al. 1996
	O_2+CoM	TCE-epoxide, CO, glyoxylic acid, trichloroethanol, trichloroacetaldhyde, trichloroacetate, dichloroacetate, glyoxylate, formic acid, carbon dioxide	Semprini et al. 1992; Dolfing et al. 1993
	NO_3	Not identified	Leahy et al. 1996
	SO_4	cDCE	Bagley and Gossett 1989
	CO_2	cDCE, tDCE, VC, carbon dioxide	Cox et al. 1994
	CO	cDCE, tDCE, 1,1-DCE, VC, ethene	Jablonski and Ferry 1992
cis-1,2-dichloroethene (cDCE)	O_2	Carbon dioxide	Bradley and Chapelle 1998; Klier et al. 1999
	O_2+CoM	cDCE-epoxide, tDCE-epoxide	Hopkins and McCarty 1995
	Fe^{3+}	Carbon dioxide	Bradley and Chapelle 1997
	CO_2	VC, ethene, ethane	Komatsu et al. 1994
	Humic acid	Carbon dioxide	Bradley et al. 1998
trans-1,2-dichloroethene (tDCE)	O_2+CoM	tDCE-epoxide	Hopkins and McCarty 1995
	O_2	Carbon dioxide	Klier et al. 1999
	CO_2	VC	Wilson et al. 1986
1,1-dichloroethene (1DCE)	O_2+CoM	Not identified	Dolan and McCarty 1994
	O_2	not identified	Klier et al. 1999
	CO_2	VC	Wilson et al. 1986
Vinyl chloride (VC)	O_2	Carbon dioxide	Davis and Carpenter 1990; Bradley and Chapelle 1998
	O_2+CoM	Not identified	Dolan and McCarty 1994
	Fe^{3+}	Carbon dioxide	Bradley and Chapelle 1996
	CO_2	Ethene, ethane, carbon dioxide, methane, chloromethane	Barrio-Lage et al. 1990; de Bruin et al. 1992; Bradley and Chapelle 1997
	Humic acid	Carbon dioxide	Bradley et al. 1998

Table 8-3 continued

Compound	Electron acceptor[a]	Products	Reference
Chlorinated ethanes			
Hexachloroethane (HCA)	NO_3	Not identified	Lew and Bouwer 1988
	SO_4	PCE, PCA, TCE	Miller et al. 1998
	CO_2	Not identified	Long et al. 1993
	An	PCE	
Pentachloroethane (PCA)	SO_4	PCE, TCE	Miller et al. 1998
	CO_2	Not identified	Boucquey et al. 1995
1,1,2,2-tetrachloroethane (TeCA)	CO_2	1,1,2-TCA, 1,2-DCA, CA, ethane, TCE, cDCE, tDCE, VC, ethene	Chen et al. 1996
1,1,1-trichloroethane (1TCA)	SO_4	1,1-DCA, CA, 1,1-DCE, carbon dioxide, acetate	Klecka et al. 1990
	CO_2	1,1-DCA, CA, carbon dioxide, nonvolatiles, acetate	Vogel and McCarty 1987; Klecka et al.. 1990
1,1,2-trichloroethane (2TCA)	CO_2	1,2-DCA, VC, ethene	Chen et al. 1996
1,2-dichloroethane (2DCA)	O_2	2-chloroethanol, chloroacetate, gylcolate, carbon dioxide	van den Wijngaard et al. 1992; Dolfing et al. 1993; Klecka et al. 1998
	O_2+CoM	Not identified	Chang and Alvarez-Cohen 1996
	CO_2	VC, CA, ethene, ethane, carbon dioxide	Bouwer and McCarty 1983a; Bosma et al. 1997
Chloroethane (CA)	O_2+CoM	Not identified	Chang and Alvarez-Cohen 1996
	CO_2	Ethane	Holliger et al. 1990
1,1,1-trifluoro-2,2-dichloroethane (HCFC-123)	CO_2	1,1,1-trifluro-2-chloroethane	Oremland et al. 1996
1,1,2-trichloro-1,2,2-trifluoro-ethane CFC-113)	NO_3	Not identified	Semprini et al. 1992
	SO_4+CO_2	1,2-dichloro-1,2,2-trifluoroethane, 1,1,2-trichloro-1,2-difluoroethane, 1-chloro-1,2,2-trifluoroethene	Jackson et al. 1990
1,2-dichloro-1,2,2-trifluoroethane (HCFC-123a)	CO_2	1-chloro-1,2,2-trifluoroethane	Lesage et al. 1990

Table 8-3 continued

Compound	Electron acceptor[a]	Products	Reference
Chlorinated methanes			
Carbon tetrachloride (CT)	NO_3	CF, carbon dioxide, cells, unknown	Bouwer and McCarty 1983b
	Fe^{3+}	Not identified	Collins et al. 1997
	SO_4+B_{12}	CF, DCM, carbon disulfide, carbon monoxide, carbon dioxide, others	Freedman et al. 1995
	CO_2	CF, DCM, CM, carbon dioxide, acetate, pyruvate, others	Egli et al. 1988
Chloroform (CF)	O_2+CoM	CO_2, soluble, particulate fractions	McClay et al. 1996
	SO_4+B_{12}	DCM, carbon dioxide, carbon monoxide	Becker and Freedman 1994
	CO_2	Carbon dioxide, unknown	Bouwer and McCarty 1983a
Dichloromethane (DCM)	O_2	Carbon dioxide, formaldehyde	Davis and Madsen 1991; Dolfing et al. 1993
	O_2+CoM	Not identified	Chang and Alvarez-Cohen 1996
	NO_3	Not identified	Freedman et al. 1997
	CO_2	Acetate, carbon dioxide, formate, pyruvate, acetate	Freedman and Gossett 1991; Magli et al. 1996
Chloromethane (CM)	O_2	Not identified	Freedman et al. 1997
	O_2+CoM	Not identified	Chang and Alvarez-Cohen 1996
	NO_3	Not identified	Freedman et al. 1997
	Acetogenic	Methanethiol, dimethyl sulfide	Braus-Stromeyer et al. 1993
Bromodichloromethane	NO_3	Not identified	Bouwer and McCarty 1983b
	CO_2	Not identified	Bouwer and McCarty 1983a
Dibromochloromethane	NO_3	Not identified	Bouwer and McCarty 1983b
	CO_2	Not identified	Bouwer and McCarty 1983a
Trichlorofluoromethane (CFC-11)	NO_3	Not identified	Semprini et al. 1992
	SO_4	Dichlorofluoromethane	Sonier et al. 1994
Dichlorofluromethane (HCFC-21)	CO_2	Not identified	Oremland et al. 1996
Chloroacetates			
Dichloroacetate	O_2	Not identified	Pries et al. 1994
Chloroacetate	O_2	Glycolic acid, Bicarbonate	Castro et al. 1996

Table 8-3 continued

Compound	Electron acceptor[a]	Products	Reference
Chloropropanes, chloroprenes, and chloroproponols, and chloropropylene oxides			
1,2-dichloropropane (2DCP)	O_2	Not identified	Adam and Veenstra 1993
	CO_2	1-chloropropane, 2-chloropropane, propene	Löffler et al. 1997
1,3-dichloropropane	O_2	Not identified	Pries et al. 1994
1-chloropropane (1-CP)	O_2	Not identified	Pries et al. 1994
2,3-dichloro-1-propanol	O_2	Not identified	Pries et al. 1994
1,3-dichloro-2-propanol	O_2	Not identified	Pries et al. 1994
cis-1,3-dichloropropene	O_2	cis-3-chloroallyl alcohol, cis-3-chloroacrolein, 3-chloroacyrlic acid, malonic acid semialdehyde	Poelarends et al. 1998
trans-1,3-dichloropropene	O_2	*trans*-3-chloroallyl alcohol, *trans*-3-chloroacrolein, 3-chloroacrylic acid, malonic acid semialdehyde	Poelarends et al. 1998
3-chloropropylene (allyl chloride)	O_2	3-chloropropylene oxide, 3-chloroacetone, acetone	Small et al. 1995
3-chloropropylene oxide (epichlorohydrin)	O_2	3-chloro-1,2-propanediol, glycerol, 3-chloroacetone, acetone	Small et al. 1995
Chlorinated ethers			
bis(2-chloroethyl)ether	O_2	Not identified	van den Wijngaard et al. 1993
2-chloroethylvinylether	O_2	2-hydroxyethylvinylether, vinyloxyacetaldehyde, vinyloxyacetic acid, 2-chloroethanol, chloroacetic acid, acetaldehyde	van den Wijngaard et al. 1993
4,4'-dichlorodiphenyl ether	O_2	4-chlorophenol, 4-chlorocatechol	Schmidt et al. 1993
2,4-dichlorodiphenyl ether	O_2	2,4-dichlorophenol, cathechol, 3,5-dichlorocatechol, phenol	Schmidt et al. 1993
Chlorinated benzenes			
Hexachlorobenzene	An	Pentachlorobenzene, 1,2,3,5- and 1,2,4,5-tetrachlorobenzene, 1,2,3-, 1,3,5- and 1,2,4-trichlorobenzene, 1,2-, 1,4-, 1,3-dichlorobenzene	Fathepure and Vogel 1991; Holliger et al. 1992; van der Meer et al. 1992; Chang et al. 1998
	SO_4	Not identified	van der Meer et al. 1992
	CO_2	1,3-dichlorobenzene, 1,4-dichlorobenzene	Pardue 1996

Table 8-3 continued

Compound	Electron acceptor[a]	Products	Reference
Chlorinated benzenes			
Pentachlorobenzene	SO_4	Not identified	van der Meer et al. 1992
	CO_2	1,2,3,4-tetrachlorobenzene, 1,2,4-trichlorobenzene, 1,2,3-trichlorobenzene, 1,2-dichlorobenzene, 1,4-dichlorobenzene, chlorobenzene	Ramanand, Balba, Duffy 1993
1,2,4,5-tetrachlorobenzene	O_2	1,3,4,6-tetrachloro-1,2-dihydroxycyclohexa-3,5-diene, 3,4,6-triclorocatechol, 2,3,5-trichloromuconate, 2-chloro-3-oxoadipate	Sander et al. 1991; Spain 1997
	SO_4	Not identified	van der Meer et al. 1992
1,2,3,4-tetrachlorobenzene	O_2	1,2,3,4-tetrachlorocatechol, 1,2,3,4-tetrachloromuconic acid, 2,3,5-trichlorodienelactone, 2,3,5-trichloromaleyl acetic acid, 2,4-dichloro-3-oxoadipic acid, tricarboxylic acid	Potrawfke et al. 1998
	SO_4	Not identified	van der Meer et al. 1992
1,2,3,5-tetrachlorobenzene	An	1,3,5-trichlorobenzene	van der Meer et al. 1992
1,3,5-trichlorobenzene	SO_4	Not identified	van der Meer et al. 1992
	An	1,3-dichlorobenzene, chlorobenzene	van der Meer et al. 1992
1,2,3-trichlorobenzene	O_2	CO_2	Marinucci and Bartha 1979
	SO_4	Not identified	van der Meer et al. 1992
	An	1,3-dichlorobenzene, chlorobenzene	van der Meer et al. 1992
1,2,4-trichlorobenzene	O_2	3,4,6-triclorocatechol, 2,3,5-trichloromucnate, 2-chloro-3-oxoadipate, carbon dioxide	Marinucci and Bartha 1979; Sander et al. 1991; Spain 1997
	SO_4	Not identified	van der Meer et al. 1992
	An	1,4-dichlorobenzene, chlorobenzene	van der Meer et al. 1992
1,2-dichlorobenzene	O_2	Not identified	van der Meer et al. 1992; Nishino et al. 1994
	SO_4	Not identified	van der Meer et al. 1992
	An	Chlorobenzene	van der Meer et al. 1992

Table 8-3 continued

Compound	Electron acceptor[a]	Products	Reference
Chlorinated benzenes *(continued)*			
1,3-dichlorobenzene	O_2	Not identified	Zaitsev et al 1995
	SO_4	Not identified	van der Meer et al. 1992
	An	Chlorobenzene	van der Meer et al. 1992
1,4-dichlorobenzene	O_2	1,4-dichlorobenzene dihydrodiol, 3,6-dichlorocatechol, 2,5-dichloro-*cis,cis*-muconic, acid; 2-chloro-4-carboxy-methylene-but-2-en-4-olide, 2-chloromaleylactic acid	Spiess et al. 1995; Spain 1997
	SO_4	Not identified	van der Meer et al. 1992
	An	Chlorobenzene	van der Meer et al. 1992
Chlorobenzene	O_2	Chlorobenzene dihydrodiol, 3-chlorocatechol, 2-chlorocatechol, 2-chloromuconic acid, 4-carboxymethylenebutenolide, maleylacetic acid, β-ketoadipic acid	Nishino et al. 1992; Spain 1997
Chlorinated toluenes			
2,3,6-trichlorotoluene	CO_2	2,5-dichlorotoluene, 2-chlorotoluene, toluene	Ramanand, Balba, Duffy 1993
2,4,5-trichlorotoluene	CO_2	2,5-dichlorotoluene, 3,4-dichlorotoluene, 4-chlorotoluene, 3-chlorotoluene, 2-chlorotoluene, toluene	Ramanand, Balba, Duffy 1993
2,4-dichlorotoluene	CO_2	4-chlorotoluene, 2-chlorotoluene, toluene	Ramanand, Balba, Duffy 1993
2,5-dichlorotoluene	CO_2	2-chlorotoluene, toluene	Ramanand, Balba, Duffy 1993
3,4-dichlorotoluene	CO_2	3-chlorotoluene, 4-chlorotoluene	Ramanand, Balba, Duffy 1993
Chlorinated phenols			
Pentachlorophenol	O_2	Carbon dioxide	Miethling and Karlson 1996
	An	2,3,4,5-tetrachlorophenol, 3,4,5-trichlorophenol, 3,5-dichlorophenol, 3-chlorophenol, phenol, methylphenol, benzoate	Chang et al. 1995
	SO_4	2,3,4,5-tetrachlorophenol, 2,4,5-trichlorophenol, 3,4,5-trichlorophenol, 2,3,5-trichlorophenol, 3,4-dichlorophenol, 3,5-dichlorophenol, 2,4-dichlorophenol, 2,5-dichlorophenol, 3,5-dichlorophenol, 3-chlorophenol, 4-chlorophenol	Masunaga et al. 1996

Table 8-3 continued

Compound	Electron acceptor[a]	Products	Reference
Chlorinated phenols			
Pentachlorophenol (*continued*)	CO_2	2,3,5,6-tetrachlorophenol, 2,3,4,5-tetrachlorophenol, 2,3,4,6-tetra chlorophenol, 2,3,6-trichlorophenol, 2,3,5-trichlorophenol, 3,4,5-trichlorophenol, 2,4,5-trichlorophenol, 2,4,6-trichlorophenol, 2,3-dichlorophenol, 3,5-trichlorophenol, 3,4-dichlorophenol, 2,4-dichlorophenol, 2-chlorophenol, 3-chlorophenol, 4-chlorophenol, carbon dioxide, methane	Nicholson et al. 1992; Hale et al. 1994
2,3,4,5-tetrachlorophenol	SO_4	3,4,5-trichlorophenol, 2,4,5-trichlorophenol, 2,3,5-trichlorophenol, 3,4-dichlorophenol, 3,5-dichlorophenol, 2,4-dichlorophenol, 2,5-dichlorophenol, 3,5-dichlorophenol, 3-chlorophenol, 4-chlorophenol	Masunaga et al. 1996
3,4,5-trichlorphenol	SO_4	3,4-dichlorophenol, 3,5-dichlorophenol, 3-chlorophenol, 4-chlorophenol	Masunaga et al. 1996
2,3,4-trichlorophenol	SO_4	3,4-dichlorophenol, 2,3-dichlorophenol, 3-chlorophenol, 4-chlorophenol	Masunaga et al. 1996
2,3,6-trichlorophenol	CO_2	2,5-dichlorophenol, 2,3-dichlorophenol, 3-chlorophenol	Mohn and Kennedy 1992
2,4,5-trichlorophenol	O_2	Carbon dioxide	Matus et al. 1996
2,4,6-trichlorophenol	An	2,4-dichlorophenol, 4-chlorophenol, phenol, methylphenol, benzoate	Armenante et al. 1992; Chang et al. 1995
2,3-dichlorophenol	SO_4	3-chlorophenol	Masunaga et al. 1996
2,4-dichlorophenol	O_2	Not identified	Zaitsev et al 1995
	SO_4	4-chlorophenol, 2-chlorophenol	Masunaga et al. 1996
2,5-dichlorophenol	SO_4	3-chlorophenol, 2-chlorophenol	Masunaga et al. 1996
2,6-dichlorophenol	SO_4	2-chlorophenol, phenol	Häggblom and Young 1991
	CO_2	4-chlorophenol	Häggblom et al. 1993b
3,4-dichlorophenol	SO_4	3-chlorophenol, 4-chlorophenol	Masunaga et al. 1996

Table 8-3 continued

Compound	Electron acceptor[a]	Products	Reference
Chlorinated phenols *(continued)*			
3,5-dichlorophenol	O_2	3-chlorophenol	Mohn and Kennedy 1992
	SO_4	3-chlorophenol	Masunaga et al. 1996
2-chlorophenol	O_2	Not identified	Zaitsev et al. 1995
	NO_3	None	Häggblom et al. 1993a
	Fe^{3+}	Not identified	Kazumi et al. 1995
	SO_4	None	Häggblom et al. 1993a
	CO_2	None	Häggblom et al. 1993a
3-chlorophenol	O_2	Not identified	Zaitsev et al. 1995
	Fe^{3+}	Not identified	Kazumi et al. 1995
	SO_4	None	Häggblom et al. 1993a
	CO_2	None	Häggblom et al. 1993a
4-chlorophenol	O_2	Not identified	Armenante et al. 1992
	Fe^{3+}	Not identified	Kazumi et al. 1995
	SO_4	None	Häggblom et al. 1993a
	CO_2	None	Häggblom et al. 1993a
Chloroanilines			
2,3,4,5-tetrachloraniline	SO_4	2,4,5-trichloroaniline	Hale et al. 1994
	CO_2	2,3,5-trichloroaniline, 2,4,5-trichloroaniline, 3,5-dichloroaniline	Kuhn et al. 1990
2,3,4-trichloroaniline	CO_2	2,4-dichloroaniline, 3,4-dichloroaniline, 4-chloroaniline	Kuhn et al. 1990
3,4-dichloroaniline	CO_2	4-chloroaniline	Kuhn et al. 1990
Chlorobenzoates			
2-chlorobenzoate	CO_2	None	Häggblom et al. 1993b
3-chlorobenzoate	O_2	Not identified	Zaitsev et al. 1995
	NO_3	None	Häggblom et al. 1993b
	Fe^{3+}	Not identified	Kazumi et al. 1995
	SO_4	None	Häggblom et al. 1993b
	CO_2	None	Häggblom et al. 1993b

Table 8-3 continued

Compound	Electron acceptor[a]	Products	Reference
Chlorobenzoates *(continued)*			
4-chlorobenzoate	NO_3	None	Häggblom et al. 1993b
3,6-Dichloro-2-methoxybenzoic acid (dicamba)	O_2	3,6-dichloro-2-salicylate, 6-chlorosalicylate	Milligan and Häggblom 1997
	CO_2	3,6-dichloro-2-salicylate, 6-chlorosalicylate	Taraban et al. 1993
Chlorinated benzidines			
3,3'-dichlorobenzidine	An	3-chlorobenzidine, benzadiene	Nyman et al. 1997
Pesticides			
α-hexachlorohexane	O_2	Pentachlorocyclohexane, 1,2,3,5-tetrachlorobenzene, 1,2,3,4-tetrachlorobenzene	Bachmann et al. 1988
ß-hexachlorocyclohexane	CO_2	Chlorobenzene, benzene	Van Eekert et al. 1998
2,4,5-trichlorophenoxyacetic acid	O_2	Carbon dioxide 2,4,5-trichlorophenol, 2,5-dichlorohydroquinone, 5-chloro-2-hydroxyhydroquinone	McCall et al. 1981 Haugland et al. 1990
(2,4,5-t)	CO_2	2,4-dichlorophenoxyacetic acid, 2,5-dichlorophenoxyacetic acid	Gibson and Suflita 1990
	An	2,4,5-trichlorophenol, 3,4-dichlorophenol, 4-chlorophenol, phenol, 2,4-cichlorophenoxy-acetic acid, 2,5-dichlorophenoxyacetic acid	Hale et al. 1994
2,4-dichlorophenoxyacetic acid	O_2	2,4-dichlorophenol, chlorohydroquinone, 3,5-dichlorocatechol, carbon dioxide	McCall et al. 1981; Haugland et al. 1990
(2,4-d)	An	2,4-dichlorophenol, 4-chlorophenol, phenol, (4-chorophenoxy)acetate	Hale et al. 1994
1,1,1-trichloro-2,2-bis(*p*-chloro-phenyl)ethane (DDT)	O_2	2,3-dihydroxy-DDT, 4-chlorobenzoic acid	Nadeau et al. 1994
	An	1,1-dichloro-2,2-bis(*p*-chlorphenyl)ethane (DDD), 1,1-dichloro-2,2-bis(*p*-chlorphenyl)ethylene (DDE), 1-chloro-2,2-bis(*p*-chlorophenyl)ethane, 1-chloro-2,2-bis(*p*-chloroethylene, 2,2-bis(*p*-chlorophenyl)ethylene, *p*-chlorobenzoate	Hale et al. 1994

Table 8-3 continued

Compound	Electron acceptor[a]	Products	Reference
Pesticides *(continued)*			
3,5,6-trichloro-4-amino-2-pyridine carboxylic acid (Picloram)	CO_2	3,6-dichloro-4-amino-2-pyridinecarboxylic acid, unidentified metabolite	Ramanand, Nagarajan, Suflita 1993
Chloroacetophenones			
4-chloroacetophenone	O_2	4-chloro-1-phenyl ethanol, 4-chlorophenyl acetate, 4-chlorophenol, 4-chlorocathecol	Havel and Reineke 1993
2,3-dichloroacetophenone	O_2	2,3-dichlorophenol, 3,4-dichlorocathecol	Havel and Reineke 1993
2,4-dichloroacetophenone	O_2	2,4-dichlorophenol, 3,5-dichlorocathecol	Havel and Reineke 1993
2,5-dichloroacetophenone	O_2	2,5-dichlorophenol, 3,6-dichlorocathecol	Havel and Reineke 1993
3,4-dichloroacetophenone	O_2	3,4-dichlorophenol, 4,5-dichlorocathecol	Havel and Reineke 1993
2,3,4-dichloroacetophenone	O_2	2,3,4-trichlorophenol, 3,4,5-dichlorocathecol	Havel and Reineke 1993
Others			
Hexachloro-1,3-butadiene	CO_2	1,1,2,3,4-pentachloro-1,3-butadiene, 1,2,3,4-tetrachloro-1,3-butadiene, trichloro-1,3-butadiene	Bosma et al. 1994
Chlorooctane	O_2	Not identified	Pries et al. 1994
1,9-dichlorononane	O_2	Not identified	Pries et al. 1994
1,3-dichloropropene	O_2	Not identified	Pries et al. 1994
2-chlorovinylethene	O_2	Not identified	Pries et al. 1994
2-chloroethanol	O_2	Not identified	Pries et al. 1994
2-chloroallyalcohol	O_2	Not identified	Pries et al. 1994
3-chloroallyalcohol	O_2	Not identified	Pries et al. 1994
2,2-dichloroproprionate	O_2	Not identified	Pries et al. 1994
3-chloroacrylate	O_2	Not identified	Pries et al. 1994
3-chlorocrotonate	O_2	Not identified	Pries et al. 1994
Chloroacetonitrile	O_2	Formate, hydrogen cyanide, carbon monoxide, hydrochloric acid	Castro et al. 1996
Chloroacetamide	O_2	Hydroxyacetamide, bicarbonate	Castro et al. 1996

[a] An = anaerobic, electron acceptor not specified; B_{12} = vitamin B_{12} added; CoM = cometabolic substrate; CO_2 = methanogenic; Fe^{3+} = iron reducing; NO_3 = nitrate reducing; O_2 = aerobic; SO_4 = sulfate reducing

pounds such as PCE, carbon tetrachloride (CT), hexachlorobenzene (HCB), PCBs, and PCDDs are thought to be degraded only under reducing conditions. Less chlorinated compounds (such as trichloroethene [TCE] or DCE; chloroform [CF]; tetra-, tri-, di-, or monochlorobenzenes; and other lower chlorinated congeners of PCBs) can be degraded aerobically or anaerobically when a cometabolic substrate is present. Many chlorinated compounds can be attacked directly under aerobic conditions. Chlorobenzenes up to and including tetrachlorobenzene are readily biodegraded under aerobic conditions (Spain 1997). Aerobic bacteria that are able to grow on chlorobenzene have been detected at many chlorobenzene-contaminated sites. Pure cultures of aerobic bacteria that metabolize lower chlorinated benzenes have been isolated, and their degradation pathways have been identified. These results show that chlorobenzenes are very good candidates for natural attenuation under both aerobic and anaerobic environments. Similarly, lower-chlorinated PCB congeners are degraded by aerobic bacteria (Flanagan and May 1993). Under denitrifying conditions, compounds such as CT, dichloromethane (DCM), hexachloroethane (HCA), and 3-chlorobenzoate are degraded. Chlorinated compounds such as *cis*-1,2-dichloroethene (cDCE), vinyl chloride (VC), and CT are degraded under Fe-reducing conditions. In some environments, sulfate may inhibit complete biodegradation of chlorinated compounds (Lee et al. 1998), while in other sulfate-reducing environments, mineralization (complete conversion of a compound to carbon dioxide, methane, water, and salts) of compounds such as CT and CF has been observed. Highly reducing environments such as those found under methanogenic conditions generally provide more favorable conditions for the anaerobic biodegradation of chlorinated compounds.

Energy-yielding oxidations

Many chlorinated compounds, including chloromethanes, chloroethanes, chloroalcohols, chloroalkanoic acids, chloroaromatics, and chloroalkenoic acids may act as the sole source of C and energy to support microbial growth (Lee et al. 1998). When present, oxygen is the primary electron acceptor under which energy-yielding oxidation occurs. However, other compounds such as iron (Fe^{3+}), sulfate, and CO_2 also serve as terminal electron acceptors (TEAs). For example, recent reports by Bradley and Chapelle (1996 and 1997) suggest that Fe-reducers can also oxidize VC and DCE.

Co-oxidations

Co-oxidation occurs as inducible mono- and di-oxygenase enzymes involved in the metabolism of light hydrocarbon gases (methane, butane, propane, etc.), aromatic compounds (toluene, phenol, xylenes, etc.), or other compounds such as ammonia fortuitously break down less chlorinated compounds such as TCE, DCE, or VC (Lee et al. 1998). The fully chlorinated compounds PCE, CT, and HCB and the highly chlorinated congeners of PCBs and PCCDs are resistant to degradation by this mechanism. The co-oxidation reaction requires both the presence of the metabolic primary substrate and oxygen, and such processes may occur at the edges of anaerobic plumes. Other chlorinated compounds that are co-oxidized include 2-chlorophenol, dichlorobiphenyls, and tetrachlorobenzene (Sander et al. 1991; Fetzner and Lingens 1994; Spain 1997). The microorganisms involved in these reactions do not appear to directly benefit from co-degradation.

Cometabolic reductive dehalogenation processes

Cometabolic reactions have been observed under nitrate-reducing, Fe-reducing, sulfate-reducing, fermentative, acetogenic, and methanogenic conditions (Mohn and Tiedje 1992; Fetzner and Lingens 1994; Fathepure and Tiedje 1998; Fetzner 1998). Chlorinated solvents such as PCE, TCA, and other chlorinated hydrocarbons such as HCB and PCBs can be metabolized through a cometabolic reaction mechanism (Vogel et al. 1987; Lee et al. 1998). The cometabolic degradation may be a minor side-reaction, and organisms do not appear to gain energy from such reactions. Biologically important molecules such as vitamin B_{12}, hemes, F_{430}, and other redox carrier molecules have been implicated in the catalysis of reductive dechlorination of halogenated aromatic and aliphatic compounds. For example, Klecka and Gonsior (1984) reported that naturally occurring iron porphyrins catalyze the reductive dehalogenation of several chlorinated ethanes and methanes. Gantzer and Wackett (1991) elegantly demonstrated the role of certain transition-metal coenzymes in the reductive dechlorination of chlorinated organic compounds. They reported that vitamin B_{12} (Co), coenzyme F_{430} (Ni), and hematin (Fe) were responsible for the reductive dechlorination of chlorinated benzenes and aliphatic hydrocarbons. Although the rates of cometabolic dechlorination reactions may be slower than those of biologically mediated reactions, they can play a significant role in the removal process in wetlands, landfills, and other environments where high levels of organics and methanogenic or sulfidogenic respiration are found.

Energy-yielding reductions: dehalorespiration

Microorganisms capable of generating energy using chlorinated compounds such as chlorophenols, chlorinated aliphatics (PCE and TCE), and chlorobenzoates have been isolated (Mohn and Tiedje 1992; Holliger et al. 1999). These organisms are known as "dehalorespirers" or "halorespirers" (Fathepure and Tiedje 1998; Lee et al. 1998). Halorespiration is an anaerobic respiration process in which certain microorganisms use chloroaliphatic or aromatic hydrocarbons as TEAs in the generation of energy for growth (McCarty 1997). Dehalogenating strains include *Desulfomonile tiedjei*, *Dehalococcoides ethenogenes* strain-195, *Desulfitobacterium dehalogenans*,

Desulfitobacterium chlororespirans, *Dehalobacter restrictus*, and *Dehalospirillum multivorans* (Fantroussi et al. 1998; Fathepure and Tiedje 1998; Lee et al 1998; Middeldorp et al. 1999). Some of the PCE-degrading bacteria produce *cis*-DCE as a final end product, while other strains such as *Dehalococcoides ethenogenes* strain-195 may carry out complete dechlorination to ethene (Maymo-Gatell et al. 1997).

Chlorinated aromatic compounds can be reductively dechlorinated in environments as diverse as marine and freshwater sediments, sewage sludges, soil, and groundwater systems (Fathepure and Tiedje 1998). Chlorinated aromatic hydrocarbons including chlorobenzenes, chlorotoluenes, chlorophenols, chloroanilines, chlorobenzoates, chloroguaiacols, chlorophenoxyacetates, chlorobiphenyls, and chlorinated dioxins undergo reductive dehalogenation under denitrifying, Fe-reducing, sulfate-reducing, fermentative, and methanogenic conditions. Reductive and hydrolytic dehalogenation seem to be the principle mechanisms for the metabolism of chloroaromatic compounds. In reductive dehalogenation, the halogen substituents are removed in a stepwise manner. In the hydrolytic degradation pathway, a halogen substituent is typically replaced by a hydroxyl group from water. Haloaromatic hydrocarbons are generally dehalogenated by the reductive mechanisms, while heterocyclic compounds typically undergo anaerobic hydrolysis. Chlorobenzenes can also serve as TEAs for some naturally occurring dehalorespiring organisms, although the organisms responsible for the reductive dechlorination have not been identified (Mohn and Tiedje 1992; Fantroussi et al. 1998).

ENVIRONMENTAL CONDITIONS FAVORING NATURAL REMEDIATION OF CHLORINATED ORGANIC COMPOUNDS

The NR of organic contaminants occurs in all natural environments, including soil, water, and sediments. However, the nature of the chemical, the physical and chemical characteristics of the environment, and the presence of the appropriate microorganisms are key factors in determining whether degradation will occur. Environmental conditions that influence NR of chlorinated compounds are presented in Table 8-4. Many of the environmental conditions are determined by microbial degradation of the chlorinated organic compounds or other organics. For example, dissolved oxygen may be depleted through biodegradation of the contaminants or dissolved organic C, allowing anaerobic processes to predominate and generating lower redox potentials. Microbial activity can also alter the pH of the environment, resulting in neutral pHs, or with an excess of certain organic compounds, resulting in production of organics acids and lowering the pH. Microbial activity can also alter the soil permeability if biomass is produced near an injection well or if soluble ferrous iron produced by Fe-reducing bacteria is oxidized to insoluble ferric iron.

Table 8-4 Environmental conditions that affect natural remediation of chlorinated compounds

Environmental condition	Effects
Dissolved oxygen	High dissolved oxygen levels increase degradation of directly aerobically degradable compounds or aerobically cometabolically degradable compounds, but may inhibit reductive dechlorination and anaerobic degradation.
Other electron acceptors	Some chlorinated compounds are degraded with nitrate, sulfate, iron, or carbon dioxide as terminal electron acceptors (TEAs). Degradation of other chlorinated compounds may be inhibited by the presence of nitrate and sulfate.
pH	Extreme pH (< 4 or > 10) may promote abiotic reactions. Neutral pHs are favorable to most microbial processes.
Dissolved organic carbon (DOC)	High concentrations of co-contaminants or naturally occurring organic carbon promotes anaerobic conditions and reductive dehalogenation and may lead to acclimation for complete degradation. High concentrations of DOC promote sorption.
Cometabolizable substrates	Under aerobic conditions, cometabolic substrates such as methane, phenol, benzene, toluene, ammonia, and other compounds can promote the biodegradation of chlorinated solvents by fortuitous reaction with enzymes produced for the breakdown of cometabolic substrates. Under anaerobic conditions, the presence of co-contaminants or naturally occurring organic material (NOM) may lead to the development of a microbial population capable of incomplete or complete degradation of the chlorinated compounds.
Water content	Low water content limits transport of organics with the water phase and reduces microbial growth. Saturated conditions generally are required to generate an anaerobic environment.
Soil permeability	High soil permeability increases transport of organics with water phase and availability of nutrients and TEAs. Compounds may be sequestered in low permeability materials and may not be bioavailable.
Temperature	Higher temperatures increase microbial growth and degradation rates.
Nutrient (N, P, trace minerals) availability	Low concentrations of nutrients and trace minerals may limit microbial growth and degradation rates.
Salinity	Highly saline environments may inhibit the growth of microbes. Marine environments are generally under sulfate-reducing conditions, which may inhibit reductive dechlorination of some compounds.
Redox potential	Some compounds are degraded best under aerobic conditions (positive redox potentials). Others degrade most efficiently under reducing conditions (generally negative redox conditions) with nitrate, sulfate, iron, or carbon dioxide as electron acceptors. Sites with multiple electron acceptors may promote complete degradation of many compounds.
Contaminant concentration	Low contaminant concentrations may not lead to acclimation. High concentrations of contaminants may be toxic. Free-phase liquids or solids may be difficult to degrade because of their limited solubility and availability.
Acclimated microbial population	The microbial population must have the enzymes necessary to degrade contaminants or acquire them through the process of acclimation. If environmental conditions are favorable. continued exposure to contaminants will often lead to acclimation.

NATURAL REMEDIATION IN AQUIFERS

Over the last decade, NR has developed as an acceptable alternative for the remediation of chlorinated organic compounds, in particular for chlorinated ethanes and ethenes. In order to support NR as a remedial option, 3 lines of evidence need to be provided:

1) Loss of contaminant in the field
2) Contaminant and geochemical analysis (presence and distribution)
3) Microbiological evidence including laboratory microcosm studies.

Wiedemeier et al. (1998) describes in detail the approaches that can be used to document NR of chlorinated solvents in groundwater. Monitored NR is one of the remedial options that should be evaluated for selection of a remedy at a hazardous waste–contaminated site. Removal, treatment, or containment of nonaqueous phase liquids (NAPLs) may be required before monitored NR is an acceptable remedial alternative. The USEPA has developed a protocol for evaluating NR of chlorinated solvents (USEPA 1998b), which is applicable to both chlorinated solvents and other chlorinated compounds. A site screening process is detailed, along with the site characterization data required to understand the site geochemistry and aquifer properties. A process for developing a site conceptual model is also described.

Reductive dehalogenation of chlorinated organics is generally dependent on the presence of alternate electron donor sources. Wiedemeier, Swanson et al. (1996) have documented NR of chlorinated compounds at sites with reductive dechlorination, using various sources of organic C, including co-contaminants (e.g., fuel hydrocarbons) and naturally occurring organic C. At some sites, reductive dechlorination is followed by cometabolic or direct oxidation as the contaminated groundwater moves from reduced to oxidized conditions.

First-order half-lives for a number of chlorinated compounds in groundwater can be estimated based on field studies. Field-derived half-lives can be estimated by several methods, including mass fluxes along transects (Weaver et al. 1996a), modeling efforts using the USEPA Multimedia Exposure Assessment Model (Lee et al. 1995) or other models, estimates made using chloride or trimethylbenzene (TMB) as tracers (Wiedemeier, Wilson, Kampbell 1996), a method with a semilog plot of individual well analyses versus time of transport, and a graphical extrapolation method that assumes the plume is at steady-state and calculates the time for the water to move between points along the center line of the plume (Ellis 1996).

Table 8-5 presents selected field-derived half-lives for a number of chlorinated compounds. This table includes the contaminant concentrations at the first monitoring point and the furthest downgradient monitoring point, as well as a description of the geology, estimated transport time, electron donors, electron acceptors, and references.

Table 8-5 Field demonstrations of natural remediation of chlorinated compounds[a]

Site	Initial average concentration (μM)	Final average concentration (μM)	Residence time (y)	Ist-order rate half-life (y)	Geology	Electron donors	Electron acceptors	References
Sites driven by primary contaminants or co-contaminants								
St. Joseph MI, 2nd transect to lake	56 TCE	< 0.011 TCE	18	3.2 net	Medium, fine, coarse-grained grained	190 mg/L undefined COD	CO_2	McCarty and Wilson 1992; Weaver et al. 1996a, 1996b
	94 cDCE	< 0.008 cDCE	18	4.1 net				
	7.4 tDCE	< 0.014 tDCE	18					
	3.5 1DCE	ND 1DCE						
	16 VC	< 0.0026 VC	18	5.8 net				
Necco Park NY, wells 146F to 156F	27 PCE	2.4 PCE			Fractured bedrock	120 mg/L TOC	SO_4, Fe^{3+}, CO_2	Lee et al. 1995; Wilson, Kampbell, Weaver 1996
	220 TCE	21 TCE						
	100 cDCE	51 cDCE						
	22 tDCE	7 tDCE						
	10 1DCE	2 1DCE						
	22 VC	40 VC						
	3.5 ethene	9.3 ethene						
	14 TeCA	1.1 TeCA						
	47 2TCA	9.8 2TCA						
		1.8 2DCA						
	140 CF	3.1 CF						
	73 mg/L TVOC	14 mg/L TVOC	1.6	1.0				
Eielson Air Force Base AK, site 45/57 to end of plume	680 TCE	1.9	6.3	0.7 (0.3 to 3.2)	Alluvial sands and gravels	Hydrocarbon	NA	Dupont et al. 1996
Plattsburgh Air Force Base NY, MW-02-108 to35PLTW13	190 TCE	0.008 TCE	18	1.9 to 3.6	Sand	17 mg/L BTEX	Fe(III), CO_2, SO_4, O_2	Wiedemeier, Wilson et al. 1996
	530 DCE	2.3 DCE		0.8 to 22				
	23 VC	0.08 VC		1.6 to 6.3				

Table 8-5 continued

Site	Initial average concentration (μM)	Final average concentration (μM)	Residence time (y)	Ist-order rate half-life (y)	Geology	Electron donors	Electron acceptors	References
Sites driven by primary contaminants or co-contaminants (continued)								
Tibbetts Road NH, segment C	5.4 TCE 2.3 DCE	0.02 TCE 0.03 DCE	10	1.3 1.6	Glacial till	6.8 mg/L BTEX	Fe(III)	Wilson, Wilson, Luce 1996
Cape Canaveral Naval Air Station	TCE CDCE		?	2.4 to 3.2 2.4 to 3.2	NA	BTEX, OM	O_2, Fe, SO_4, CO_2	Swanson et al. 1996
Naval Air Station Cecil Field FL	0.16 PCE 0.34 cDCE 0.05 VC	ND PCE ND cDCE ND VC	1.2	0.1 to 0.2 0.1 to 0.2	NA	fuels	Fe(III), SO_4, CO_2	Chapelle 1996; Wilson, Kampbell, Weaver 1996
Rotterdam, the Netherlands	726 2DCA 24 VC	2.1 DCA 0 VC	2.0	0.9-31	Layered sand	2DCA VC	SO_4	Bosma et al. 1997
Southern U.S.	52 VC	1.2 VC		1.1	Fine and very fine sand	NA	anaerobic	Leethem and Larson 1997
Palo Alto CA	0.17 CF 0.029 CClBr2 0.075 1TCA 0.015 PCE 0.061 TCE	0.0063 CF < 0.00048 CClBr2 0.026 1TCA 0.072 PCE 0.024 TCE	1.0	0.063 0.063 0.063 0.063 0.063	Layered sand, gravel, silt, and clay	3.4 mg/L TOC	O_2, ?	Roberts et al. 1982
Naturally occurring organic matter drives reactions								
Pictatinny Arsenal NJ, source to discharge	14 TCE TCE DCE	5.8 TCE	3.1 4.2	2.2 0.6 > 1.8	Fine to coarse sand	5 to 14 mg/L DOC	Fe(III), SO_4	Imbrigiotta et al. 1996; Wilson, Kampbell, Weaver 1996
Mixed sites with both reducing and aerobic conditions								
Dover Air Force Base DE, entire plume	150 TCE 100 DCE 16 VC	ND TCE ? ?	51	2.8 to 4.3 TCE>DCE 1.4 to 2.8 DCE>VC 1.8 to 22 VC>ETE	Sand	NA	CO_2, SO_4, Fe(III), NO_3, O_2	Klecka et al. 1996

Table 8-5 continued

Site	Initial average concentration (μM)	Final average concentration (μM)	Residence time (y)	Ist-order rate half-life (y)	Geology	Electron donors	Electron acceptors	References
Mixed sites with both reducing and aerobic conditions								
Sacramento CA, wells 1727 to 3054	11 TCE 29 DCE 22 VC 14 1DCA 1.3 CA	0.06 TCE 0.7 DCE < 0.16 VC 0.06 1DCA < 0.16 CA	1.8	0.6 0.8 0.2	Alluvial sediments	4 mg/L BOD	SO_4, CO_2, O_2	Cox et al. 1995; Wilson, Kampbell, Weaver 1996
Aerobic or unknown electron acceptor sites								
Canadian Forces Borden, Ontario	? HCA	? PCE		0.11	NA	NA	O_2	Criddle et al. 1986
Brookhaven National Lab, Long Island NY	1×10^{-10} MF CT	0.1×10^{-10} MF CT	50	14 ± 4	Fine to very coarse sand and gravel	NA	O_2	Happell and Wallace 1998
Santa Clara CA	1TCA	1DCE, 1DCA	NA	2.3	Silt and silty sand	acetic acid(abiotic degradation product)	NA	Wing 1997
Livermore CA	1TCA	1DCE	47	2	Flood plain deposits	none	O_2	McNab and Naraslmhan 1994
Otis Air Force Base MA	5.4 PCE 1.3 TCE	0.0006 PCE 0.00076 TCE	47	10 11	Sand and gravel	Surfactants sewage	NO_3	Ala and Domenico 1992

Table 8-5 continued

Site	Initial average concentration (µM)	Final average concentration (µM)	Residence time (y)	Ist-order rate half-life (y)	Geology	Electron donors	Electron acceptors	References
Aerobic or unknown electron acceptor sites								
Luddington, UK	7.1×10^{-4} HpCDD 8.0×10^{-3} OCDF 1.3×10^{-4} HpCDF 3.2×10^{-4} OCDF	3.9×10^{-4} HpCDD 4.3×10^{-3} OCDF 6.3×10^{-5} HpCDF 1.8×10^{-4} OCDF	18 18 18 18	20 20 19 22	79% sand, 6% silt, 15% clay, 1.8% organic matter content	Sewage sludge	NA	McLachlan et al. 1996
Jutland, Denmark	1.0 oDCB 1.0 pDCB 0.9 2,4-DCP 0.9 2,6-DCP 0.8 4,6-DCC	0.1 oDCB 0.2 pDCB ND 2,4-DCP 0.02 2,6-DCP 0.08 4,6-DCC	0.27 0.27 0.27 0.27 0.27	0.05 0.04 0.02 0.06 0.01	Unconfined	Various hydrocarbons and chlorinated compounds	O_2, NO_3	Nielsen et al. 1996
Rhine River	0.3 to 1 HCB	0.1 HCB		7	Not defined	Natural organic	CO_2	Beurskens et al. 1993

[a] BOD = biochemical oxygen demand; BTEX = benzene, toluene, ethylbenzene, and xylenes; CA = chloroethane; CClBr2 = dibromochloromethane; cDCE = cis-1,2-dichloroethen;, CF = chloroform; CO_2 = Carbon dioxide; COD = chemical oxygen demand; CT = carbon tetrachloride1DCA = 1,1-dichloroethane; 2DCA = 1,2-dichloroethane; DCC = -o-dichlorocreosol; 1DCE = 1,1-dichloroethen; DCP = dichlorophenol; DOC = dissolved organic carbon; ETE = ethene; Fe^{III} = iron reducing; HCA = hexachloroethane; HpCDD = 1,2,3,4,6,7,8-heptachlorodibenzo-*p*-dioxin; HpCDF = 1,2,3,4,6,7,8-heptachlorodibenzofuran; MF = mole fraction; NA = not available; ND = non-detectible; NO_3 = Nitrate; oDCB = ortho-dichlorobenzene; OCDD = octachlorodibenzo-*p*-dioxin; OCDF = octachlorodibenzofuran; O_2 = oxygen; pDCB = *para*-dichlorobenzene; PCE = tetrachloroethene; Ref. = references, RT = residence time, SO_4 = sulfate; 1TCA = 1,1,1-trichloroethane; 2TCA = 1,1,2-trichloroethane; TCE = trichloroethene; tDCE = *trans*-1,2-dichloroethene; TeCA = 1,1,2,2-tetrachloroethane; TVOC = total volatile organic compounds; VC = vinyl chloride; > x = removal to below detection limit.

Natural remediation with co-contaminants

There are a number of sites where fuels or other organic co-contaminants supported reductive dehalogenation, including Palo Alto, California (Roberts et al. 1982); St. Joseph, Michigan, Superfund site (McCarty and Wilson 1992; Weaver et al. 1996a; Weaver et al. 1996b; Adriaens et al. 1997); Necco Park, New York (Lee et al 1995; Wilson, Kampbell, Weaver 1996); Eielsen Air Force Base, Alaska (Dupont et al. 1996); Cape Canaveral Air Station, Florida (Swanson et al. 1996); Plattsburgh Air Force Base, New York (Wiedemeier, Wilson, Kampbell 1996); Naval Air Station Cecil Field, Florida (Chapelle 1996; Wilson, Kampbell, Weaver 1996); Tibbetts Road, New Hampshire (Wilson, Wilson, Luce 1996); Rotterdam, the Netherlands (Bosma et al. 1997); and a site in the southern U.S. (Leethem et al. 1997). The compounds serving as electron donors for these sites included benzene, toluene, ethylbenzene, and xylenes (BTEX), other fuel hydrocarbons, tertiary treated wastewater, 2-dichloroethane (2DCA), and VC. As much as 18% of the reduction in COD across the St. Joseph, Michigan, site was estimated due to TCE dechlorination and another 25% to methanogenesis (McCarty and Wilson 1992). The highest estimated degradation rates were 0.1 to 0.2 years at the Naval Air Station Cecil Field, Florida, where unknown fuels and warm temperatures supported rapid reductive dechlorination rates (Chapelle 1996). Rapid half-lives were also observed for a site in Palo Alto, California, where trace levels of several chlorinated organics were disposed with tertiary treated wastewater. The primary electron acceptors at these sites were oxygen, nitrate, iron (Fe^{3+}), sulfate, and carbon dioxide.

Natural remediation with naturally occurring organics

Natural remediation of TCE at Picatinny Arsenal, New Jersey, was supported by between 5 and 14 mg/L of dissolved organic C, consisting primarily of humic and fulvic acids (Imbrigiotta et al. 1996). The plume at the Picatinny Arsenal site discharges to a brook and releases an estimated 50 kg TCE per year. Dissolution of TCE into the groundwater was estimated to contribute 550 kg TCE per year to the plume, with an unknown quantity of TCE released from denser-than-water nonaqueous phase liquid (DNAPL) dissolution. The fate of the TCE in the plume was attributed to the following mechanisms: 360 kg TCE is anaerobically biotransformed per year; 50 kg TCE per year is volatilized during advection; and lateral dispersion, diffusion-driven volatilization, and sorption are each thought to remove less than 1 kg of TCE per year. Humic acids were shown to support reductive dechlorination in laboratory studies (Lyon et al. 1995). At the Picatinny Arsenal site, half-lives for TCE ranged from 0.6 years (Wilson, Kampbell, Weaver 1996) to 2.2 years (Imbrigiotta et al. 1996), and the half-life for DCE was more than 1.8 years. Both Fe^{3+} and sulfate were used as TEAs.

Natural remediation with sequential anaerobic and aerobic behaviors

Sequential anaerobic and aerobic biodegradation of chlorinated organic compounds was thought to occur at sites in Sacramento, California (Cox et al. 1995) and Dover Air Force Base, Delaware (Klecka et al. 1996). Contamination levels decreased by 99.4% for TCE, 98% for DCE, more than 99.3% for VC, 99.6% for 1DCA, and more than 88% for CA in groundwater at the Sacramento site, with an estimated residence time of 1.8 years (Cox et al. 1995). During this same interval, a reduction in 4 mg/L biochemical oxygen demand (BOD) to less than 1 mg/L was noted. An undefined organic represented by a reduction of 3 mg/L BOD may be supporting some reductive dechlorination. In the downgradient portion of the plume, dissolved oxygen may be used for cometabolic biodegradation of the TCE, DCE, and VC. First-order biodegradation half-lives of 0.6, 0.8, and 0.2 years were estimated for TCE, DCE, and VC, respectively (Wilson, Kampbell, Weaver 1996).

The chlorinated solvent plume at Dover Air Force Base, which has multiple sources of solvent contamination, extends for approximately 2,700 m and is 910 m wide (Klecka et al. 1996). The plumes of TCE, DCE, VC, and ethene are spatially segregated, which suggests that the more mobile daughter products are degraded before they move away. Dissolved oxygen, nitrate, sulfate, iron, and carbon dioxide are all present and potentially could be used as TEAs at the site. Microcosm studies with soils from this site indicate the direct aerobic mineralization of ^{14}C-DCE (Klecka et al. 1996). Half-lives for TCE to DCE, DCE to VC, and VC to ethene were estimated to be 2.8 to 4.2 years, 1.4 to 2.8 years, and 1.8 to 2.2 years, respectively.

Sites where electron acceptor is not defined

At some sites, the nature of the TEA facilitating the dechlorination reaction could not be accurately defined. Criddle et al. (1986) documented the transformation of HCA to tetrachloroethene, with both biological and abiotic mechanisms believed to be involved. At the Brookhaven National Laboratory, soil gas levels of CT were reduced approximately 90% during biological transformation in the vadose zone (Happell and Wallace 1998). 1TCA was transformed to 1DCE and 1DCA at 2 sites in California where the TEA was thought to be oxygen or was not known (McNab and Naraslmhan 1994; Wing 1997). Acetic acid, an abiotic degradation product of 1TCA may have supported reductive dechlorination of the TCA. PCE and TCE were degraded slowly at a site in Massachusetts where surfactants and sewage support nitrate-reducing conditions (Ala and Domenico 1992). Various dioxins applied with sewage sludge were slowly biodegraded at a site in Luggington, United Kingdom (McLachlan et al. 1996).

NATURAL REMEDIATION OF SEDIMENTS AND OTHER ENVIRONMENTS

Most of the cases where NR of chlorinated organic compounds has been documented are in aquifers, because of the concerns over impacts of these compounds to the drinking water (Table 8-5). Similar types of degradative processes have been shown to occur in aquatic sediments. For example, highly chlorinated PCBs have been shown to undergo reductive dechlorination in anoxic river sediment systems, resulting in lower chlorinated homologs. Support for the naturally occurring anaerobic PCB dechlorination was provided by analysis of PCB congener patterns in contaminated river sediments (Brown et al. 1987). The resulting lower chlorinated PCBs were subsequently shown to undergo naturally aerobic biodegradation, as evidenced by the presence of intermediate metabolites of PCB degradation (e.g., chlorobenzoates) in sediment cores (Flanagan and May 1993).

Recently, NR as a technology has begun to be considered as an effective remedial strategy for aquatic sediments. USEPA's Contaminated Sediments Management Strategy (USEPA 1998a) has identified NR as a potential remedy:

> The Agency will consider whether a combination of pollution prevention and source control will allow contaminated sediments to recover naturally without unacceptable impacts to human health or the environment. On a site-specific basis, cleanup programs intend to consider natural attenuation as a remedy.

Although NR processes are known to occur in sediments, the feasibility for applying these processes to contaminated sediments as a remedial alternative has not been well established. Unlike the situation for contaminated groundwater, there are no well-recognized protocols for implementing NR in sediments. As a result of this deficiency, there is not a consensus between the regulatory community and practitioners on how to implement NR or on what type of data needs to be developed to successfully support the application of NR to sediments. However, over the next decade, the principles and practices for implementing NR in sediments are expected to become more established.

The application of NR to sediments as a risk-based corrective action (RBCA) will rely on many of the same physical, chemical, and biological processes that have been outlined for contaminated aquifers. However, unlike the situation for contaminated groundwater, the physical process of sediment deposition, mixing, and burial will likely be the dominant process impacting exposure. That is, the prevailing mechanism that controls exposure in sediments is contact with the sediment. Most of the chemicals of concern in aquatic ecosystems, including many chlorinated organics, are hydrophobic and associated with the solid phase (e.g., sediment). Sediment deposition processes cap or bury contaminants and reduce the overall net flux of these compounds to the overlying water, thus reducing exposure to the biota.

In the final analysis, the physical, chemical, and biological processes involved in the application of NR as a remedial strategy must result in a decrease in the overall bioavailability and toxicity of the organic contaminant to levels that are protective of human health and the ecosystem.

Since the application of NR to aquatic sediments is still in its infancy, there are very few published examples. Probably one of the first applications of NR as a risk management strategy for contaminated sediments was the result of wide-scale contamination of the James River by the chlorinated pesticide Kepone. It was estimated that ~40,000 lbs of Kepone were present in the river sediment (Brossman 1978). Although removal by dredging was considered as a remedial option, studies indicated that dredging of the contaminated sediment might increase the risk of exposure of Kepone to the biota (Lunsford 1987). Since much of the contamination occurred in zones of high sedimentation, it was concluded that these deposition processes likely reduced the exposure of indigenous organisms to the pesticide. Eventually, NR was selected as the final remedial option (Bender 1984).

Several authors have reported the underlying principles of NR in sediments and other environmental matrices. Happell and Wallace (1998) documented the loss of CT relative to atmospheric concentrations in the vadose zone and aerobic groundwater from a site in Long Island, New York. Lorah et al. (1997) observed NR in a freshwater wetland as the groundwater moved from an aerobic aquifer into a peat unit. Naturally occurring bioremediation of HCB to lower chlorinated benzenes was observed at a wetlands in Louisiana (Pardue 1996). Adriaens et al. (1997) observed aerobic degradation of reduced degradation products as groundwater moved from an anaerobic aquifer into the aerobic Lake Michigan at the St. Joseph site.

Half-lives of between 40 and more than 100 years were estimated for toxaphene found in the sediments of the Great Lakes (Pearson et al. 1997). Toxaphene, a broad-spectrum pesticide comprised of multichlorinated bornenes and boranes, was extensively used from the 1950s until it was banned in the United States in 1982. The primary input of toxaphene into the Great Lakes is from atmospheric deposition. Present-day concentrations of toxaphene are lower than the levels deposited in the sediments from the early 1970s to early 1980s by between 0% and 64%. In some core samples from the Great Lakes, but not all, there was an increasing contribution of the hexa and hepta homologs and decreasing contribution of octa and nona homologs with increasing depth. This suggests that biodegradation of the toxaphene is occurring.

Beurskens and coworkers (1993) examined the fate of HCB in the anaerobic sediments from a lake in the Netherlands. Numerous reports have demonstrated that, under anaerobic conditions, HCB is reductively dechlorinated to di-, tri-, and tetra-chlorinated benzenes (Fathepure et al. 1988; Cozza and Woods 1992; Nowak et al. 1996; Susarla et al. 1997). Beurskens et al. (1993) reported the in situ concentrations of HCB from anaerobic lake sediments from freshly taken cores in 1988 as compared to cores taken in 1972, which had been preserved for approximately 16 years. Heat and desiccation had inhibited biological activity in the historical cores. The concentration of HCB was significantly higher in the historical versus the recent cores, and they concluded that HCB had disappeared (> 80%) in the river sediments. The decreased concentration of HCB in the fresh cores was likely due to the in situ natural degradative activities occurring in the lake sediments, since there was a concomitant increase in di- and tri-chlorobenzenes in the recent versus historical cores. These results demonstrate a clear loss of HCB in the anaerobic portions of the lake sediment from 1972 to 1988, with a reported half-life for HCB of ~7 years.

A recent study by Quensen and coworkers (1998) indicated that natural degradative processes play important roles in the recovery of estuarine and marine sediments contaminated with DDT and 1,1-dichloro-2,2,bis(*p*-chlorophenyl)-ethylene (DDE), a toxic degradation product of DDT. During laboratory studies with marine sediments collected from a Superfund site off the coast of southern California on the Palos Verdes Shelf, they observed that DDE was biologically dechlorinated to 1-chloro-2,2,bis(*p*-chlorophenyl)ethylene (DDMU) under both methanogenic and sulfidogenic conditions.

The relevance of the laboratory observations to the in situ fate of DDE has yet to be clearly established. However, analysis of field data from the DDT-contaminated marine sediment cores collected from the Palos Verdes Shelf has demonstrated a decrease in the magnitude of DDE concentrations over the last 2 decades, with the corresponding accumulation of DDMU. The in situ half-life of DDE has been estimated to be in the range of 6 to 10 years, although some locations show constant DDE concentrations (Renner 1998).

CONCLUSIONS

The NR of chlorinated compounds is well documented. Chlorinated compounds are clearly being naturally remediated at a number of sites where the reduction in the mass, concentration, toxicity, bioavailability, or mobility of these compounds has been documented.

Natural remediation has been identified as a feasible remedial approach because of its effectiveness and lower cost as compared to alternatives. Quinton et al. (1997) recently compared remediation cost using several different technologies (pump-and-treat with air sparging and activated C adsorption, substrate-enhanced anaerobic bioremediation as a biobarrier or an in situ recirculating system, in situ permeable reactive barrier with zero-valent Fe, and NR) for a hypothetical PCE-contaminated aquifer. Over a period of 30 years, they found that NR was the least expensive alternative of the 5 technologies evaluated, with NR being less than one-tenth the cost of a groundwater pump-and-treat remediation system. NR may not be the least expensive alternative for risk reduction. The costs for conducting an evaluation of NR will be site specific and will vary based on the complexity of site conditions, the types and concentrations of contaminants, the extent of contamination, and the understanding of the in situ conditions. Cost components for an NR assessment include site assessment, short- and long-term monitoring, laboratory studies, modeling, literature searches, and evaluation and reporting of data. Long-term monitoring will be needed to ensure protection of human health and the environment.

REFERENCES

Adam AM, Veenstra JN. 1993. Effects of selected environmental parameters on the biodegradation of 1,2-dichloropropane. In: 48th Purdue Industrial Waste Conference Proceedings. Chelsea MI: Lewis. p 261–278.

Adriaens P, Lendvay J, McCormick ML, Dean SM. 1997. Biogeochemistry and dechlorination potential at the St. Joseph Aquifer-Lake Michigan interface. In: Volume 3, In situ and on site bioremediation. Papers from the Fourth International In Situ and On Site Bioremediation Symposium; 1997 April 28–May 1; New Orleans LA. Columbus OH: Battelle Pr. p 173–178.

Ala NK, Domenico PA. 1992. Inverse analytical techniques applied to coincident contaminant distributions at Otis Air Force Base, Massachusetts. *Ground Wat* 30:212–218.

Alexander M. 1995. How toxic are toxic chemicals in soil? *Environ Sci Technol* 29:2713–2717.

Armenante PM, Kafkewitz D, Lewandowski G, Kung C-M. 1992. Integrated anaerobic-aerobic process for the biodegradation of chlorinated aromatic compounds. *Environ Prog* 1:113–122.

Bachmann A, de Bruin W, Jumelet JC, Rijnaarts HN, Zehnder AJB. 1988. Aerobic biomineralization of *alpha*-hexachlorocyclohexane in contaminated soil. *Appl Environ Microbiol* 54:548–554.

Bagley DM, Gossett J. 1989. Tetrachloroethene transformation to trichloroethene and cis-1,2-dichloroethene by sulfate-reducing enrichment culture. *Appl Environ Microbiol* 56:2511–2516.

Barrio-Lage GA, Parsons FZ, Narbaitz RM, Lorenzo PA, Archer HE. 1990. Enhanced anaerobic biodegradation of vinyl chloride in ground water. *Environ Toxicol Chem* 9:403–415.

Becker JG, Freedman DL. 1994. Use of cyanocobalamin to enhance anaerobic biodegradation of chloroform. *Environ Sci Technol* 28:1942–1949.

Bender ME, Huggett RJ. 1984. Fate and effects of Kepone in the James River. *Rev Environ Toxicol* 1:5–50.

Beurskens JEM, Dekker CGC, Jonkhoff J, Pompstra L. 1993. Microbial dechlorination of hexachlorobenzene in a sedimentation area of the Rhine River. *Biogeochem* 19:61–81.

Beyerle-Pfnur R, Lay JP. 1990. Adsorption and desorption of 3,4-dichloroaniline on soil. *Chemosphere* 21:1087–1094.

Boucquey J-B, Renard P, Amerlynck P, Filho PM, Agathos GN, Naveau H, Nyns E-J. 1995. High rate continuous biodegradation of concentrated chlorinated aliphatics by a durable enrichment of methanogenic origin under carrier-dependent conditions. *Biotech Bioengineer* 47:298–307.

Bosma TNP, Cottaar FHM, Posthumus MA, Teunis CJ, van Veldhuizen A, Schraa G, Zehnder AJB. 1994. Comparison of reductive dechlorination of hexachloro-1,3-butadiene in Rhine sediment and model systems with hydroxocobalamin. *Environ Sci Technol* 28:1124–1128.

Bosma TNP, van Aalst MA, Rijnaarts HHM, Tatt J, Bovendeur J. 1997. Intrinsic dechlorination of 1,2-dichloroethane at an industrial site. In: Volume 3, In situ and on site bioremediation. Papers from the Fourth International In Situ and On Site Bioremediation Symposium; 1997 April 28–May 1; New Orleans LA. Columbus OH: Battelle Pr. p 155–159.

Bouwer EJ, McCarty PL. 1983a. Transformations of 1- and 2-carbon halogenated aliphatic organic compounds under methanogenic conditions. *Appl Environ Microbiol* 45:1286–1294.

Bouwer EJ, McCarty PL. 1983b. Transformations of halogenated organic compounds under denitrification conditions. *Appl Environ Microbiol* 45:1295–1299.

Bradley PM, Chapelle FH. 1996. Anaerobic mineralization of vinyl chloride in Fe(III)-reducing aquifer sediments. *Environ Sci Technol* 30:2084–2086.

Bradley PM, Chapelle FH. 1997. Kinetics of DCE and VC mineralization under methanogenic and Fe(III) reducing conditions. *Environ Sci Technol* 31:2692–2696.

Bradley PM, Chapelle FH. 1998. Effect of contaminant concentration on aerobic microbial mineralization of DCE and VC in stream-bed sediments. *Environ Sci Technol* 32:553–557.

Bradley PM, Chapelle FH, Lovely DR. 1998. Humic acids as electron acceptors for anaerobic microbial oxidation of vinyl chloride and dichloroethene. *Appl Environ Microbiol* 64:3102–3105.

Braus-Stromeyer SA, Cook AM, Leisinger T. 1993. Biotransformation of chloromethane to methanothiol. *Environ Sci Technol* 27:1557–1559.

Brossman MW, Mackenthun KM, Kohler JA, Terrell CR. 1978. Kepone mitigation feasibility project. Kepone-contaminated Hopewell/James River areas. Report. EPA/440/5-78/004.

Brown JF, Bedard DL, Brennan MJ, Carnahan JC, Feng H, Wagner RE. 1987. Polychlorinated biphenyl dechlorination in aquatic sediments. *Science* 236:709–712.

Castro CE, O'Shea SK, Wand W, Bartnicki EW. 1996. Biodehalogenation: Oxidative and hydrolytic pathways in the transformation of acetonitrile, chloroacetonitrile, chloracetic acid, and chloroacetamide by *Methylosinus trichosporium* OB-3b. *Environ Sci Technol* 30:1180–1184.

Chang B-V, Liao K-T, Pan T-M, Yuan S-Y. 1995. Biotransformation of chlorophenols in river sediments. *Chem Ecolog* 10:105–114.

Chang B-V, Su C-J, Yuan S-Y. 1998. Microbial hexachlorobenzene dechlorination under three reducing conditions. *Chemosphere* 13:2721–2730.

Chang H-L, Alvarez-Cohen L. 1996. Biodegradation of individual and multiple chlorinated aliphatic hydrocarbons by methane-oxidizing cultures. *Appl Environ Microbiol* 62:3371–3377.

Chapelle FH. 1996. Identifying redox conditions that favor the natural attenuation of chlorinated ethenes in contaminated ground-water systems. U.S. Environmental Protection Agency Symposium on Natural Attenuation of Chlorinated Organics in Ground Water; 1996 Sept 11–13; Dallas TX. Washington DC: Office of Research and Development. EPA/540/R-96/509. p 17–20.

Chen C, Puhakka JA, Ferguson JF. 1996. Transformation of 1,1,2,2-tetrachloroethane under methanogenic conditions. *Environ Sci Technol* 30:542–547.

Collins R, Kim S, Picardal F. 1997. Humic substance and iron mineral effects on reductive dechlorination reactions. In: Volume 4, In situ and on site bioremediation. Papers from the Fourth International In Situ and On Site Bioremediation Symposium; 1997 April 28–May 1; New Orleans LA. Columbus OH: Battelle Pr. p 41.

Cox EE, Acton DW, Major DW. 1994. Evaluating in situ trichloroethene biotransformation using in situ microcosms. In: Hinchee RE, Leeson A, Semprini L, Ong SK, editors. Bioremediation of chlorinated and polycyclic aromatic hydrocarbon compounds. Boca Raton FL: Lewis. p 314–319.

Cox E, Edwards E, Lehmicke L, Major D. 1995. Intrinsic biodegradation of trichloroethene and trichloroethane in a sequential anaerobic-aerobic aquifer. In: Hinchee RE, Wilson JT, Downey DC, editors. Intrinsic bioremediation. Columbus OH: Battelle Pre. p. 223–231.

Cozza CL, Woods SL. 1992. Reductive dechlorination pathways for substituted benzenes: A correlation with electronic properties. *Biodegrad* 2:265–278

Criddle CS, McCarty PL, Elliott MC, Barker JF. 1986. Reduction of hexachloroethane to tetrachloroethylene in groundwater. *J Contamin Hydrol* 1:133–142.

Davis JW, Carpenter CL. 1990. The aerobic biodegradation of vinyl chloride in ground water. *Appl Environ Microbiol* 56:3878–3880.

Davis JW, Madsen SS. 1991. The biodegradation of methylene chloride in soils. *Environ Toxicol Chem* 10: 463–474

de Bruin WP, Kotterman MJJ, Posthumus MA, Schraa G, Zehnder AJB. 1992. Complete biological reductive transformation of tetrachloroethene to ethane. *Appl Environ Microbiol* 58:1996–2000.

De Leer EWB. 1985. The identification of highly chlorinated ethers and diethers in river sediment near an epichlorohydrin plant. *Wat Res* 19:1411–1419.

Dolan ME, McCarty PL. 1994. Factors affecting transformation of chlorinated aliphatic hydrocarbons by methanotrophs. In: Hinchee RE, Leeson A, Semprini L, Ong SK, editors. Bioremediation of chlorinated and polycyclic aromatic hydrocarbon compounds. Boca Raton FL: Lewis. p 303–308.

Dolfing J, van den Wyngaard AJ, Jannsen DB. 1993. Microbiological aspects of the removal of chlorinated solvents from air. *Biodegradation* 4:261–282.

Dupont RR, Gorder K, Sorensen DL, Kemblowski MW, Haas P. 1996. Case study: Eielson Air Force Base, Alaska. U.S. Environmental Protection Agency. Symposium on Natural

Attenuation of Chlorinated Organics in Ground Water; 1996 Sept 11–13; Dallas TX. Washington DC: Office of Research and Development. EPA/540/R-96/509. p. 104–109.

Durham RW, Oliver BG. 1983. History of Lake Ontario contamination from the Niagara River by sediment radiodating and chlorinated hydrocarbon analysis. *J Great Lakes Res* 9:160–168.

Egli C, Tschan T, Scholtz R, Cook AM, Leisinger T. 1988. Transformation of tetrachloromethane to dichloromethane and carbon dioxide by *Acetobacterium woodii*. *Appl Environ Microbiol* 54:2819–2824.

Ellis DE. 1996. Intrinsic remediation in the industrial marketplace. U.S. Environmental Protection Agency Symposium on Natural Attenuation of Chlorinated Organics in Ground Water; 1996 Sept 11–13; Dallas TX. Washington DC: Office of Research and Development. EPA/540/R-96/509. p. 120–123.

Fantroussi SE, Naveau H, Agathos SN. 1998. Anaerobic dechlorinating bacteria. *Biotechnol Prog* 14:167–188.

Fathepure BZ, Tiedje JM. 1998. Anaerobic bioremediation: Microbiology, principles and applications. In: Adriano DC, Bollag J-M, Frankenberger WT, Sims RC, editors. Bioremediation of contaminated soils. Madison WI: American Society of Agronomy p 339–396.

Fathepure BZ, Tiedje JM, Boyd SA. 1988. Reductive dechlorination of hexachlorobenzene to tri- and dichlorobenzenes in anaerobic sewage sludge. *Appl Environ Microbiol* 54:327–330.

Fathepure BZ, Vogel TM. 1991. Complete degradation of polychlorinated hydrocarbons by a two-stage biofilm reactor. *Appl Environ Microbiol* 57:3418–3422.

Feenstra S. 1994. Chlorinated organic substances in groundwater. *Environ Sci Pollut Control Ser* 11(Groundwater Contamination and Control):337–353.

Fetzner S. 1998. Bacterial dehalogenase. *Appl Microbiol Biotechnol* 50: 633–657.

Fetzner S, Lingens F. 1994. Bacterial dehalogenases: Biochemistry, genetics, and biotechnological applications. *Microbiol Rev* 58:641–685.

Flanagan WP, May RJ. 1993. Metabolite detection as evidences for the naturally occurring aerobic biodegradation in the Hudson River. *Environ Sci Technol* 27:2207–2212.

Focht R, Vogan J, O'Hannesin S. 1996. Field application of reactive iron walls for in situ degradation of volatile organic compounds in groundwater. *Remediation*. Summer:81–94.

Freedman DL, Gossett JM. 1991. Biodegradation of dichloromethane in a fixed film reactor under methanogenic conditions. In: Hinchee RE, Olfenbuttel RF, editors. On-site bioreclamation processes for xenobiotic and hydrocarbon treatment. Stoneham MA: Butterworth-Heinemannn. p. 113–133.

Freedman DL, Lasecki M, Hashsham S, Scholze R. 1995. Accelerated biotransformation of carbon tetrachloride and chloroform by sulfate-reducing enrichment cultures. In: Hinchee RE, Leeson A, Semprini L, editors. Bioremediation of chlorinated solvents. Columbus OH: Battelle Pr. p. 123–138.

Freedman DL, Smith CR, Noguera DR. 1997. Dichloromethane biodegradation under nitrate-reducing conditions. *Water Environ Res* 69:115–122.

Freedman DL, Woertz J, Verce MF, Reis IML, Barbosa RAM. 1997. Biodegradation of chloromethane under anoxic conditions. In: Volume 3, In situ and on site bioremediation. Papers from the Fourth International In Situ and On Site Bioremediation Symposium; 1997 April 28–May 1; New Orleans LA. Columbus OH: Battelle Pr. p. 1–6.

Freeze RA, Cherry JA. 1979. Groundwater. Englewood Cliffs NJ: Prentice-Hall. p 75.

Freitag D, Ballhorn L, Behechti A, Feischer K, Thumm W. 1994. Structural configuration and toxicity of chlorinated alkanes. *Chemosphere* 28:253–259.

Gantzer CJ, Wackett LP. 1991. Reductive dechlorination catalyzed by bacterial transition-metal coenzymes. *Environ Sci Technol* 25:715–722.

Gibson SA, Suflita JM. 1990. Anaerobic biodegradation of 2,4,5-trichlorophenoxyacetic acid in samples from a methanogenic aquifer: Stimulation by short-chain organic acids and alcohols. *Appl Environ Microbiol* 56:1825–1832.

Golden KA, Wong CS, Jeremiason JD, Eisenreich SJ, Hallgren J, Swackhamer DL, Engstrom DR, Long DT. 1993. Accumulation and preliminary inventory of organochlorines in Great Lakes sediments. *Water Sci Technol* 28:19–31.

Gribble GW. 1994. The natural production of chlorinated compounds. *Environ Sci Technol* 28:310A–319A.

Gribble GW. 1996. Naturally occurring organohalogen compounds: A comprehensive survey. *Prog Org Nat Prod* 68:1–498

Häggblom MH, Rivera MD, Young LY. 1993a. Effects of auxiliary carbon sources and electron acceptors on methanogenic degradation of chlorophenols. *Environ Toxicol Chem* 12:1395–1403.

Häggblom MH, Rivera MD, Young LY. 1993b. Influence of alternative electron acceptors on the anaerobic biodegradability of chlorinated phenols and benzoic acids. *Appl Environ Microbiol* 59:1162–1167.

Häggblom MH, Young LY. 1991. Chlorophenol degradation coupled to sulfate reduction. *Appl Environ Microbiol* 56:3255–3260.

Hale DD, Jones WJ, Rogers JE. 1994. Biodegradation of chlorinated homocyclic and heterocyclic compounds in anaerobic environments. In: Chaudhry CR, editor. Biological degradation and bioremediation of toxic chemicals. Portland OR: Dioscorides Pr. p 148–168.

Happell JD, Wallace DWR. 1998. Removal of atmospheric CCl_4 under bulk aerobic conditions in groundwater and soils. *Environ Sci Technol* 32:1244–1252.

Haugland RA, Schlemm DJ, Lyons RP III, Sferra PR, Chakrabarty AM. 1990. Degradation of the chlorinated phenoxyacetate herbicides 2,4-dichlorophenoxyacetic acid and 2,4,5-trichlorophenoxyacetic acid by pure and mixed bacterial cultures. *Appl Environ Microbiol* 56:1357–1362.

Havel J, Reineke W. 1993. Microbial degradation of chlorinated acetophenones. *Appl Environ Microbiol* 59:2706–2712.

Henschler D. 1994. Toxicity of chlorinated organic compounds: Effects of the introduction of chlorine in organic molecules. *Angew Chem Int Ed Engl* 33:1920–1935.

Hileman B. 1993. Concerns broaden over chlorine and chlorinated hydrocarbons. *Chem Engin News* 71.

Holliger C, Schraa G, Stams AJM, Zehnder AJB. 1990. Reductive dechlorination of 1,2-dichloroethane and chloroethane by cell suspensions of methanogenic bacteria. *Biodegrad* 1:253–261.

Holliger C, Schraa G, Stams AJM, Zehnder AJB. 1992. Enrichment and properties of an anaerobic mixed culture reductively dechlorinating 1,2,3-trichlorobenzene to 1,3-dichlorobenzene. *Appl Environ Microbiol* 58:1636–1644.

Holliger C, Wohlfarth, Diekert G. 1999. Reductive dechlorination in the energy metabolism of anaerobic bacteria. *FEMS Microbiol Rev* 22:383–398.

Hopkins GD, McCarty PL. 1995. Field evaluation of in situ aerobic cometabolism of trichloroethylene and three dichloroethylene isomers using phenol and toluene as the primary substrates. *Environ Sci Technol* 29:1628–1637.

Howard PH, Boethling RS, Jarvis WF, Meylan WM, Michalenk EM. 1991. Handbook of environmental degradation rates. Chelsea MI: Lewis.

Imbrigiotta TE, Ehlke TA, Wilson BH, Wilson JT. 1996. Case study: Natural attenuation of a trichloroethene plume at Picatinny Arsenal, New Jersey. U.S. Environmental Protection Agency Symposium on Natural Attenuation of Chlorinated Organics in Ground Water; 1996 Sept 11–13; Dallas TX. Washington DC: Office of Research and Development. EPA/540/R-96/509. p 83–89.

Isenesse AR. 1991. Bioaccumulation and food chain accumulation. In: Grover I, Raj R, editors. Environmental chemistry of herbicides. Boca Raton FL: CRC Pr. p 187–197.

Jablonski PE, Ferry JG. 1992. Reductive dechlorination of trichloroethylene by the CO-reduced CO dehydrogenase enzyme complex from *Methanoscarcina thermophila. FEMS Microbiol Letters* 96:55–60.

Jackson RE, Priddle MW, Lesage S. 1990. Transport and fate of CFC-113 in groundwater. Proceedings of NWWA/API Conference on Petroleum Hydrocarbons and Organic Chemicals in Ground Water: Prevention, Detection, and Restoration; 1990 Oct 31–Nov 2; Houston TX. Dublin OH: Water Well Journal Publishing Co. p. 129–142.

Jafvert CT, Wolfe NL. 1987. Degradation of selected halogenated ethanes in anoxic sediment-water systems. *Environ Toxicol Chem* 6:827–837.

Jeffers PM, Ward LM, Woytowitch LM, Wolfe NL. 1989. Homogeneous hydrolysis rate constants for selected chlorinated methanes, ethanes, ethenes, and propanes. *Environ Sci Technol* 23:965–969.

Kazumi J, Häggblom MM, Young LY. 1995. Diversity of anaerobic microbial processes in chlorobenzoate degradation: Nitrate, iron, sulfate, and carbonate as electron acceptors. *Appl Microbiol Biotechnol* 43:929–936.

Klecka GM, Carpenter CL, Gonsior, SJ. 1998. Biological transformations of 1,2-dichloroethane in subsurface soil and groundwater. *J Contam Hydrol* 34:139–154.

Klecka GM, Gonsior SJ. 1984. Reductive dechlorination of chlorinated methanes and ethanes by reduced iron (II) porphyrins. *Chemosphere* 13:391–402.

Klecka GM, Gonsior SG, Markham DA. 1990. Biological transformation of 1,1,1-trichloroethane in subsurface soils and ground water. *Environ Toxicol Chem* 9:1437–1451.

Klecka, GM, Wilson JT, Lutz E, Klier N, West R, Davis JW, Weaver J, Kampbell V, Wilson B. 1996. Intrinsic remediation of chlorinated solvents in groundwater. Proceedings of Intrinsic Bioremediation Conference; 1996 Mar 18–19; London UK.

Klier NJ, West RJ, Donberg PA. 1999. Aerobic biodegradation of dichloroethylenes in surface and subsurface soils. *Chemosphere* 38:1175–1188.

Knabel DB, Federle T, MacVoy DC, Vestal RJ. 1996. Microbial mineralization of organic compounds in acidic agricultural soil: Effect of preadsorption to various soil components. *Environ Toxicol Chem* 15:1865–1875.

Komatsu T, Momonoi K, Matsuo T, Hanaki K. 1994. Biotransformation of *cis* 1,2-dichloroethylene to ethylene under anaerobic conditions. *Water Sci Technol* 30:75–84.

Kriegman-King MR, Reinhard M. 1994. Transformation of carbon tetrachloride by pyrite in aqueous solution. *Environ Sci Technol* 28:692–700.

Kuhn EP, Townsend GT, Suflita JM. 1990. Effect of sulfate and organic carbon supplements on reductive dehalogenation of chloroanilines in anaerobic aquifer slurries. *Appl Environ Microbiol* 56:2630–2637.

Leahy JG, Byrne AM, Olsen RH. 1996. Comparison of factors influencing trichloroethylene degradation by toluene-oxidizing bacteria. *Appl Environ Microbiol* 62:825–833.

Lee MD, Mazierski PF, Buchanan Jr RJ, Ellis DE, Sehayek LS. 1995. Intrinsic in situ anaerobic biodegradation of chlorinated solvents at an industrial landfill. In: Hinchee RE, Wilson JT, Downey DC, editors. Intrinsic bioremediation. Columbus OH: Battelle Pr. p 205–222.

Lee MD, Odom JM, Buchanan Jr RJ. 1998. New perspectives on microbial dehalogenation of chlorinated solvents. Insights from the field. *Ann Rev Microbiol* 52:423–452.

Leethem JT, Larson JR. 1997. Intrinsic bioremediation of vinyl chloride in groundwater at an industrial site. In: Volume 3, In situ and on site bioremediation. Papers from the Fourth International In Situ and On Site Bioremediation Symposium; 1997 April 28–May 1; New Orleans LA. Columbus OH: Battelle Pr. p 167–172.

Lesage S, Jackson RE, Priddle MW, Riemann PG. 1990. Occurrence and fate of organic solvents residues in anoxic groundwater at the Gloucester Landfill, Canada. *Environ Sci Technol* 24:559–266.

Lew DJ, Bouwer EJ. 1988. Anoxic biotransformation of halogenated aliphatic micropollutants. Proceedings Water Pollution Control Federation Conference; 1988 Oct 3–6; Dallas TX.

Löffler FE, Champine JE, Ritalahti KM, Sprague SJ, Tiedje JM. 1997. Complete reductive dechlorination of 1,2-dichloropropane by anaerobic bacteria. *Appl Environ Microbiol* 63:2870–2873.

Long JL, Stensel HD, Ferguson FF, Strand SE, Ongerth JE. 1993. Anaerobic and aerobic treatment of chlorinated aliphatic compounds. *J Environ Engin* 119:300–320.

Lorah MM, Olsen LD, Smith BL. 1997. Natural attenuation of chlorinated hydrocarbons in a freshwater wetland. In: Volume 3, In situ and on site bioremediation. Papers from the Fourth International In Situ and On Site Bioremediation Symposium; 1997 April 28–May 1; New Orleans LA. Columbus OH: Battelle Pr. p 207–212.

Lunsford CA, Weinstein MP, Scott L. 1987. Uptake of Kepone by the estuarine bivalve *Rangia cuneata* during the dredging of contaminated sediments in the James River. *Water Res* 21:411–416.

Lyon WG, West CC, Osborn ML. Sewell GW. 1995. Microbial utilization of vadose zone organic carbon for reductive dechlorination of tetrachloroethene. *J Environ Sci Health* A30:1629–1939.

Mackay D, Guardo AD, Patterson S, Kacsi G, Cowan CE. 1996. Assessing the fate of new and existing chemicals: A five stage process. *Environ Toxicol Chem* 15:1618–1626.

Mackay D, Patterson S, Shiu WY. 1992. Generic models for the evaluating the regional fate of chemicals. *Chemosphere* 24:695–717.

Mackay D, Shiu WY, Ma KC, editors. 1992. Illustrated handbook of physical–chemical properties and environmental fate for organic chemicals. Volume 1, Monoaromatic hydrocarbons, chlorobenzenes and PCBs. Boca Raton FL: Lewis.

Magli A, Wendt M, Leisinger T. 1996. Isolation and characterization of *Dehalobacterium formicoaceticum* gen. nov. sp. nov., a strictly anaerobic bacterium utilizing dichloromethane as source of carbon and energy. *Arch Microbiol* 166:101–108.

MAK-Werte-List-Maximale Arbeitsplatzkonzentrationen. Deutsche Forschungsgemeinschaft. Since Part 10VCH, Weinheim, 1974.

Manahan SE. 1994. Environmental chemistry. Boca Raton FL: Lewis.

Marinucci AC, Bartha R. 1979. Biodegradation of 1,2,3- and 1,2,4-trichlorobenzene in soil and in liquid enrichment culture. *Appl Environ Microbiol* 38:811–817.

Masunaga S, Susarla S, Gundersen J, Yonezawa Y. 1996. Pathway and rate of chlorophenol transformation in anaerobic estuarine sediment. *Environ Sci Technol* 30:1253–1260.

Matus V, Vàsquez M, Vincente M, Gonzàlez B. 1996. Microbial mineralization of 2,4,5-trichlorophenol in soil. *Environ Sci Technol* 30:1472–1476.

Maymo-Gatell X, Chien Y-t, Gossett JM, Zinder SH. 1997. Isolation of a bacterium that reductively dechlorinates tetrachloroethene to ethene. *Science* 276:1568–1571.

McCall PJ, Vrona SA, Kelley SS. 1981. Fate of uniformly carbon-14 ring labeled 2,4,5-trichlorophenoxyacetic acid and 2,4-drichlorophenoxyacetic acid. *J Agric Food Chem* 29:100–107.

McCarty PL. 1997. Breathing with chlorinated solvents. *Science 16*:1521–1522.

McCarty PL, Wilson JT. 1992. Natural anaerobic treatment of a TCE plume, St. Joseph, Michigan, NPL site. U.S. Environmental Protection Agency. Bioremediation of hazardous wastes. Cincinnati OH: USEPA Center for Environmental Research Information. EPA/600/R-92/126. p. 47–50.

McClay K, Fox BG, Steffan RJ. 1996. Chloroform mineralization by toluene-oxidizers. *Appl Environ Microbiol* 62:2716–2722.

McLachlan MS, Sewart AP, Bacon JR, Jones KC. 1996. Persistence of PCDD/Fs in a sludge-amended soil. *Environ Sci Technol* 30:2567–2571.

McNab Jr WW, Naraslmhan TN. 1994. Degradation of chlorinated hydrocarbons and groundwater geochemistry: A field study. *Environ Sci Technol* 28:769–775.

Middeldorp PJM, Luijten MLGC, van de Pas LBA, van Eekert MHA, Kengen SWM, Schraa G, Stams AJM. 1999. Anaerobic microbial reductive dehalogenation of chlorinated ethenes. *Bioremed J* 3:151–169.

Miethling R, Karlson U. 1996. Accelerated mineralization of pentachlorophenol in soil upon inoculation with *Mycobacterium chlorophenolicum* PCP1 and *Spingomonas chlorophenolica* RA2. *Appl Environ Microbiol* 62:4361–6366.

Miller PL, Vasudeavan D, Gschwend PM, Roberts AL. 1998. Transformation of hexachloroethane in a sulfidic natural water. *Environ Sci Technol* 32:1269–1275.

Milligan PW, Häggblom MM. 1997. Anaerobic biotransformation of dicamba and chlorosalicylic acids under denitrifying, iron reducing, and methanogenic conditions. In: Volume 2, In situ and on site bioremediation. Papers from the Fourth International In Situ and On Site Bioremediation Symposium; 1997 April 28–May 1; New Orleans LA. Columbus OH: Battelle Pr. p. 293–298.

Mohn WR, Kennedy KJ. 1992. Limited degradation of chlorophenols by anaerobic sludge granules. *Appl Environ Microbiol* 58:2131–2136.

Mohn WW, Tiedje JM. 1992. Microbial reductive dechlorination. *Microbiol Rev* 56:482–507.

Nadau LJ, Menn F-M, Breen A, Sayler GS. 1994. Aerobic degradation of 1,1,1-trichloro-2,2-*bis*(4-chlorophenyl)ethane (DDT) by *Alcaligenes eutrophus* A5. *Appl Environ Microbiol* 60:51–55.

Nicholson DK, Woods SL, Istok JD, Peek DC. 1992. Reductive dechlorination of chlorophenols by a pentachlorophenol-acclimated methanogenic consortium. *Appl Environ Microbiol* 58:2280–2286.

Nielsen PH, Bjerg PL, Nielsen P, Smith P, Christensen TH. 1996. In situ and laboratory determined first-order degradation rate constants of specific organic compounds in an aerobic aquifer. *Environ Sci Technol* 30:31–37.

Nishino SF, Spain JC, Belcher LA, Litchfield CD. 1992. Chlorobenzene degradation by bacteria isolated from contaminated groundwater. *Appl Environ Microbiol* 58:1719–1726.

Nishino SF, Spain JC, Pettigrew CA. 1994. Biodegradation of chlorobenzene by indigenous bacteria. *Environ Toxicol Chem* 13:871–877.

Nowak J, Kirsch NH, Hegemann W, Stan HJ. 1996. Total reductive dechlorination of chlorobenzenes to benzene by a methanogenic mixed culture enriched from Saale River sediment. *Appl Microbiol Biotechnol* 45:700–709.

Nyman MC, Nyman AK, Lee LS, Nies LF, Blatchley III ER. 1997. 3,3'-Dichlorobenzidine transformation processes in natural sediments. *Environ Sci Technol* 31:1068–1073.

Oliver BG, Nicol KD. 1982. Chlorobenzenes in sediments, water, and selected fish from Lakes Superior, Huron, Erie, and Ontario. *Environ Sci Technol* 6:532–536.

Oremland RS, Lonergan DL, Culbertson CW, Lovley DR. 1996. Microbial degradation of hydrochlorofluorocarbons ($CHCl_2F$ and $CHCl_2CF_3$) in soils and sediments. *Appl Environ Microbiol* 62:1818–1821.

Palm H, Lammi R. 1995. Fate of pulp mill organochlorines in the Gulf of Bothnia sediments. *Environ Sci Technol* 29:1722–1727.

Pardue JH. 1996. Reductive dechlorination and the intrinsic bioremediation of contaminated sediments in the Devil's Swamp Wetlands. Proceedings WEPTEC 96, 69th Annual Conference and Exposition; 1996; Dallas TX. Volume 3, Part 1, Remediation of soil and groundwater. p 253–263.

Pearson RF, Swackhamer DL, Eisenreich SI, Long DT. 1997. Concentrations, accumulations, and inventories of toxaphene in sediments of the Great Lakes. *Environ Sci Technol* 31:3523–3529.

Peijnenburg WJGM, 'T Hart MJ, den Hollander HA, van de Meent D, Verboom HH, Wolfe NL. 1992. Reductive transformation of halogenated aromatic hydrocarbons in anaerobic water-sediment systems: Kinetics, mechanisms, and products. *Environ Toxicol Chem* 11:289–300.

Poelarends GJ, Wilkens M, Larkin MJ, van Elsas JD, Janssen DB. 1998. Degradation of 1,3-dichloropropene by *Pseduomonas cichorii* 170. *Appl Environ Microbiol* 64:2931–2936.

Potrawfke T, Timmis KN, Wittich R-M. 1998. Degradation of 1,2,3,4-tetrachlorobenene by *Pseudomonas chloraphis* RW71. *Appl Environ Microbiol* 64(10):3798–3806.

Pries F, van der Ploeg JR, Dolfing J, Janssen DB. 1994. Degradation of halogenated aliphatic compounds: The role of adaptation. *FEMS Microbiol Rev* 15:279–295.

Quensen JF, Mueller SA, Jain MK, Tiedje JM. 1998. Reductive dechlorination of DDE to DDMU in marine sediment microcosms. *Science* 280:733–735.

Quinton GE, Buchanan Jr RJ, Ellis DE, Shoemaker SH. 1997. A method to compare groundwater cleanup technologies. *Remediation* Autumn: 7–16.

Ramanand R, Balba MT, Duffy J. 1993. Reductive dehalogenation of chlorinated benzenes and toluene under methanogenic conditions. *Appl Environ Microbiol* 59:3266–3272.

Ramanand K, Nagarajan A, Suflita JM. 1993. Reductive dechlorination of the nitrogen heterocyclic herbicide picloram. *Appl Environ Microbiol* 59:2251–2256.

Reinhard M, Semadeni M, Sharma PK, Haag WR. 1998. Dehalogenation of haloaliphatic hydrocarbon compounds in the aquatic and terrestrial environment. In: Adriano DC, Bollag J-M, Frankenberger WT, Sims RC, editors. Bioremediation of contaminated soils. Madison WI: American Soc Agronomy. p 133–174.

Renner R. 1998. "Natural" remediation of DDT, PCBs debated. *Environ Sci Technol* 32:360A–363A.

Roberston BK, Alexander M. 1998. Sequestration of DDT and Dieldrin in soil: Disappearance of acute toxicity but not the compound. *Environ Toxicol Chem* 17:1034–1038.

Roberts PV, Schreiner J, Hopkins GD. 1982. Field study of organic water quality changes during groundwater recharge in the Palo Alto Baylands. *Wat Res* 16:1025–1035.

Sarkar JM, Malcolm RL, Bollag JM. 1988. Enzymic coupling of 2,4-dichlorophenol to stream fulvic acid in the presence of oxidoreductases. *Soil Sci Soc Am J* 52:688–694.

Sander P, Wittich R-M, Fortnagel P, Wilkes H, Francke W. 1991. Degradation of 1,2,4-trichlorobenzene and 1,2,4,5-tetrachlorobenzene by *Pseudomonas* strains. *Appl Environ Microbiol* 57:1430–1440.

Scharp S, Marea V, Pieter J, Opperhuizen A. 1994. Experimental problems in determining sorption coefficients of organic chemicals: An example for chlorobenzenes. *Chemosphere* 28:931–945.

Schmidt S, Fortnagel P, Wittich R-M. 1993. Biodegradation and transformation of 4,4'- and 2,4-dihalodiphenyl ethers by *Sphingomonas* sp. Strain SS33. *Appl Environ Microbiol* 59:3931–3933.

Semprini L, Hopkins GD, Roberts PV, McCarty PL. 1992. In situ transformation of carbon tetrachloride and other halogenated compounds resulting from biostimulation under anoxic conditions. *Environ Sci Technol* 26:2454–2461.

Small FJ, Tilley JK, Ensign SA. 1995. Characterization of a new pathway for Epichlorohydrin degradation by whole cells of *Xanthobacter* Strain Py2. *Appl Environ Microbiol* 61:1507–1513.

Sonier DN, Duran NL, Smith GB. 1994. Dechlorination of trichlorofluoromethane (CFC-11) by sulfate-reducing bacteria from an aquifer contaminated with halogenated aliphatic compounds. *Appl Environ Microbiol* 60:4567–4572.

Spain J. 1997. Synthetic chemicals with potential for natural attenuation. *Bioremed J* 1:1–9.

Spiess E, Sommer C, Gorish H. 1995. Degradation of 1,4-dichlorobenzene by *Xanthobacter flavus*. *Appl Environ Microbiol* 61:3884–3888.

Susarla S, Masunaga S, Yonezawa Y. 1997. Transformation kinetics and pathways of chlorophenols and hexachlorobenzene in fresh water lake sediment under anaerobic conditions. *Environ Tech* 18: 903–911.

Swanson M, Wiedemeier TH, Moutoux DE, Kampbell DH, Hansen JE. 1996. Patterns of natural attenuation of chlorinated aliphatic hydrocarbons at Cape Canaveral Air Station, Florida. U.S. Environmental Protection Agency Symposium on Natural Attenuation of Chlorinated Organics in Ground Water; 1996 Sept 11–13; Dallas TX. Washington DC: Office of Research and Development. EPA/540/R-96/509. p. 166.

Taraban RH, Berry DF, Berry DA, Walker Jr HL. 1993. Degradation of dicamba by an anaerobic consortium enriched from wetland soil. *Appl Environ Microbiol* 59:2332–2334.

Toro DM, Zarba CS, Hansen DJ, Berry WJ, Schwartz RC, Cowan CE, Pavlou SP, Allen HE, Thomas NA, Paquin PR. 1991. Technical basis for establishing sediment quality criteria for nonionic organic chemicals using equilibrium partitioning method. *Environ Toxicol Chem* 10:1541–1583.

Ulrich HJ, Stone AT. 1989. Oxidation of chlorophenols adsorbed to manganese oxide surfaces. *Environ Sci Technol* 23:421–428.

[USEPA] U.S. Environmental Protection Agency. 1986. Groundwater contamination. In: Protecting groundwater: Pesticides and agricultural practices. Washington DC: USEPA. EPA 440/6-88-001. p 9.

[USEPA] U.S. Environmental Protection Agency. 1996a. Toxics release inventory public data release. EPA 745-R-98-005.

[USEPA] U.S. Environmental Protection Agency. 1996b. Toxics release inventory state fact sheets. EPA 745-F-98-001. www.epa.gov/opptintr/tri/pdr96/

[USEPA] U.S. Environmental Protection Agency. 1997 (Nov). Directive 9200.4-17. Use of monitored natural attenuation at Superfund, RCRA corrective action and underground storage tank sites. Washington DC: Office of Solid Waste and Emergency Response.

[USEPA] U.S. Environmental Protection Agency. 1998a. EPA's contaminated sediment management strategy. EPA-823-F-98-004.

[USEPA] U.S. Environmental Protection Agency. 1998b. Technical protocol for evaluating natural attenuation of chlorinated solvents in ground water. Washington DC: Office of Research and Development. EPA/600/R-98-128.

U.S. International Trade Commission. 1995. Investigation 332-135. Report on domestic production and sales of synthetic organic chemicals and the raw materials from which they are made. Publication 2933.

Van den Wijngaard AJ, Prins J, Smal AJAC, Janssen DB. 1992. Degradation of 2-chlorethylvinylether by *Ancylobacter aquaticus* AD25 and AD27. *Appl Environ Microbiol* 59:2777–2783.

van der Meer JR, Bosma TNP, de Bruin WP, Harms H, Holliger C, Rynaarts HHM, Tros MC, Schraa G, Zehnder AJB. 1992. Versatility of soil column experiments to study biodegradation of halogenated compounds under environmental conditions. *Biodegrad* 3:265–284.

Van Eekert MHA, van Ras NJP, Mentink GH, Rijnaarts HHM, Stams AJM, Field JA, Schraa G. 1998. Anaerobic transformation of ß-hexachlorocyclohexane by methanogenic sludge and soil microflora. *Environ Sci Technol* 32:3299–3304.

Vogel TM, Criddle CS, McCarty PL. 1987. Transformation of halogenated aliphatic hydrocarbons. *Environ Sci Technol* 21:722–736.

Vogel TM, McCarty PL. 1985. Biotransformation of tetrachloroethylene to trichloroethylene, dichloroethylene, vinyl chloride and carbon dioxide under methanogenic conditions. *Appl Environ Microbiol* 49:1080–1083.

Vogel TM, McCarty PL. 1987. Abiotic and biotic transformation of 1,1,1-trichloroethane under methanogenic conditions. *Environ Sci Technol* 21:1208–1213.

Weaver JW, Wilson JT, Kampbell DH. 1996a. Case study of natural attenuation of trichloroethene at St. Joseph, MI. U.S. Environmental Protection Agency Symposium on Natural Attenuation of Chlorinated Organics in Ground Water; 1996 Sept 11–13; Dallas TX. Washington DC: Office of Research and Development. EPA/540/R-96/509. p 65–68.

Weaver JW, Wilson JT, Kampbell DH. 1996b. Extraction of degradation rate constants from the St. Joseph, Michigan, trichloroethene site. U.S. Environmental Protection Agency Symposium on Natural Attenuation of Chlorinated Organics in Ground Water; 1996 Sept 11–13; Dallas TX. Washington DC: Office of Research and Development. EPA/540/R-96/509. p. 69–73.

Wiedemeier TH, Swanson MA, Moutoux DE, Gordon EK, Wilson JT, Wilson BH, Kampbell DH, Haas PE, Miller RN, Hansen JE, Chapelle FH. 1998. Technical protocol for evaluating natural attenuation of chlorinated solvents in ground water. Washington DC: Office of Research and Development. EPA/600/R-98/128.

Wiedemeier TH, Swanson MA, Moutoux DE, Wilson JT, Kampbell DH, Hansen JE, Haas P. 1996. Overview of the technical protocol for natural attenuation of chlorinated aliphatic hydrocarbons in ground water under development for the U.S. Air Force Center for Environmental Excellence. U.S. Environmental Protection Agency Symposium on Natural Attenuation of Chlorinated Organics in Ground Water; 1996 Sept 11–13; Dallas TX. Washington DC: Office of Research and Development. EPA/540/R-96/509. p. 35–59.

Wiedemeier TH, Wilson JT, Kampbell DH. 1996. Natural attenuation of chlorinated aliphatic hydrocarbons at Plattsburgh Air Force Base, New York. U.S. Environmental Protection Agency Symposium on Natural Attenuation of Chlorinated Organics in Ground Water; 1996 Sept 11–13; Dallas TX. Washington DC: Office of Research and Development. EPA/540/R-96/509. p. 74–82.

Wilson BH, Smith GB, Rees JF. 1986. Biotransformation of selected alkylbenzenes and halogenated aliphatic hydrocarbons in methanogenic aquifer material: A microcosm study. *Environ Sci Technol* 20:997–1002.

Wilson BH, Wilson JT, Luce D. 1996. Design and interpretation of microcosm studies for chlorinated compounds. U.S. Environmental Protection Agency Symposium on Natural Attenuation of Chlorinated Organics in Ground Water; 1996 Sept 11–13; Dallas TX. Washington DC: Office of Research and Development. EPA/540/R-96/509. p. 21–28.

Wilson JT, Kampbell DH, Weaver JW. 1996. Environmental chemistry and kinetics of biotransformation of chlorinated organic compounds in ground water. U.S. Environmental Protection Agency Symposium on Natural Attenuation of Chlorinated Organics in Ground Water; 1996 Sept 11–13; Dallas TX. Washington DC: Office of Research and Development. EPA/540/R-96/509. p. 124–127.

Wing MR. 1997. Apparent first-order kinetics in the transformation of 1,1,1-trichloroethane in groundwater following a transient release. *Chemosphere* 34:771–781.

Zaitsev GM, Uotila JS, Tsitko TV, Lobanok AG, Salkinoja-Salonen MS. 1995. Utilization of halogenated benzenes, phenols, and benzoates by *Rhodococcus opacus* GM-14. *Appl Environ Microbiol* 61:4191–4201.

9

Natural Remediation of Metals

Yujun Yin and Herbert E. Allen

CONTAMINATION OF THE ENVIRONMENT by metals has been of great concern worldwide. The detected concentrations of metals in the environment span several orders of magnitude from site to site (Moore 1990; Alloway 1995). Aside from input of natural origin, anthropogenic input has been a major source and remains staggering (Nriagu 1988; Nriagu and Pacyna 1988). Adverse effects of metals to both wildlife and human health have been extensively documented. The metals most often cited as problems are As, Cd, Cr, Cu, Hg, Ni, Pb, and Zn (U.S. Environmental Protection Agency [USEPA] 1986a; Förstner 1995). Other metals, including Ag, Ba, Fe, Mn, Ra, Se, Th, and U, also are cited on the U.S. National Priority List of hazardous waste sites (USEPA 1986a). Many efforts have been made to established environmental quality criteria to protect biota and human life. In many countries, environmental quality criteria are based on the total concentration alone because of the ease of analysis (Beyer 1990; Wu et al. 1991; Canadian Counsel of Ministers of the Environment [CCME] 1991). When the total concentration is extremely high, it is deemed that adverse effects will occur. Likewise, if the concentration is below a certain value, no effect is expected. In most cases, the total concentration falls somewhere in between, and the effects are not directly related to the total concentration of a metal. Instead, the toxicity and bioavailability of a metal are attenuated by many environmental processes, which vary from site to site. Examples have been reported that a low concentration caused adverse effects at one site and a high concentration caused no adverse effects at another (Allen 1996).

Natural Remediation of Environmental Contaminants. Michael Swindoll et al., editors.
 ISBN 1-880611-33-3.

Without consideration of the site-specific situation, the environmental quality criteria established could either under- or overestimate the biological effects. Overestimates of the effects would result in unnecessary cost for cleanup, while underestimates would render the criteria unprotective. Several environmental processes, including aqueous speciation, adsorption or desorption, chemical transformation, precipitation or dissolution, and biological alteration, could alter the fate and toxicity of a metal in a particular environment. It is therefore not surprising that no adverse effects are observed when the concentrations of a metal at many sites are well above the total concentration criteria, which are frequently established based on bioassay tests using relatively simple systems in soluble metal salts. The reduction of the toxicity of pollutants to the biota by these processes is referred as "natural remediation" (NR). The objective of this chapter is to address how these processes affect the environmental fate, bioavailability, and toxicity of metals. It is not our goal to review all the details related to these processes; rather, we will focus on some of the issues important to understanding and predicting the bioavailability and toxicity of metals. Following a brief discussion of the importance of clean techniques in sampling and chemical analysis, we discuss the environmental processes that control metal fate and the issues related to the measurement and prediction of these processes. We then examine the influences of these processes on bioavailability and toxicity of metals as well as their importance with respect to environmental quality criteria. Finally, we provide conclusions and recommendations for future work.

CLEAN TECHNIQUE IN SAMPLING AND ANALYSIS

Typically, the metals of environmental concern are present in trace quantities (Adriano 1992). To obtain valid data with such low concentrations, it is necessary to take special caution, that is, "clean technique," in sampling and analysis (Bruland 1983; Nriagu et al. 1993; USEPA 1994b). Although the need to use clean techniques for environmental sampling and analysis had been proposed in early 1976 (Patterson and Settle 1976), it did not receive wide attention until the 1990s (EA Engineering 1994). The oceanographic community quickly realized the importance of using these clean techniques, while those working on freshwater and effluents were slower to respond. This attitude changed after Windom and his coworkers challenged the adequacy of the U.S. Geological Survey's (USGS) NASQAN data for assessing metal trends in the nation's rivers (Windom et al. 1991). By using a clean technique, they determined Cu concentrations about 3-fold lower, Cd concentrations about 30-fold lower, Pb concentrations 100-fold lower, and Zn concentrations almost 20-fold lower than those obtained by the USGS network.

The biased high metal concentrations resulting from contamination in sampling and analysis could invoke both regulatory and economic issues. The argument on the water quality of New York Harbor is a perfect example in this regard. A summary of the analysis results for waters of New York Harbor is presented in Table 9-1 (Battelle Ocean Sciences 1991). The harbor water had been routinely sampled and analyzed by the New York City Department of Environmental Protection (NYCDEP). Of the 6 metals studied, the only metal whose annual average concentration for 1987 was below the value of the water quality criteria (WQC) was Cd. The reported 1987 annual average Zn concentration exceeded the criteria more than 3-fold. In the following year, a series of cruises was carried out by the USEPA on the Research Vessel *Anderson*. Sampling and analyses were carried out by Battelle scientists using low-level techniques. The concentrations of all 6 elements obtained in 1988 decreased greatly, as much as 40-fold in the case of Cd, compared to those of 1987. The average zinc concentration for the 1987 data was more than 25-fold greater than was the result obtained using the clean techniques in 1988. As a result of evaluation of the newer data, the potential ecological impacts were reassessed and an unnecessary, costly upgrade of sewage treatment plants discharging to the lower Hudson River and the New York Harbor was not instituted.

Table 9-1 Water quality criteria and annual average concentrations using conventional sampling and analytical methods and clean sampling and analytical methods for New York Harbor

		Annual average concentration (μg/L)	
Metal	Water quality criteria (μg/L) 1987	New York City Department of Environmental Protection (conventional) 1987	Research Vessel *Anderson* (clean) 1988
Cd	9.3	4.3	0.11
Cu	2.9	13	4.6
Pb	8.6	70	3
Hg	0.025	0.2	0.015
Ni	8.3	15.6	2.6
Zn	86	264	10

Source: Battelle Ocean Sciences 1991.

As the importance of clean techniques for trace metal studies became increasingly apparent, USEPA proposed guidelines for trace metal sampling and analysis in order to obtain reliable data (Prothro 1993; USEPA 1994b). These guidelines address the basic components of a measurement program. The apparatus, materials, operating protocols, and quality assurance/quality control (QA/QC) procedures that are used during sampling and chemical analysis are specified.

ENVIRONMENTAL FATE OF METALS

Metals are present in the environment in a variety of forms and are subject to alteration by several environmental processes. Depending on environmental conditions, one or more processes dominate the metal state and behavior at a specific location. What follows is a detailed discussion of these processes, namely aqueous speciation, adsorption and desorption, precipitation or dissolution, and redox reaction.

Aqueous speciation

Metals in aqueous solutions can exist either as free metal ions or as complexes with inorganic and organic ligands. Free metal cations are hydrated in the form of $M(H_2O)_n^{x+}$. They behave as Lewis acids and therefore tend to form complexes with Lewis bases containing electron donor atoms, such as N, O, and S. Based on the molecular size, charge, electropositivity, polarizability, excitable outer electrons on donor atoms, and the nature of chemical bonds, metals are classified as hard or soft acids. Hard acids include Li, Na, K, Be, Mg, Ca, Sr, Mn, Al, Sc, Ga, In, La, Gd, Lu, Cr, Co, Fe, As, Mo, Ti, Ce, U, Th, Pu, Hf, while soft acids include Cu, Ag, Au, Ti, Hg, Pd, Cd, Pt, and Te (Pearson 1963). Hard acids prefer to react with hard bases, while soft acids prefer to react with soft bases. The preference sequences of hard and soft acids for electron donor atoms are listed in Table 9-2.

Table 9-2 Preference sequences of hard and soft acids for electron donor atoms[a]

Hard acid	Soft acid
N >> P > As > Sb	N << P > As > Sb
O >> S > Se > Te	O << S < Se < Te
F >> Cl > Br > I	F << Cl < Br < I

[a] Source: after Pearson 1963.

The inorganic ligand OH^- has been most studied by scientists because of its ubiquitous presence. The hydrolysis reactions of metal ions can be expressed by Equation 9-1.

$$M^{x+} + yH_2O = M(OH)_y^{(x-y)+} + yH^+ \qquad K_{OH} \qquad \text{(Equation 9-1)}$$

The reaction in Equation 9-1 depends on the solution pH and the acidic nature of the metal ion. The stronger the acid nature of a metal (the larger the hydrolysis constant [log K_{OH}]), the lower the pH value at which metal hydrolysis occurs. Other important inorganic ligands include Cl^-, HCO_3^-, CO_3^{2-}, NH_3, SO_4^{2-}, and PO_4^{3-}. The reactions of metals with these ligands are expressed by Equation 9-2.

$$M^{x+} + nL_i^{j-} = ML_{i,n}^{(x-nj)+} \qquad K_i \qquad \text{(Equation 9-2)}$$

where

i = i^{th} ligand

K_i = its stability constant

The carbonate and bicarbonate ion concentrations can be easily determined by alkalinity analysis, and others by ion chromatographic measurements. The stability constants between these ligands and metals are well established (Martell and Smith 1993), and the fraction of each species among the total inorganic metal concentration can be readily computed, as indicated by Equation 9-3 in the case of free metal ion:

$$f_{M^{x+}} = \frac{1}{1 + \sum_{y=1}^{m} K_{OH,y}/[H^+]^y + \sum_{i=1}^{k} K_i[L_i^{j-}]n} \qquad \text{(Equation 9-3)}$$

where

m, k = integers

$f_{M^{x+}}$ = fraction that free metal ion comprises of the total metal concentration

Several computer programs, such as MINEQL$^+$ (Schecher and McAvoy 1991) and MINTEQA2 (Allison et al. 1991), have been developed to perform the computation.

In natural systems, inorganic forms of metals usually account for only limited percentages of total aqueous metals (Mantoura et al. 1978; Vuceta and Morgan 1978; Sigg and Xue 1994). A majority of metals are present in the forms of organic complexes. Both anthropogenic organic ligands and natural organic matter are responsible for metal binding. Nitrilotriacetic acid (NTA) and ethylenediaminetetraacetic acid (EDTA) are two of the most extensively investigated anthropogenic ligands because of their wide application and very strong binding affinity for metals (Means and Crerar 1978; Morel and Hering 1993; Kari et al. 1995; Xue et al. 1995). Most EDTA-metal complexes except Fe-EDTA are neither biologically nor photochemically degradable (Natarajan and Endicott 1973; Alder et al. 1990). It has been reported that most of a metal could be bound by EDTA in certain natural environments where EDTA is present at high concentration levels (Means and Crerar 1978). Recent research by Sedlak et al. (1997) also suggests that a large portion of Ni and Cu in wastewater may be bound by anthropogenic ligands. The binding of metals with synthetic ligands is well documented, and the most comprehensive binding information can be found from the National Institute of Standards and Technology (NIST) database (Martell and Smith 1993). It needs to be pointed out that the thermodynamically favored metal-chelate complexes may not always be the dominant species in natural systems (Hering and Morrel 1989). The speciation also depends on the initial forms of ligands and the kinetics of metal exchange reactions.

The dominant role of naturally occurring organic matter (NOM) in complexing metals has been well documented (Schnitzer 1978; Stevenson 1982; Sposito 1986). A number of approaches, including ultrafiltration, ion exchange, equilibrium dialysis, ion selective electrode potentiometry, cathodic and anodic striping voltametry, lanthanide ion probe spectroscopy, and nuclear magnetic resonance spectroscopy, have been used to characterize the binding of NOM with metals (Christman and Gjessing 1983; van den Berg et al. 1990; Yin et al. 1997a). As a mixture of multiligands, NOM behaves quite differently from simple ligands in binding metals. Carboxylic and phenolic groups are the dominant functional groups on NOM for metal binding (Buffle 1990). Thiol and amine functional groups account for a small fraction of the total functional group; however, they have very strong binding affinity for metals (Reimers and Krenkel 1974; Yin et al. 1997b). Generally speaking, higher valent cations have stronger binding affinity for NOM than do lower valent cations. For divalent cations, Tipping and Hurley (1992) reported a binding strength sequence of $Mg^{2+} < Ca^{2+} < Mn^{2+} < Cd^{2+} < Co^{2+} < Ni^{2+} \sim Zn^{2+} < Pb^{2+} < Cu^{2+} < VO^{2+}$ based on evaluation of a large set of data. Decreases in pH increase the surface potential and proton competition, thus decrease metal binding.

As an example, Figure 9-1 illustrates the effect of NOM on Cu speciation in river water. For the sake of easy illustration, 1:1 two-site binding between Cu and NOM is assumed. Because in many cases conditional stability constants between inorganic metals and NOM were determined (Van den Berg et al. 1990), rather than thermodynamic constants between free metal ion and NOM, the reaction of inorganic Cu and NOM is considered in this illustration. The reactions of inorganic Cu with 2 binding sites (L_1 and L_2) are expressed as:

$$Cu + L_1 = CuL_1 \qquad K_{CuL1} \qquad \text{(Equation 9-4)}$$

$$Cu + L_2 = CuL_2 \qquad K_{CuL2} \qquad \text{(Equation 9-5)}$$

where

K_{CuL1}, K_{CuL2} = conditional stability constants for inorganic Cu with NOM at the given conditions

The fractions of equilibrium inorganic Cu and the organically bound Cu were calculated in the same fashion as expressed in Equation 9-3. The values of the 2 conditional stability constants were adopted from McKnight et al. (1983), who reported studies for 9 rivers, 4 lakes, and 6 wetlands. One log stability constant is about 6, with a binding density of 1 µmol/mg C, and the other is about 8, with a binding density of 0.26 µmol/mg C. An NOM concentration range of 0 to 8.75 mg/L, a typical range found in natural waters, was employed in the calculation. The

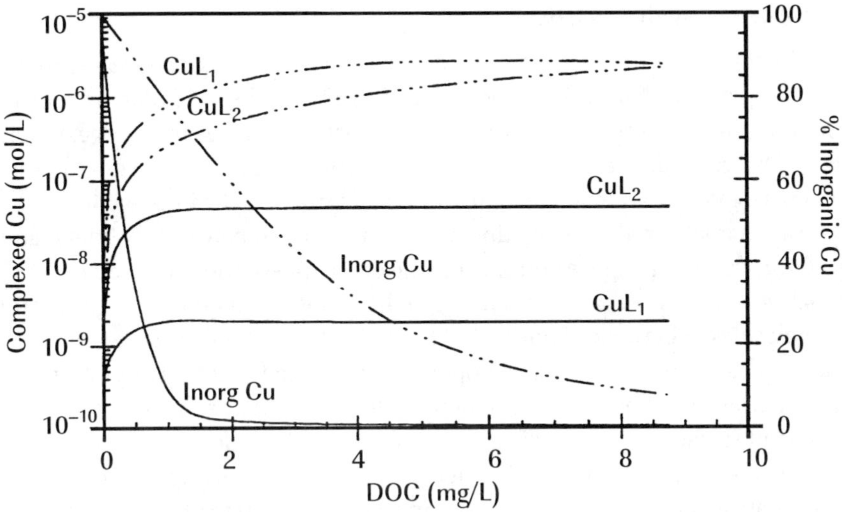

Figure 9-1 Complexation of Cu by dissolved organic matter, expressed as dissolved organic carbon (DOC) (CuL_1 and CuL_2 = Cu bound by sites 1 and 2; solid line = total Cu concentratlon of 5×10^{-8} M; dashed line = total Cu concentration of 5×10^{-6} M)

concentrations of Cu bound to both sites and the fraction of inorganic Cu are plotted against the total organic matter concentration.

The fraction of inorganic Cu depends on concentrations of both NOM and the total Cu present. At a low total Cu concentration of 5×10^{-8} M, almost all Cu is bound by the strong ligand on the NOM at an NOM concentration above 1 mg C/L. Increasing Cu to a higher concentration of 5×10^{-6} M results in Cu being bound by both strong and weak ligands on NOM. The fraction of inorganic Cu at a given NOM concentration increases because of the decrease in overall binding strength of NOM for Cu.

Although in many cases 1-site or 2-site models can fit the experimental data very well, they oversimplify the complex nature of NOM, thus the stability constants obtained are conditional (Perdue and Charles 1983). These constants depend on experimental conditions such as pH, metal to NOM ratio, and ionic strength. Several multisite models have been developed to describe the complexation reactions. Representative of these models are the normal distribution model (Perdue and Charles 1983), the simple mixture model (Mattigod and Sposito 1979), and the WHAM discrete site model (Tipping 1994). A review of these models was given by Westall et al. (1995) recently. The calculation of metal–NOM complexation using these models relies on the availability of stability constants. At present, the WHAM has the largest database and was shown to best predict the speciation in natural rivers (Warwick et al. 1994).

Adsorption and desorption

Adsorption of metals onto environmental particulate matter can be treated as a surface complexation formation process (Stumm 1992). The solids can be considered as inorganic and organic polymers, whose surfaces can be viewed as extending structures bearing surface functional groups (Stumm 1992). Both inner-sphere and outer-sphere complexation are responsible for adsorption. The inner-sphere complex is characterized by formation of a chemical bond between the metal and the electron-donating surface atom, while the outer-sphere complexation occurs by approaches of a metal of opposite charge to the surface with typically 1 (or more) water molecules in between (Sposito 1984; Stumm 1992). Inner-sphere complexes are more stable than outer-sphere complexes. Differentiation of adsorption mechanisms can be achieved by using fast kinetic techniques, such as the pressure jump method, and surface spectroscopic methods (Sparks 1995). Selenite, for example, has been shown to form both inner-sphere and outer-sphere complexes on geothite, while selenate adsorption on geothite occurred only via outer-sphere complexes (Sparks 1992).

The extent of adsorption of metals on solids is strongly pH dependent. A recent review of adsorption of metals on mineral surfaces was given by Stumm (1992). The pH affects adsorption by affecting both metal speciation and surface charge properties. Decreases in pH increase surface potential and proton competition, and thus do not favor cation binding. At pH values much higher than the point of zero charge, the surface is highly negatively charged and metal anion adsorption decreases. Adsorption of metal cations on mineral surfaces generally increases with increasing pH (Dzombak and Morel 1990). Maximum adsorption is typically achieved within a pH increase of 1 to 2 units, which is the so-called "adsorption edge." Stronger acidity of a metal results in the adsorption edge shifting to a lower pH range. Adsorption of metal anions on mineral surfaces either decreases or follows an inverted U-shape as pH increases, depending on the nature of solid surfaces (Stumm 1992). Adsorption of metals on natural materials, such as soils, sediments, and suspended particles, is much more complicated because of the interactions among the particulates. In natural systems, not all minerals are necessarily important in adsorption. Some of the mineral surfaces may be coated by other components. For example, clay minerals are generally coated by metal oxides and organic matter (Hart 1982; Davis 1984; Jenne 1988), while the metal oxides can also be coated by organic matter (Stumm 1992).

Adsorption of 8 divalent heavy metals on 15 noncontaminated agricultural soils of varying properties was recently investigated in our laboratory. An initial metal concentration of 1×10^{-7} M and an ionic strength of 0.01 M were employed. The fraction of the total metal adsorbed was calculated based on the equilibrium soluble metal concentration and the added metal concentration as well as the metal naturally present in the soil. As an example, Figure 9-2 shows adsorption of the metals

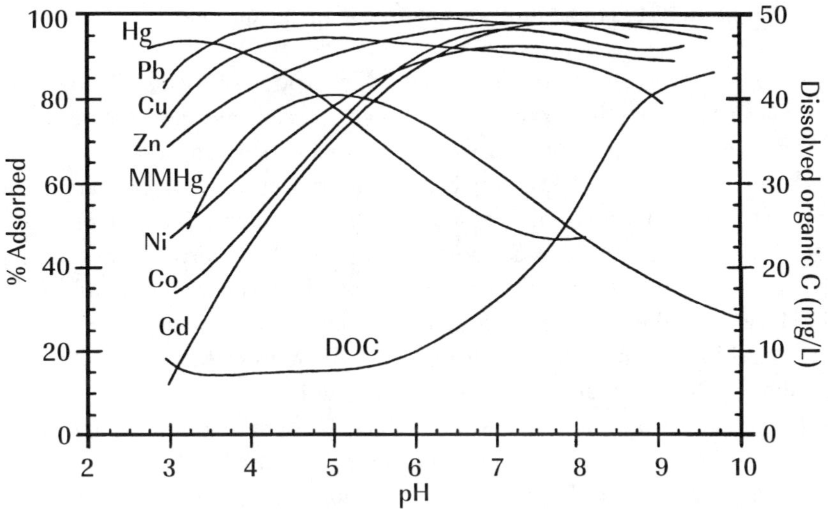

Figure 9-2 Adsorption of metals on a Freehold sandy soil subsurface (DOC = dissolved organic C; soil:solution = 1 g/100 mL; I = 0.01 M $NaNO_3$; T = 25 °C)

on a subsurface sandy loam. At pH 3, adsorption followed a sequence of Hg > Pb > Cu > Zn > Ni > Co > Cd, which is similar to the sequence of the first hydrolysis constants ($\log K_{OH}$) of the metals, that is, Hg (10.6) > Pb (6) ~ Cu (6) > Zn (5) > Ni (4.1) ~ Co (4.3) > Cd (3.9). The stronger Lewis acids not only have stronger binding affinity for OH^- but also for the surface sites that can be considered as bases. Thus, at low pH values, while the adsorption of metals of weaker acids was small because of the competition of protons, the adsorption of metals of stronger acids was large because of their stronger binding affinity for the surfaces.

Increases in pH increased adsorption principally because the surface potential decreases. Further increases in pH decreased adsorption. The complexation of metals by dissolved organic matter (DOM), whose concentration increased with increasing pH, was found to be the major reason responsible for adsorption decreases (Yin et al. 1996). The pH values at which adsorption starts decreasing depended on the acidity of the metal. For metals that are stronger Lewis acids, adsorption reached maxima at lower pH values and then decreased as complexation by the DOM increased with increasing pH. Mercury exhibited the strongest affinity for the DOM, Cu was next, and the rest were affected only slightly.

Besides dissolved humic substances, the presence of inorganic ligands and anthropogenic organic ligands in the solution phase can also decrease adsorption by formation of nonadsorbable complexes with metals. Chloride and CO_3^{2-}, which have a strong affinity for metals and could be present at high concentrations in the solution, are two of the most important inorganic ligands (Yin et al. 1996). EDTA, on the other hand, is the most important anthropogenic organic ligand (Means et al. 1980).

The background electrolyte can also affect adsorption of metals. Zachara et al. (1993) investigated Cd (II) sorption on specimen and soil smectites in Na and Ca electrolytes and found that ion exchange dominated Cd sorption at low ionic strength in Na^+ electrolyte. Increasing Na^+ concentration to 0.1 M, or changing the electrolyte to Ca^{2+} at ionic strength of 0.003 to 0.006 M, suppressed Cd ion exchange. Yin and Allen (1997) also reported greater Hg(II) adsorption on soils in Ca electrolyte than in Na electrolyte. The presence of a high concentration of Ca^{2+} (0.003 M) reduced the solubility of organic matter, either by coagulation or by serving as a bridge connecting organic matter to the solid surfaces, thus decreasing the amount of DOM available to affect Hg adsorption.

Adsorption of heavy metals on solid surfaces is characterized by 2 temporal regimes: a fast step followed by a slow step. Typically, more than 90% of adsorption occurs during the fast step. The fast step is commonly associated with surface reactions, while the slow step results from intraparticle diffusion and/or surface precipitation (Sparks 1989; Fuller et al. 1993). The slow adsorption could last for days, weeks, months, or even years.

No method is available to allow direct measurement of the quantity of metals bound to different particulate components. Sequential extraction, initially proposed by Tessier et al. (1979), has been extensively used to characterize the particulate trace metals. By using increasingly strong extractants, the trace metals associated with exchangeable carbonate metal oxide or reducible organics and sulfides and residual mineral phases are sequentially released. This method has been criticized for not being able to provide accurate information regarding metal binding (Martin et al. 1987).

Perhaps the most important adsorbing phase in many cases is the NOM. Many researchers have reported that organic matter was the major component responsible for metal binding to oxalic sediments (Fu et al. 1992). Metal partitioning to suspended particles collected from receiving waters has also been reported to be dominated by the particulate organic matter (Shi 1996). In recent studies on adsorption of metals on 15 New Jersey soils, Lee et al. (1996) and Yin et al. (1997c) also found that adsorption correlated very well with the soil organic matter content.

Adsorbed metals could remobilize upon environmental alteration. Not all adsorption reactions are reversible (Di Toro et al. 1986). Diffusion of metals into organic matter and clay mineral micropores or structure seems to be the major reason for the reaction irreversibility (Bruemmer et al. 1988; Yin, Allen, Huang, Sparks, Sanders 1997). The irreversibility increases with increased aging. For example, more metal could be released from freshly contaminated soils than from sludge-amended soils. The introduction of chelating ligands would increase both the amount of metals released and the desorption rate. The decomposition of particulate organic matter and dissolution of minerals can also release metals to solution. Batch extraction has frequently been used to determine the potential for metal mobility with water, electrolyte, or chelating solution as extractants. Methodological problems related to artifacts, including sampling procedures, sample preparation and extraction techniques, have been reviewed (Wenzel and Blum 1995). Due to the inhibition of desorption resulting from the released metal remaining in the system, the batch method may underestimate the potential mobility of metals (Yin 1996). Flow techniques provide a better estimate of metal mobility.

A number of models have been proposed to describe metal adsorption. The dependence of adsorption on metal loading levels is commonly described with the following generalized Equation 9-6 (Sposito 1985):

$$q(c)=\sum_{i=1}^{n}\frac{b_i k_i^{\beta_i} C^{\beta_i}}{1=B_i C_i^{\beta_i}} \qquad \text{(Equation 9-6)}$$

where

$q(c)$ = amount of solute adsorbed
C = concentration of solute in the solution
b_i, B_i, k_i, β_i = adjustable parameters

To describe surface complexation formation reactions as a function of pH, both one-site and multisite models have been proposed. One-site models include constant capacitance, double layer, and triple layer models. The adsorption of H^+, OH^-, metal cations, and anions on the surface sites can be represented by the following reactions (Equations 9-7 through 9-15; Stumm 1992):

Acid–base equilibria

$$S-OH+H^+ \Leftrightarrow S-OH_2^+ \qquad \text{(Equation 9-7)}$$

$$S-OH+OH^- \Leftrightarrow S-O^- +H_2O \qquad \text{(Equation 9-8)}$$

Metal cation binding

$$S-OH+M^{z+} \Leftrightarrow S-OM^{(z-1)+}+H^+ \qquad \text{(Equation 9-9)}$$

$$2S-OH+M^{z+} \Leftrightarrow (S-O)_2M^{(z-2)}+2H^+ \qquad \text{(Equation 9-10)}$$

$$S-OH+M^{z+}+H_2O \Leftrightarrow S-OMOH^{(z-2)+}+2H^+ \qquad \text{(Equation 9-11)}$$

$$S-OH+L^- +M^{z+} \Leftrightarrow S-L-M^{(z-1)+}+OH^- \qquad \text{(Equation 9-12)}$$

$$S-OH+L^- +M^{z+} \Leftrightarrow S-OM-L^{(z-2)+}+H^+ \qquad \text{(Equation 9-13)}$$

Metal anion binding

$$S-OH+L^- \Leftrightarrow S-L+OH^- \qquad \text{(Equation 9-14)}$$

$$2S-OH+L^{-} \Leftrightarrow S_2-L^{+}+2OH^{-} \qquad \text{(Equation 9-15)}$$

Only the metal ion of interest is considered in the constant capacitance model. Both the metal of interest and background electrolyte ions are considered in double layer and triple layer models. Mass balance and surface charge balance form the bases for model calculation (Dzombak and Morel 1990). A representative of multisite models is MUSIC (MUlti SIte Complexation) (Van Riemsdijk and Hiemstra 1993). Based on the results of infrared analysis of metal oxides, this model considers 3 different surface sites in which oxygen groups are singly, doubly, or triply coordinated to an underlying metal ion.

The surface complexation models have been used to describe metal adsorption on minerals. The FITEQL (Herbelin and Westall 1994) computer program is ideal for fitting the experimental data based on these models. In some cases, different models can fit the same experimental data equally well (Westall and Hohl 1980). Thus, a better fit of the experimental data to a model does not imply any mechanism. Some researchers have used these surface complexation models to describe metal adsorption to natural materials. The model-fitting parameters obtained vary from one study to another. To make the models useful in predicting metal partitioning in natural systems, a large number of samples of different sources must be studied under different conditions. The model-fitting parameters must be related to environmental variables, so the metal partitioning can be predicted based on major environmental variables.

As an alternative to surface complexation modeling, the adsorption of metals has also been directly related to the properties of soils, sediments, and suspended particles (Jenne et al. 1986). An example is given in Figure 9-3. Lee et al. (1996) investigated Cd adsorption on 15 soils and found that adsorption correlated very well with the soil organic matter content at given pH values. By normalizing the partition coefficients with soil organic matter content, the authors obtained a very good correlation between the normalized partition coefficient and pH.

PRECIPITATION AND DISSOLUTION

Metal ions in solutions can be precipitated by ligands such as hydroxide, carbonate, chloride, silicate, phosphate, and in anoxic environments, sulfide. The solubility products of metals with these ligands are well defined (Martell and Smith 1993), and the solubility of metals therefore can be easily calculated. Metal ions would compete for the ligands present. The smaller the solubility product, the easier the precipitation. For example, many divalent metal ions M^{2+} (e.g., Cd^{2+}, Cu^{2+}, Hg^{2+}, Ni^{2+}, Pb^{2+}, and Zn^{2+}) have stronger binding affinities for S^{2-} than for Fe^{2+}, and thus can replace Fe^{2+} from FeS to precipitate according to the reaction in Equation 9-16.

$$FeS+M^{2+}=MS+S^{2-} \qquad \text{(Equation 9-16)}$$

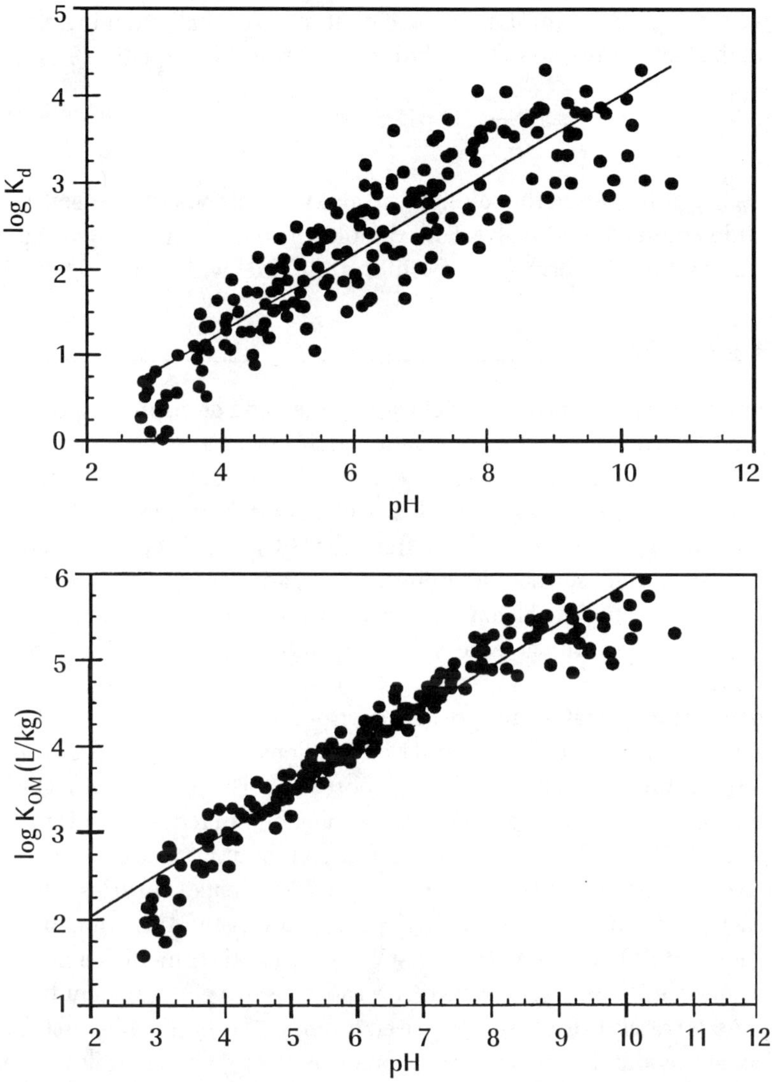

Figure 9-3 Soil–water partitioning of Cd as a function of pH (K_d = partitioning coefficient; K_{OM} = soil organic matter normalized partitioning coeficient; data from Lee et al. 1996)

Therefore, if a large quantity of FeS is present in a sediment, all the metals between the parentheses could be precipitated and rendered nontoxic. In addition, metal hydroxides can precipitate at solid surfaces at concentrations at which no precipitation would occur in the bulk solution (Fuller et al. 1993). However, it is impossible to differentiate precipitation from adsorption based on macroscopic measurements.

Under acidic conditions, most metal precipitates and crystallized minerals are unstable and thus release metals to the solution as expressed by Equation 9-17:

$$MS + 2H^{+} = H_2S + M^{2+} \qquad \text{(Equation 9-17)}$$

The presence of ligands in the solution can also facilitate adsorbed metal and metal-containing minerals to dissolve. For example, Cl^- and EDTA can significantly increase metal solubility by formation of soluble complexes with metals.

REDOX REACTIONS

Redox reactions are very important to metal adsorption and desorption, speciation, and toxicity in the environment. Recent reviews on this subject are given by Fish (1993), Grundl (1995), and Pardue and Patrick (1995). Redox reactions can occur as a result of changes in the environmental Eh and pH, or can be induced by introduction of redox-active materials. In aerobic waters, the H_2O/O_2 redox couple dominates the Eh. In subsurface soils, underground waters, and sediments, oxygen is usually depleted or absent. The Eh and pH are often altered by the microbially mediated reduction of electron acceptors including nitrate, sulfate, and oxidized forms of Fe and Mn.

Three redox-active metals of the greatest environmental concern are Cr, As, and Se. Chromium exists in 3 states: Cr(VI) and Cr(III) anions and Cr(III) hydroxide. Cr(VI) is the most soluble, least sorbable, and most toxic. The presence of NOM and anthropogenic hydroxylated compounds, such as phenol, can reduce Cr(VI) to Cr(III) (Pardue and Patrick 1995). In oxalic surface soils and surface waters, Cr(III) is expected to be oxidized to Cr(VI) by MnO_2 or Fe- and Mn-coated particles. The oxidation reaction would be hindered at high pH because of the formation of $Cr(OH)_3$ (Fendorf 1993). Both As(III) and As(V), as well as their methylation products, are present in the environment. As(V) adsorbs to a greater extent than does As(III). As(V) rarely causes any adverse environmental effect, while As(III) has been listed as a potential carcinogen. Oxidation of As(III) to As(V) in soils has been observed, but the reverse reaction has not (Dias 1997). Selenium can exist in 4 oxidation states: Se(II), Se(O), Se(IV), and Se(VI). Higher oxidation states of Se have greater solubility. It has been reported that more than 50% of the total Se was released when a sediment was oxidized (Pardue and Patrick 1995).

NATURAL REMEDIATION EFFECTS ON BIOAVAILABILITY AND TOXICITY OF METALS

The environmental processes discussed previously have dramatic effects on the bioavailability and toxicity of metals (Allen 1993). It is not surprising that there is no correlation between total metal concentrations and biological effects. An example is given in Figure 9-4. Except for very low values, the Cu concentration distribution in sediments, with and without adverse effects, overlaps.

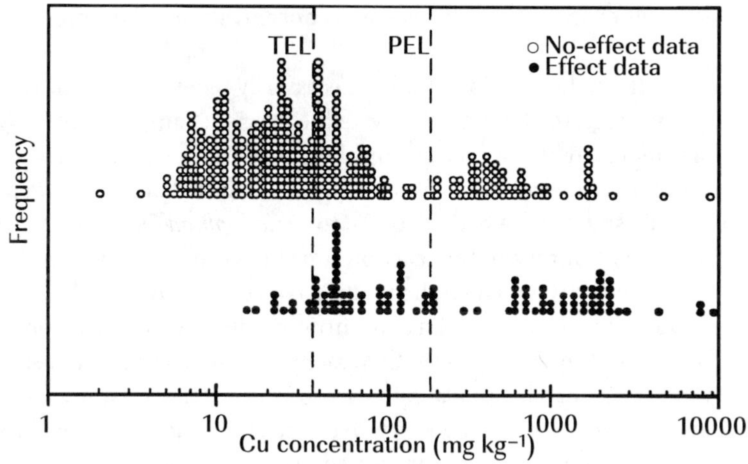

Figure 9-4 Correlation of total Cu concentration with biological effects (TEL = threshold effect level; PEL = probable effect level; data from Allen 1996)

The ratio of acid-volatile sulfide (AVS)/simultaneously extracted metal (SEM), including Cd^{2+}, Cu^{2+}, Hg^{2+}, Ni^{2+}, Pb^{2+} and Zn^{2+}, on a mole-to-mole basis, has been proposed as an index for metal toxicity in sediments (Di Toro et al. 1990; Allen et al. 1993). Based on Equation 9-17, acidification of a sediment would dissolve metal sulfide precipitates, releasing hydrogen sulfide and metal ions to the solution phase. Thus, if the measured AVS/SEM > 1, all the extracted metals could be present in sediment as sulfide precipitates because of their high stability, and consequently could become biologically unavailable. This hypothesis was supported by laboratory bioassay tests (Di Toro et al. 1990). In cases where AVS/SEM < 1, the adverse effect may or may not occur, depending on the forms of the metals present.

Extractable metals have frequently been correlated with biological uptake and toxicities in sediments and soils. This method assumes that the metal associated with certain solid phases is bioavailable without consideration of the competition of organisms and the effects of environmental conditions such as pH. Thus, only a statistical relationship is given with respect to bioavailability. The correlation obtained for a single metal therefore varies from one study to another. For example, some researchers (Whitten and Ritchie 1991; Novozamsky et al. 1993) reported that uptake of Cd by plants correlated well with $CaCl_2$-extractable Cd, while others (Korcak and Fanning 1978; Huang 1997) found that Cd uptake correlated well with DTPA-extractable Cd.

All plants and many soft-bodied animals take up metals from aqueous solutions. The partitioning of metals between solid and solution phases is therefore primarily important in determining the bioavailability of metals. For example, plant uptake for metals from soils can be increased by lowering soil pH and adding chelating agents, which increase aqueous metal concentrations (Cunningham and Ow 1996).

Not all the metal in the solution phase is necessarily bioavailable. The free metal ion has often been suggested to be the bioavailable species (Campbell 1995). Many researchers have reported linear relationships between the free metal ion activity and biological uptake. An example is given in Figure 9-5 based on data from Ma et al. (1999). They investigated the toxicity of Cu to *Ceriodaphnia dubia* as a function of the humic acid concentration and the reaction time between humic acid and Cu. The survival rate of *C. dubia* depended on the total Cu concentration as well as on aqueous speciation. For a given total Cu concentration, increases in reaction time and humic acid concentration decreased Cu toxicity. No correlation between total Cu concentration and *C. dubia* survival rate was observed. The measured free Cu^{2+} ion concentration, however, correlated very well with the *C. dubia* survival rate, indicating that only the free Cu^{2+} ion was bioavailable.

In a recent study, McLaughlin et al. (1997) investigated plant uptake of Cd and Zn from nutrient solutions, with and without the presence of chelating ligands. They found that for a fixed, free metal ion activity, plant uptake of Cd and Zn was much greater in the presence of chelating ligands than in the absence of ligands. This was attributed to plant uptake of intact metal–ligand complexes and increases in the diffusion rate of metal ions to the plant surface driven by complexation formation.

It is impossible to quantitatively define the bioavailable metal in a system. Rather, it is more appropriate to consider that the introduction of an organism constitutes another receptor site that can compete for metal ions with the soluble ligands and adsorption sites. The amount of metal taken up by an organism depends on its binding constant for the metal, competing ligand concentrations and formation constants, as well as environmental conditions. For example, the calculated free metal concentration could be very small in the presence of ligands; however, all of the metal could be taken up by a plant if the plant has much stronger binding affinity for the metal than do other ligands. To predict the bioavailability of a metal to organisms, one needs to consider the equilibrium competition among the various binding sites in the system. An example is the gill complexation model, which treats the receptor sites on the gill as competing ligands. With this model, Playle et al. (1993a, 1993b) described the competition of gills with DOM for metals.

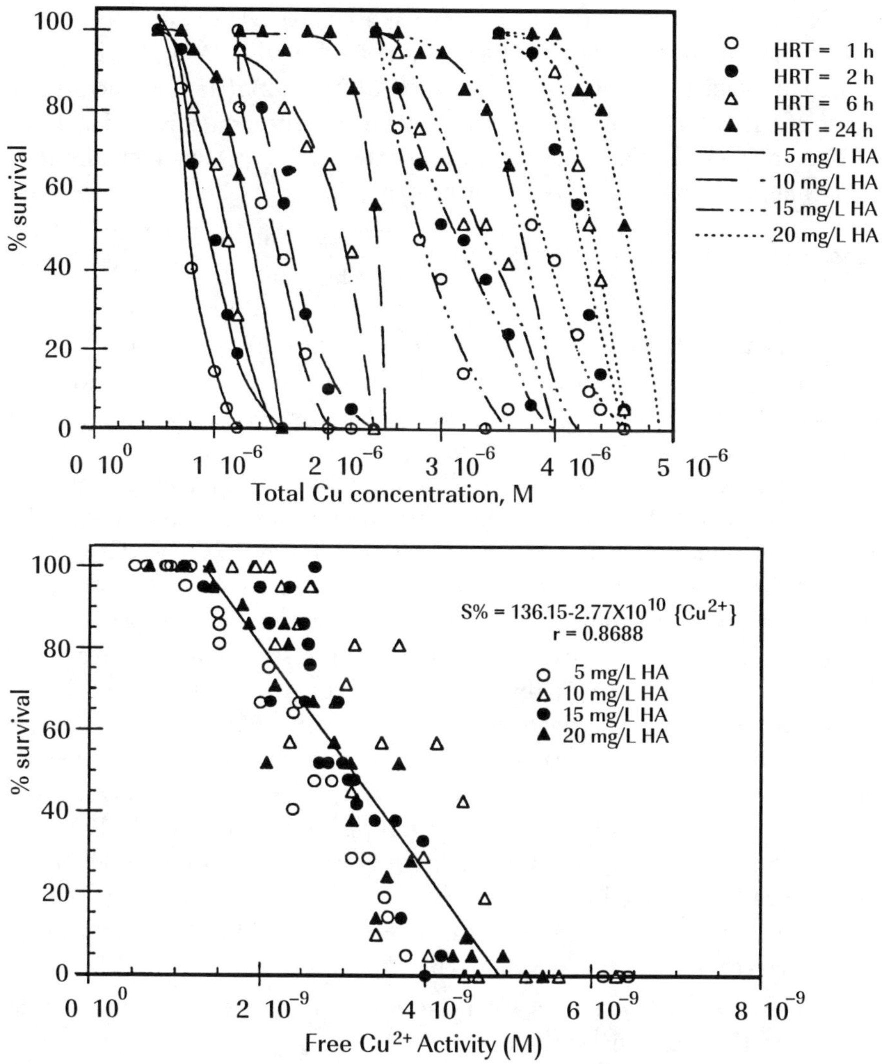

Figure 9-5 Aqueous toxicity of Cu to *Ceriodaphnia dubia* as a function of Cu concentration, speciation, and reaction kinetics with humic acid (HA) (data from Ma et al. 1999)

ENVIRONMENTAL QUALITY CRITERIA

The importance of chemical speciation to biological effects has been well established; however, this is not fully reflected in the current environmental quality criteria. The establishment of environmental quality criteria based on bioavailability is difficult because of many uncertainties, which need to be investigated further. Water hardness has been incorporated into WQC (USEPA 1986b). Besides hardness, bioavailability of metals in the aquatic system is also affected by many other factors. To account for the site-specific situation, the water effect ratio (WER) has been proposed to calculate the WQC at a given site (USEPA 1992). The WER, defined as the ratio of the LC50 for site water to the LC50 for reference water, is used as a multiplier to adjust the national water quality criteria (NWQC) to account for the effect of matrix on bioavailability at a specific site, that is:

$$\text{Site-specific WQC} = \text{NWQC} \times \text{WER} = \text{NWQC} \times \frac{\text{site water LC50}}{\text{reference water LC50}} \quad \text{(Equation 9-18)}$$

Caution should be taken in using WER to calculate a site-specific WQC. The WER value depends on organism sensitivity to metals, which in turn depends on competition ability of organism with ligands for the available metal. For a given site, the WER value usually increases as the organism sensitivity to metals increases because more metals would be bound by ligands and fewer would be taken up by insensitive organisms. Similarly, the changes in matrix characteristics among different sites would make less difference to the WER for the less sensitive organisms than for the more sensitive ones. Consequently, the use of an insensitive organism for toxicity tests would result in a lower WER and a conservative site-specific criterion, while the use of a sensitive organism for the tests would result in a larger WER and a site-specific criterion that is not sufficiently protective. A detailed treatment of this subject was given by Allen and Hansen (1996).

Equilibrium partitioning–based estimates of metal concentrations in interstitial water have been proposed to derive sediment quality criteria (SQC; USEPA 1994a). The technical basis for this approach has been reviewed by Ankley et al. (1997). Briefly, the SQC for 5 metals (Cd, Cu, Ni, Pb, and Zn) collectively can be derived based on the following 4 criteria:

1) The ratio of the sum of the molar concentrations of 5 metals to the molar concentration of the AVS should be < 1.
2) The measured interstitial water concentrations of the metals should be less than the WQC final chronic values (FCVs).
3) The interstitial water concentrations of metals calculated based on organic carbon partitioning coefficients, in addition to the AVS and SEM relationships, should be less than the FCVs.
4) The calculated interstitial water concentrations of metals based on the minimum partitioning coefficients (e.g., generated from chromatographic sand) should be less than the FCVs.

The first criterion is supported by both short-term and long-term laboratory and field toxicity tests of sediments (Ankley et al. 1997), and is experimentally easily determined (Allen and Hansen 1996). This criterion does not predict toxicity when SEM/AVS > 1. The second criterion would ensure no adverse effects. The direct determination of interstitial water concentrations, however, is still not routinely operational. The third and fourth criteria are still questionable because the partitioning coefficients depend on pH and metal loading levels. More research is needed to predict partitioning coefficients as a function of environmental conditions in order to safely adopt these environmental criteria.

The cation exchange capacity (CEC) has been used to develop guidelines to limit the application of metals to soils (USEPA 1979). The guidelines thus formulated, however, are not supported by long-term field experimentation (Sommers et al. 1987). In some cases, the results related to the effect of CEC on the bioavailability of metals are in conflict. For example, Korcak and Fanning (1985) found that, as soil CEC increased, plant uptake for metals decreased in metal salt–treated soils but increased in sludge-treated soils.

Lee et al. (1993) propose a methodology for the development of metal criteria that will be protective of groundwater quality. According to this methodology, the maximum metal concentrations in soils are calculated based on partitioning coefficients and pore water concentrations, which are adopted from drinking water standards. The ability to accurately predict partitioning coefficients under different environmental conditions is the key to using this method to formulate soil standards.

CONCLUSIONS

This review indicates that proper clean techniques must be employed in environmental sampling and analysis in order to obtain reliable data. Caution needs to be taken with data obtained using alternative techniques. In aquatic systems, adverse biological effects are not always directly correlated to total metal concentrations, since the metals are attenuated by environmental processes. Solid–water partitioning of metals, as affected by sorption and desorption and by precipitation or dissolution, is the primary factor that determines the amount of metal available in the aqueous solution for potential leaching to groundwater and for biological uptake. Not all the metal in the solution phase is necessarily bioavailable; instead, organism uptake of a metal depends on competition for the metal between the binding sites on the organism and the ligands present in the system.

As a long-term goal, environmental quality criteria should be based on the whole system chemistry, which accounts for site-specific characteristics. More research needs to be conducted using natural materials, that is, soils, sediments, and suspended particles, with emphasis on the prediction of metal partitioning based on environmental conditions and metal loading levels. To enable models to have wide application, large numbers of samples of different sources need to be studied, and the model-fitting parameters should be correlated to the environmental properties. Special environmental entities with unique characteristics may need to be treated separately. The metal-binding sites of an organism need to be identified and the binding constants determined so that the competitive binding processes can be predicted.

REFERENCES

Adriano DC. 1992. Biogeochemistry of trace metals. Boca Raton FL: Lewis.

Alder AC, Siegrist H, Gujer W, Giger W. 1990. Behavior of NTA and EDTA in biological wastewater treatment. *Water Res* 24:733–742.

Allen HE. 1993. The significance of trace metal speciation for water, sediment and soil quality standards. *Sci Total Environ* 134 (Supplement Part 1):23–45.

Allen HE. 1996. Standards for metals should not be based on total concentration. *SETAC News* 16:18–19.

Allen HE, Fu G, Deng B. 1993. Determination of acid-volatile sulfide (AVS) and simultaneously extracted metals (SEM) for the estimation of potential toxicity in aquatic sediments. *Environ Toxicol Chem* 12:1441–1453.

Allen HE, Hansen DJ. 1996. The importance of trace metal speciation to water quality criteria. *Water Environ Res* 68:42–54.

Allison JD, Brown DS, Novo-Gradac KJ. 1991. MINTEQA2/PRODEFA2, A geochemical assessment model for environmental systems. Version 3.0. Athens GA: U.S. Environmental Protection Agency (USEPA).

Alloway BJ. 1995. Soil processes and the behavior of metals. In: Alloway BJ, editor. Heavy metals in soils. New York NY: Blackie Academic & Professional. p 11–37.

Ankley GT, Di Toro DM, Hansen DJ, Berry WJ. 1997. Technical basis and proposal for deriving sediment quality criteria for metals. *Environ Toxicol Chem* 15:2056–2066.

Battelle Ocean Sciences. 1991. Final results of the ambient and municipal sample interlaboratory comparison study in ambient waters and discharges to New York/New Jersey Harbor. Report submitted to the U.S. Environmental Protection Agency (USEPA). Duxsbury MA.

Beyer WN. 1990. Evaluating soil contamination. Washington DC: U.S. Department of Interior (USDOI).

Bruemmer GW, Gerth J, Tiller KG. 1988. Reaction kinetics of the adsorption and desorption of nickel, zinc and cadmium by goethite. I. Adsorption and diffusion of metals. *J Soil Sci* 39:37–52.

Bruland KW. 1983. Trace elements in seawater. *Chem Oceanography* 8:157–220.

Buffle J. 1990. Complexation reactions in aquatic systems: An analytical approach. Chichester UK: Ellis Horwood.

Campbell PGC. 1995. Interactions between trace metals and aquatic organisms: A critique of the free-ion activity model. In: Tessier A, Turner DR, editors. Metal speciation and bioavailability in aquatic systems. New York NY: Wiley. p 45–102.

[CCME] Canadian Council of Ministers of the Environment. 1991. Interim Canadian environmental quality standards for contaminated sites. Winnipeg: CCME.

Christman RF, Gjessing ET, editors. 1983. Aquatic and terrestrial humic materials. Ann Arbor MI: Ann Arbor Science.

Cunningham SD, Ow DW. 1996. Promises and prospects of phytoremediation. *Plant Physiol* 110:715–719.

Davis JA. 1984. Complexation of trace metals by adsorbed natural organic matter. *Geochim Cosmochim Acta* 48:679–69.

Dias F. 1997. Adsorption and desorption of arsenic in fifteen soils from New Jersey [M.S. thesis]. Newark DE: Univ Delaware.

Di Toro DM, Mahony JD, Hansen DJ, Scott KJ, Hicks MB, Mayr SM, Redmond MS. 1990. Toxicity of cadmium in sediments: The role of acid volatile sulfide. *Environ Toxicol Chem* 9:1487–1502.

Di Toro DM, Mahony JD, Kirchgraber RR, O'Byrne AL, Pasquale LR, Piccirilli DC. 1986. Effects of nonreversibilty, particle concentration, and ionic strength on heavy metal sorption. *Environ Sci Technol* 20:55–61.

Dzombak DA, Morel FMM. 1990. Surface complexation modeling; Hydrous ferric oxide. New York NY: Wiley-Interscience.

EA Engineering, Science, and Technology, Inc. 1994. Chemical translator guidance manual. Hunt Valley MD: EA Eng Sci Tech.

Fendorf SE. 1993. Oxidation and sorption mechanisms of hydrolyzable metal ions on oxide surface [PhD thesis]. Newark DE: Univ Delaware.

Fish W. 1993. Sub-surface redox chemistry: A comparison of equilibrium and reaction-based approaches. In: Allen HE, Perdue EM, Brown DS, editors. Metals in groundwater. Chelsea MI: Lewis. p 73–102.

Förstner U. 1995. Land contamination by metals: Global scope and magnitude of problem. In: Allen HE, Huang CP, Bailey GW, Bowers AR, editors. Metal speciation and contamination of soil. Boca Raton FL: Lewis. p 1–34.

Fu G, Allen HE, Cao Y. 1992. The importance of humic acids to proton and cadmium binding in sediments. *Environ Toxicol Chem* 11:1363–1372.

Fuller CC, Davis JA, Waychunas GA. 1993. Surface chemistry of ferrihydrite: Part 2. Kinetics of arsenate adsorption and coprecipitation. *Geochim Cosmochem Acta* 57:2271–2282.

Grundl T. 1995. Determination of redox status in sediments. In: Allen HE, editor. Metal contaminated aqueous sediments. Chelsea MI: Ann Arbor. p 149–159.

Hart BT. 1982. Uptake of trace metals by sediments and suspended particles: A review. In: Sly PG, editor. Sediment/freshwater interaction. Boston MA: Dr. W. Junk Publ. p 293–298.

Herbelin A, Westall J. 1994. FITEQL: A computer program for determination of chemical equilibrium constants from experimental data. Version 3.1. Corvallis OR: Oregon State Univ.

Hering JG, Morrel FMM. 1989. Slow coordination reactions in seawater. *Geochim Cosmochim Acta* 53:611–618.

Huang PM. 1997. Chemical speciation and bioavailability of soil cadmium in the Canadian Prairies: An overview. In: Iskandar IK, Hardy SE, Chang AC, Pierzynski GM, editors. Proceedings of extended abstracts from the Fourth International Conference on the Biogeochemistry of Trace Elements; Berkeley CA. p 97–98.

Jenne EA. 1988. Controls on Mn, Fe, Co, Ni, Cu, and Zn concentrations in soils and water: The significant role of hydrous Mn and Fe oxides. In: Baker RA, editor. Trace inorganics in water. Washington DC: American Chemical Society (ACS). Adv Chem Ser 73. p 337–387.

Jenne EA, Di Toro DM, Allen HE, Zarba CS. 1986. An activity-based model for developing sediment criteria for metals: Part I. A new approach. Proceedings of the International Conference Chemicals in the Environment; Lisbon, Portugal. p 560–568.

Kari FG, Hilger S, Canonica S. 1995. Determination of the reaction quantum yield for the photochemical degradation of Fe(III)-EDTA: Implications for the environmental fate of EDTA in surface waters. *Environ Sci Technol* 29:1008–1017.

Korcak RF, Fanning DS. 1978. Extraction of cadmium, copper, nickel, and zinc by double acid versus DTPA and plant content at excessive soil levels. *J Environ Qual* 7:506–512.

Korcak RF, Fanning DS. 1985. Availability of applied heavy metals as a function of type of soil materials and metal source. *Soil Sci* 140:23.

Lee SZ, Allen HE, Huang CP, Sparks DL, Sanders PF, Peijnenberg WJGM. 1996. Predicting soil-water partition coefficients for cadmium. *Environ Sci Technol* 30:3418–3424.

Lee SZ, Chen B, Allen HE, Huang CP, Sparks DL, Sanders P. 1993. Soil metal standards for groundwater protection. Proceedings of the 24th Mid-Atlantic Industrial and Hazardous Waste Conference. Lancaster PA: Technomic.

Ma H, Allen HE, Kim S, Cha D. 1999. Effect of kinetics of complexation by humic acid on the toxicity of copper to *Ceriodaphnia dubia*. *Environ Toxicol Chem* 18:828–837.

Mantoura RFC, Dickson A, Riley JP. 1978. The complexation of metals with humic materials in natural waters. *Estuar Coastal Mar Sci* 6:387–408.

Martell AE, Smith RM. 1993. NIST critical stability constants of metal complexes database. Gaithersberg MD: National Institute of Standards and Technology (NIST).

Martin JP, Nirel P, Thomas AJ. 1987. Sequential extraction techniques: Promises and problems. *Mar Chem* 22:313.

Mattigod SV, Sposito G. 1979. Chemical modeling of trace metal equilibria in contaminated soil solutions using the computer program GEOCHEM. In: Jenne EA, editor. Chemical modeling in aqueous systems. Washington DC: American Chemical Society (ACS). ACS Symp Ser 93. p 837–855.

McKnight DM, Feder GL, Thurman EM, Wershaw RL, Westall JC. 1983. Complexation of copper by aquatic humic substances from different environments. *Sci Total Environ* 28:65–76.

McLaughlin MJ, Smolders E, Merckx R, Maes A. 1997. Plant uptake of Cd and Zn in chelator-buffered nutrient solution depends on ligand type. In: Ando T, Fujita K, Mae T, Matsumoto H, Mori S, Sekija J, editors. Plant nutrition for sustainable food production and environment, XIIIth International Plant Nutrition Colloquium. Dordrecht, the Netherlands: Kluwer Academic. p 113–118.

Means JL, Crerar DA. 1978. Migration of radioactive wastes: Radionuclide mobilization by complexing agents. *Science* 200:1477–1481.

Means JL, Kucak T, Crerar DA. 1980. Relative degradation rates of NTA, EDTA, and DTPA and environmental implications. *Environ Pollut (Series B)* 1:45–60.

Moore JW. 1990. Inorganic contaminants of surface water. Research and monitoring priorities. New York NY: Springer-Verlag.

Morel FMM, Hering JG. 1993. Principles and applications of aquatic chemistry. New York NY: Wiley.

Natarajan P, Endicott JF. 1973. Photoredox behavior of transition metal-ethylenediamine-tetraacetate complexes. A comparison of some group VIII metals. *J Physicol Chem* 77:2049–2054.

Novozamsky I, Lexmond Th M, Houba VJG. 1993. A single extraction procedure of soil for evaluation of uptake of some heavy metals by plants. *Intern J Environ Anal Chem* 51:47–58.

Nriagu JO. 1988. A silent epidemic of environmental metal poisoning? *Environ Pollut* 50:139–161.

Nriagu JO, Larson G, Wong HKT, Azcue JM. 1993. A protocol for minimizing contamination in the analysis of trace metals in Great Lakes waters. *J Great Lakes Res* 19:175.

Nriagu JO, Pacyna JM. 1988. Quantitative assessment of worldwide contamination of air, water and soils by trace metals. *Nature* 333:134–139.

Pardue JH, Patrick Jr WH. 1995. Changes in metal speciation following alteration of sediment redox status. In: Allen HE, editor. Metal contaminated aqueous sediments. Chelsea MI: Ann Arbor. p 160–186.

Patterson CC, Settle DM. 1976. The reduction of orders of magnitude errors in lead analyses of biological materials and natural waters by evaluating and controlling the extent and sources of industrial lead contamination introduced during sample collecting and analysis. In: LaFleur PD, editor. Accuracy in trace analysis: Sampling, sample handling, analysis. National Bureau of Standards. NBS Special Publication 422:321.

Pearson RG. 1963. Hard and soft acids and bases. *J Am Chem Soc* 85:3533–3539.

Perdue EM, Charles RL. 1983. Distribution model for binding of protons and metal ions by humic substances. *Environ Sci Technol* 17:654–660.

Playle RC, Dixon DG, Burnison DG. 1993a. Copper and cadmium binding to fish gills: Modification by dissolved organic carbon and synthetic ligands. *Can J Fish Aquat Sci* 50:2667–2677.

Playle RC, Dixon DG, Burnison DG. 1993b. Copper and cadmium binding to fish gills: Estimates of metal-gill stability constants and modeling of metal accumulation. *Can J Fish Aquat Sci* 50:2678–2687.

Prothro MG. 1993. Office of water policy and technical guidance on interpretation and implementation of aquatic life metals criteria. Washington DC: U.S. Environmental Protection Agency (USEPA).

Reimers RS, Krenkel PA. 1974. Kinetics of mercury adsorption and desorption in sediments. *J Water Pollut Control Fed* 46:352–365.

Schecher WD, McAvoy DC. 1991. MINEQL$^+$: A chemical equilibrium program for personal computers. Edgewater MD: Environmental Research Software.

Schnitzer M. 1978. Humic substance: Chemistry and reactions. Amsterdam, the Netherlands: Elsevier/North Holland.

Sedlak DL, Phinney JT, Bedworth WW. 1997. Strongly complexed Cu and Ni in wastewater effluents and surface runoff. *Environ Sci Technol* 31:3010–3016.

Shi B. 1996. Fate and effects of metals in aquatic environment: Evaluation for POTW effluent discharges and metal partitioning in surface water [PhD thesis]. Newark DE: Univ Delaware.

Sigg L, Xue HB. 1994. Metal speciation: Concepts, analysis and effects. In: Bidoglio G, Stumm W, editors. Chemistry of aquatic system: Local and global perspectives. Boston MA: Kluwer Academic. p 153–181.

Sommers L, Volk VV, Giordano PM, Sopper WE, Bastian R. 1987. Effect of soil properties on accumulation of trace elements by crops. In: Page AL, Logan TJ, Ryan JA, editors. Land application of sludge: Food chain implications. Chelsea MI: Lewis.

Sparks D. 1995. Kinetics of metal sorption reactions. In: Allen HE, Huang CP, Bailey GW, Bowers AR, editors. Metal speciation and contamination of soil. Boca Raton FL: Lewis.

Sparks DL. 1989. Kinetics of soil chemical processes. New York NY: Academic Pr.

Sparks DL. 1992. Soil kinetics. *Encyclopedia Earth Sys Sci* 4:219–227.

Sposito G. 1984. The surface chemistry of soils. New York NY: Oxford Univ Pr.

Sposito G. 1985. Chemical models of inorganic pollutants in soils. *CRC Crit Rev Environ Control* 15:1–14.

Sposito G. 1986. Sorption of trace metals by humic materials in soils and natural waters. *CRC Crit Rev Environ Control* 16:193–229.

Stevenson FJ. 1982. Humus chemistry. New York NY: Wiley.

Stumm W. 1992. Chemistry of the solid-water interface: Processes at the mineral-water and particle-water interface in natural systems. New York NY: Wiley.

Tessier A, Campbell PGC, Bisson M. 1979. Sequential extraction procedure for the speciation of particulate trace metals. *Anal Chem* 51:844–851.

Tipping E. 1994. WHAM: A chemical equilibrium model and computer code for waters, sediments, and soils incorporating a discrete site/electrostatic model of ion-binding by humic substances. *Comput Geosci* 20:973–1023.

Tipping E, Hurley MA. 1992. A unifying model of cation binding by humic substances. *Geochem Cosmochim Acta* 56:3627–3641.

[USEPA] U.S. Environmental Protection Agency. 1979. 40 CFR 257. *Federal Register* 44:53460.

[USEPA] U.S. Environmental Protection Agency. 1986a. National priority list fact book, HW 7.3. Washington DC: USEPA.

[USEPA] U.S. Environmental Protection Agency. 1986b. Quality criteria for water. Washington DC: Office of Water. EPA 440/5-86-001.

[USEPA] U.S. Environmental Protection Agency. 1992. Interim guidance on interpretation and implementation of aquatic life criteria for metals. Washington DC: USEPA.

[USEPA] U.S. Environmental Protection Agency. 1994a. Equilibrium partitioning approach to predicting metal bioavailability in sediments and the derivation of sediment quality criteria for metals, Volume 1, Briefing report to the Science Advisory Board. Washington DC: Office of Water. EPA 822-D-94-002.

[USEPA] U.S. Environmental Protection Agency. 1994b. Method 1669, Sampling ambient water for determination of trace metals at EPA water quality criteria levels, Washington DC: Engineering and Analysis Division (4303).

van den Berg CMG, Nimmo M, Daly P, Turner DR. 1990. Effects of the detection window on the determination of organic copper speciation in estuarine waters. *Anal Chim Acta* 232:149–159.

Van Riemsdijk WH, Hiemstra T. 1993. Adsorption to heterogeneous surfaces. In: Allen HE, Perdue EM, Brown DS, editors. Metals in groundwater. Chelsea MI: Lewis. p 1–36.

Vuceta J, Morgan JJ. 1978. Chemical modeling of metals in fresh waters: Role of complexation and adsorption. *Environ Sci Technol* 12:1302–1308.

Warwick P, Hall A, Read D. 1994. Proceedings of an NEA Workshop on Binding Models Concerning Natural Organic Substances in Performance Assessment; Bad Zurzach, Switzerland. p 109–136.

Wenzel WW, Blum WEH. 1995. Assessment of metal mobility in soil: Methodological problems. In: Allen HE, Huang CP, Bailey GW, Bowers, AR, editors. Metal speciation and contamination of soil. Boca Raton FL: Lewis.

Westall JC, Hohl H. 1980. A Comparison of electrostatic models for the oxide solution interface. *Adv Colloid Interf Sci* 12: 265–294.

Westall JC, Jones JD, Turner GD, Zachara JM. 1995. Models for association of metal ions with heterogeneous environmental sorbents. 1. Complexation of Co(II) by Leonardite humic acid as a function of pH and $NaClO_4$ concentration. *Environ Sci Technol* 29:951–959.

Whitten MC, Ritchie GSP. 1991. Calcium chloride extractable cadmium as an estimate of cadmium uptake by subterranean clover. *Aust J Soil Res* 29:215–221.

Windom HL, Byrd JT, Smith Jr. RG, Huan F. 1991. Inadequacy of NASQAN data for assessing metal trends in the nation's rivers. *Environ Sci Technol* 25:1137–1142.

Wu Y, Zhou Q, Adriano DC. 1991. Interim environmental guidelines for cadmium and mercury in soils of China. *Water Air Soil Pollut* 57–58:733–743.

Xue H, Sigg L, Kari FG. 1995. Speciation of EDTA in natural waters: Exchange kinetics of Fe-EDTA in river water. *Environ Sci Technol* 9:1035–1038.

Yin Y. 1996. Adsorption and desorption reactions of mercury(II) and methylmercury at the soil-solution interface [PhD thesis]. Newark DE: Univ Delaware.

Yin Y, Allen HE. 1997. Adsorption of mercury(II) and methylmercury in sodium and calcium electrolytes. In: Iskandar IK, Hardy SE, Chang AC, Pierzynski GM, editors. Proceedings of extended abstracts from the Fourth International Conference on the Biogeochemistry of Trace Elements; Berkeley CA. p 271–272.

Yin Y, Allen HE, Huang CP, Li Y, Sanders PF. 1996. Adsorption of Hg(II) by soil: Effects of pH, chloride concentration, and organic matter. *J Environ Qual* 25:837–844.

Yin Y, Allen HE, Huang CP, Sanders PF. 1997a. Interaction of Hg(II) with soil-derived humic substances. *Anal Chim Acta* 341:73–82.

Yin Y, Allen HE, Huang CP, Sanders PF. 1997b. Effects of pH, chloride, and calcium(II) on adsorption of methylmercury by soils. *Environ Toxicol Chem* 16:2457–2462.

Yin Y, Allen HE, Huang CP, Sanders PF. 1997c. Adsorption/desorption isotherms of Hg(II) by soil. *Soil Sci* 162:35–45.

Yin Y, Allen HE, Huang CP, Sparks DL, Sanders PF. 1997. Kinetics of Hg(II) adsorption and desorption on soil. *Environ Sci Technol* 31:496–503.

Zachara JM, Smith SC, McKinley JP, Resch CT. 1993. Cadmium sorption on specimen and soil smectites in sodium and calcium electrolytes. *Soil Sci Soc Am J* 57:1491–1501.

10

Natural Remediation of Radionuclides

C. Marjorie Aelion

RADIONUCLIDES ARE ELEMENTS that emit energy, or radiation, during their natural decay. Of the 340 natural and manmade elements, or nuclides, approximately 70 are radionuclides. Radionuclides have been present on the earth since it was formed, occur in geologic formations, and are naturally produced in the atmosphere. With the nuclear age, anthropogenic radionuclides have been produced for military, energy, and research and development applications. Radionuclides occur in the environment as the result of releases from natural sources, nuclear fuel and energy production including mining activities and oil and gas production, nuclear testing and nuclear accidents, and nearly a century of disposal of radioactive wastes.

Radionuclides are unique among environmental contaminants because their natural spontaneous decay is accompanied by a release of energy proportional to the mass of the particle released. This energy is in the form of ionizing radiation. As ionizing radiation passes through biological tissues, high-energy free radicals are created. Tissue interaction with free radicals results in chemical reactions that can damage living cells. Functional and morphological changes then occur, induced directly or indirectly by the radiation exposure (Severa and Bár 1991).

Ionizing radiation has enough energy to change the atomic balance of substances (e.g., tissues) through which it moves. These changes are manifested in different ways and can cause several types of damage at all biological levels:

- For example, at the molecular level, enzymes, ribonucleic acids (RNAs) and deoxyribonucleic acids (DNAs) can be altered and impaired, which in turn can impact metabolic processes.

Natural Remediation of Environmental Contaminants. Michael Swindoll et al., editors.
 ISBN 1-880611-33-3.

- At the subcellular level, damage to cellular membranes, cell nuclei, chromosomes, mitochondria, and lysosomes can occur.
- At the cellular level, blocking of cell division, cell death, and malignant transformations of cells can occur.
- At the tissue and organ level, systemic breakdown of the central nervous system, blood, and gastrointestinal tract can occur, and malignancy can be induced.
- At the level of the whole organism, death and shortening of life span can occur.
- Finally, at the population level, alterations in genetic characteristics and genetic and chromosomal mutations at the individual level can occur (Severa and Bár 1991).

Scientific inquiry into the biological impacts of radionuclides has focused primarily on human receptors, with much less attention paid to ecological effects. Those environmental studies that have been conducted generally fall into 2 categories: those that monitor the concentrations of radionuclides in environmental media (e.g., air, land, water, and biota) and those that investigate the chemistry of radionuclides with respect to their fate and transport in the environment. Although it is apparent that ecological impacts may result from both natural and anthropogenic radionuclides, much remains to be understood, and the picture for environmental contamination and ecological risk of radionuclides is far from complete.

Decades of radionuclide production, radiological waste disposal, and nuclear testing and accidents have resulted in widespread radioactive contamination. In many cases, active remediation is not technically or economically feasible, or is not necessary from a risk perspective. Management of radiological materials and contamination is complicated by several factors. First, many of the radionuclides emit sufficient energy that they are of danger to biota, even at very low levels. Second, radionuclides with long half-lives represent potential long-term environmental problems. In addition, radionuclides and radioactive wastes are not readily destroyed by conventional treatment methods such as biodegradation, chemical complexation, or incineration. Presently, management alternatives for radioactive materials are limited primarily to time, isolation, or natural attenuation. Time, as a natural remediation (NR) factor, is unique to the radionuclides and is significant for risk management only for those radionuclides that decay relatively quickly, that is, have short half-lives. Isolation shields and distances biological resources and humans from radioactive material, and it is typically accomplished through land burial or vitrification. Natural attenuation as a management alternative has been practiced for decades with the disposal of radioactive materials into the environment, primarily into air and oceans. This management alternative takes advantage of NR processes, primarily dilution, which can contribute to the reduction of potential acute risks posed by these contaminants. However, dilution does not remove or convert the radionuclide to a nonhazardous form.

This chapter addresses the processes that affect the fate and ecological impact of radionuclides. Once released into the environment, chemical, physical, and biological interactions and reactions between the radionuclide and the environment all play a role in the fate, transport, and biological impact of radionuclides. Because these impacts are affected by properties of both the radionuclides and the environment, environmental conditions at the time of release often determine the ultimate distribution of radionuclides in the environment. Atmospheric conditions and hydrology and hydraulics are the major factors that control atmospheric and aquatic transport. Atmospheric, aquatic, and soil and sediment chemistry are the major controls over chemical reactions and bioavailability. Interactions with forests, plants, microbes, and animals impact biological interception of deposited particles and biological uptake and bioaccumulation of radionuclides. With an understanding of these interactions between radionuclides and the environment, risk management decisions can be tailored to provide the greatest protection of ecological resources from the potential impacts of this unique group of contaminants.

CHEMISTRY OF RADIONUCLIDES

To assess the fate and potential ecological impact of radionuclides, we must understand their chemistry. As with all elements, the nucleus of the atom consists in part of neutrons and protons that are present in a ratio specific to the element. Usually when the neutron-to-proton ratio is too large, the nucleus changes its state by emitting either a particle or electromagnetic radiation. This nuclear disintegration of the parent compound is termed "radioactivity," and the isotope that changes its nuclear state is called a "radioisotope" or "radionuclide." Isotopes, also known as "nuclides," are any of 2 or more species of atoms of a chemical element with the same atomic number and position in the periodic table and nearly identical chemical behavior, but different atomic mass or mass number and physical properties. Some isotopes are radioactive, while others are not. The terms "radioisotope" and "radionuclide" are used to distinguish radioactive from nonradioactive (or stable) forms of an isotope. For example, ^{12}C and ^{13}C are the stable and most abundant forms of the C isotope, and ^{14}C (the radiocarbon often used for archeological dating) is the radioactive form of C. All have atomic number of 6 but differ in their atomic mass (12, 13, and 14 atomic mass units, respectively).

The number of protons in an atom corresponds to the atomic number. The number of neutrons equals the atomic weight expressed as the nearest whole number, minus the number of protons (or atomic number). All nuclides with atomic numbers greater than 83, or atomic weights greater than 200, are unstable and prone to radioactive decay. In addition, stable atoms can be converted, either naturally or artificially, to radioactive ones by bombardment with neutrons. In the

International Union of Pure and Applied Chemists' (IUPAC) nomenclature, the atomic mass (neutrons plus protons) is the superscript used to distinguish between isotopes of an atom, and the atomic number (number of protons) is the subscript. For example, lead has 4 isotopes, $^{204}_{82}Pb$, $^{206}_{82}Pb$, $^{207}_{82}Pb$ and $^{208}_{82}Pb$, each of which has 82 protons but a different number of neutrons, 122, 124, 125, and 126, respectively.

For different radioisotopes, the half-life ($t_{1/2}$), or the amount of time for the radioisotope to reach one-half its concentration by radioactive decay, varies considerably (Table 10-1). There are both short- and long-lived naturally occurring radionuclides and artificially created radionuclides. The energy emitted by short-lived radionuclides attenuates so quickly that they are typically not a threat to ecological receptors, although under certain circumstances they can present a risk to human health. In contrast, the long-lived radionuclides emit energy for sufficient duration to potentially pose a threat of adverse impact to ecological receptors. As such, the long-lived radionuclides are the primary focus of this chapter.

Types of emitters

As radionuclides decay, they can emit 3 types of ionizing radiation, termed "alpha" (α), "beta" (β), and "gamma" (γ) radiation (Table 10-1). Alpha particles are positively charged matter, have considerable mass, and move at 10% of the speed of light. Because they are big and slow, they do not penetrate material and are stopped by an ordinary piece of paper. The α particle is the same as a nucleus of a He atom, containing 2 protons and 2 neutrons. The product formed from α particle decay is a new element containing 2 fewer protons than the original parent atom (Davis and Cornwell 1991). Alpha particles are considered the most benign of the emitters because they are relatively easy to stop physically. However, if taken internally through ingestion or respiration, the particles can cause severe biological damage because they have a large amount of ionizing power due to their large mass. A typical example is U decay to Th and the release of an α particle.

$$^{238}_{92}U \rightarrow\ ^{234}_{90}Th +\ ^{4}_{2}He\ (\alpha \text{ particle})$$

Beta particles are negatively charged, have small mass, and move at 33% to 99% the speed of light. Emission of the β particle also changes the parent atom into a new element in which the number of protons is increased by 1, relative to the parent atom. Thus, when an element emits a β particle, the resulting product is a different element that is one place to the right of the parent element on the periodic table. Beta particles are electrons and travel several hundred feet through air; they can be stopped by Al that is a few millimeters thick. Because of the lower mass, β particles have less ionizing power than α particles but require greater shielding to protect biota. Thus, they have a greater impact from external exposure but less impact from internal ingestion than do α particles. A typical example is K decay to Ca and the release of a β particle.

Table 10-1 Example radioisotopes and their half-lives

Radioisotope	Half-life[a]	Major emission[b]	Major source
^{212}Polonium	3.04×10^{-7} s	α	Natural, Th-series
^{222}Radon	3.825 d	α, < 1% γ	Natural, U-Series
^{131}Iodine	8.04 d	β, γ	Fission, weapons
^{32}Phosphorus	14.3 d	β	Natural, cosmic
^{210}Polonium	138.4 d	α, < 1% γ	Natural, U-series
^{60}Cobalt	5.27 y	β, γ	Nuclear weapons, reactors
^{3}Hydrogen	12.33 y	β	Natural, cosmic, fission
^{90}Strontium	28.8 y	β	Weapons
^{137}Cesium	30.17 y	β, γ	Fission
^{241}Americium	432.2 y	α, γ	Decay of 241-Pu
^{226}Radium	1,622 y	α, γ	Natural, U-series
^{14}Carbon	5,730 y	β	Natural, cosmic, nuclear weapons
^{239}Plutonium	2.4×10^{4} y	α	Fission, nuclear weapons
^{234}Uranium	2.5×10^{5} y	α, < 1% γ	Natural, U-series
^{235}Uranium	7.04×10^{8} y	α, γ	Natural, Ac-series
^{40}Potassium	1.28×10^{9} y	β, γ	Natural, primordial, non-series
^{238}Uranium	1.25×10^{10} y	α, < 1% γ	Natural, U-series

[a] Sources: Davis and Cornwell 1991; Blackman 1996.
[b] Source: Eisenbud and Gesell 1997.

$$^{40}_{19}\mathrm{K} \rightarrow {}^{40}_{20}\mathrm{Ca} + {}^{0}_{-1}\mathrm{e}\ (\beta \text{ particle})$$

Gamma radiation is true electromagnetic radiation (not particles) and travels at the speed of light. The unit of γ radiation is termed a "photon," and γ radiation has no electrical charge. When γ rays are emitted, the element's nucleus retains its original composition. There is no change in the size of the nucleus or in the number of protons or neutrons. Gamma rays originate when the nucleus transfers from a higher to a lower atomic energy state, and they have properties, such as frequencies and amplitudes, that are associated with other rays (e.g., x-rays, radio waves, and light waves). The penetrating power of rays increases as the wavelength decreases. Compared to other rays, γ rays have the greatest impact on ecological systems because of their short wavelength and great penetrating power. Several centimeters of Pb are required to shield and protect biota from γ radiation. A typical example is the decay of Co and the release of a γ ray.

$$^{60}_{27}Co \rightarrow ^{60}_{27}Co + \gamma\,(\gamma \text{ ray})$$

The natural decay of a radionuclide may result in the release of more than one type of emitter. Two or more emissions are possible from the decay of a radionuclide, or several particles of the same type but different energies can be emitted (Davis and Cornwell 1991). Many of the products of nuclear and military industries are γ emitters as well as β emitters. A typical example of a multiple emission is the decay of Co to Ni and the release of both a β particle and a γ ray.

$$^{60}_{27}Co \rightarrow ^{60}_{28}Ni + ^{0}_{-1}e\,(\beta \text{ particle}) + (\gamma \text{ ray})$$

Radioactive elements spontaneously emit α and β particles or γ rays by the disintegration of the nuclei of atoms, and a measure of their radioactivity is the number of disintegrations that occurs per unit time (minute or second). One curie (Ci) was defined in 1912 as the number of radioactive disintegrations per second undergone by 1 gm of Ra (equals 3.7×10^{10} disintegrations per second). One becquerel (Bq), the SI unit for radioactivity, equals 1 disintegration per second.

Of all the types of ionizing radiation, γ and β radiation represent a significant potential risk to ecological resources from external exposure. Alpha particles, which have the lowest energy, are of little environmental concern from external exposure because they are essentially unable to penetrate tissue. Beta particles can penetrate tissue, and γ rays have extensive penetrating power. However, to assess potential impact of radionuclides to the biosphere, we must know the type of emitter, the energy associated with the emission, the half-life, and how the exposure occurs, whether internal (respiration or ingestion) or external.

SOURCES OF RADIATION

There are 2 major sources of ionizing radiation in the environment, those that are naturally occurring and those that are anthropogenic (Table 10-1). Naturally occurring radionuclides are present in the atmosphere and in geological formations. Radioisotopes naturally produced in the atmosphere by cosmic ray bombardment of N, O, and Ar include, among others, tritium (^{3}H), beryllium (^{7}Be), carbon (^{14}C), sodium (^{22}Na), phosphorus (^{32}P and ^{33}P), and sulfur (^{33}S) (Appleby et al. 1993). Although cosmic radiation is highly penetrating and is represented by radionuclides with half-lives that vary from relatively short (12.3 y for ^{3}H) to long (5,730 y for ^{14}C), its impacts are restricted primarily to the stratosphere above an altitude of about 25 km (Eisenbud and Gesell 1997). Thus, the ecological impact of cosmic radiation is relatively minor.

Natural radionuclides in the earth's crust are either primordial, with half-lives comparable to the age of the earth, or were generated by the decay of long-lived precursors (Appleby et al. 1993). Primordial radionuclides are either nonseries (creating a stable daughter product during decay) or series (creating a chain of short-lived radioactive daughter products during decay). The 3 series radionuclides include ^{238}U, ^{232}Th, and ^{235}U for the uranium series, the thorium series, and the actinium series, respectively. The largest geologic sources of radiation are ^{40}K, which is a natural nonseries radioisotope, and the radioactive decay chain of ^{238}U and ^{232}Th (Bou-Rabee 1997). Their world average concentrations in the soil are estimated to be 370 Bq/kg (^{40}K), 40 Bq/kg (^{238}U), and 40 Bq/kg (^{232}Th) (United Nations Scientific Committee on the Effects of Atomic Radiation [UNSCEAR] 1993). Relative percentages of the U radionuclides found in nature are 0.0058% (^{234}U), 0.71% (^{235}U), and 99.28% (^{238}U).

Certain naturally occurring radioisotopes are emitted from soil as gaseous decay products. Radon (^{222}Rn) is one of the most prevalent naturally occurring radioactive isotopes. ^{222}Rn is the gaseous decay product of radium (^{222}Ra), which is a member of the ^{238}U series. ^{226}Radium and its decay products are responsible for a great deal of human exposure to radioactivity from natural sources (National Council on Radiation Protection and Measurements [NCRP] 1987; Muller et al. 1989; Ottoboni 1991; Harley 1992). Many of these gaseous radionuclides are α emitters with great ionizing energy. This produces an internal respiratory hazard important to human health (Harley 1992) and may produce lung cancer, but it poses less of an ecological risk from external exposure.

The release of natural sources of radioactivity is enhanced by human activities such as mining and energy production. Such activities remove radionuclides from geologic formations and place them on land or in the ocean, often chemically concentrating them. Mining, concentrating, and purifying the radioactive fuel for nuclear power facilities produce radioactive wastes that are deposited on the earth's surface. Offshore production of oil and gas may lead to removal of radionuclides from geological formations that subsequently are deposited in the world's oceans. Most of the world's oil fields have formed over organic shales, whose uraniferous phosphate content can raise U concentrations to 1000 mg/kg or more (Appleby et al. 1993). ^{226}Ra and its daughter products are generated by the U in shale and can co-precipitate after temperature and pressure changes associated with the oil production and the brine water. Radioactive deposits (scale) on equipment can contain from 1 to more than 15 Bq/kg of radioactivity. In instances where active waste disposal is required for this equipment, disposal is typically at sea. With offshore oil production, natural attenuation (i.e., dilution and subsequent sequestration in sediments) is the primary treatment for these radioactive materials. Studies on the adverse impacts of radionuclides have tended to focus on human health, with less focus on ecological impacts. In some of the cases mentioned above, as in offshore oil production, there is limited human interaction because these areas are remote. Because of this, the ecological consequences tend to be ignored and are not well studied. However, biological receptors in the vicinity may be at potential risk.

Artificial radionuclides are created by the nuclear power and military industries. Nuclear energy production is based on fission in which the nucleus of an atom is bombarded by neutrons, which splits the atom into 2 smaller atoms. The fission process releases radioactive products, additional neutrons, and energy. The additional neutrons are available for further bombardment of the nuclear fuel. The majority of artificial radionuclides found in the environment occurs as a result of nuclear weapons production and testing, nuclear accidents, nuclear processing from mining, fuel refining, power production and spent fuel reprocessing, and from radioactive waste disposal to oceans, rivers, land, and the atmosphere. Some of these practices have become less frequent, as is the case with nuclear testing. Other releases are unpredictable, such as nuclear accidents. Several countries have left environmental legacies regarding radiological contamination from accidental releases (Shcherbak 1996; Segerstahl et al. 1997), nuclear weapons production (Luttrell 1993), and waste disposal to oceans (Chung et al. 1986; Luttrell 1993), rivers and lakes (Zorpette 1996), and land (Luttrell 1993).

Technologies have not been developed to deal with planned and accidental releases, and these artificial radionuclides often have long half-lives, making waste management by radioactive decay impossible. This contrasts to radionuclides used in nuclear medicine, which have short half-lives (^{99}Mo, $t_{1/2}$ = 67 h, and iodine ^{131}I, $t_{1/2}$ = 8 d), can be stored properly, allowed to decay 10 half-lives to background levels, and then disposed of as nonradioactive waste. Long-lived artificial radionuclides measured in the environment include plutonium (^{239}Pu, $t_{1/2}$ = 24,000 y) and uranium (^{234}U, $t_{1/2}$ = 250,000 y), as well as several other radionuclides of varying half-lives, such as ^{3}H, strontium (^{90}Sr), rubidium (^{103}Ru, ^{106}Ru), tin (^{113}Sn), iodine (^{129}I), cesium (^{137}Cs), promethium (^{147}Pm), U (total), ^{238}Pu, ^{239,240}Pu, americium (^{241}Am), and gross α and β radiation (Luttrell 1993). The occurrence of long-lived radionuclides in the environment because of nuclear weapons production and testing, nuclear power generation, nuclear accidents and nuclear waste disposal, and direct placement of radioactive wastes in oceans, rivers, and lakes contributes to global radioactive contamination and represents a serious threat to the biosphere.

FATE AND TRANSPORT OF RADIONUCLIDES

Radionuclides are found in all parts of the environment, including the atmosphere, aquatic and terrestrial systems, and throughout the biosphere. As with other chemicals, the fate and transport of radionuclides in the environment results from the interaction of physical, chemical, and biological processes. Similarly to metals in general, and unlike organic compounds, these processes do not change the chemical into a benign end product. The radionuclides are still radioactive.

Physical processes such as advection and dispersion can transport radionuclides long distances in the atmosphere and oceans and can dilute radioactive contamination over large areas. Transport also occurs on a smaller scale. For example, U mining activities have contributed to radiological contamination of the land surface and the atmosphere. The radioactive sediment of tailings piles, the crushed waste rock from the mining process, can disperse particles into the atmosphere or be eroded by water or wind action.

The environmental transport of radionuclides is a function of their physical form (i.e., complexed, particle, vapor) at the time they are introduced into the environment. Artificial radionuclides are emitted to the atmosphere as inert gases, as reactive species, as small particles, or as attachments to coarse particles. For example, the artificial radionuclides resulting from nuclear weapons testing are made up of relatively water soluble ^{137}Cs and ^{90}Sr. However, ^{137}Cs and ^{90}Sr emanating from the 1986 accident at the Russian nuclear power plant in Chernobyl were present in nonexchangeable chemical forms, and some were present as fuel particles that were insoluble in water (Bell et al. 1993). Thus the same ^{137}Cs and ^{90}Sr radionuclides will be transported differently in these 2 cases because the chemical form from which they originated was different.

Soils and aquatic systems become contaminated with radionuclides through direct application of radioactive waste to land and water, through sedimentation of radionuclides from water to sediment, and as a result of atmospheric fallout. Chemical processes such as sorption may immobilize the contaminants in soils and sediments, and chemical properties may be most important for migration in sediment. This includes chemical environmental conditions such as soil pH, redox conditions, types of minerals and colloids present for sorption reactions to occur, and the reversibility of sorption–desorption reactions. Complexation of many radioactive elements occurs with humic substances and colloids.

Biological processes can also impact radionuclide fate and transport. Biota may intercept radionuclides from atmospheric fallout, and biological processes such as uptake by plants may remove radionuclides from the soil and redeposit them on the land surface when the plants die. Biological sequestration from sediment may depend on how strongly the radionuclides are sorbed or whether they are chelated and complexed to soluble organic matter in aquatic systems. In contrast to the high-energy environment of coastal zones for which radionuclide transport is dominated by physical processes, transport of radionuclides in the ocean is more impacted by biological processes. As radionuclides are adsorbed and sequestered in oceanic particles and biogenic debris, they are consumed by zooplankton and transferred to sediment as fecal pellets in a few months, or they are transferred by aggregation of phytoplankton and subsequent fallout (Aarkrog et al. 1993).

Physical transport

The atmosphere is a large reservoir for artificial radionuclides. The atmosphere can provide rapid dispersal of radionuclides long distances from their release, primarily by advection and diffusion. Advection is a function of winds at the earth's surface. Mixing characteristics are influenced by vertical temperature gradients, turbulent eddies, and wind shear. Near-surface radionuclide dust may be transported and diffused over distances of tens to hundreds of kilometers, with a mean residence time in the troposphere that ranges from 5 days for lower rain-bearing regions to 40 days in higher altitudes (Eisenbud and Gesell 1997). Mean residence time of stratospheric aerosols ranges from less than 0.5 to 3 years in the middle of the stratosphere and from 5 to 10 years at 100 km (ApSimon et al. 1993). The use of nuclear weapons and nuclear testing has expanded the radionuclide occurrence to the upper atmosphere over 30 km (Eisenbud and Gesell 1997). Since 1962 and 1963 when atmospheric ^{14}C, one of the radionuclides released during nuclear testing, hit an all-time high concentration, natural attenuation has resulted in a slow decline to the current levels because of dilution in this large-scale system. Deposition to land and oceans also has contributed to ^{14}C decline in the atmosphere.

The ocean represents a second large reservoir for radionuclide contamination. Waterborne radionuclides can be dispersed by the ocean current systems (Aarkrog et al. 1993). For example, radionuclides from northern Europe deposited in the northeast Atlantic Ocean are dispersed by currents from the Irish Sea and the Baltic Sea, which move north along the west coast of Norway and then in the direction of the Siberian north coast, finally entering the Arctic Ocean. Since the 1960s, ^{90}Sr and ^{137}Cs have been measured at stations in the northeast Atlantic at the Faroe Islands and the coast of East Greenland. Similarly, in the Gulf of Mexico, waste discharges from offshore drilling have increased the amount of ^{226}Ra, ^{228}Ra, and ^{210}Pb in the water and biota (Meinhold and Holtzman 1997). However, radionuclides released were found to be dispersed quickly to ambient levels within 5 m from discharge.

Radionuclides can be transported from the atmosphere and the sea to land, and subsequent to deposition on land, wind action can cause resuspension of the particles into the atmosphere. The solubility, molecular weight, and reactivity of each chemical in the atmosphere and the ocean will help define its fate and transport in the environment. Radionuclides originally discharged to sea from the nuclear fuel reprocessing plant in Sellafield, West Cumbria, England, were transported to land in sea spray from surf breaking on shore and were subsequently deposited (Pattenden et al. 1983). $^{238, 239, 240}Pu$ present in the sea spray were readily adsorbed to fine particles, which were scavenged by the bubbles of the sea spray, incorporated into droplets, and ultimately carried on-shore by winds. In contrast, Cs remained predominantly in solution in the seawater and was not transported to land.

Aerosols in the troposphere can be distributed by winds and deposited back to the earth's surface. Particulate matter and gases are removed and deposited on the earth's surface by a variety of processes, including nucleation scavenging; gas and particle diffusion to cloud and rain droplets; fallout from rain, sleet, and snow; and dry deposition from diffusion and gravitational settling (ApSimon et al. 1993). For atmospheric radionuclide releases in arid environments with limited vegetation, soil and sediments are the major repository of radionuclides such as Pu, Cs, and Sr. Fallout from atmospheric inputs of ^{90}Sr and ^{137}Cs into the upper layers of soil in western European pastures has been investigated for 30 to 40 years. It is estimated that about 66% of the fallout radioactivity remains in the upper 10 cm of the soil (Frissel and Pennders 1983). For releases in more temperate, lush environments, rainfall can contribute to total deposition, and the vegetation is an important environmental medium to intercept and take up radionuclides. The type of surface of the vegetation is an important factor to radionuclide fate, as is the season in which the radionuclides are released. Forests are one of the main filters for atmospheric contamination by radionuclides.

Chemical reactions

Chemical processes affect the fate and transport of radionuclides, often decrease a radionuclide's mobility and bioavailability to plants and animals, and thus can affect potential ecological risk. Similarly to metals, radionuclides are not destroyed but are transformed by chemical reactions including sorption, complexation, isotopic and ion exchange reactions, precipitation and co-precipitation (Lu et al. 1998). Migration of radionuclides in soil is primarily a function of how tightly they are sorbed onto solids in sediment, water, or air. Sorption may be the ultimate fate of even atmospheric contamination by radionuclides. For example, more than 95% of the ^{137}Cs and $^{239,\ 240}Pu$ that has entered Lake Michigan as atmospheric fallout is estimated to be sorbed to sediments (Aarkrog et al. 1993).

Sorption in sediment and aquatic systems, which generally decreases a radionuclide's mobility in the environment (Beneš 1982), is often expressed through an experimentally derived distribution coefficient (K_d), defined as the equilibrium distribution of a chemical between the solid and solution phases (Miettinen et al. 1982). In general, Pu and Am essentially do not migrate in soil because of sorption, that is, they have a high K_d value of approximately 10^3 to 10^4 in clay soils. ^{137}Cs migration in general is also minimal, with a K_d value of approximately 10^3 in clay soils. Neptunium and ^{90}Sr migrate slowly over short distances (1 cm/y), with a K_d value of approximately 5×10^1 to 5×10^2 in clay soils (Bell et al. 1993).

In sediment and water, most uncomplexed radionuclides are sorbed onto surfaces. The strength of this association is determined by the type of solid surface to which radionuclides are chemically and physically bound (e.g., sand, clay, organic matter, colloids, or larger natural particles). Thus K_d values can be used to estimate and compare relative sorption and subsequent migration of different radionuclides, but environmental factors such as pH or salinity also must be considered.

Based on measurements in the environment, Cs ions were found to be strongly adsorbed within the clay mineral platelets of illite, a group of clay minerals having a crystalline structure. In general, the mobility and bioavailability of ^{137}Cs decreased with increasing soil organic matter and increasing clay content. Because Cs was more strongly absorbed, it migrated less than Sr, although Sr ions also were incorporated into the crystal lattice of soil minerals, forming a strong chemical bond (Bell et al. 1993). Following Chernobyl, a 30% to 90% decrease in Cs bioavailability from soil to plants was reported after 1 year of deposition. After 10 years, only approximately 10% of the original Cs in soil remained bioavailable.

Once bound to sediment, the release of sediment-bound radionuclides is due to adsorption–desorption reactions. The kinetics of adsorption–desorption reactions determine whether the overall outcome is radionuclide release or immobilization. Adsorption–desorption reactions, and thus the K_d value, are impacted by environmental conditions such as type of sediments, salinity, pH, and redox of the environment, which affects the speciation and oxidation state of the specific radionuclides. For example, in laboratory leaching experiments, ^{90}Sr adsorption and desorption were reversible. Desorption of ^{137}Cs was a function of the type of material to which it was sorbed. Desorption was greatest for humin (mineral-free), lower for the intact sediment, and lowest for the humin plus mineral (or clay) fraction (Rigol et al. 1998). So, strong complexes are formed with humic and fulvic acids, which can act as slow release sources of radionuclides, including actinide cations (elements with atomic number between 89 [actinium] and 103 [lawrencium]). In contrast to humic acids, the occurrence of clay minerals inhibits desorption.

The characteristics of soil, including pH, clay content, organic content, and moisture content, are all important to radionuclide transport and mobility. In natural soils, pH may be important to radionuclide mobility, particularly in areas of sandy, low organic content. As with other metals, low pH increases the mobility of radionuclides if limited sorption sites are available. A great deal of research has been carried out in agricultural soils because of the interest in the transfer of radionuclides from soil to plants that are then consumed by humans. Agricultural soils differ from natural soils because they usually have a high organic matter content, are irrigated, and receive fertilizers and pesticides. For agricultural soils, soil type and organic matter content are the main determinants of complexation, and pH is not as important as other soil characteristics (Bell et al. 1993). In wet soils rich in organic matter, as is typical of agricultural soils, Cs is more available in solution than for clayey soils (Bell et al. 1993). This is because its sorption to particulate organic matter may not be as strong as that within the clay mineral platelets, and there may be ion exchange processes occurring with the excess cations in the fertilizer.

In aquatic systems, sorption is also important to the fate of radionuclides and is impacted by whether the system is oxidized or reduced, which affects the oxidation state of the radionuclide. For example, although Cs exists in only 1 oxidation state (i.e., is monovalent), Pu exists in aqueous solution in any of 4 oxidation states: Pu (III) and Pu(IV), which are the reduced forms, and Pu(V) and Pu(VI), which are the oxidized forms (Aarkrog et al. 1993). In general, adsorption of radionuclides to natural particles in water occurs in reducing environments or with lower oxidation states (reduced states) of radionuclides (Aarkrog et al.1993). However, natural particles themselves take a role in redox reactions, as has been found in the reduction of Pu(V) to Pu(III, IV). So, even in a naturally oxidized aquatic system with a great deal of natural particles, the existence of reduced actinides promotes sorption of the radionuclides to the natural particles. Also, the sediment–water interface in lakes and reservoirs is often reduced, and many aquatic systems are oxidized but may have seasonal anoxia (change from oxidized to reducing conditions during hot summer months), which may change the mobility, bioavailability, and thus, the ecological risk associated with the radionuclides at different times of the year.

Radionuclides have strong affinities not only for mineral surfaces with irreversible sorption (Moulin et al. 1996) but also for colloids. The majority of the ^{137}Cs deposited in Danube River sediment by contamination from the Chernobyl accident was associated with small particles (< 20 µ) (Marginer et al. 1989). Sorption to colloids is determined in part by the quantity of colloidal particles and the concentration of suspended sediment (Sanchez et al. 1982), as well as the geologic matrix and chemical reactions associated with this matrix. As with other natural particles, colloidal organic matter that is present in high concentrations will play a role in reducing oxidized actinides, that is, Pu(V, VI) to Pu(III , IV), and in increasing sorption. Sorption to small colloids as well as to humic substances can increase the mobility of radionuclides in aquatic systems and in the soil–water interface, since colloids are often mobile. Conversely, sorption to large particles and subsequent precipitation out of the water column may decrease mobility and bioavailability.

Complexation, loosely defined as the reaction of 2 soluble species to form a third soluble species, occurs between radionuclides and organic materials through chelation, hydration, oxygenation, and halogenation. Chelation is a specific type of complex in which a radionuclide or metal combines with an organic compound (a complexing agent) with several reactive functional groups. Complexation is an important process that can enhance the mobility and bioavailability of radionuclides because radionuclides are more soluble in chelated form, and therefore, less sorption to solid surfaces occurs. Low-molecular-weight complexing agents like citrate, phosphonate, and hydroxamic acids occur naturally in aquatic environments and may complex radionuclides such as Pu(IV).

If the concentration of radionuclides and complexing agents in aquatic environments is low, mobile complexed radionuclides may not be present in significant amounts and therefore may not be ecologically important (Bulman 1983). If the concentration of radionuclides and complexing agents in aquatic environments is high, as may occur with anthropogenic addition of actinides, complexation reactions are important and increase solubility and mobility of radionuclides. For example, in oxidized aquatic environments with limited natural particles and colloids, higher oxidation states of the actinides exist, and they typically have higher water solubilities. In natural waters, Pu(V) in the form of PuO_2^+ is the stable oxidized form (Aarkrog et al. 1993). Under aerobic conditions in the natural environment including a freshwater pond, groundwater, and soils, Pu(V) and organically complexed Pu(IV) were the most mobile forms of Pu. Pu(V) did not interact strongly with inorganic carbonates (Bondietti 1982) that can cause precipitation. Similarly, in laboratory studies, the oxidized form of cobalt (CO(III)) was more mobile and stable compared to the reduced form (CO(II)), which can dissociate at pH below 6.5 and become immobilized or sorbed in the presence of manganese (IV) oxide minerals (Gorby et al. 1998).

Precipitation may occur if chemical reactions such as co-precipitation take place or if the solubility of the radionuclide or radionuclide complex is exceeded. Co-precipitation reactions, commonly with ferric hydroxide and barium sulfate, can occur in surface waters to form colloids and larger particles. The larger the particle that is formed, the less mobile it will be and precipitation out of the water column can occur. These processes limit transport and bioavailability of the radionuclides. For example, at a uranium deposit in Australia, U was enriched in the iron nodules 10^6 times more than in the groundwater because of a combination of adsorption and fixation by precipitation of copper uranyl phosphate microcrystals. This resulted in long-term retardation of U in this system and reduced dispersion of the U in the environment (Sato et al. 1997).

Biological reactions

Biological interactions may alter the mobility, transport, and availability of radionuclides in the environment. However, unlike interactions with organic contaminants, biological processes do not destroy radionuclides. In general, biota serve as storage compartments or sorptive surfaces for radionuclides. Therefore, the biological interactions are perhaps better categorized as chemical or physical reactions carried out by biological organisms.

Recently, interest has grown in the use of biological organisms as a removal mechanism for radionuclides from the environment, with subsequent controlled disposal or treatment. Many of the interactions between radionuclides and biota remove radionuclides from the surrounding media and concentrate them in biota. The specific mechanisms of biological uptake of radionuclides are not always identified. After the radionuclides are concentrated, whether internally or externally by uptake by plants, bacteria, fungi, or lichens, they must be collected, managed, and disposed of properly.

Plants have 3 general mechanisms for interception and uptake of radionuclides:

1) Deposition on foliage with large surface areas
2) Foliar absorption
3) Root absorption.

After absorption by a plant, the radionuclides can be translocated within the plant. This process ultimately impacts the bioavailability of the radionuclides and increases their potential transfer to grazing animals. Animals consuming the foliage may come into contact with the contaminants, while soil invertebrates may have more exposure to radionuclides translocated to the plant roots.

Monitoring of forests, grasses, lichens, native plants, crops, and wild animals was carried out following the nuclear accident at Chernobyl (Savchenko 1995). It was estimated that 60% to 90% of the radionuclides deposited from the atmosphere were trapped by the crown of trees (Bell et al. 1993). Sorption of radionuclides by deciduous trees was less likely than sorption to evergreens because at the time of the year of the accidental release (spring), there was little foliage and less surface area to intercept radionuclides. The greater the surface area of the foliage, the greater the absorption. Also, the longer a contaminant remains on the plant surface, the greater the potential that it will be absorbed into the plant tissue. If the radionuclides are soluble, then more uptake occurs. Radionuclides in solution can be taken up by foliage through the stomate or across the cuticle of the plant.

Once in the foliage of deciduous trees, the radionuclides ultimately migrate to the litter layer when the leaves fall to the ground. After the foliage falls to the ground, the radionuclides migrate to the upper layer of sediment when the leaves decompose. Because deciduous leaves are more readily decomposed than pine needles, in the years following the Chernobyl accident, most (55% to 96%) of the radionuclide contamination intercepted in the pine forests was subsequently transported to the litter layer, which had limited further decomposition. Conversely, radionuclide contamination intercepted by birch forests was present in the upper layer of sediment, as the leaves decomposed more quickly.

Once radionuclides reach the ground and penetrate soil, roots can come into contact with them. Rhizofiltration is a process in which roots can absorb, concentrate, and precipitate radionuclides and heavy metals. Root uptake can be complicated not only by the chemical and physical properties of the radionuclide but also by such factors as the presence or absence of bacteria associated with the roots of a plant and the partitioning of the radionuclide between nonavailable and bioavailable fractions. Soil characteristics of pH, cation exchange capacity (CEC), organic matter content, and clay content affect chemical bonding or sorption of radionuclides. The strength of the chemical sorption of radionuclides, primarily to clay minerals, directly impacts transport and bioavailability. The more strongly the sorption occurs (irreversible), the less available the radionuclides are for plant uptake. Conversely, complexation and the presence of humic substances can increase the solubility, mobility, and transport of actinides in groundwater (Moulin et al. 1996) and therefore potentially contribute to their uptake into plants (Bulman 1983).

The presence of exchangeable ions in soils affects radionuclide bioavailability. Increasing K^+ concentration in solution decreased the root uptake of ^{137}Cs in plant roots, as did increasing concentrations of Ca^{2+} and Mg^{2+} (Smolders et al. 1997). The bioavailability of ^{137}Cs decreased with increasing exchangeable K content of the soil because the K^+ was exchanged with the Cs, and Cs was readily sorbed to the soil. However, excess K in the form of K-fertilizers applied to agricultural soils may cause desorption of the fixed Cs, which may have the opposite effect and increase the uptake of previously sorbed Cs by crops (Bell et al. 1993).

In aqueous waste streams, roots can remove low concentrations of radionuclides that cannot be efficiently removed using conventional technologies. In growth-chamber experiments with sunflowers, 95% of U was removed in 24 hours. Of this, more than 99% of the plant-associated ^{238}U was in the roots of the plant, and ^{238}U was removed more quickly than Cs or Sr (Dushenkov et al. 1997). Bioconcentration of radionuclides in water by hydroponically grown plants in growth chambers is expected to be greater and more efficient than bioconcentration of radionuclides in sediments by plants grown in natural environments. In natural environments, bioconcentration may take longer, and there may be greater exchange of radionuclides between plant roots and shoots that also concentrate radionuclides. For example, when the sunflower (*Helianthus annuus* L.) was tested in a pond contaminated with Cs and Sr from Chernobyl, Cs was found to concentrate more in the sunflower roots and Sr, which is generally more soluble, was found to concentrate more in the sunflower shoots (Cooney 1996).

Uptake of radionuclides by plants varies not only by radionuclides and plant tissue but also by plant species. The uptake of ^{137}Cs in 3 plant species, Indian mustard (*Brassica juncea* L. Czern), red root pig weed (*Amaranthus retroflexus* L.), and tepary bean (*Phaseolus acutifolius* A. Gray), occurred primarily in plant shoots (Lasat et al. 1998). Red root pig weed removed approximately 3% of the total Cs from the top 15 cm of soil in shoots of a 3-month-old plant, approximately 4 times the accumulation of Cs than the other 2 species examined. If removal of less tightly bound Cs occurred first, then the efficiency of the removal would be expected to decrease with time. Although the specific mechanisms of uptake often are not examined, factors such as general resistance, plant growth rate, transpiration rate, and size and structure of the leaves may impact the distribution of radionuclides between different parts of the plant and between different plant species.

Biosorption by bacteria (Huang et al. 1998) and fungi (Guibal et al. 1995) has been examined for removal of uranyl ions from the aqueous phase. Uranium uptake by bacteria could occur by sorption of uranyl ions on the surface of bacterial cells, sorption through the formation of surface precipitates with bacteria as nuclei, and cell internalization. Internal accumulation of U was not detected after bacterial cells were washed with EDTA and digested with acid, suggesting removal of uranyl ions was primarily through extracellular adsorption (Huang et al. 1998). Similarly, in fungi *Aspergillus niger, Penicillium chrysogenum,* and *Mucor miehei,* uranyl ions were biosorbed to amino ligands present in chitin or chitosan of the cell wall. At lower pH, however, there was reduced complexation and biosorption of the uranyl ions (Guibal et al. 1995).

In terrestrial systems, lichens are efficient bioaccumulators of radionuclides and are relatively resistant to γ radiation (Bell et al. 1993; Savchenko 1995). Lichens absorb radionuclides from precipitation in the form of rain, as well as from soils, and thus are highly susceptible to accidental releases of radionuclides. ^{137}Cs and ^{90}Sr accumulation in lichens and native Alaskans was investigated in the 1960s and 1970s (Hanson 1973; Hanson and Thomas 1982). Concentrations of ^{137}Cs in caribou and reindeer flesh increased during winter months when their diet consisted predominantly of lichens, demonstrating effective transfer of the radionuclides from the lichens to the grazing animals.

In aquatic organisms, bioaccumulation of radionuclides is dependent on the specific radionuclide and the amount of contact between contaminated areas and biota. In a pond containing 5 Ci of $^{239,240}Pu$ and other actinides ($^{233,234}U$, ^{238}U, ^{238}Pu, ^{241}Am, and ^{244}Cm), the relative uptake by aquatic and terrestrial organisms was U > Cm ≥ Am > Pu (Garten et al. 1982). Because the chemical oxidation state of radionuclides impacts sorption, and thus bioavailability, trivalent forms of Am and Cm were bioaccumulated more than were the tetravalent Pu in aquatic and terrestrial studies. In addition, the association of a particular organism with the sediment–water interface was the primary factor controlling bioaccumulation in aquatic biota. Thus, if a radionuclide is strongly sorbed to sediment, sedimentary particles may be the primary source of exposure to transuranic actinides. If organisms have extensive contact with these sedimentary particles, then a greater chance for bioaccumulation exists.

Not all studies have demonstrated biomagnification of radionuclides. Field studies from U- and Ra-contaminated areas in Slovenia suggested that algae were efficient bioaccumulators of U, and larval organisms (*Trichoptera* sp. and *Gammarus fossaurm*) accumulated ^{226}Ra, but neither of these was biomagnified in fish tissues (Stegnar and Kobal 1982).

RISK MANAGEMENT

The uncertainties associated with radionuclides and their health implications cause many people to fear radioactivity. With chemicals, one often hears that "the dose makes the poison." With radionuclides, acute impacts as a function of dose are known, but it is not clear that there is any safe level for chronic exposure. Radionuclides are a class of environmental pollutant for which public perception is often divided into perceived versus real risk. Because cancer is a potential outcome of radiation exposure, it becomes complicated to formulate policy regarding the handling and disposal of radioactive materials. In addition, general mistrust is rampant in the area of radioactive waste (Renner 1997). These issues impact not only environmental regulations but also decision-making on the selection and use of remedial actions to address radionuclide contamination.

The U.S. Environmental Protection Agency (USEPA) sets standards for exposure of the general public to radiation and reviews impact statements for major projects. The USEPA also has some responsibility toward remedial activities of radiologically contaminated sites under programs such as the Comprehensive Environmental

Response, Compensation and Liability Act (CERCLA, also known as the Superfund Act), and the Resource Conservation and Recovery Act (RCRA). The USEPA has the authority to clean up most releases of radioactive material from private and federal sites. There are exceptions for sites designated under the Uranium Mill Tailings Radiation Control Act (UMTRCA) and for sites subject to the Nuclear Regulatory Commission (NRC) where there has been a nuclear incident (USEPA 1990). The NRC regulates and licenses all radioactive waste handling, processing, and disposal activities.

The history and regulatory system for radiological wastes and contamination are complicated, in part, by the complexity of this type of contamination relative to other inorganic or organic contamination. In the past, the regulatory framework increased radionuclide contamination in the environment (i.e., no reprocessing of spent fuel rods as an issue of national security). The current regulatory framework has focused on decreasing radionuclides in the environment, particularly those associated with localized environmental radiological contamination from nuclear weapons and energy production and from leaking of radioactive waste disposed at these facilities.

The current price tags for these programs are great and have come under great scrutiny. In July 1996, the U.S. Department of Energy (USDOE) announced a 10-year plan to clean up radioactive waste on their sites. They have been criticized for their slow pace of cleanup and technology development and for the large estimated amounts of money to be expended. The USDOE has received $23 billion for environmental management since 1989 (Renner 1997). Although NR has been relied on in the past because of the lack of suitable disposal and treatment technologies for radionuclides, the time has come to use NR principles to reduce ecological risk, until more efficient disposal and treatment technologies are developed.

CONCLUSIONS

Radionuclides represent a class of chemicals that are distinct from other environmental contaminants because of their energy-emitting properties. Some radionuclides can cause severe ecological and human health effects. This group encompasses many chemicals of varying half-lives and differing strengths of energy emitters. Radionuclides in the environment include elements with half-lives that range from microseconds to millions of years. Many of the human-generated radionuclides have very long half-lives. Given the long half-lives of some radioactive elements, extremely long time periods may be required to affect environmental concentrations of radionuclides. In contrast, NR may be viable for radionuclides of short half-life through radioactive decay. For example, tritium ($t_{1/2}$ = 12.3 years) in groundwater may be effectively treated by physically delaying transport and allowing the natural decay of the radioactivity to occur.

Both naturally occurring and artificially produced radionuclides from the nuclear and military industries contribute to environmental contamination. The environment has become contaminated by radionuclides from natural geologic sources and from historical controlled and accidental releases to land and ocean. This includes waste disposal, introduction into the atmosphere from nuclear testing, controlled and accidental nuclear-energy releases, and releases through the production of nuclear weapons. Many, although not all, radioactive elements are a special class of metals, and they behave chemically, physically, and biologically in the environment in a manner that is similar to metals. Therefore, the processes that impact metals also affect radionuclides.

Radionuclides in the atmosphere are rapidly attenuated by atmospheric transport, primarily advection and dispersion. Once in the atmosphere, these contaminants may travel great distances, essentially contaminating the global environment. They are diluted and dispersed to low concentrations in the atmosphere, and they fall out by wet and dry deposition onto oceans and land. A similar situation occurs in oceanic systems in which dilution occurs in the large-scale aquatic environment, and the radionuclides subsequently settle out as insoluble particles to the ocean floor.

Many radioactive elements are readily immobilized in soils and sediments, (i.e., naturally attenuated through sorption processes) because radionuclides have strong affinities for sediments and soils. Adsorption of dissolved or uncomplexed species on suspended solids and bottom sediments, co-precipitation with solids in natural waters, and sedimentation of particulate forms in aquatic systems reduce the bioavailability of radionuclides to aquatic organisms, assuming the organisms are not in constant contact with the contaminated sediments. Increased solubility due to complexation with organic ligands, resuspension of bottom sediments, and dissolution from suspended solids or bottom sediments increases bioavailability of radionuclides to aquatic organisms. Conditions in soil and aquatic systems, such as amount of suspended solids and colloids, and soil pH and redox impact radionuclide adsorption in soils and sediment. Sorption to sediment also is impacted by chemical characteristics of the radionuclide, including chemical speciation and complex oxidation states. For example, the reduced state of the radionuclide is favored for sorption. In general, sorption in sediment is a primary process that keeps radionuclides from being transported to groundwater, and if it isolates the contamination from aquatic organisms, it is a form of "distancing and shielding."

Biosorption and bioaccumulation increase biological sequestration and uptake of radionuclides. Mechanisms that account for the biological–radionuclide interactions are not well understood. Bioaccumulation occurs in foliage and root systems of plants. These processes have the potential to concentrate the radionuclides from sediment and aquatic systems into biota. Other organisms such as bacteria sequester radionuclides through external sorption to cell walls, but their contribution may be more important in active remediation than in natural remediation.

For the case of radionuclides, unlike other elements, time, shielding, and distance are the 3 greatest forms of protection. Currently, radioactive chemicals are disposed of or remediated passively by isolating them in a protective structure so that long-lived isotopes can decay without exposing plants and animals. Other active remedial technologies that remove radiation from concentrated liquid effluents or solid surfaces are based on essentially the same physical and chemical processes as those that occur naturally in the environment: desorption, dissolution, and chelation. All these processes create radioactive waste that must be disposed of properly. For certain highly radioactive wastes, technologies are currently being developed and treatment is being carried out on-site. Many of these technologies are not natural methods; on the contrary, they involve significant inputs of energy, such as vitrification at high temperatures of liquid, high-level radioactive waste.

Natural remediation of gross radiological contamination is possible for those elements with short half-lives or through the process of isolation. Unfortunately, the long half-lives associated with many of the artificially produced radionuclides make their past and continued generation a problem that will not go away any time soon. Many NR processes reduce potential adverse ecological impacts of radionuclides in the environment by altering the potential for exposure: time, distance, shielding, and dilution. Natural remediation has been used to address globally dispersed radiation and radioactive waste generation since the introduction of artificially produced radioactivity in the 1930s. In the past, NR has been considered in many cases the “solution” to the problem of radioactive waste, albeit based on the economic perspective of the cheapest method of disposal and not necessarily the best method. While we have no “solution” to decades of radioactive waste generation and environmental contamination by natural and artificial radionuclides, maximizing NR can certainly contribute to resolving the problem.

REFERENCES

Aarkrog A, Calmet DP, Charmasson S, Carpenter R, Fowler S, Ivanovich M, Kuznetsov G, Luttrell SP, Malcolm SJ, Mitchell PI, Nies H, Patel B, Polikarpov GG, Rjabov I, Swift D, Tveten U. 1993. Radionuclide aquatic pathways. In: Warner FE, Harrison RM, editors. Radioecology after Chernobyl. Chichester UK: Wiley. p 177–274.

Appleby LJ, Devell L, Mishra UC, Voice EH. 1993. Sources. In: Warner FE, Harrison RM, editors. Radioecology after Chernobyl. Chichester UK: Wiley. p 1–31.

ApSimon HM, Beck H, Devell L, Dickerson M, Garland JA, Graziani G, Gudiksen P, McKay WAS, Mishra UC, Nicholson KW, Smith FB, Shapiro CS. 1993. Atmospheric pathways. In: Warner FE, Harrison RM, editors. Radioecology after Chernobyl. Chichester UK: Wiley. p 55–100.

Bell JNB, Coughtrey PJ, Frissel M, Hakonson TE, Hanson WC, Horrill D, Howard BJ, Lane LJ, Myttenaere C, Robison WL, Ronneau C, Shaw G, Schell WR, Van Den Hoek J,

Konoplyov A, Zezina N. 1993. Terrestrial pathways. In: Warner FE, Harrison RM, editors. Radioecology after Chernobyl. Chichester UK: Wiley. p 101–176.

Beneš P. 1982. Physico-chemical forms and migration in continental waters of radium from uranium mining and milling. In: Environmental migration of long-lived radionuclides. Vienna, Austria: International Atomic Energy Agency (IAEA). STI/PUB/597. p 3–23.

Blackman Jr WC. 1996. Basic hazardous waste management. 2nd edition. Boca Raton FL: Lewis. 397 p.

Bondietti EA. 1982. Mobile species of Pu, Am, Cm, Np, and Tc in the environment. In: Environmental migration of long-lived radionuclides. Vienna, Austria: International Atomic Energy Agency (IAEA). STI/PUB/597. p 81–96.

Bou-Rabee F. 1997. Soil radioactivity atlas of Kuwait. *Environ Int* 23:5–15.

Bulman RA. 1983. Complexation of transuranic elements: A look at factors which may enhance their biological availability. In: Coughtrey, PJ, Bell, JNB, Roberts, TM, editors. Ecological aspects of radionuclide release. Oxford UK: British Ecological Society, Blackwell Scientific. p 105–113.

Chung C, Pan L-K, Yeh C-S, Tsai CM. 1986. Ocean transport of low-level radioactive waste off Taiwan. *Nucl Chem Waste Manage* 6:183–191.

Cooney CM. 1996. Sunflowers remove radionuclides from water in ongoing phytoremediation field tests. *Environ Sci Technol News* 5:194A.

Davis ML, Cornwell DA. 1991. Introduction to environmental engineering. 2nd ed. New York NY: McGraw-Hill. 822 p.

Dushenkov S, Vasudev D, Kapulnik Y, Gleba D, Fleisher D, Ting KC, Ensley B. 1997. Removal of uranium from water using terrestrial plants. *Environ Sci Technol* 31:3468–3474.

Eisenbud M, Gesell T.1997. Environmental radioactivity. 4th ed. New York NY: Academic Pr. 656 p.

Frissel MJ, Pennders R. 1983. Models for the accumulation and migration of ^{90}Sr, ^{137}Cs, $^{239,240}Pu$ and ^{241}Am in the upper layers of soil. In: Coughtrey PJ, Bell JNB, Roberts TM, editors. Ecological aspects of radionuclide release. Oxford UK: British Ecological Society, Blackwell Scientific. p 63–71.

Garten Jr CT, Trabalka JR, Bogle MA. 1982. Comparative food chain behaviour and distribution of actinide elements in and around a contaminated fresh-water pond. In: Environmental migration of long-lived radionuclides. Vienna, Austria: International Atomic Energy Agency (IAEA). STI/PUB/597. p 299–312.

Gorby YA, Caccavo Jr F, Bolton Jr. H. 1998. Microbial reduction of cobalt^{III}EDTA^{-} in the presence and absence of manganese (IV) oxide. *Environ Sci Technol* 32:244–250.

Guibal E, Roulph C, le Cloirec P. 1995. Infrared spectroscopic study of Uranyl biosorption by fungal biomass and materials of biological origin. *Environ Sci Technol* 29:2496–2503.

Hanson WC. 1973. Fallout ^{90}Sr and ^{137}Cs in Northern Alaskan ecosystems during 1959–1970 [dissertation]. Fort Collins CO: Colorado State Univ.

Hanson WC, Thomas JM. 1982. Prediction of ^{90}Sr body burdens and radiation dose in Anaktuvuk Pass Alaskan Eskimos due to fallout. *Health Phys* 43:323–333.

Harley NH. 1992. Radon and daughters. In: Lippmann M., editor. Environmental toxicants: Human exposures and their health effects. New York NY: Van Nostrand Reinhold. p 520–542.

Huang FYC, Brady PV, Lindgren ER, Guerra P. 1998. Biodegradation of uranium-citrate complexes: Implications for extraction of uranium from soils. *Environ Sci Technol* 32:379–382.

Lasat MM, Fuhrmass M, Ebbs SD, Cornish JE, Kochian LV. 1998. Phytoremediation of a radiocesium-contaminated soil: Evaluation of Cesium-137 bioaccumulation in the shoots of three plant species. *J Environ Qual* 27:165–169.

Lu N, Kung KS, Mason CFV, Triay IR, Cotter CR, Pappas AJ, Pappas MEG. 1998. Removal of plutonium-239 and americium-241 from Rocky Flats soil by leaching. *Environ Sci Technol* 32:370–374.

Luttrell SP. 1993. Case-studies of significant radioactive releases. In: Warner FE, Harrison RM, editors. Radioecology after Chernobyl. Chichester UK: Wiley. p 33–53.

Marginer FJ, Rank D, Terlunen J. 1989. Radioactivity in the River Danube: Special investigations. In: Feldt W, editor. The radioecology of natural and artificial radionuclides. Fachverb Strahllenschutz e. V. 22. p 332–337.

Meinhold AF, Holtzman S. 1997. Radiation risk and dose to recreational fishermen from ingestion of fish caught near eight oil platforms in the Gulf of Mexico. Draft report to the U.S. Department of Energy, Energy Science and Technology Division, Department of Applied Science, Brookhaven National Laboratory, Upton NY. 62 p.

Miettinen JK, Nikula A, Leskiinen S. 1982. Distribution coefficients or radionuclides between Finnish soils and groundwater. In: Environmental migration of long-lived radionuclides. Vienna, Austria: International Atomic Energy Agency (IAEA). STI/PUB/ 597. p 153–158.

Misra M, Mehta RK. 1995. Remediating heavy metals, radionuclides, and organic pollutants from soils. *JOM* September:45–53.

Moulin VM, Moulin CM, Dran J-C. 1996. Role of humic substances and colloids in the behavior of radiotoxic elements in relation to nuclear waste disposal. Chapter 16. In: Gaffney JS, Marley NA, Clark SB, editors. Humic and fulvic acids. Washington DC: American Chemical Society (ACS). p 259–271.

Muller J, Kusiak R, Ritchis AC. 1989. Factors modifying lung cancer risk in Ontario uranium miners, 1955–1981. Ontario Ministry of Labour Report. Toronto, Canada: Ministry of Labour.

Ottoboni MA. 1991. The dose makes the poison. 2nd ed. New York NY: Van Nostrand Reinhold. 244 p.

[NCRP] National Council on Radiation Protection and Measurements. 1987. Exposure of the population of the United States and Canada from natural background radiation. Bethesda MD: NCRP. Report No. 94.

Pattenden NJ, Cambray RS, Eakins JD. 1983. Radionuclide transfer processes from sea to land. In: Coughtrey, PJ, Bell, JNB, Roberts, TM, editors. Ecological aspects of radionuclide release. Oxford UK: British Ecological Society, Blackwell Scientific. p 259–271.

Renner R. 1997. U.S. nuclear cleanup shows signs of progress. *Environ Sci Technol* 31:134–137.

Rigol A, Vidal M, Rauret G, Shand CA, Cheshire MV. 1998. Competition of organic and mineral phases in radiocesium partitioning in organic soils of Scotland and the area near Chernobyl. *Environ Sci Technol* 32:663–669.

Sanchez AL, Schell WR, Sibley TH. 1982. Distribution coefficients for plutonium and americium on particulates in aquatic environments. In: Environmental migration of long-lived radionuclides. Vienna, Austria: International Atomic Energy Agency (IAEA). STI/PUB/597. p 188–203.

Sato T, Murakami T, Yanase N, Isobe H, Payne TE, Airey PL. 1997. Iron nodules scavenging uranium from groundwater. *Environ Sci Technol* 31:2854–2858.

Savchenko VK. 1995. The ecology of the Chernobyl catastrophe: Scientific outlines of an international programme of collaborative research. In: Jeffers JNR, editor. Volume 16, Man and the biosphere series. Paris, France: United Nations Educational, Scientific and Cultural Organization (UNESCO). p 25–42.

Segerstahl B, Akleyev A, Novikov V. 1997. The long shadow of Soviet plutonium. *Environment* 39:12–20.

Severa J, Bár J. 1991. Handbook of radioactive contamination and decontamination. Studies in Environmental Science 47. Amsterdam, the Netherlands: Elsevier. 251 p.

Shcherbak YM. 1996. Ten years of the Chernobyl era. *Sci Am* April, 49:44–49.

Smolders E, van den Brande K, Merckx R. 1997. Concentrations of ^{137}Cs and K in soil solution predict the plant availability of ^{137}Cs in soils. *Environ Sci Technol* 31:3423–3438.

Stegnar P, Kobal I. 1982. Uptake and distribution of radium and uranium in the aquatic food chain. In: Environmental migration of long-lived radionuclides. Vienna, Austria: International Atomic Energy Agency (IAEA). STI/PUB/597. p 364–373.

[UNSCEAR] United Nations Scientific Committee on the Effects of Atomic Radiation. 1993. UNSCEAR 1993 Report to the General Assembly, with Annexes. New York NY: United Nations.

[USEPA] U.S. Environmental Protection Agency. 1990. Assessment of technologies for the remediation of radioactively contaminated Superfund sites. Washington DC: Office of Solid Waste and Emergency Response, Office of Radiation Programs. EPA/540/2-90/001. 28 p.

Zorpette G. 1996. Down the drain. *Sci Am* December. p 20–24.

11

Natural Remediation of Sediments

Teresa Michelsen

SEDIMENTS IN LAKES, rivers, bays, and other surfacewater bodies are at the receiving end of many environmental transport pathways. Contaminants released to the environment are transported in groundwater, surfacewater runoff, atmospheric deposition, and direct discharges to surfacewater bodies, where metals and persistent organic chemicals partition into organic C and/or mineral surfaces in sediments. As a result of all of these processes, sediments in many urban areas of North America have become contaminated above levels that may be harmful to human health and the environment (U.S. Environmental Protection Agency [USEPA] 1994b; Canadian Council of Ministers of the Environment [CCME] 1995; National Research Council [NRC] 1997).

Sediments serve not only as a repository for contaminants but also may release contaminants to the environment. Contaminants in sediments interact with pore water, overlying water, and aquatic biota living in or ingesting sediments. Contaminated sediments are of concern because they can be directly toxic to organisms that live in sediments, reducing or eliminating benthic communities that serve as prey for fish and other organisms higher in the food web. Benthic organisms, shellfish, and fish that come into contact with sediments may themselves become contaminated, in turn exposing fish, mammals, and birds to contaminants in their food (USEPA 1994b; CCME 1995). Exposure of some fish, such as salmon, to contaminated sediments during early life stages has been shown to reduce their growth, immunity to diseases, and survival rates (Varanasi et al. 1993).

Natural Remediation of Environmental Contaminants. Michael Swindoll et al., editors.
 ISBN 1-880611-33-3.

Contaminants in sediments, particularly those that bioaccumulate in the food chain, can also pose risks to human health. These risks are particularly acute for populations that consume large amounts of seafood caught in contaminated areas, such as Native Americans, recreational fishermen, and Asian immigrants (Landolt et al. 1987; Toy et al. 1996). In addition, contaminated sediments interfere with port navigation and development projects and waterfront construction projects, by increasing the cost and difficulty of obtaining permits and disposing of sediments. Contaminated sediments may also create legal liabilities for property owners, degrade habitat, and impact commercial, recreational, and tribal fisheries (NRC 1997).

States and provinces in North America have begun developing and implementing regulatory programs to address contaminated sediments, including programs to accomplish source control, conduct cleanup of contaminated sites, and dispose of contaminated dredged material. A common problem faced by cleanup programs is the extensive area and volume of sediments that are contaminated. A large urban area may have many millions of cubic yards of contaminated sediments to address (Washington Department of Ecology [WDOE], also known as "Ecology," 1991a). Costs of remediation range from tens to hundreds of dollars per cubic yard (USEPA 1994a), making complete cleanup or removal of contaminated sediments problematic or cost prohibitive for many jurisdictions. Entire watersheds may exceed levels designed to be protective of human health and the environment, particularly in the case of chlorinated compounds that bioaccumulate through the food chain (WDOE 1997). Finally, sediments may contain complex mixtures of contaminants from a variety of urban and industrial sources, decreasing the feasibility of treatment technologies and complicating the resolution of liability for cleanup. As a result of the magnitude of the problem, natural remediation (NR) has emerged as an alternative to active cleanup measures for contaminated sediments. In areas where conditions are appropriate for NR, this approach allows active cleanup activities to be focused on areas of highest risk, conserving economic and agency resources.

Although often attractive from a cost and feasibility standpoint, NR of sediments raises many environmental, policy, and technical issues that need to be addressed to ensure successful use of this approach as part of a comprehensive sediment management program. This chapter presents an overview of regulatory programs for sediments that include NR, reviews the environmental processes of NR in sediments, introduces field and modeling methods for estimating the rates of these processes, and discusses policy considerations and technical issues that should be considered when evaluating the benefits and costs of NR for sediments at a specific site.

Regulations, guidance documents, and literature pertaining to sediments often refer to NR as "natural recovery." The definition of natural recovery in this context is the same as that used for NR in this book: naturally occurring physical, chemical, or biological processes that reduce the risk of ecological effects from contaminants in sediments. In the context of this chapter, "natural recovery" and "natural remediation" are synonymous.

REGULATORY FRAMEWORKS FOR NATURAL REMEDIATION IN SEDIMENTS

Natural remediation of sediments is normally discussed in the context of a sediment cleanup program and can be considered a natural, or passive, cleanup alternative (as opposed to active cleanup alternatives such as capping, dredging, treatment, and disposal). Cleanup of contaminated sediments can be accomplished under a variety of different regulatory authorities. Principal among these are state, provincial, and federal Clean Water Acts and site cleanup regulations, such as the Comprehensive Environmental Response, Compensation, and Liability Act (CERCLA; also known as "Superfund"), or the Resource Conservation and Recovery Act (RCRA), along with similar state laws. Some sediment cleanup actions have also been conducted pursuant to Natural Resource Damage Assessment (NRDA) settlements, and are viewed as partial restoration of natural resource damages (e.g., Elliott Bay/ Duwamish Restoration Board [EBDRP] 1994).

As part of the decision process used in these cleanup programs, NR of sediments is considered one possible cleanup alternative among many and is normally evaluated along with a range of active cleanup alternatives as part of a feasibility study. Natural remediation is similar in concept to the National Environmental Policy Act (NEPA) "no action" alternative, but with a few key differences. First, NR processes should result in a reduction of ecological and human health risks over time, whereas the "no action" alternative often assumes no change in the site conditions. Second, NR, when used as a cleanup alternative, is not without costs. Site-specific field studies and/or models are typically required by regulatory agencies to determine that NR is an alternative that will meet the project objectives, and monitoring costs are incurred to verify its success (WDOE 1991b). In addition, a contingency plan may be required in the event that NR does not proceed as predicted. In contrast, "no action" under NEPA is usually interpreted to mean that no studies, monitoring, or other activities or expenditures will occur.

Although sediment quality can be addressed under all these authorities, relatively few states or provinces have promulgated regulations that specifically address cleanup or NR of contaminated sediments. Sediment quality standards have been developed by Washington (WDOE 1991c), Florida (MacDonald 1994), New York (New York Department of Environmental Conservation [NYDEC] 1994), Ontario (Persaud et al. 1992), and British Columbia (British Columbia Ministry of the Environment, Lands and Parks [BCMOELP] 1994, 1995). At a national level, USEPA has draft criteria for 2 pesticides (USEPA 1993b, 1993c), and Canada has developed interim sediment quality guidelines for several metals, polynuclear aromatic hydrocarbons (PAHs), and pesticides (Environment Canada [EC] 1997a–i,

1998). These sediment quality standards provide a numeric goal or target for NR and can be used in conjunction with a model to estimate the time required for NR.

Contaminants in sediments are generally only of concern if they are present in the surface layer in which biological organisms live, termed the "biologically active zone." In marine sediments on the West Coast, for example, 95% of benthic organisms inhabit the top 10 cm of sediment (the depth of the biologically active zone may vary significantly, particularly in freshwater environments) (WDOE 1995). This information provides additional refinement to the success measures for NR by defining, in addition to a numeric sediment quality criterion (SQC), a depth over which that criterion should be met to be protective of benthic organisms.

Cleanup of contaminated sediments in a watershed can be legally and technically quite complex, as there may be a variety of sources commingled in the receiving environment and a variety of different types of legal actions occurring at different facilities. An ongoing need for navigational dredging, pier maintenance, and waterfront construction further complicates project planning and resolution of liability (NRC 1997; Michelsen et al. 1998). Before an in-water cleanup can be conducted, a variety of local, state, and federal permits are required, and proprietary and tribal treaty rights may also be affected (WDOE 1990). These complications provide added incentive to determine, on a watershed basis or a site-specific basis, which areas may recover on their own if left to natural processes.

In 1991, the State of Washington published the first set of state regulations specifically addressing cleanup of contaminated sediments (Sediment Management Standards, Chapter 173-204 WAC). Currently, this is the only set of promulgated rules that provides guidance on the use of natural recovery for sediments. WAC 173-204-560(4)(f)(ii) provides that "the proposed site cleanup alternatives may include establishment of a sediment recovery zone" but cautions that "establishment or expansion of a sediment recovery zone shall not be used as a substitute for active cleanup actions, when such actions are practicable and meet the [numeric sediment quality] standards..." In general, the WDOE considers natural recovery to be a viable alternative if it will result in sediments in the biologically active zone meeting the site-specific sediment cleanup standards within a reasonable time frame, defined as a maximum of 10 years (WDOE 1991b). If viable, natural recovery must then be weighed against active cleanup alternatives as part of the feasibility study.

In addition, the rule provides that "cleanup standards shall be defined in consideration of the net environmental effects (including the potential for natural recovery of the sediments over time), cost, and engineering feasibility..." Net environmental effects include consideration of both benefits and impacts associated with natural recovery and active cleanup alternatives. This includes habitat disturbance during active remediation and time for the ecosystem to recover from that disturbance, time required to achieve cleanup standards under both active and passive cleanup alternatives, and continued ecological effects that may occur during the natural recovery period. The regulations also specify approved models for predicting natural recovery, monitoring requirements for natural recovery zones, and the criteria that should be evaluated in determining whether to authorize a sediment recovery zone.

In addition to incorporating natural recovery into its sediment cleanup regulations, WDOE has published a number of guidance documents and reports on natural recovery and natural recovery models. These reports include reviews of natural recovery and sediment impact zone models (PTI Environmental Services [PTI] 1989, 1991a), further guidance on the selected natural recovery model WASP4 (PTI 1993), and guidance on integrating natural recovery considerations into investigations and evaluations of alternatives for cleanup (WDOE 1991b). WDOE also considers the potential for a site to naturally recover as part of its site ranking formula (PTI 1991b).

In 1994, USEPA incorporated the concept of NR into its agency-wide Contaminated Sediment Management Strategy (USEPA 1994b). The strategy states, "In certain circumstances, the best strategy will be to implement pollution prevention measures as well as point and nonpoint source controls to allow natural recovery processes such as biodegradation, chemical degradation, and the deposition of clean sediments to diminish risks associated with the sites." USEPA lists factors to consider in determining whether NR is appropriate: "the specific contaminants present and their associated risks, the designated uses impaired during recovery, the size of the affected area, the feasibility of remediation, site hydrodynamics, including the potential for downstream transport, the time required for natural recovery, and the liability associated with active remediation."

The strategy discusses some of the tradeoffs in determining the net environmental benefit of NR at a specific site. It states that natural recovery is the best option when cleaning up contaminated sediments would do more harm than good. For example, dredging or capping may harm sensitive habitats such as eelgrass beds that would take a very long time to recover from physical disturbance. On the other hand, USEPA concludes that "natural recovery is not acceptable where contamination poses severe and substantial risks to aquatic life, wildlife, and human health. In addition, natural recovery may not be the method of choice for contaminants that biodegrade or transform into more persistent, toxic compounds."

The Oregon Department of Environmental Quality (DEQ) has also recently developed a draft strategy on cleaning up contaminated sediments, in which it discusses the role of natural recovery (DEQ 1998). It contains many of the same ideas expressed in the USEPA Contaminated Sediment Management Strategy and, in addition, provides a useful definition of natural recovery as follows: "The term 'natural recovery' implies both 'attenuation' (those processes that reduce contaminant concentrations in all media, and do not result in crossmedia or offsite transport of contaminants) and 'recovery' (reestablishment of viable benthic and pelagic communities through time, as well as restoration of other beneficial uses)."

NATURAL REMEDIATION PROCESSES IN SEDIMENTS

Contaminant concentrations may be reduced in sediments in a variety of ways (Figure 11-1), including burial, mixing, degradation, dissolution into water, transfer of contaminants into biota (bioaccumulation), and advection. These may be considered NR processes if the overall risks to the environment are reduced. However, not all processes that reduce concentrations in sediments contribute to risk reduction in the overall aquatic environment. In these cases, the processes may be more appropriately considered contaminant transfer mechanisms and may result in undesirable increases in risks to aquatic life. It is important when assessing the potential for NR at a site to clearly distinguish between and be aware of all these processes, to better understand the net environmental benefit or cost associated with leaving sediments to be influenced by natural processes alone.

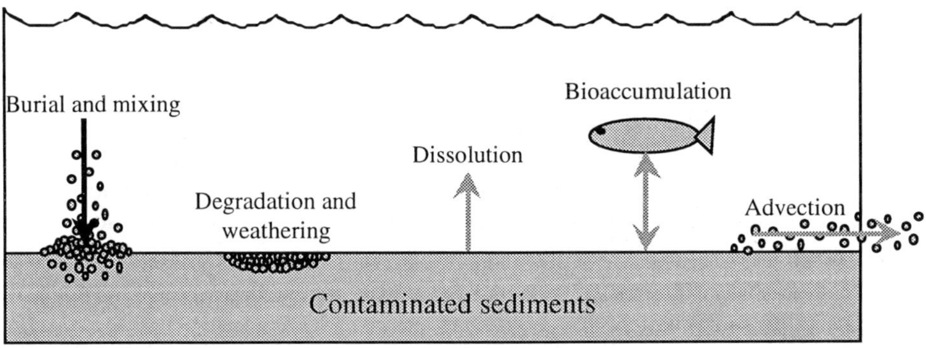

Figure 11-1 Processes that affect natural recovery. Processes shown in black, including burial and mixing and degradation and weathering, enhance natural recovery and reduce surface concentrations of contaminants. Processes shown in gray, including dissolution, bioaccumulation, and advection, may also occur during natural recovery but result in greater exposure of aquatic life to contaminants.

The processes discussed below are listed roughly in order of the degree to which they influence NR, although the importance of actual mechanisms will vary from site to site. The summary that follows is based on the author's involvement in more than 70 sediment site investigations in the Pacific Northwest and on compilations of site information by USEPA, EC, and university researchers and consultants for more than 40 additional sites in California, the Great Lakes, and East Coast regions (International Joint Commission [IJC] 1997; USEPA 1998; WDOE 1998; see also specific citations below). In these areas, physical processes of sediment deposition, mixing, and transport have by far the greatest influence on NR. At sites located in active hydrologic environments, contaminant transfer pathways are also important and may serve to increase the bioavailability of contaminants during the NR process. Finally, biodegradation is an important secondary process at some sites and can be quite significant in certain cases.

Burial

The process that most often predominates during NR is burial by clean sediments. This process can rapidly reduce contaminant concentrations in the surface sediments while also limiting bioavailability (through sequestration) of the remaining contaminants. Burial may occur in areas with significant inputs of clean sediments, such as near a river mouth or eroding bluffs. In such areas, gradual deposition of clean sediments may eventually form a cap and isolate deeper layers of contaminated sediments, reducing their bioavailability. In many areas, however, there is little or no natural sedimentation, and consequently, it is important to assess the potential for burial on a site-specific basis.

Sedimentation rates in nearshore areas are highly variable, depending on proximity to a source of sediments, erosional or resuspension processes operating in the area, and other hydrologic characteristics of the water body. For example, a recent compilation of sedimentation rates measured at various sites in Puget Sound identified a range of 0.1 to 5.7 g/cm^2-y, and found more than an order of magnitude variation even within waterways or study areas (Norton and Michelsen 1995). In river systems, sedimentation and erosion rates can be much more variable and may change with river flow. For example, in the Fox River, sedimentation ranges from 0 to 6 g/cm^2-y at median flow rates, while sedimentation rates ranged from erosional to 15 g/cm^2-y during a high-flow event (Gailani et al. 1991).

The amount of sedimentation necessary to affect NR depends entirely on the initial concentrations in sediments compared to the target cleanup standards, the time frame considered acceptable for NR, and the degree of mixing occurring in surface sediments. Because the sedimentation rate is one of the critical parameters for predicting the rate of NR, and because it is spatially and temporally variable, site-specific measurements of this parameter are needed, often from several different locations. This NR process is most likely to be effective in quiescent areas where erosion or resuspension of sediment is not likely and in areas where there is a significant source of clean sediments.

If input sources of contamination are not sufficiently reduced, NR may be slowed by contaminants in the particulates contributing to sedimentation; if source loading is great enough, increased contamination of bottom sediments may even occur. Any point sources, such as combined sewer overflows and storm drains that remain in the system must be evaluated and their continuing impact on the remediation area assessed. In addition, it cannot be assumed that control of point sources is sufficient to prevent recontamination of an area to be cleaned up. Nonpoint sources that are difficult to identify, quantify, and control still remain in many urban waterfronts and agricultural areas. Contaminant concentrations in water column particulates can be measured to ensure that contaminant inputs from nonpoint sources are known and factored into NR assessments (Norton and Michelsen 1995).

Mixing

Where a benthic community is present or when other disturbances exist (e.g., propeller wash, wave resuspension), recently deposited sediments may be mixed into the underlying contaminated sediments. If newly deposited sediments are clean, this mixing process will gradually reduce overall concentrations in surface sediments through dilution. Bioturbation may also introduce oxygen into the sediments, facilitating biodegradation. Mixing and ingestion of sediments by benthic organisms may increase the bioavailability of certain compounds by bringing buried contaminants to the surface and allowing their transfer to biota (Drake et al. 1994). Oxidation reactions can also occur when heavily polluted sediments begin to recover and become aerobic. These reactions may release metals that have been bound to sulfides under anaerobic conditions (National Oceanic and Atmospheric Administration [NOAA] 1995; USEPA 1995), particularly in marine environments where sulfur is abundant.

Advection and erosion

Sediments and associated contaminants may be transported offsite through erosion, advection, or bedload transport during the NR period. These processes may result in additional downstream or offshore areas becoming contaminated and may complicate remediation of sediments by replacing an easily removed, highly concentrated smaller area of contamination with a diffuse, high-volume problem (USEPA 1994b). In river systems, it will be particularly critical to evaluate the potential for large-scale sediment transport events such as infrequent floods to remove a clean cap and re-expose or mobilize underlying contaminated sediments (Cardenas et al. 1995). In large lakes and coastal areas, sediment transport events can also be caused by large storms (Lick et al. 1994). Understanding the sediment transport, erosional, and depositional characteristics of the river or other water body is important for accurate predictions of potential for long-term burial.

Once disturbed, bottom sediments may be resuspended into the water column where they may become more susceptible to bioaccumulation and biodegradation. Contaminants on particles may dissolve into the water, and particles may be ingested by fish or filter-feeding invertebrates. Storms and other resuspension mechanisms are believed to be the primary mechanisms for continuing exposure of the food chain to historically contaminated sediments in the Great Lakes (Lick et al. 1994). For this reason, NR may not be a recommended approach in areas that are erosional or subject to storms, heavy propeller wash, or other disturbances (WDOE 1991b; USEPA 1994b; NRC 1997).

Weathering

Many organic contaminants in sediments, such as petroleum compounds, undergo a process called "weathering" over time. The term "weathering" refers to aging of sediments, organic matter, and chemicals partitioned into the organic matter. As the sediments age, the organic matter changes structure, and chemicals within it may become more tightly bound (McCall and Agin 1985; Connaughton et al. 1993; Pignatello et al. 1993). In addition, contaminants (including metals) may diffuse into interior interstices or binding sites in mineral grains from which the contaminant is not easily released (Steinberg et al. 1987). With reference to petroleum and other mixtures, the term "weathering" is also used to refer to the gradual loss of the more soluble and volatile components of the mixture, leaving behind more refractory and closely bound components (Wilcock et al. 1996). Because of the decrease in solubility that occurs, all of these processes of weathering reduce the bioavailability of the contaminant or mixture and, consequently, its toxicity as well as its susceptibility to biodegradation (Harms and Zehnder 1995; Hatzinger and Alexander 1995; Sandoli et al. 1996).

Degradation

Biodegradation and chemical transformation of contaminants occur in sediments, and these processes can be important contributors to NR. Biodegradation in the environment is most rapid and likely to occur for chemicals that are naturally occurring, such as petroleum hydrocarbons (PHs), as there are natural populations of microorganisms that can use these substances as a source of energy.

An important consideration in assessing the environmental effect of biodegradation is the intermediate and end products of biodegradation and their relative toxicity and bioavailability compared to the parent compounds. Many organic compounds, such as low-molecular-weight PHs, can be completely broken down or mineralized into water and carbon dioxide. Other compounds, such as the chlorinated hydrocarbons (e.g., dioxins and polychlorinated biphenyls [PCBs]), may undergo dechlorination to less chlorinated compounds but may not be completely mineralized (Liu et al. 1996). In the case of dioxins, lower-chlorinated degradation products are generally more toxic than the parent compounds (Safe 1990), while lower-chlorinated PCB congeners are generally less toxic than the parent compounds (Safe 1994; Mousa et al. 1996). Other tradeoffs may occur; for example, anaerobic dechlorination of PCBs generates lower-chlorinated congeners that are less toxic (Quensen et al. 1988) but more soluble and volatile, and hence, more bioavailable (Bushart et al. 1988) than the parent compounds.

Over time, weathering processes reduce bioavailability and the rate of biodegradation. For example, researchers have determined that biodegradation of petroleum compounds in sediments and soils weathered for only 1 year occurs up to 10 times more slowly than in freshly contaminated sediments (Hatzinger and Alexander 1995; Carpenter et al. 1996; Krieger-Brockett et al. 1997). However, most reported chemical half-lives are derived from studies using chemicals in dissolved form spiked into sediments. This form of the chemical is much more bioavailable and degradable than in weathered sediments (Renner 1998). In addition, many contaminated sediments are anaerobic because of an excess of organic matter (e.g., petroleum, wood waste, sewage). Biodegradation rates in anaerobic sediments will generally be much slower than aerobic degradation rates (McCarty et al. 1984; USEPA 1994a; Madsen and Bilotta 1996).

In addition to environmental factors, sediments often contain complex mixtures of contaminants. Typically, more soluble or degradable components of mixtures such as PAHs and PCBs will be removed first, leaving residuals that are relatively resistant to biodegradation (Wilcock et al. 1996). Some components of a mixture may be toxic to the biodegrading organisms, reducing or eliminating the potential for biodegradation even of the relatively degradable components of the mixture (USEPA 1994a). Finally, other rate-limiting factors may be present. The concentration of a chemical may itself affect biodegradation rates. At high concentrations, the chemical may be toxic to microorganisms (USEPA 1994a). At lower concentrations, a threshold may be reached below which the microorganisms find other sources of energy that are more efficiently utilized, and biodegradation of the target chemical may slow or cease (Sokol et al. 1998).

For the reasons described above, biodegradation rates reported in the literature should be used with caution in long-term NR modeling and should be considered most applicable to fresh spills in aerobic environments.

Dissolution and bioaccumulation

Contaminants in sediments may partition into other environmental media, including pore water and overlying surface water, and to aquatic organisms in contact with sediments or suspended particulates. These potential transfer mechanisms should be evaluated as part of the assessment of the net environmental effect of NR. For many sediment contaminants, transfers to the overlying water and to biota will decrease over time because of the burial, weathering, and biodegradation processes described above. Similarly, where contaminants have been in place for decades, contaminant transfer mechanisms may be of less concern than they would be where contaminant inputs are new or relatively recent. However, some organisms may continue to be exposed to contaminants bound to sediments by direct contact or ingestion of sediments or by filtering out suspended sediments from the water column. Benthic flux studies (e.g., Bothner et al. 1980) may be used to supplement partitioning models to determine whether sediments are a source or a sink with respect to the overlying water. Tissue analyses or bioaccumulation studies may be used to determine whether sediments continue to be a source of contaminants to biota and the food chain.

NATURAL REMEDIATION MODELING AND FIELD INVESTIGATIONS

Natural remediation models for sediments generally try to take into account all of the processes shown in Figure 11-1 and include simple equations such as SEDCAM (WDOE 1991b), spreadsheet models (e.g., adapted from Officer and Lynch 1989), and complex hydrodynamic models capable of detailed assessment of multiple sources and a complex receiving environment (USEPA 1993a; Drake et al. 1994). A combination of CORMIX (nearfield model; Akar and Jirka 1991) and WASP4 (farfield model; Ambrose et al. 1992) is a preferred 3-dimensional model used by USEPA and WDOE, and both agencies have developed guidance manuals for its use in NR and source control evaluations for sediments (USEPA 1993a; PTI 1993).

Use of these models requires a great deal of site-specific information; even the simple models require data that are not likely to be available without site-specific field studies (NRC 1997). Table 11-1 lists the minimum amount of site-specific data needed and the field studies that are typically used to acquire these data. It is important to assess early in a site investigation whether NR is likely to be seriously considered during the feasibility study so that the necessary data collection effort can be incorporated into the study design. It is interesting to note that the same types of studies and models are also used in assessing the potential for recontamination after cleanup. In some cases, it may not be known whether recontamination will occur until the studies are complete. The following sections review each of the types of data that can be used to assess NR.

Deposition rate

To assess the potential for NR, it is critical to know the site-specific rate of sediment deposition. Sediment deposition rates vary greatly from area to area and may even vary significantly within a site. Therefore, it is necessary to collect these data directly from the area proposed for NR, and several locations may be needed if site characteristics vary. There are 2 primary methods of determining the sedimentation rate: 1) analyze core samples by radioisotope dating methods and 2) deploy sediment traps. These methods are often used in combination.

Radioisotope dating is usually accomplished by dividing a core sample into small (e.g., 2-cm) vertical sections and analyzing for ^{210}Pb and/or ^{137}Cs in each section. ^{210}Pb is a naturally occurring radionuclide with a half-life of approximately 22.3 years. ^{210}Pb is continually produced from radon in the atmosphere and has a

Table 11-1 Required data for natural recovery models and associated field studies

Model input parameter	Field studies
Deposition rate	Sediment trap accumulation Core dating techniques
Source loading	Sediment trap particulate analyses Point source loading estimates coupled with hydrodynamic models
Mixing depth	Field observations Core dating curves
Initial sediment concentrations	Surface sediment samples
Biodegradation rates	None (literature-derived)
Partitioning coefficients	Sediment and porewater samples or literature-derived

constant concentration in newly deposited sediments. As the sediments are buried and removed from atmospheric sources, the ^{210}Pb decays, causing a logarithmic decline in concentration with depth that can be modeled (Krishnaswamy et al. 1971; Koide et al. 1973; Christensen 1982). Because this decline occurs at a known rate that is relevant to the time scale we are interested in, the age of each core section can be determined. ^{137}Cs is an isotope associated with aboveground nuclear testing and had a worldwide maximum atmospheric concentration from 1963 to 1965. It is often used as a confirmatory analysis for the ^{210}Pb dating method.

Although generally considered the most definitive method of obtaining sedimentation rates, radioisotope dating can be confounded by several factors, including abrupt or gradual changes in sedimentation rates with time and major sediment disturbances such as dredging, dumping, installation of subsurface utilities, and scouring from propeller wash. Where dredging has occurred, a general sense of sedimentation rates can sometimes be obtained from examining dredging records and bathymetric surveys. However, past surveys and dredging records often are not accurate enough to develop a useful estimate. For these reasons, an attempt should be made to locate core stations in areas without large-scale anthropogenic or natural disturbances (such as slumping of sediments in areas with steep slopes).

Use of sediment traps is another common method for determining sedimentation and resuspension rates. Generally straight-sided glass cylinders of about 0.5 m in height, sediment traps are deployed in the water column and collect particulates in the water as they deposit out (Norton and Michelsen 1995). It is important to recognize that sediment traps collect both newly depositing material and resuspended bottom sediments. For this reason, they are commonly paired with radioisotope coring stations to allow the sediment trap accumulation rate to be divided into resuspension and deposition components. In addition, chemical fingerprinting and ^{210}Pb ratios can be used to estimate the relative quantities of bottom sediments and water column particulates in the trap (Michelsen et al. 1999). Information on the resuspension rate is a useful component of an NR evaluation, as it assists in determining whether resuspension, erosion, or transport of contaminants to downstream areas is likely to be an issue.

Source loading

Source loading information is needed to determine the input of contaminants to the system being modeled. This information can be obtained in various ways. Particulates collected in sediment traps can be analyzed for conventional parameters and contaminants. Coupled with the deposition rate, this analysis provides an estimate of mass loading to the surface sediments. Sediment traps are particularly helpful in quantifying the effects of nonpoint sources on a system, even when the nonpoint sources cannot be identified, although care must be taken to differentiate between water column particulates falling into the trap and resuspension of bottom sediments. Alternatively, a source loading and hydrodynamic model can be used in conjunction with effluent data to estimate loading from point sources such as combined sewer overflows or stormwater outfalls. This is particularly useful when source loading is likely to be spatially variable and certain areas will receive greater contaminant inputs than others.

It is important to avoid the common mistake of setting incoming chemical concentrations to 0 (i.e., assuming 100% clean sediment deposition). Even uncontaminated sediments will have a baseline metals concentration from crustal sources, and certain organic contaminants (e.g., PAHs, dioxins/furans, and PCBs) are ubiquitous at low levels due to atmospheric sources. Finally, there may be contaminants in the system from nonpoint or unknown sources. Failing to quantify chemical concentrations of depositing sediments will result in an overestimate of natural recovery rates when a burial and mixing model is used. Initial concentrations in sediments at the site also need to be well characterized, including any large-scale spatial variations that may be present.

Mixing depth

Particulates that settle out onto sediment may be mixed into underlying sediments through bioturbation and other processes (e.g., prop wash, resuspension by wind and storms). A key parameter that affects the outcome of NR predictions is the depth in surface sediments over which this mixing takes place. If mixing is largely due to bioturbation, the mixing depth is generally the same as the biologically active zone (the depth to which organisms colonize and actively work the sediments).

Mixing depth can often be determined through visual inspection of the sediment profile, which can be done in the field or using a sediment profile camera

(Rhoads and Germano 1986). The mixing depth may appear as a lighter-colored oxygenated zone, with darker anaerobic sediments below. A more quantitative estimate can be obtained from ^{210}Pb dating (Krishnaswamy et al. 1971; Koide et al. 1973; Christensen 1982). The mixing depth in the sediment ^{210}Pb profile normally appears as a region where ^{210}Pb concentrations are constant because of continual mixing of recently deposited material into the sediments. Mixing depths in marine waters typically range from 0 to 20 cm (WDOE 1995) and should be quantified to the nearest centimeter. The mixing depth can be 0 if the site is very quiescent and there is no biological activity, such as in highly contaminated anaerobic sediments.

Other site-specific parameters

Many complex hydrodynamic models such as WASP4 will require additional information on the sources and receiving environment, including bathymetry and shoreline configuration, source locations, stratification (if any) in the water column, density of water layers, current vectors, advection and dispersion rates, particle sizes and settling rates, and particle densities (PTI 1993). Additional information may be needed for sources with special characteristics, such as diffusers. Tidal variations are sometimes incorporated into regional hydrodynamic models (e.g., King County (Washington) Department of Natural Resources [KCDNR] 1998). Complex hydrodynamic modeling typically is appropriate when contaminant transport is being evaluated on a regional scale, in areas with multiple sources, or in complex hydrodynamic regimes such as rivers, but is generally not necessary for a screening-level evaluation of the potential for NR at less complex sites.

Chemical-specific parameters

Chemical-specific parameters may need to be obtained if processes such as diffusion and biodegradation are included in the model. These may include partitioning coefficients between sediment and pore water, degradation rates under aerobic or anaerobic conditions, and volatilization rates. Care must be taken that degradation rates are selected from studies that accurately reflect conditions in sediments, including aerobic or anaerobic conditions and weathering effects on bioavailability. Because laboratory-derived degradation rates reported in the literature are often higher than actual rates in weathered sediments, NR models are often run with degradation rates set to 0 as a conservative screening step. This avoids overoptimistic NR predictions that may result in eventual implementation of a costly contingency plan if monitoring shows that NR is not proceeding as expected. Degradation rates of 0 will also be appropriate for metals and most persistent chlorinated organics.

Empirical evidence

Evidence for NR is sometimes available if sediments at the site have been sampled several times over a period that is long enough to observe significant concentration changes (e.g., 10 years). At these sites, historical data may be sufficient to demonstrate that NR is occurring and to estimate future trends. Figure 11-2 shows NR of Hg in sediments at a site in Bellingham Bay, Washington, where a large source of clean sediments (the Nooksack River) is nearby (Hart Crowser 1997). These data were sufficient to project probable trends in remediation over time. Another method that may provide recovery information for sites that have not been previously sampled is analysis of core samples in conjunction with the core dating techniques described above. Multiple cores may need to be composited to allow for chemical analyses because very small sections are typically taken for core dating. Therefore, it is helpful to have already determined the specific chemicals of interest to limit the volume needed for chemical analyses as much as possible.

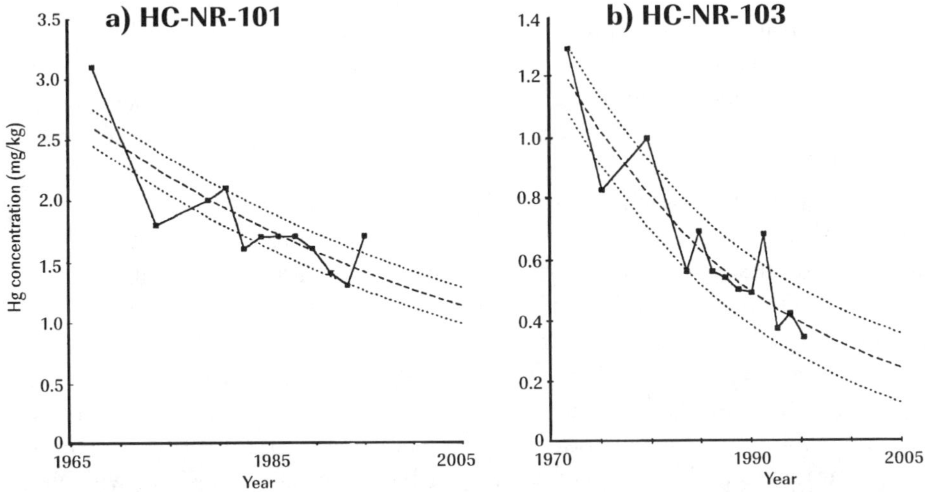

Figure 11-2 Natural recovery in Bellingham Bay, Washington. a) Natural recovery occurs more slowly in a core sample close to the source, subject to disturbance and recontamination by shipping traffic. b) Natural recovery occurs more rapidly in a core sample farther from the source, in a deeper area receiving significant clean sedimentation from the Nooksack River. (adapted from Hart Crowser 1997, with permission from Georgia Pacific West, Inc.)

MANAGEMENT AND POLICY CONSIDERATIONS RELATED TO NATURAL REMEDIATION

Once it has been determined that NR is possible or already occurring at a site, it may be added to the list of sediment management alternatives for detailed evaluation of net environmental benefits and costs. Although it might be assumed that NR should always be selected as the management alternative where it is occurring, there are several policy and management considerations that must be evaluated carefully prior to selecting the management alternative.

Among the first issues that should be considered is that of potential net environmental benefits of the various alternatives. The goals of sediment management programs typically include improving conditions in the environment, protecting public health, and ensuring suitable habitat for wildlife and fisheries resources. These goals should be incorporated into the criteria for the acceptability of management alternatives. The assessment should also consider the extent of risk reduction expected for each alternative. In some cases, there may be little net environmental benefit obtained from active remediation versus NR. The assessment of management alternatives should evaluate whether a protracted NR period interferes with designated beneficial uses of the water body or whether unacceptable risks to human health and the environment will continue to exist. Such impairments may have economic, environmental, and health consequences that outweigh benefits.

The potential for adverse impacts of the active remediation and NR alternatives should be considered. For instance, active remediation may impair the habitat through physical disturbance. This is an especially important consideration for certain sensitive environments, such as wetlands or eelgrass beds. Resuspension of contaminated sediments is a concern, whether from natural causes or as a result of remediation activities such as dredging. In some cases, these active measures may have a greater detrimental impact than allowing contamination to remain. In addition, there are situations in which active cleanup alternatives are not technically feasible. In these situations, existing risks should be managed to the extent practical, possibly by limiting human and/or wildlife access to the site and minimizing exposures as much as possible.

Cost considerations are often important. Natural remediation alternatives often appear to be the lowest cost alternative considered in feasibility studies because NR does not require construction or dredging. However, consideration of construction costs alone may not take into account all of the long-term costs associated with NR. There is often a significant up-front cost just to determine whether NR is feasible at a specific site. In addition, monitoring of the site throughout the NR period is often required by regulatory agencies. Monitoring plans developed for past sites typically include physical, chemical, and biological components. For smaller sites, the cumulative cost of long-term monitoring may exceed the cost of a one-time removal action. Finally, contingency plans may be required, in case recovery does not proceed as planned. If NR does not occur as predicted, additional expense may have been incurred for NR modeling and monitoring when an active remediation has to be implemented in the end.

If the presence of contamination affects other site uses, other economic costs may be incurred. For example, NR may not be the best option if a property transaction or development action is likely in the near future. The presence of contamination can complicate property transactions, lower property values, and result in costly changes to dredging and development plans to address contamination issues during permitting. If property transactions or development projects are likely to occur at the site in the foreseeable future, quick resolution of cleanup liability through an active one-time cleanup action may be desirable in order to obtain financing and assure potential buyers that cleanup has been completed. Certain site uses may also be incompatible with the presence of contamination, particularly those that result in resuspension and transport of contaminants offsite, such as waterfront construction or heavy vessel traffic. Natural remediation obviously cannot be selected for areas that may need to be dredged to maintain navigability.

Finally, there are potential costs associated with natural resource damages. Most state and federal cleanup regulations have provisions that allow natural resource trustees to recover the cumulative cost of past and continuing damages to natural resources, in addition to cleanup costs. Natural resource trustees generally include state and federal wildlife agencies and Native American tribes. Damages may include reductions in populations of fish, shellfish, or other aquatic life; closure of fisheries because of contamination; and impacts on habitat. Natural resource damages are often calculated using simple formulas related to the length of the exposure and the size of the contaminated area. If it can be demonstrated that contamination at the site is resulting in ongoing environmental risks or damages to natural resources such as fisheries, selecting NR as a cleanup method may increase one's liability for damages because of the increased time frame over which the contaminants remain at concentrations that may have adverse effects.

ENHANCED NATURAL REMEDIATION

In some areas, traditional active cleanup alternatives (e.g., capping, dredging, or disposal) may be impracticable, yet NR alone does not appear likely to reduce concentrations in sediments to levels that are sufficiently protective of human health and the environment within an acceptable time frame. This is often the case in areas that lack a significant source of clean sedimentation. Recently, a new approach has been taken to remediation that is designed to overcome these shortcomings and provide a low-cost enhancement or replacement for recovery processes such as sedimentation and subsequent mixing that are not naturally present in an area. This approach is variously referred to as "thin-layer capping" or "enhanced natural recovery."

Unlike full-scale, or "thick" capping, thin-layer caps are not designed to isolate underlying contaminated sediments from the environment. Rather, their purpose is to provide a source of clean sediment so that natural processes of mixing and ecosystem recovery can take place. The effect of this clean sediment can be modeled using NR models in the same way as natural sedimentation. The clean sediment is normally obtained from other dredging projects in the vicinity, and is relatively low-cost for that reason. If the remediation area is closer to the clean sediment dredging location than the usual disposal site, the dredger may be able to dispose of the material for little or no charge. In addition, the target thickness may be only 10 to 30 cm, requiring relatively little material to cover a large area. Repeated applications can be performed over a period of years, corresponding with local dredging cycles.

Various methods of application can be employed to obtain a thin-layer cap. At the Eagle Harbor site in Washington State, sediment was initially planned to be placed in "windrows" of material that would be thinnest (10 cm) between rows and thickest (30 to 45 cm) at the apex of each row (USACE 1992; USEPA 1992). The goal was for the layer to be thin over 50% of the area, to allow organisms to recolonize the surface of the sediment and work the clean material into the underlying contaminated sediments. An additional technique was tested during actual implementation, which consisted of slowly washing sediments of the back of a barge, rather than bottom-dumping or other methods that result in thicker placement or clumps of material. This method of placement is also effective in allowing some limited remediation of sensitive habitat areas such as eelgrass without smothering or otherwise damaging the plants.

Enhanced natural recovery offers a low-cost middle ground between traditional active remediation alternatives and NR, which may not be effective at all sites. It is also part of an emerging trend toward less costly remedial alternatives that are better integrated with ongoing maintenance dredging and other waterfront activities. Finally, it offers a method for remediation of sediment in habitats that could be damaged by more intrusive alternatives.

CONCLUSIONS

Natural remediation of sediments may be a viable management alternative for some areas. The sites that are most likely to be successfully remediated through natural processes include those where contaminant input sources are controlled, a significant source of clean sediment is nearby, and sediments are quiescent and fine grained with relatively low-level, widespread contamination that will recover in a reasonably short time frame without substantial losses to other media. Presence of even a limited benthic community may also assist NR by mixing and oxygenating the sediments.

Natural remediation may be selected when active remediation of sediments is technically infeasible, when sensitive habitats would be adversely impacted by active remediation, or when the cost of active cleanup is so high that it outweighs the benefits of remediation. Natural remediation may also be appropriate as a remedy for lower-risk areas within a site or lower-risk sites within a region, at least until all higher-priority areas have been addressed. Once remediation of higher-priority areas is completed, the lower-risk areas can be monitored to determine whether NR is occurring at an acceptable rate, and if not, enhanced NR and/or active remediation can then be initiated.

Natural remediation may be less appropriate for high-energy areas in which sediments may be resuspended and continually introduce contaminants into the food chain, areas where source control is not adequate to allow recovery, areas where natural resource damages or human health risks are significant and ongoing, and areas for which NR modeling predicts little likelihood of success. In these areas, active measures to reduce ecological and human health risks and decrease contaminant mobility and bioavailability will be appropriate.

REFERENCES

Akar PJ, Jirka GH. 1991. CORMIX2: An expert system for hydrodynamic mixing zone analysis of conventional and toxic multiport diffuser discharges. Washington DC: U.S. Environmental Protection Agency (USEPA). EPA-600/3-91/073.

Ambrose RB, Wool TA, Martin JM, Connolly JP, Schanz RW. 1992. WASP4, a hydrodynamic and water quality model: Model theory, user's manual, and programmer's guide. Athens GA: U.S. Environmental Protection Agency (USEPA), Environmental Research Laboratory.

[BCMOELP] British Columbia Ministry of the Environment, Lands and Parks. 1994. Approved and working criteria for water quality: 1994. Victoria BC: BCMOELP, Water Quality Branch, Water Management Division.

[BCMOELP] British Columbia Ministry of the Environment, Lands and Parks. 1995. Approved and working criteria for water quality: 1995. Victoria BC: BCMOELP, Water Quality Branch, Water Management Division.

Bothner MH, Jahnke RA, Peterson ML, Carpenter R. 1980. Rate of mercury loss from contaminated estuarine sediments. *Geochimt Cosmo Acta* 44:273–285.

Bushart SP, Bush B, Barnard EL, Bott A. 1988. Volatilization of extensively dechlorinated polychlorinated biphenyls from historically contaminated sediments. *Environ Toxicol Chem* 17(10):1927–1933.

Cardenas M, Gailani J, Ziegler CK, Lick W. 1995. Sediment transport in the lower Saginaw River. *Mar Freshwater Res* 46:317–347.

Carpenter SD, Smith W, Krieger-Brockett BB, Deming JW. 1996. An approach to estimating in situ PAH degradation rates in Eagle Harbor sediments. WA Sea Grant Workshop on Marine Extremozymes and Bioremediation; 1996 Sept; Seattle WA.

[CCME] Canadian Council of Ministers of the Environment. 1995. Protocol for the derivation of Canadian sediment quality guidelines for the protection of aquatic life. Ottawa ON: Environment Canada. CCME EPC-98E.

Christensen ER. 1982. A model for radionuclides in sediments influenced by mixing and compaction. *J Geophys Res* 87:566–572.

Connaughton DF, Stedinger JR, Lion LW, Shuler ML. 1993. Description of time-varying kinetics: Release of naphthalene from contaminated soils. *Environ Sci Technol* 27:2397–2403.

[DEQ] Oregon Department of Environmental Quality. 1998. Contaminated sediment management strategy. 9/10/98 Revised Draft. Portland OR: Waste Management and Cleanup Division.

Drake DE, Sherwood CR, Wilberg PL. 1994. Predictive modeling of the natural recovery of the contaminated effluent-affected sediment, Palos Verdes margin, Southern California. Menlo Park CA: U.S. Geological Survey (USGS).

[EBDRP] Elliott Bay/Duwamish Restoration Panel. 1994. Concept document, Elliott Bay/ Duwamish Restoration Program. Seattle WA: National Oceanic and Atmospheric Administration (NOAA) Damage Assessment and Restoration Center.

[EC] Environment Canada. 1997a. Canadian sediment quality guidelines for arsenic: Supporting document. Ottawa ON: Guidelines and Standards Division. Draft report.

[EC] Environment Canada. 1997b. Canadian sediment quality guidelines for cadmium: Supporting document. Ottawa ON: Guidelines and Standards Division.

[EC] Environment Canada. 1997c. Canadian sediment quality guidelines for chlordane, dieldrin, endrin, heptachlor epoxide, and lindane: Supporting document. Ottawa ON: Guidelines and Standards Division. Draft report.

[EC] Environment Canada. 1997d. Canadian sediment quality guidelines for chromium: Supporting document. Ottawa ON: Guidelines and Standards Division. Draft report.

[EC] Environment Canada. 1997e. Canadian sediment quality guidelines for copper: Supporting document. Ottawa ON: Guidelines and Standards Division. Draft report.

[EC] Environment Canada. 1997f. Canadian sediment quality guidelines for lead: Supporting document. Ottawa ON: Guidelines and Standards Division. Draft report.

[EC] Environment Canada. 1997g. Canadian sediment quality guidelines for mercury: Supporting document. Ottawa ON: Guidelines and Standards Division.

[EC] Environment Canada. 1997h. Canadian sediment quality guidelines for polychlorinated biphenyls. Ottawa ON: Guidelines and Standards Division. Draft report.

[EC] Environment Canada. 1997i. Canadian sediment quality guidelines for polycyclic aromatic hydrocarbons: Supporting document. Ottawa ON: Guidelines and Standards Division. Draft report.

[EC] Environment Canada. 1998. Canadian sediment quality guidelines for DDTs: Supporting document. Ottawa ON: Guidelines and Standards Division. Draft report.

Gailani J, Ziegler CK, Lick W. 1991. The transport of sediments in the Fox River. *J Great Lakes Res* 17:479–494.

Harms H, Zehnder AJB. 1995. Bioavailability of sorbed 3-chlorodibenzofuran. *Appl Environ Microbiol* 61:27–33.

Hart Crowser. 1997. Remedial investigation report, Whatcom Waterway site, Bellingham, Washington. Draft report. Prepared for Georgia-Pacific West, Inc., Bellingham WA.

Hatzinger PB, Alexander M. 1995. Effect of aging of chemicals in soil on their biodegradability and extractability. *Environ Sci Techol* 29:537–545.

[IJC] International Joint Commission. 1997. Overcoming obstacles to sediment remediation in the Great Lakes Basin. Great Lakes Water Quality Board. Internet document: http://www.ijc.org/boards/wqb/sedrem.html

[KCDNR] King County Department of Natural Resources. 1998. King County combined sewer overflow water quality assessment for the Duwamish River and Elliott Bay. Volume IV, Hydrodynamic and fate and transport numerical model. Draft report. Seattle WA: Duwamish River and Elliott Bay Water Quality Assessment Team.

Koide M, Bruland KE, Boldberg ED. 1973. Th-228/Th-232 and Pb-210 geochronologies in marine and lake sediments. *Geochimt Cosmo Acta* 37:1171–1183.

Krieger-Brockett BB, Herwig RP, Deming JW. 1997. An assessment of organic contaminant biodegradation rates in marine environments. Proceedings of the Fourth International Symposium: In Situ and On-Site Bioremediation; 1997 April 28–May 1. 4(4):427–432.

Krishnaswamy D, Lal D, Martin JM, Meybeck M. 1971. Geochronology of lake sediments. *Earth Plant Science Letters* 11:407–414.

Landolt M, Kalman D, Nevissi A, van Belle G, Van Ness K, Hafer F. 1987. Potential toxicant exposure among consumers of recreationally caught fish from urban embayments of Puget Sound: Final report. Rockville MD: National Oceanographic and Atmospheric Administration (NOAA). NOS-AMO-33.

Lick W, Lick J, Ziegler CK. 1994. The resuspension and transport of fine-grained sediments in Lake Erie. *J Great Lakes Res* 20(4):599–612.

Liu X, Sokol RC, Kwon OS, Bethoney CM, Rhee GY. 1996. An investigation of factors limiting the reductive dechlorination of polychlorinated biphenyls. *Environ Toxicol Chem* 15(10):1738–1744.

MacDonald DD. 1994. Approach to the assessment of sediment quality in Florida coastal waters. Volume 1, Development and evaluation of the sediment quality assessment guidelines. Tallahassee FL: Florida Department of Environmental Protection.

Madsen EL, Bilotta SE. 1996. Oxygen limitations and aging as explanations for the field persistence of naphthalene in coal tar–contaminated surface sediments. *Environ Toxicol Chem* 15:1876–1893.

McCall PJ, Agin GL. 1985. Desorption kinetics of picloram as effected (sic) by residence time in the soil. *Environ Toxicol Chem* 4:37–44.

McCarty PL, Rittmann BE, Bouwer EJ. 1984. Microbiological processes affecting chemical transformations in groundwater. In: Bitton G, Gerba CP, editors. Groundwater pollution microbiology. New York NY: Wiley.

Michelsen TC, Boatman CD, Norton D, Ebbesmeyer CC, Floyd T, Francisco MD. 1999. Resuspension and transport of contaminated sediments along the Seattle waterfront. Part 1: Field investigations and conceptual model. *J Mar Environ Engin* 5:35–65.

Michelsen TC, Boatman CD, Norton D, Ebbesmeyer CC, Francisco MD. 1998. Transport of contaminants along the Seattle waterfront: Effects of vessel traffic and waterfront construction activities. *Water Sci Tech* 37(6–7):9–15.

Mousa MA, Quenson JF III, Chou K, Boyd SA. 1996. Microbial dechlorination alleviates inhibitory effects of PCBs on mouse gamete fertilization in vitro. *Environ Sci Technol* 30:2087–2092.

[NOAA] National Oceanic and Atmospheric Administration. 1995. The utility of AVS/EqP in hazardous waste site evaluations. Seattle WA: Office of Ocean Resources Conservation and Assessment. NOAA Technical Memorandum NOS ORCA 87.

Norton D, Michelsen TC. 1995. Elliott Bay waterfront recontamination study. Olympia WA: Washington Department of Ecology, Environmental Investigations and Laboratory Services. Pub. No. 95–335.

[NRC] National Research Council. 1997. Contaminated sediments in ports and waterways. Washington DC: National Academy Pr. 295 p.

[NYDEC] New York Department of Environmental Conservation. 1994. Interim guidance: Freshwater navigational dredging. Division of Water.

Officer CB, Lynch DR. 1989. Bioturbation, sedimentation, and sediment-water exchanges. *Estuar Coast Shelf Sci* 28:1–12.

Persaud D, Jaagumagi R, Hayton A. 1992. Guidelines for the protection and management of aquatic sediment quality in Ontario. Toronto ON: Ontario Ministry of the Environment, Water Resources Branch.

Pignatello JJ, Ferrandino FJ, Huang LQ. 1993. Elution of aged and freshly added herbicides from a soil. *Environ Sci Technol* 27:1563–1571.

[PTI] PTI Environmental Services. 1989. Puget Sound contaminated sediments impact and recovery zones workshop summary. Olympia WA: Washington Department of Ecology.

[PTI] PTI Environmental Services. 1991a. Recommended sediment impact and recovery zone models and case study analysis. Olympia WA: Washington Department of Ecology.

[PTI] PTI Environmental Services. 1991b. Sediment Management Standards sediment site ranking: SEDRANK guidance document. Olympia WA: Washington Department of Ecology.

[PTI] PTI Environmental Services. 1993. WASP application guidance manual. Olympia WA: Washington Department of Ecology.

Quenson JF III, Mousa MA, Boyd SA, Sanderson JT, Froese KL, Giesy JP. 1988. Reduction of aryl hydrocarbon receptor-mediated activity of polychlorinated biphenyl mixtures due to anaerobic microbial dechlorination. *Environ Toxicol Chem* 17(5):806–813.

Renner R. 1998. Natural remediation of DDT, PCBs debated. *Environ Sci Technol* 32(15):360A–363A.

Rhoads DC, Germano JD. 1986. Interpreting long-term changes in benthic community structure: A new protocol. *Hydrobiologia* 142:291–308.

Safe SH. 1990. Polychlorinated biphenyls (PCBs), dibenzo-p-dioxins (PCDDs), dibenzofurans (PCDFs), and related compounds: Environmental and mechanistic considerations that support the development of toxic equivalency factors (TEFs). *Crit Rev Toxicol* 21(1):51–88.

Safe SH. 1994. Polychlorinated biphenyls: Environmental impact, biochemical and toxic responses, and implications for risk assessment. *Crit Rev Toxicol* 24:87–149.

Sandoli RL, Ghiorse WC, Madsen EL. 1996. Regulation of microbial phenanthrene mineralization in sediment samples by sorbent-sorbate contact time, inocula, and gamma irradiation-induced sterilization artifacts. *Environ Toxicol Chem* 15(11):1901–1907.

Sokol RC, Bethoney CM, Rhee GY. 1998. Effect of Aroclor 1248 concentration on the rate and extent of polychorinated biphenyl dechlorination. *Environ Toxicol Chem* 17(10):1922–1926.

Steinberg SM, Pignatello JJ, Sawhney BL. 1987. Persistence of 1,2-dibromoethane in soils: Entrapment in intraparticle micropores. *Environ Sci Technol* 21:1201–1208.

Toy KA, Polissar NL, Liao S, Mittelstaedt GD. 1996. A fish consumption survey of the Tulalip and Squaxin Island Tribes of the Puget Sound region. Prepared by Mountian-Whisper-Light Statistical Consulting for the Squaxin Island and Tulalip Tribes.

[USACE] U.S. Army Corps of Engineers. 1992. Eagle Harbor, Washington, thin-layer placement of dredged material, feasibility analysis. Seattle WA: Seattle District.

[USEPA] U.S. Environmental Protection Agency. 1992. West Harbor Operable Unit, Wyckoff/Eagle Harbor Superfund site. Record of decision. Seattle WA: Region 10.

[USEPA] U.S. Environmental Protection Agency. 1993a. Assessment and remediation of contaminated sediments (ARCS) program: Risk assessment and modeling overview document. Chicago IL: Great Lakes National Program Office. EPA 905-R93-007.

[USEPA] U.S. Environmental Protection Agency. 1993b. Sediment quality criteria for the protection of benthic organisms: Dieldrin. Washington DC: Office of Water. EPA-822-R-93-015.

[USEPA] U.S. Environmental Protection Agency. 1993c. Sediment quality criteria for the protection of benthic organisms: Endrin. Washington DC: Office of Water. EPA-822-R-93-016.

[USEPA] U.S. Environmental Protection Agency. 1994a. Assessment and remediation of contaminated sediments (ARCS) program, remediation guidance document. Chicago IL: Great Lakes National Program Office. EPA 905-R94-003.

[USEPA] U.S. Environmental Protection Agency. 1994b. EPA's contaminated sediment management strategy. Washington DC: Office of Water.

[USEPA] U.S. Environmental Protection Agency. 1995. An SAB report: Review of the agency's approach for developing sediment criteria for five metals. Washington DC: Science Advisory Committee. EPA-SAB-EPEC-95-0XX.

[USEPA] U.S. Environmental Protection Agency. 1998. Realizing remediation: A summary of contaminated sediment remediation activities in the Great Lakes Basin. Chicago IL: Great Lakes National Program Office.

Varanasi U, Casillas E, Arkoosh MR, Hom T, Misitano DA, Brown DW, Chan S-L, Collier TK, McCain BB, Stein JE. 1993. Contaminant exposure and associated biological effects in juvenile Chinook salmon (*Oncorrhyncus tshawytscha*) from urban and nonurban estuaries of Puget Sound. Seattle WA: National Oceanic and Atmospheric Administration (NOAA), National Marine Fisheries Service. NMFS-NWFSC-8.

[WDOE] Washington Department of Ecology. 1990. Commonly required environmental permits for Washington State. Olympia WA: WDOE. Pub. No. 90-29.

[WDOE] Washington Department of Ecology. 1991a. Final environmental impact statement for the Washington State Sediment Management Standards, Chapter 173-204 WAC. Olympia WA: Sediment Management Unit.

[WDOE] Washington Department of Ecology. 1991b. Sediment cleanup standards user's manual. Olympia WA: Sediment Management Unit.

[WDOE] Washington Department of Ecology. 1991c. Sediment management standards, Chapter 173-204 WAC. Olympia WA: Sediment Management Unit.

[WDOE] Washington Department of Ecology. 1995. Sediment Sampling and Analysis Plan Appendix: Guidance on the development of sediment sampling and analysis plans meeting the requirements of the Sediment Management Standards: Chapter 173-204 WAC. Olympia WA: Sediment Management Unit.

[WDOE] Washington Department of Ecology. 1997. Developing health-based sediment quality criteria for cleanup sites: A case study report. Olympia WA: Sediment Management Unit. Pub. No. 97-114.

[WDOE] Washington Department of Ecology. 1998. Toxics Cleanup Program and Sediment Management Unit site files and reports. Unpublished. Bellevue WA and Olympia WA.

Wilcock RJ, Corban GA, Northcott GL, Wilkins AL, Langdon AG. 1996. Persistence of polycyclic aromatic compounds of different molecular size and water volubility in surficial sediment of an intertidal sandflat. *Environ Toxicol Chem* 15(5):670–676.

12

Natural Remediation in Wetlands

Steven M. Jones, Daniel W. Smith, Alan F. Weston

WETLANDS ARE A DIVERSE GROUP of natural ecosystems that range from salt marshes to prairie potholes to riparian forests to forested swamps. Despite the diversity of biological, chemical, and physical attributes exhibited by wetland ecosystems, all have 3 characteristics in common:

1) Standing water (inundation) or high water table (saturation) for some period of time during the biological growing season
2) Hydric soils that provide a predominantly anaerobic environment for chemical and biological processes
3) Specialized plants (hydrophytes) adapted to periodic or permanent inundation or saturation.

Under U.S. regulations, wetlands are defined as "those areas that are inundated or saturated by surface water or groundwater at a frequency and duration sufficient to support, and under normal circumstances, do support vegetation adapted typically for life in saturated soil conditions" (33 Code of Federal Regulations [CFR] 328.3(b) 1984).

Historically, wetlands have often been perceived as areas with little or no value, or worse, as breeding grounds for mosquitoes and disease. As such, many wetland areas were filled or drained to provide land for agriculture or other beneficial uses. Prior to wetland protection in the U.S., it was common practice to use wetlands for disposal of wastes. In attempts to manage mosquito-borne diseases, pesticides were intentionally introduced into wetlands.

Natural Remediation of Environmental Contaminants. Michael Swindoll et al., editors.

Today, we recognize the many ecological and human-use benefits provided by wetlands. Among the benefits, wetlands provide storage capacity for storm and flood water, and enhance water quality by trapping sediments, assimilating nutrients, and chemically transforming pollutants. A great diversity of plants and wildlife, including many threatened or endangered species, live, breed, or forage in wetlands. Wetlands serve as nursery areas for many valuable recreational fish species. Wetlands are often quite rich in nutrients and organic matter and are among the most productive ecosystems (Odum 1971; Mitsch and Gosselink 1993), producing the biomass that forms the base of complex food webs. Our current appreciation of the value of wetlands is reflected in the protection afforded them under the federal Clean Water Act (1977) and regulations enacted by several individual states (e.g., New Jersey).

The ability of wetlands to cleanse contaminated waters has long been recognized by the engineering community. Wetlands are used in stormwater management systems to reduce sediment loads and nutrients in surface runoff (Reddy and Smith 1987; Reed et al. 1995). Constructed wetlands are used to treat acid mine drainage (Wildeman et al. 1992) and wastewater (Hammer 1990; Kadlec and Knight 1996). Both constructed and natural wetlands are used to treat groundwater and surface water from sites where discharges of waste and chemicals, such as heavy metals, petroleum hydrocarbons (PHs), and chlorinated solvents, have occurred.

Although there is an abundance of scientific literature on the use of wetlands as water treatment systems, there is relatively little that addresses remediation and risk reduction in wetlands containing contaminants. The paucity of this literature is due in part to the priority given by regulators and the regulated community to conventional remedial measures, such as excavation, to address contamination in wetlands. Unfortunately, these conventional remediation approaches often destroy the integrity of the wetlands ecosystem by removing vegetation and hydric soils, altering hydrology, and displacing wildlife and other biota that use the wetlands. Because wetlands systems represent a delicate balance of hydrology, soils, and vegetation, it may be impossible to fully, or even partially, re-create wetland functions following remediation. Moreover, even under the best of conditions and the most optimistic scenarios, restoring ecological structure and functions, such as habitat for wildlife, to "remediated" forested wetlands can be very long term—a minimum of 30 to 50 years. Conventional methods of remediation are also problematic because even if basic functions can be restored, the resulting wetlands may vary significantly from natural wetlands. With the recognition of the adverse impacts associated with active remediation, natural remediation (NR) in wetlands is becoming an increasingly important risk management consideration in the evaluation of remedial alternatives.

This chapter identifies and discusses the significance of those processes in wetlands that contribute to NR of contaminants. As used throughout this chapter, natural remediation refers to nonanthropogenic processes and mechanisms that reduce the potential for risk to ecological receptors. We discuss the characteristics of wetlands and how these attributes can contribute to the reduction of risk posed by contaminants to biota. Further, we discuss the biological, chemical, and physical processes that occur in wetlands and how these processes contribute to NR. We present several case histories in which NR has been documented or is part of a selected remedy to illustrate the viability of NR as a risk management alternative for wetlands. We also present the tradeoffs between NR and active remediation of contaminants in wetlands and we offer conclusions.

CHARACTERISTICS OF WETLANDS

The co-occurrence of 3 characteristics distinguishes wetlands from other ecosystems:

1) Saturation or inundation for some period of time during the biological growing season
2) Hydric soils
3) Hydrophytic vegetation.

The combination of these wetland characteristics produces a suite of physical, biological, and chemical conditions that affect the fate and ecological impact of contaminants in wetlands and that determine the nature and extent of NR. Wetland characteristics and how they affect contaminants are discussed in this section.

Wetlands hydrology

Saturation or inundation for some period during the growing season is the most important attribute of wetlands, creating conditions that distinguish wetlands from other ecosystems. Accordingly, the hydrology of a wetland is the major determinant of its environmental conditions. Hydrology is defined by the source of water, the quantity of water, and the temporal pattern of inundation or saturation. Hydrology determines the spatial and temporal extent of the anaerobic and reducing conditions, which are characteristic of wetlands. The characteristics of hydric soil, as well as the species of hydrophytic vegetation inhabiting wetlands, depend on the hydrology of the specific wetlands. Fate, transport, and ecological impact of contaminants are strongly influenced by hydrology. The primary routes of contaminant transport into and out of wetlands are a function of hydrology and are summarized in Figure 12-1.

Water enters wetlands by surface runoff (i.e., storm water), groundwater inflow, and precipitation. The relative contribution of these sources can influence the kinds of contaminants and the NR processes occurring in a particular wetland. For instance, surface runoff can introduce contaminants dissolved in water or adsorbed to particulates suspended in storm water. Therefore, a variety of contaminants with

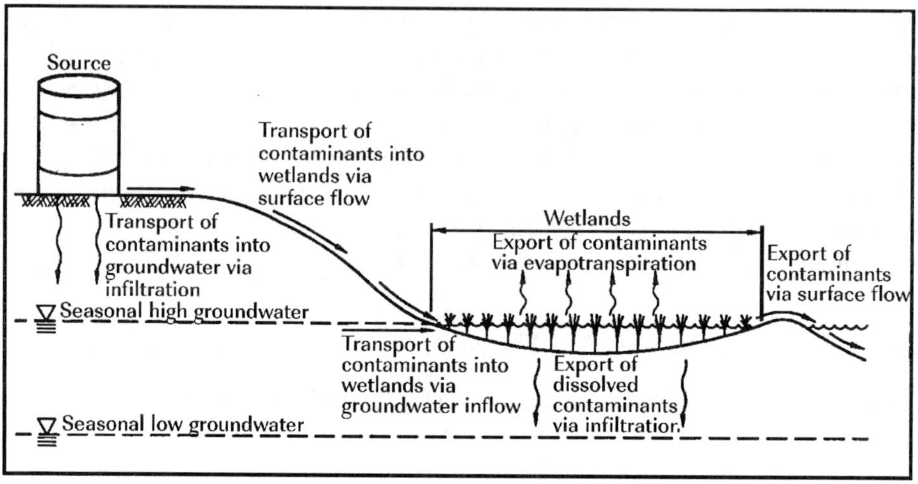

Figure 12-1 Routes of contaminant transport in and out of wetlands

a wide range of properties (e.g., solubilities) may be transported into wetlands by surface runoff. However, contaminants transported by groundwater must be in the dissolved phase, which minimizes transport and potential impacts of less soluble and hydrophobic contaminants.

Water is transported out of wetlands primarily by 4 mechanisms:

1) Volatilization
2) Transpiration
3) Infiltration
4) Surface outflow.

Volatilization occurs as water vaporizes directly from wetlands into the atmosphere. Transpiration occurs as water taken into the tissue of vascular plants is released to the atmosphere. The collective loss of water through volatilization (evaporation) and transpiration, termed "evapotranspiration," can result in a significant loss of water from wetlands. Water transport by infiltration occurs as water moves downward through the soil. Infiltrating water can be held in the underlying soil or continue to flow downward to the groundwater table. The relative importance of transport mechanisms often changes on a seasonal basis. For instance, infiltration, or recharge, generally occurs during the drier portions of the year when the water table recedes. During periods of high water table, the direction of flow can switch, and groundwater flows, or discharges, into a wetland.

Outflows also influence the types and effectiveness of NR processes. Wetlands with well-defined outlets can transport both dissolved and adsorbed chemicals out of the wetlands. Water-soluble contaminants can also be transported out of wetlands through evapotranspiration and through infiltration to the groundwater. Depending on the solubility of the contaminant and the properties of the subsurface soil (e.g., organic content, cation exchange capacity [CEC], grain size), attenuation of the contaminant in the infiltrating water can occur because of adsorption to soil.

In general, because of the typically minimal relief and abundant vegetation of most wetlands, water currents are negligible. Consequently, wetlands more typically function as traps for particles and particle-bound contaminants. Export of contaminants adsorbed to particulate matter occurs only when flows are of sufficient velocity to transport the particles. Tidal wetlands are an exception because they tend to have higher energy flows sufficient to transport sediments and particulate matter.

Based on the source of water and position in the landscape, there are 4 general categories of nontidal freshwater wetlands (Novitski 1989):

1) Surfacewater-depression wetlands
2) Surfacewater-slope wetlands
3) Groundwater-depression wetlands
4) Groundwater-slope wetlands.

Each of these wetlands types has unique properties that contribute to the fate and transport of environmental contaminants. Novitski's classification system and mechanisms of NR for each type of wetlands are summarized in Table 12-1.

Table 12-1 Nontidal wetlands defined by hydrology

Type of wetlands	Primary inflows	Primary outflows	Primary mechanisms of natural remediation
Surfacewater-depression	Surface runoff	Evaporation, infiltration	Deposition and burial Biological transformations Chemical transformations
Surfacewater-slope	Surface runoff	Surface flow	Limited deposition and burial Export of biologically or chemically transformed chemicals
Groundwater-depression	Groundwater, surface water	Evaporation, infiltration	Deposition and burial Biological transformations Chemical transformations
Groundwater-slope	Groundwater, surface water	Surface flow	Limited deposition and burial Export of biologically or chemically transformed chemicals

Hydric soils

Wetland soils, commonly referred to as "hydric soils," are saturated or flooded during the growing season for a time sufficient to allow the development of anaerobic conditions that favor growth and regeneration of hydrophytic vegetation (U.S. Drug Administration [USDA] 1985). During periods of saturation, 2 distinct chemical environments occur within wetlands: an anaerobic or anoxic zone and an aerobic or oxygenated zone. Through most of the wetlands soil profile, oxygen is depleted and anaerobic conditions prevail. These anaerobic conditions contrast to the aerobic conditions typical of most nonsaturated (e.g., nonwetland) soils. In addition to the primarily anaerobic environment in wetland soils, a thin aerobic layer occurs at the surface of the wetlands. This aerobic layer can occur at the air–water interface (inundated wetlands) or the air–soil interface (saturated wetlands). Aerobic conditions can also occur at the root–soil interface (rhizosphere) where oxygen is released from plant roots.

Hydric soils tend to be very high in organic matter because of the often abundant growth of vegetation and relatively slow rates of decomposition associated with anaerobic conditions. Hydric soils often contain 20% or more organic C, which is an order of magnitude higher than many upland soils and aquatic sediments. The high organic content of wetlands soil provides an abundance of sites for adsorption of metals (e.g., Powell et al. 1997) and hydrophobic organic chemicals (Detenbeck et al. 1996), thus contributing to the sequestering and immobilization of many contaminants.

In general, most chemical reactions that are critical to the fate and transport of contaminants in wetlands occur in soils. These chemical processes are a function of both the physical and chemical characteristics of the soil and the microbial community contributing to the chemical environment of wetlands (Mitsch and Gosselink 1993). Chemical processes within wetlands soils that contribute to NR of contaminants include oxidation–reduction (redox) reactions, hydrolysis, adsorption, precipitation, ion exchange, and chelation.

Hydrophytic vegetation

Hydrophytes are specialized plants adapted to the periodic or permanent inundation or saturation that exists in wetlands. Hydrophytes come in a variety of growth forms including trees, shrubs, grasses, forbs, submerged aquatic vegetation, and periphyton. When compared to the many thousands of plant species that live in terrestrial habitats, relatively few plants are adapted for life in wetlands. The primary challenge faced by plants inhabiting wetlands is a reduced supply of oxygen in soil available for root uptake. Limited availability of oxygen to plant roots results in a series of physiological and biochemical changes within the plant that affect the ability of the plant to survive. For example, a switch to anaerobic metabolism, in response to a reduced supply of oxygen, not only decreases energy, in the form of

ATP, produced by the plant, but also results in production of end products such as ethanol, which can be toxic to the plant (Mitsch and Gosselink 1993).

Hydrophytes promote a number of processes that facilitate NR of contaminants. They slow the velocity of water transported through the wetlands, allowing suspended particles with adsorbed contaminants to settle out of the water column. Wetland plants can take up contaminants dissolved in water or desorbed from particulate matter, and they may store, chemically transform, or release those contaminants to the atmosphere through transpiration. These and other remediation processes are discussed in more detail the following section.

WETLANDS NATURAL REMEDIATION PROCESSES

Contaminants are introduced into wetlands by 4 routes:

1) Surfacewater flows
2) Groundwater
3) Atmospheric deposition
4) Direct discharge or dumping.

Contaminants that enter wetlands may undergo NR through one or more biological, chemical, or physical processes, as summarized in Table 12-2. These processes are not unique to wetlands; however, some may be enhanced or accelerated in wetlands, while others (e.g., oxidation and biodegradation) tend to be slowed or retarded in the mostly anaerobic conditions prevalent in wetlands (e.g., see Madsen et al. 1996).

It should be noted that not all of the processes identified in Table 12-2 occur in all wetlands, nor do they necessarily result in benefits to the environment or detoxification of chemicals. The anoxic conditions that prevail in many wetlands can retard or prevent degradation of chemicals that are labile to oxidation (Madsen et al. 1996; Vink and Van Der Zee 1997). In some cases, transformation processes in wetlands may even produce more toxic forms of contaminants. For example, microbial transformation of relatively nontoxic inorganic Hg to more toxic, more bioavailable forms of Hg may be facilitated in wetland environments under anaerobic conditions.

Biological processes

Biological NR processes are those mechanisms in which biota (flora and fauna) are directly and primarily responsible for reducing potential risks attributable to contaminants. Typically, plants and microbes are the major biotic contributors to NR in wetlands, although other biota may be important. The 2 primary biological mechanisms that contribute to NR are transformations (e.g., biodegradation) and uptake.

Table 12-2 Summary of natural remediation processes in wetlands

Category	Process	Mechanisms for remediation
Biological	Plant uptake	Removal of contaminants from water and soil Temporary or permanent storage of contaminants in plant tissue Release of contaminants to atmosphere through transpiration Biochemical transformation to less toxic forms
	Microbial biodegradation	Degradation and transformation of contaminants to less toxic forms Facilitation of reduction reactions Facilitation of chelation of metals Decomposition of plant litter and organic matter
Chemical	Oxidation, reduction	Formation of insoluble metal sulfides Dechlorination Enhanced degradation of some organic contaminants
	Adsorption	Removal of hydrophobic contaminants from solution Trapping of metals on soil particles and decomposing litter Cation exchange
	Precipitation	Transformation of contaminants to less soluble forms
	Chelation	Binding of metals and other contamination and reduced availability to biota
Physical	Photolysis	Chemical transformation of contaminants to less toxic forms
	Dilution	Reduced contaminant concentrations in surface water
	Deposition, burial	Reduced availability of contaminants adsorbed to sediment particles to biota
	Export	Transport of contaminants out of wetlands
	Evaporation, volatilization	Loss of contaminants to atmosphere

Vascular plants may contribute to the NR of contaminants through sequestration, biodegradation, and evapotranspiration processes that occur following uptake of the contaminants into plant tissue. Cattail (*Typha* sp.), bulrush (*Scirpus* sp.), and duckweed (*Lemna* sp., *Spirodela* sp., *Wolfiella* sp.), are common wetlands plants that have well-documented abilities to take up a variety of contaminants (Reed et al. 1995; Kadlec and Knight 1996). In general, chemicals with a log K_{OW} between 0.5 and 3.0—such as benzene, toluene, ethylbenzene, xylenes, chlorinated solvents (e.g., trichloroethylene [TCE], tetrachloroethylene [PCE]), and nitrotoluene (from ammunition wastes, e.g., 2,2,6-trinitrotoluene [TNT])—can be transported into plant tissue (Schnoor et al. 1995; Best et al. 1997; Price et al. 1997). Plant uptake of metals, including As, Cd, Cu, Fe, and Pb, has also been reported (Chan et al. 1982; Pratt and Pluto 1989; Otte and Ernst 1994). Some hydrophytes are able to metabolize and sequester chemicals into various tissue types, allowing for additional uptake and storage of chemicals. For example, *Potamogeton* (pond weed) differentially concentrates some heavy metals in stem and leaf tissues and others in roots and rhizomes (Chan et al. 1982). Plant uptake of contaminants may lower the potential for ecological risk by reducing contaminant mobility and bioavailability or by transforming the contaminants to less hazardous forms. The latter not only reduces toxicity to the plant but also reduces potential adverse effects that may occur as the plant matter is consumed by herbivores and detritivores. In assessing potential risks, it should be recognized that the sequestration and concentration of contaminants in plants can pose additional risk to wildlife species (e.g., waterfowl and muskrats) that forage on plant materials.

Biodegradation of contaminants is a major mechanism of NR in wetlands (Kadlec and Knight 1996). A variety of wetland organisms, including bacteria and fungi, are able to degrade and transform contaminants. Given the presence of both aerobic and anaerobic zones, along with an abundance of organic C, wetlands typically support a high diversity and abundance of microorganisms, providing a wide range of potential degradation pathways (Tiedje 1993). Some bacteria and fungi secrete extracellular enzymes that can facilitate NR. Amino acids secreted by bacteria also facilitate chelation of trace metals in wetland soils. Bacteria also play an important role in the abiotic chemical reactions that occur in wetlands, such as redox reactions, by altering the chemical environment.

Hydrophytes can also facilitate biodegradation processes. Plants provide surface area for sorption of contaminants as well as attachment of bacteria and fungi, thus bringing contaminants and biodegraders in close proximity. Mycorhyzae, fungi in symbiotic relationships with plant root systems, can secrete enzymes that facilitate degradation of organic materials. Some hydrophytes also release enzymes that break down contaminants (Beelen and Burris 1995). Schnoor and colleagues (1995) have isolated and identified several plant enzymes specific to some degradation processes. For example, the enzymes nitroreductase and laccase facilitate the breakdown of TNT in organic detritus. The plant enzyme dehalogenase facilitates degradation of TCE and other chlorinated solvents. Peroxidase and nitrilase are other enzymes excreted by hydrophytes that have been isolated and demonstrated to be important in the degradation of organic chemicals (Schnoor et al. 1995). The ability of plants to metabolize chemicals to less toxic forms not only reduces toxicity to the plant but also reduces potential adverse effects that may occur as the plant matter is consumed by herbivores and detritivores.

Chemical processes

Naturally occurring chemical processes can result in changes in the structure, properties, and toxicity of contaminants. Chemical processes, including redox reactions, adsorption, precipitation, chelation, and photolysis, can contribute to reduction of ecological risk in wetlands. The chemical environments found in a typical wetland and associated NR processes are presented in Figure 12-2. Several processes may operate simultaneously or sequentially to facilitate NR.

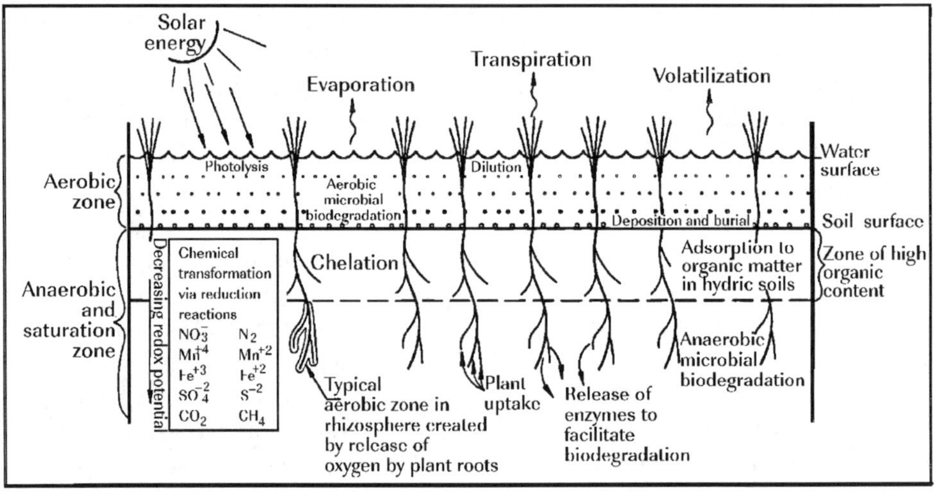

Figure 12-2 Cross-section of a typical wetlands and the zones where processes of natural attenuation occur

Redox reactions are common in wetlands (Mitsch and Gosselink 1993) and are often facilitated either directly or indirectly by microbial activity. Oxidation typically occurs in the aerobic zones of wetlands, whereas reduction occurs in the more extensive anaerobic zones. In wetlands that are characterized by periodic inundation and saturation, the chemical environment may alternate between anaerobic and aerobic, sometimes favoring reduction and at other times favoring oxidation. Although reduction is the predominant reaction in wetlands, it should be noted that oxidation and other oxygen-dependent chemical reactions that occur within the aerobic zones of wetlands can also be important mechanisms for NR. For example, Fe, which is toxic to many plants, can be immobilized in the rhizosphere through the oxidation of ferrous iron to insoluble iron oxide. Table 12-3 identifies the major elements in wetland soils, the oxidized and reduced forms, and the redox potential at which the switch from oxidation to reduction occurs. Examples of contaminants that are reduced to less toxic forms under anaerobic conditions are also provided.

Several NR mechanisms are favored in the anaerobic environment that is characteristic of wetlands. Denitrification and methanogenesis are favored under the reducing conditions found in saturated soil. Cd, Cu, Pb, Hg, Ag, and Zn form insoluble sulfides in reducing environments (Debusk et al. 1996). Dechlorination of

Table 12-3 Redox potential in wetlands

Electron acceptor	Oxidized form	Reduced form	eV for reduction (mV)	Contaminants reduced
Nitrogen	NO_3^-	N_2O, N^2, NH_4^+	250 to 350	Nitrobenzene, aniline
Manganese	Mn^{4+}	Mn^{2+}	220 to 225	Hydrocarbons
Iron	Fe^{3+}	Fe^{++}	120 to 00	Tetrachloroethene, trichloroethene, chromium VI
Sulfur	SO_4^{2-}	S^{2-}	–75 to –150	Nitrobenzene
Carbon	CO_2	CH_4	–250 to –350	Nitrobenzene, trichloroethylene (TCE)

some chlorinated hydrocarbons, polychlorinated biphenyls (PCBs), and organophosphates occurs more rapidly under anaerobic conditions than under aerobic conditions (Pratt and Pluto 1989; Abramowitz 1990; Quesin et al. 1990; McCarty and Semprini 1994; Lorah and Olsen 1997; Vink and Van der Zee 1997). Swindoll et al. (1995) reported attenuation of nitrobenzine and aniline under anaerobic conditions.

The predominantly anaerobic environment of wetlands leads to the formation and accumulation of soil that can be very rich in organic C. Many chemicals, especially hydrophobic organic chemicals (e.g., PCBs) and many heavy metals, bind strongly to organic matter. Given the high C content of most wetland soils, adsorption may be a significant mechanism for reducing the bioavailability of many contaminants (DeBusk et al. 1996; Kelsey and Alexander 1997; Polonsky and Clements 1999). Frequently, adsorption occurs when chemicals dissolved in surface water and groundwater come into contact with the organic matter of wetland soils. If the partitioning coefficient (K_{OC}) favors adsorption over dissolution, chemicals adsorb to soil particles or other organic material. In addition to PCBs, chemicals that typically adsorb to organic particles include P compounds, PHs, chlorinated hydrocarbons, and most metals (Powell et al. 1997).

Cation exchange, where positively charged cations bind to negatively charged clay or humus particles, is another attenuation process common in wetlands. Because certain cations have a greater tendency to attach to soil particles, there is a continual process of attachment and displacement of cations as water flows through wetland soils. Attenuation occurs as those cations with higher affinity for attachment (H and Ca) displace those with a lower affinity for attachment (e.g., Na and K). Acids released during the decomposition of organic material introduce a continual source of H ions, which tend to displace most other cations. The uptake of cations by plants also contributes to the attenuation process. When cations are removed from the soil, attachment sites for other cations become available.

Anaerobic conditions can produce anions that can combine with heavy metals to produce insoluble compounds. This process, which is a form of precipitation, effectively immobilizes heavy metals and retains the insoluble compound in the system indefinitely in a less bioavailable and, thus, less hazardous form. This NR process has been used extensively in constructed wetlands to remove hazardous heavy metals from groundwater and surface water flows. An example is the precipitation of Hg as its sulfide (Balba and Weston, unpublished data).

Chelation occurs as organic molecules (chelating agents) form complexes with metal ions, generally reducing their toxicity to biota (Polonsky and Clements 1999). Chelating agents are excreted by wetland plants to facilitate uptake of essential metals ions, and these agents may reduce the toxicity of some heavy metals. Humic acids released during the microbial decomposition of wetland litter facilitate the chelation of metals (Chan et al. 1982) and may catalyze the degradation of some compounds such as atrazine (Detenbeck et al. 1996).

Photolysis, which is the breakdown of a compound caused by absorption of radiant energy, is facilitated in the shallow water typical of wetlands. Photolysis can be a major fate process for photosensitive chemicals (e.g., the herbicide atrazine). However, direct sunlight onto the water or soil surface is often precluded by wetland foliage, and wetland waters tend to be stained with dissolved organic material, further reducing the amount of sunlight available for photolysis. Consequently, photolysis is not expected to be a primary fate process in most wetlands under most conditions. This process may be seasonally important in some emergent wetlands during early spring, when solar radiation is relatively high, but prior to emergence of dense marsh vegetation.

Physical processes

Physical NR processes reduce the potential for biota to be exposed to concentrations of contaminants that can have adverse affects. Physical processes important to NR in wetlands include dilution, deposition and burial, export of contaminants out of wetlands, and volatilization. Physical processes often work in conjunction with the chemical and biological processes discussed in previous sections to reduce exposure and risk.

Dilution is perhaps the simplest and most common of the physical NR processes occurring in wetlands. Dilution occurs as a contaminant goes from a higher to a lower concentration. Unless the only source of water for a wetland is contaminant-laden water, dilution will occur. For contaminants transported by surface water, dilution can be provided by surface water flowing into a wetland from noncontaminated portions of the watershed and from groundwater. For contaminants dissolved in groundwater flowing into the wetlands, surface water, which generally represents a greater proportion of the volume of total water, provides dilution. Dilution can reduce risk by lowering the concentration of contaminants below the threshold that induces adverse effects.

Deposition and burial of sediments can be significant mechanisms of NR in wetlands. Deposition occurs when the velocity of water in wetlands is slowed to the point where the water can no longer hold particles in suspension. The suspended particles settle out of the water column and become trapped in wetlands. Contaminants that are adsorbed to these particles are also deposited and trapped. Accordingly, wetlands are commonly referred to as nutrient and sediment "sinks." Burial occurs when multiple depositional events occur over time, with each subsequent event depositing a layer of sediments or particulate matter over previously deposited layers. The tandem processes of deposition and burial can provide a mechanism for NR if sediments deposited subsequent to the deposition of contaminants are "clean." Although these chemicals may persist, deposition and burial can be effective mechanisms for reducing risk by eliminating potential exposures to contaminants (Janssen et. al. 1997; Kelsey and Alexander 1997; Maruya et. al. 1997).

Because most soil-dwelling organisms are found in the top few centimeters of wetland soils and sediments due to the anoxic conditions that typically prevail (Mathooko 1995), contaminant burial can be a relatively rapid and effective mechanism of risk reduction. Risk to ecological receptors can be reduced or eliminated with the deposition of only a few centimeters of sediment. Deposition and burial can assist other processes of NR by holding contaminants in wetlands so that other processes, such as chemical reduction, can operate.

The rate at which deposition and burial of chemicals occurs within a wetlands is a function of the hydrology, the type and density of vegetation, and the size and density of particles suspended in water. Orson et al. (1992) report deposition rates of 0.70 to 2.22 cm/y for freshwater tidal marshes of the Delaware River. Lawson et al. (1996) report similar rates of deposition (1 to 4 cm) for an estuarine tidal marsh in Alaska.

Ecological risk in wetlands can be reduced by the export of contaminants with outflow of surface water and groundwater (see the section titled "Wetlands hydrology"). Contaminants dissolved in water that infiltrates into wetlands soils have several possible fates. They can be biologically or chemically transformed in the soil. They can be held in the soil pore water or sorbed to soil particles. They may also reach the water table, where dilution is likely to occur. Because of topography, surfacewater-slope and groundwater-slope wetlands provide the greatest opportunity for export. Tidal wetlands, both freshwater and estuarine, also have a high potential for NR through export. Conversely, wetland types with a high potential for export of chemicals have the lowest potential for NR through deposition and burial.

Volatilization of chemicals dissolved in the surface water of wetlands can also reduce the concentration of contaminants in wetlands. Because the depth of water in wetlands tends to be shallow (generally < 3 feet), the large surface-to-volume ratio favors volatilization. Where the wind blows over the surface of wetlands, the process is enhanced. However, the dense vegetation cover common to many wetlands may limit the effectiveness of this process. A further limitation, given that water in wetlands can be rich in particulate and dissolved organic carbon (DOC), is that many compounds, especially hydrophobic organic compounds, are unlikely to be dissolved and are thus unavailable for volatilization.

CASE HISTORIES

The protection and preservation of wetlands have been key considerations in developing remedies for contaminated sites. Under some regulatory programs, such as Superfund, the U.S. Environmental Protection Agency (USEPA) routinely considers NR as a potential remedy (USEPA 1997a). Natural remediation, used either alone or in conjunction with other remedial technologies, has been selected as the remedy for several sites involving contamination of wetlands. Several case studies are presented below that illustrate the use of natural processes in wetlands to reduce contaminant concentrations and/or reduce risk to ecological and human receptors.

Oak Grove Landfill

The Oak Grove Sanitary Landfill, located in Anoka County, Minnesota, is a 45-acre abandoned landfill. During active operation, the landfill accepted a variety of municipal and industrial wastes. The landfill is in the watershed of the Rum River. Cedar Creek, a tributary to the Rum River, flows within 3,000 feet of the landfill and flows through a large wetlands complex adjacent to the landfill. Contaminants from the landfill have been transported to these wetlands by 2 routes: leachate that emerges from the base of the landfill travels by surface flow into the wetlands and groundwater discharging from a shallow aquifer that flows beneath and into the landfill. The shallow aquifer, a sand outwash with a thickness of 13.5 to 54 feet, is the primary source of contaminants to the wetlands. The southern end of the shallow aquifer, near the point of discharge into the wetlands, is overlain with a peat layer. The shallow water aquifer is hydrologically separated from a deeper, water-supply aquifer by a semipermeable layer of glacial till. Consequently, contaminants from the landfill infiltrate down into the shallow aquifer but are prevented from moving into the deeper aquifer. The semi-impermeable layer directs the flow of water in the shallow aquifer to the south and eventually upward to the point of discharge into the wetlands.

A remedial investigation/feasibility study (RI/FS) identified several contaminants in the shallow water aquifer. The primary contaminants are volatile organic compounds (vinyl chloride [VC], 1,2-dichloroethene, benzene, ethylbenzene, toluene, and xylenes) and metals (As, Ba, and Ni). Most of these contaminants were believed to be associated with petroleum-based fuels and lubricating oils deposited in the landfill. Vinyl chloride and dichloroethene are possible degradation products of chlorinated solvents. Of the identified contaminants, VC, benzene, 1,2-dichloroethene, As, and Ni exceeded both federal and state groundwater criteria.

Surface water from the wetlands was analyzed as part of the RI/FS. Several of the contaminants identified in the shallow aquifer were also detected in the surface water of the wetlands. However, VC, benzene, 1,2-dichloroethene, As, and Ba, which exceeded the regulatory limits in the shallow aquifer, did not exceed these limits in the wetlands. Because groundwater from the shallow aquifer discharges directly into the wetlands, the observed reduction in the concentrations of these contaminants provided evidence that the contaminants were attenuated in the wetlands. As identified in the record of decision (ROD) for the site (USEPA 1990), 2 mechanisms of NR are occurring in the wetlands: volatilization and adsorption. Once the groundwater discharges into the wetlands, the volatile organic compounds volatilize to the atmosphere. Hydrophobic contaminants, such as the metals, are adsorbed by the organic C in the peat layer of the shallow aquifer prior to infiltrating the wetland.

Active remediation of the landfill occurred in 2 phases. In the first phase, source control of the contaminants was implemented. The landfill was capped to eliminate infiltration of water into the landfill and subsequent leaching and migration of contaminants from the landfill into the shallow aquifer. One of the primary objectives in the second phase of remediation was to prevent impacts to the surface water of the wetlands from the discharge of contaminated groundwater. Because the RI/FS provided strong evidence of NR, the selected remedy for Phase Two was NR, coupled with long-term monitoring of the shallow and deep aquifers and the surface water and sediments of the wetlands. Pump and treat as a remedial action was considered not justified because the "toxicity, mobility and volume of contaminants in the groundwater are being addressed via natural attenuation" (USEPA 1990).

Burnt Fly Bog

Burnt Fly Bog represents an example where risk management decisions concerning chemical contamination in a wetland were based on numerous considerations, not just potential risk. The Burnt Fly Bog is a 680-ha forested wetlands and freshwater marsh complex in central New Jersey. Approximately 20 ha were contaminated by a combination of oil, tar, Pb, and PCBs from past waste handling and storage practices (New Jersey Department of Environmental Protection [NJDEP] 1998). For investigation purposes, the wetlands complex was partitioned into 3 sections:

1) Tar Patch Area
2) Northerly Wetlands
3) Westerly Wetlands.

The appropriately named Tar Patch Area, about 2.2. ha in area, contained visible globules of tar and oil and little vegetation. In contrast, the Northerly

Wetlands and Westerly Wetlands supported dense wetlands vegetation. The contamination in the Westerly Wetlands and Northerly Wetlands encompassed 10 ha and 1 ha, respectively.

The USEPA and the NJDEP conducted both ecological and human health risk assessments. Sampling of soil identified concentrations of PCBs and Pb that were higher than screening-level benchmarks used by NJDEP. The ecological risk assessment (ERA) employed a combination of earthworm bioassays and tissue sampling of local rodents. Earthworm bioassays showed little evidence of direct toxicity to the earthworms themselves from either Pb or the PCBs, but significant bioaccumulation of both contaminants. According to the risk assessment, Pb accumulation in earthworms was sufficiently high to pose unacceptable risk to avian consumers of earthworms. Whole body analyses of local shrews, carnivores of local soil invertebrates, showed appreciable bioaccumulation of PCBs, with PCB body burdens high enough to potentially cause problems both to the shrews and to their predators. A human health risk assessment was also conducted, and unacceptably high human health risks were found based on a worst-case exposure scenario.

The site posed a dilemma. According to conventional worst-case risk assessments, the contaminated portions of the wetlands posed an unacceptable hazard to human and ecological receptors. This perception of unacceptable risk was based on worst-case assumptions that likely overestimated potential injury to a small number of species. In fact, as documented by field surveys, Burnt Fly Bog supported a diverse plant and animal community. The uncertain relationship between worst-case risk and actual injury was problematic because conventional remedial methods that involve soil removal would destroy valuable and ecologically functional forested areas of the bog. Under the best of conditions, active remediation of the forested areas would have required several decades to restore the habitat potential of the current bog. Remediation by excavation and offsite soil disposal followed by wetlands reconstruction was estimated to cost \$28.2 million, \$2.6 million, and \$14.0 million for the Westerly Wetlands, Northerly Wetlands, and Tar Patch Area, respectively (NJDEP 1998).

Natural remediation provided an alternative to destructive remedial methods. Based on the hydrology and vegetative structure of Burnt Fly Bog, burial and adsorption to abundant organic C in wetlands soil are likely to be significant processes of NR for both Pb and PCBs in the Westerly Wetlands and Northerly Wetlands. PCBs are amenable to anaerobic biodegradation under certain conditions: relatively warm temperatures, anoxic conditions, and large microbial populations. These conditions should occur, at least seasonally, at Burnt Fly Bog. Natural remediation would allow the wetlands to continue to provide wetlands functions. An assessment of the wetlands functions and values was conducted using the Wetlands Evaluation Technique (WET; Adamus 1983a, 1983b) and Habitat Evaluation Procedures (HEP; U.S. Fish and Wildlife Service 1980). Natural remediation also had obvious cost advantages. With institutional controls such as fencing and deed restrictions, NR was estimated to cost about $115,000, or about 0.2% of the almost $30 million estimated for soil removal and wetlands restoration.

Given the alternative, NJDEP selected a middle course, electing to remove soil and contaminants from the largely nonvegetated Tar Patch Area and the Northerly Wetlands. Active remediation of the Tar Patch area was warranted because the area is a source of chemicals to the downstream areas of the Westerly Wetlands. In addition, the existing vegetation and resulting habitat potential of this area is marginal, so the ecological damages associated with remediation are comparatively minor. NJDEP also elected to remove soil from about 1 ha of Northerly Wetlands, followed by wetlands restoration. In the Westerly Wetlands, NJDEP elected to allow NR, primarily burial and adsorption to organic C, to remediate the chemicals. A fence was erected to minimize human exposure. The rationale for allowing NR in the Westerly Wetlands versus the Northerly Wetlands was that the contaminants are more localized, and thus easier to remediate, in the Northerly Wetlands. The Northerly Wetlands could also be a source of chemicals to the Westerly Wetlands, so the former was remediated to prevent recontamination of the Westerly Wetlands.

Aberdeen Proving Grounds

Lorah and Olsen (1997) provide strong evidence that NR of chlorinated volatile organic compounds (VOCs) is occurring in wetlands at Aberdeen Proving Grounds near Baltimore, Maryland. Evidence of ongoing NR was provided by both field investigations and related laboratory studies. At Aberdeen, shallow groundwater with elevated concentrations of TCE and 1,1,2,2-tetrachloroethane (TeCA) flow into wetlands. The wetlands are characterized by 2 distinct soils units: a unit of organic peat and a lower unit of silt to sandy clay or clayey sand. TCE is biodegraded by microorganisms under anaerobic conditions through a sequential process of dechlorination, with 1,2-dichoroethene (1,2-DCE), VC, 1,1,2-trichloroethane (1,1,2-TCA), and 1,2-dichloroethane (1,2-DCA) as products of degradation. TeCA is biodegraded to 1,1,2-TCA, which is in turn reduced to 1,2-DCE and VC. Adsorption of TCE and TeCA by the highly organic peat unit is also a likely mechanism for NR.

Concentrations of total chlorinated VOCs decrease from 4 μmol/L in the shallow aquifer to 1 μmol/L in the lower peat unit, and to 0.2 μmol/L in the upper peat unit. Further evidence of NR was provided by the relative concentrations of TCE and TeCA and the degradation products in the aquifer and wetlands soils. Within the shallow aquifer, concentrations of TCE and TeCA were relatively high, whereas concentrations of the degradation products (1,2-DCE, VC, 1,1,2-TCA, and 1,2-DCA) were near or below the limits of detection. In the wetland soils, TCE and TeCA were near or below the limits of detection, while the degradation products were present at higher concentrations. The decrease in parent chemical and appearance of degradation products coincided with a transition from generally aerobic conditions in the aquifer to anaerobic conditions in the wetland soils. Evidence for aerobic conditions in the aquifer includes low concentrations of ferrous iron, sulfide, and methane. Groundwater flowing from the aquifer to the lower sandy unit to the upper peat unit is subject to an increasingly reducing environment. Iron reduction is the predominant biogeochemical process in the lower sandy unit of the wetlands, and methanogenesis is predominant in the upper peat unit. Lorah and Olsen (1997) suggest that biodegradation of TCE occurs primarily under conditions favoring methanogenesis. The authors further suggest that complete remediation of chlorinated VOCs is possible in methanogenic wetland soils.

Upper Delaware River Estuary

Orson et al. (1992) documented NR of heavy metals (Cu, Pb, Cd, Cr, Zn, and Ni) in tidal marshes of the Upper Delaware River Estuary. Processes contributing to NR include transport of chemical laden sediments, high deposition rates (0.70 to 2.22 cm/y), uptake of heavy metals by plants, and adsorption to particulate organic matter. Relatively high rates of decomposition and a high clay content ($\pm$ 30%) of marsh sediments also contribute to rapid sequestering of metals. Incoming tides transport metals adsorbed to sediments and particulates. These suspended solids, which are high in organics, are deposited in the marshes at low tide. Contributing to the transport of metals into the marshes is stormwater runoff from the highly urbanized watershed of the study area, which extended from Trenton, New Jersey to the south of Philadelphia. Heavy metals are either taken up and incorporated into plant tissue or remain adsorbed to sediments.

Concentrations of metals in sediments are highest during late summer and fall, when litter accumulation is highest. As litter decomposes, metals are adsorbed to fine organic matter released through decomposition and then incorporated into the marsh substrate. Due to the rapid rate of deposition, the particulate-bound metals are quickly buried by continuous deposition of new sediments. Because of the high rates of deposition, any mixing with the water column is limited to growth from the previous growing season. This process led the authors to conclude that plant litter acts as a "sponge" to temporarily sequester metals until adsorption and deposition incorporate them into the marsh sediments.

Several key observations related to NR in tidal marsh systems are noted by the authors. Historical accumulation of heavy metals was greatest in years with the highest storm activity. The authors attribute this to increased transport and deposition of sediments brought about by increased flow, higher burial rates of organic particulates, or a combination of both. Another observation was that incorporation of heavy metals in the sediments of freshwater tidal marshes occurs primarily on the surface of the marsh. In contrast, incorporation of heavy metals into the sediments of estuarine tidal marshes occurs primarily below ground. This is due to the relatively high density of roots and rhizomes in estuarine marshes where adsorption can occur.

Natural remediation of white phosphorus, Eagle River Flats, Alaska

Eagle River Flats is a 868-ha tidal salt marsh at the mouth of the Eagle River, north of Anchorage, Alaska. The marsh has been used as an artillery range since the 1940s. As a result, the marsh became contaminated with white P, a smoke-producing compound used by the military in training artillery shells. In the presence of air, white P spontaneously combusts. The compound persists under conditions of inundation characteristic of the tidal marsh. Most of the white P at Eagle River Flats is contained in a series of large shallow ponds that occur within the marsh. Prior to oxidation, the compound is quite toxic. Unusually high mortality of migratory waterfowl in the area has been attributed to their ingestion of the white P particles while feeding on the bottom of ponded areas in the marsh. At the same time, the vegetation and many other components of the marsh ecosystem appear relatively unstressed, arguing against destructive remedial measures such as dredging. Another complicating factor is that conventional remediation itself entails significant risk from unexploded ordnance.

Based on a series of studies concerning physical processes, such as erosion and deposition in the tidal marsh, it was determined that most of the area can be remediated naturally (Lawson et al. 1996). Like many river deltas, the marsh topography is quite dynamic. Because of ongoing deposition and erosion within the marsh, new drainage channels are continually being formed while old channels are silting in. Based on aerial photographs taken over the last several decades, Lawson et al. (1996) estimated that within 2 to 10 years, new drainage channels would be eroded and connect to the larger ponds containing much of the white P. Once these channels were completed, the ponds should drain completely during low tide, and the white P should remediate naturally (i.e., combust) when exposed to the air. Lawson et al. (1996) determined that ponds farther from the main channels were unlikely to be drained in the near future. In these areas, sediment deposition, which was measured to range from 1 to 4 cm/y, was predicted to bury the white P.

Ironically, in this case, it is the wetland characteristics—standing water and saturated sediments—that precluded natural destruction of the white P. The NR for much of the marsh—waiting till the ponds are drained—inexorably depends on the wetland becoming less wet. Thus, this instance of NR illustrates that physical and chemical conditions of wetlands can retard as well as promote remediation of anthropogenic chemicals.

Natural remediation of nitrobenzene and aniline

Natural remediation of nitrobenzene and aniline in wetlands associated with a manufacturing facility was documented by Swindoll et al. (1995). Over a period of 2 years, the concentration of nitrobenzene in wetlands soils decreased from 2,000 mg/kg to < 0.5 mg/kg, and aniline decreased from > 1,000 mg/kg to 180 mg/kg. The proposed mechanism for NR is microbial degradation under both aerobic and anaerobic conditions. Nitrobenzene can be reduced under sulfate-reducing, denitrifying, or methanogenic conditions. Aniline, which is a degradation product of nitrobenzene, can be reduced under denitrifying conditions.

Analysis of water and soils collected from the wetlands indicated that conditions were favorable for microbial degradation. N was sufficiently abundant to support microbial populations. S and Fe were available as electron acceptors for reduction reactions under anaerobic conditions. High alkalinity provided buffering capacity for conditions favorable for methanogenic bacteria. Significant amounts of organic C were available as an energy source. Laboratory studies confirmed that nitrobenzene and aniline were degraded by microbes indigenous to the site.

NATURAL REMEDIATION OR DESTRUCTIVE REMEDIATION?

Conventional active technologies for remediating contaminants often do so at the expense of ecological value. In short, the process of remediation often destroys what it intends to protect. This is not a significant problem when applied to brownfields or abandoned industrial sites with minimal ecological value. However, destructive remediation poses fundamentally different problems when applied to wetlands or other areas with ecological value. In this regard, it is important to understand that the presence of contaminants or worst-case risk does not necessarily mean that all, or even any, of the important ecological functions of a wetlands have been significantly impaired. The authors have worked in a number of contaminated wetlands, ranging from mature, forested swamps contaminated with Pb and PCBs to an estuarine marsh with chlorinated hydrocarbons to a bed of dense, submerged aquatic vegetation contaminated with Hg and chlorinated organics. In all of these cases, productive plant communities, supporting diverse wildlife communities, co-occurred with anthropogenic chemicals that, based on worst-case ERAs, posed "unacceptably high risks."

In such cases, the critical question becomes whether the existing ecological functions provided by the contaminated wetlands should be subjected to the certain effects of destructive remediation or left to the uncertain risks of leaving the contaminants in place while allowing NR to occur (e.g., Powell et al. 1997). Traditional risk assessment methods, which are based on worst-case assumptions and scenarios, are not adequate to evaluate such a tradeoff. Such methods are, by design, conservative and therefore they intentionally, and often grossly, overestimate potential impacts of contaminants on ecological resources. For example, current ecological risk methods intentionally focus upon risks to the most sensitive and most exposed individuals of a population (USEPA 1997b, 1998). Such worst-case methods should not be used to predict actual impacts to entire wetland ecosystems. Risk managers could be misled into selecting remedies that disturb largely or totally functional wetlands to eliminate the potential for damage to a small number of species or, even worse, a small number of individuals. For example, potential toxicity of PCBs to the hypersensitive mink is generally used as the basis for ERAs, making it a de facto rationale for destruction of wetlands in which less sensitive animal and plant communities are apparently healthy (e.g., as in Burnt Fly Bog in New Jersey.).

Compared to the tenuous relationship between "unacceptably high risks" and actual, significant degradation of wetlands ecosystems, impacts to wetlands from destructive remediation are certain, generally severe in the short term, and potentially significant over the long term. Consequently, destructive remediation of wetlands deserves careful consideration. Risk managers, resource agencies, and stakeholders should seek to maximize "net ecological benefit" by evaluating risk at 2 levels: risk of chemicals to human health and the environment and risk to wetlands from destructive remediation. The importance of protecting human health and the environment should not be compromised. However, conservative risk assessment methods that exaggerate the potential for likely effects are not sufficient to make risk management decisions that maximize net ecological benefit. Instead, decision-makers should recognize the benefits of a "top down" approach (i.e., ecologically based field studies that quantify population- and community-level parameters) rather than a "chemistry up" approach (i.e., sampling and analytical studies that focus on chemical parameters with little attention to existing ecological functions of wetlands). These considerations will provide a more scientifically valid basis for evaluation and selection of remedial alternatives that reduce risk to acceptable levels while minimizing impacts to functioning wetlands systems.

CONCLUSIONS

Many of the natural processes that occur in wetlands have the potential to reduce risk to both human and ecological receptors through a variety of biological, chemical, and physical mechanisms that transform contaminants to less toxic forms or reduce exposure concentrations. In addition to their potential for NR, wetlands also provide many important ecological functions. Evidence has been presented in this chapter that NR does occur in wetlands. The remediation process occurring in wetlands, as well as the ecological functions, should be considered by risk managers when they evaluate remedial technologies, as was the case for Burnt Fly Bog.

Not all processes discussed in this chapter occur in all wetlands or are effective in reducing risk for all contaminants. Toxicity of some contaminants may actually increase in wetlands. In some cases, concentrations of contaminants may exceed acceptable limits for protection of human health or the environment. A prime example is the Kesterson Natural Wildlife Refuge in the San Joaquin Valley of California. Drainage containing Se and other contaminants from surrounding agricultural land was directed into the Kesterson Marsh. Concentration of these contaminants in the ecosystem through evaporation and bioaccumulation resulted in a variety of adverse impacts to fish and waterfowl inhabiting the marsh, including disappearance of fish, reduced productive success, and a high incidence of deformities. In this example, and many others, NR is not an appropriate remedy. Where NR is not appropriate, less destructive forms of remediation and attenuation, such as in situ treatment, can potentially conserve wetland resources while simultaneously satisfying regulatory requirements.

REFERENCES

Abramowitz DA. 1990. Aerobic and anaerobic biodegradation of PCBs: A review. *Crit Rev Biotechnol* 10:241–251.

Adamus PR. 1983a. A method for wetlands functional assessment. Volume I, Criteria review of evaluation concepts. Washington DC: U.S. Department of Transportation. FHWA-IP-82-23. 176 p.

Adamus PR. 1983b. A method for wetlands functional assessment. Volume II, FHWA assessment method. Washington DC: U.S. Department of Transportation. FHWA-IP-82-24. 134 p.

Beelen PV, Burris DR. 1995. Reduction of the explosive 2,4,6-trinitrotoluene by enzymes from aquatic sediments. *Environ Toxicol Chem* 14:2115–2123.

Best E, Zappi M, Fredrickson H, Sprecher S, Larson S, Miller, J. 1997. Screening of aquatic and wetland plant species for phytoremediation of explosives-contaminated groundwater from the Iowa army ammunition plant. Vicksburg MS: U.S. Army Corps of Engineers Waterways Experiment Station. Technical Report EL-97-2.

Chan E, Bursztynsky TA, Hantzsche N, Litwin YJ. 1982. The use of wetlands for water pollution control. Cincinnati OH: U.S. Environmental Protection Agency (USEPA). EPA-600-82-086. 261 p.

Debusk TA, Laughlin Jr., RB, Schwartz LN. 1996. Retention and compartmentalization of lead and cadmium in wetland microcosms. *Water Resources* 30(11):2709–2716.

Detenbeck N, Hermanutz R, Allen K, Swift M. 1996. Fate and effects on the herbicide atrazine in flow-through wetland mesocosms. *Environ Toxicol Chem* 15:937–946.

Hammer DA. 1990. Constructed wetlands for wastewater treatment. Boca Raton FL: Lewis. 856 p.

Janssen R, Peijnenburg W, Posthuma L, Van Den Hoop M. 1997. Equilibrium partitioning of heavy metals in Dutch field soils. I. Relationship between metal partition coefficients and soil characteristics. *Environ Toxicol Chem* 16:2470–2478.

Kadlec RH, Knight RL. 1996. Treatment wetlands. Boca Raton FL: Lewis. 928 p.

Kelsey J, Alexander M. 1997. Declining bioavailability and inappropriate estimation of risk of persistent compounds. *Environ Toxicol Chem* 16:582–585.

Lawson D, Hunter L, Bigl S. 1996. Physical processes and natural attenuation alternatives for remediation of white phosphorus contamination, Eagle River Flats, Fort Richmond, Alaska. Hanover NH: U.S. Army Alaska Environmental Resources Department, Cold Regions Research and Engineering Laboratory. CRREL Report 96-13.

Lorah MM, Olsen LD. 1996. Natural attenuation of hydrocarbons in a freshwater wetland. *In Situ On-Site Bioremed* 3:207–212.

Mathooko JM. 1995. Vertical distribution of macrobenthos and coarse particulate organic matter in the sediment surface of a pool biotrope in the Njoro River Kenya. *Arch Hydrobiol* 133:95–106.

Madsen E, Mann C, Bilotta S. 1996. Oxygen limitations and aging as an explanation for the field persistence of naphthalene in coal tar-contaminated surface sediments. *Environ Toxicol Chem* 15:1876–1882.

Maruya K, Risebrough R, Horne A. 1997. The bioaccumulation of polynuclear aromatic hydrocarbons by benthic invertebrates in an intertidal marsh. *Environ Toxicol Chem* 16:1087–1097.

McCarty PL, Semprini L. 1994. Groundwater treatment for chlorinated solvents. In: Norris RD, editor. Handbook of bioremediation. Boca Raton FL: CRC Pr. p 87–116.

Mitsch WJ, Gosselink JG. 1993. Wetlands. 2nd ed. New York NY: Van Nostrand Reinhold. 772 p.

[NJDEP] New Jersey Department of Environmental Protection. 1998. Record of decision. Burnt Fly Bog site. Marlboro Township, Monmouth County, NJ.

Novitski RP. 1989. Wetland hydrology. In: Majumdar SK, Brooks RP, Brenner FJ, Tiner Jr RW, editors. Wetlands ecology and conservation: Emphasis in Pennsylvania. Easton PA: The Penn Acad Science. p 47–64.

Odum E. 1971. Fundamentals of ecology. 3rd ed. Philadelphia PA: WB Saunders. 516 p.

Orson RA, Simpson RL, Good RE. 1992. A mechanism for the accumulation and retention of heavy metals in tidal freshwater marshes of the Upper Delaware River Estuary. *Estuar Coast Shelf Sci* 34:171–186.

Otte ML, Ernst WHO. 1994. Arsenic in vegetation of wetlands. In: Nriag JO, editor. Arsenic in the environment, Part. I: cycling and characterization. New York NY: Wiley. p 365–379.

Polonsky AP, Clements, WH. 1999. Contaminant assimilation within the water column of two newly created prairie wetlands. *Environ Contam Toxicol* 36:140–145.

Powell R, Kimerle R, Coyle G, Best R. 1997. Ecological risk assessment of a wetland exposed to boron. *Environ Toxicol Chem* 16:2409–2414.

Pratt PR, Pluto T. 1989. Use of wetlands for the treatment of industrial and nonpoint source pollutants. In: Majumdar SK, Brooks RP, Brenner FJ, Tiner Jr RW, editors. Wetlands ecology and conservation: Emphasis in Pennsylvania. Easton PA: Penn Acad Science. p 315–332.

Price R, Pennington J, Larson S, Neumann D, Hayes C. 1997. Plant uptake of explosives from contaminated soil and irrigation water at the former Nebraska ordnance plant, Mead, Nebraska. Vicksburg MS: U.S. Army Corps of Engineers Waterway Experiment Station. Technical Report EL-97-11.

Quensen III JF, Boyd SA, Tiedje JM. 1990. Dechlorination of four commercial polychlorinated biphenyl mixtures (aroclors) by anaerobic microorganisms from sediments. *Appl Environ Microbiol* 56:2360–2369.

Reddy KR, Smith WH. 1987. Aquatic plants for water treatment and resource recovery. Orlando FL: Magnolia. 1032 p.

Reed SC, Crites RW, Middlebrooks EJ. 1995. Natural systems for waste management and treatment. New York NY: McGraw-Hill. 433 p.

Schnoor JL, Licht LA, McCutcheon SC, Wolfe NL, Carreira LH. 1995. Phytoremediation of organic and nutrient contaminants. *Environ Sci Technol* 29:318–323.

Swindoll CM, Perkins RE, Gannon JT, Holmes M, Fisher GA. 1995. Assessment of bioremediation of a contaminated wetland. In: Hinchee LE, Wilson JT, Downey DC, editors. Intrinsic bioremediation. Columbus OH: Battelle Pr. p 163–169.

Tiedje JM. 1993. Bioremediation from an ecological perspective. In: In situ bioremediation: When does it work? National Academy Pr. p 110–120.

[USDA] U.S. Department of Agriculture Soil Conservation Service. 1985. Hydric soils of the United States. Washington DC: USDA-SCS. National Bulletin No. 430-5-9.

U.S. Fish and Wildlife Service. 1980. Habitat evaluation procedures (HEP). Washington DC: U.S. Fish and Wildlife Service, Division of Ecological Services. ESM 102.

[USEPA] U.S. Environmental Protection Agency. 1990. Superfund record of decision, Oak Grove Sanitary Landfill, MN, second remedial action – Final. EPA/ROD/R05-91/153.

[USEPA] U.S. Environmental Protection Agency. 1997a. Use of monitored natural attenuation at Superfund, RCRA corrective action and underground storage tank sites. OSWER Directive 9200.4-17.

[USEPA] U.S. Environmental Protection Agency. 1997b. Ecological risk assessment guidance for Superfund. Process for designing and conducting ecological risk assessments. Edison NJ: Environmental Response Team. EPA/540/R-97/006.

[USEPA] U.S. Environmental Protection Agency. 1998. Guidelines for ecological risk assessment. Risk Assessment Forum. Washington DC: USEPA. EPA/630/R-95/002F.

Vink J, Van Der Zee S. 1997. Effects of oxygen status on pesticide transformation and sorption in undisturbed soil and lake sediment. *Environ Toxicol Chem* 16:608–616.

Wildeman RR, Brodie GA, Gusek JJ. 1992. Wetland design for mining operations. Vancouver BC: Bitech. 300 p.

13

Natural Remediation in Subsurface Ecosystems

Duane Graves

ECOSYSTEMS THAT EXIST immediately beneath the earth's surface represent numerous chemical and physical milieus that support a diverse biota. These subsurface ecosystems contain organisms ranging from viruses, bacteria, and protozoa to plants, invertebrates, and vertebrates. The subsurface spans a spectrum of environmental conditions that includes extremes from oxidized to reduced states and from being completely water saturated to being dry. Subsurface geology is often complex, with multiple layers of sedimentary and volcanic deposits composed of silt, sand and clay, various minerals, organic matter, basalts, consolidated (rock) and unconsolidated strata. The subsurface is a critical component of the earth's biosphere. The shallow subsurface provides the support and inorganic nutrient base for primary production from forests, prairies, and arable land. In addition to sustaining primary production, soil supports the agricultural production critical for sustaining the human population and serves as a purifying filter for fresh water.

At many locations, industrial, municipal, domestic, and agricultural practices have compromised the quality of soil and groundwater by introducing unnatural concentrations of organic compounds, xenobiotic chemicals, inorganic compounds, heavy metals, and radionuclides. These contaminants can have both acute and chronic effects on biota. Acute effects such as mortality or environmental sterility are obvious when plants and animals fail to grow or when groundwater is discolored or foul tasting. Chronic effects, including reproductive impairment and abnormal growth, may be more subtle and may not be detected for several years or several generations of exposure to the impacted soil or groundwater. Because of the

Natural Remediation of Environmental Contaminants. Michael Swindoll et al., editors.
 ISBN 1-880611-33-3.

importance of subsurface resources and the finite quantity of high-quality groundwater, it is critical that these resources be protected.

Regulatory and technical efforts to curb contamination and restore subsurface quality have had some great successes, yet subsurface contamination still remains a pervasive worldwide issue. In the past decade, we have begun to realize that technical and resource limitations impair our ability to actively restore contaminated subsurface ecosystems. Engineered solutions are often ineffective or too costly for widespread application to clean subsurface soil and groundwater. Concurrent with our understanding of these limitations, there has been an increased awareness of the important role that naturally occurring phenomena contribute to the restoration of environmental quality.

Extensive research has demonstrated natural remediation (NR) of a wide range of environmental contaminants, including petroleum hydrocarbons (PHs), chlorinated solvents, nitroaromatics (e.g., trinitrotoluene [TNT]), fuel oxygenates (e.g., methyl tertiary butyl ether [MTBE]), inorganics, radionuclides, and metals (McCormick et al. 1976; Spain 1996; Townsend and Myers 1996; Weidemeier et al. 1996; Vance 1998; Brady et al. 1999). Natural remediation is being applied at many sites and is often coupled with active remediation approaches. Yet, in many cases, the conditions necessary to support effective NR remain ill defined, and the prevalence of appropriate conditions has not been established (Preus et al. 1993; Bradley et al. 1997; Cattaneo et al. 1997; Krumholz et al. 1997).

Significant advances in the understanding and application of NR in the subsurface have occurred in the last decade. Protocols and approaches for establishing the effectiveness of NR are well defined and widely accepted. The U.S. Air Force Center for Environmental Excellence (AFCEE) and the U.S. Environmental Protection Agency (USEPA) National Groundwater Research Laboratory have set the standard for assessing the role of NR for restoring groundwater impacted with PHs and chlorinated solvents (Wiedemeier et al. 1994, 1998; USEPA 1996). Several states have adopted the technical protocols advanced by AFCEE and the USEPA largely intact. The USEPA Office of Solid Waste and Emergency Response (OSWER) has issued a directive describing the appropriate application of NR for contaminated soil and groundwater (USEPA 1999). In general, as evidence mounts regarding the effectiveness of NR in subsurface environments, the requirements and burden of proof are being relaxed.

Currently, environmental regulatory agencies have differing attitudes about the use, effectiveness, and applicability of NR. Both federal and state regulatory programs incorporate NR as an acceptable remediation alternative. However, the level of detail required to gain regulatory approval for NR as the primary or sole remedial action can vary considerably. Typically, a lack of familiarity increases the difficulty of gaining regulatory acceptance for NR. The end result is a requirement for very thorough and detailed descriptions of NR phenomena with as much quantitation as possible.

Acceptance of NR requires evidence that remediation is occurring, that subsurface restoration will occur within a reasonable time frame, and that the site owner remains responsible for site oversight and monitoring until satisfactory restoration has occurred. Natural remediation is also misperceived to be slower than other remedial processes. Although NR may be slow, it is often the primary contributor to subsurface restoration during the application of other technologies.

APPLYING NATURAL REMEDIATION FOR SUBSURFACE RESTORATION

Applying NR to the subsurface requires an understanding of site conditions, including contaminant concentration and distribution, environmental chemistry, geology, hydrogeology, and microbiology. Contrary to initial concerns that NR is a "do nothing" alternative, significant site assessment, characterization, and ongoing monitoring are necessary for the implementation of NR as a treatment alternative for the subsurface.

Defining site conditions is the first action necessary to develop a site management and remediation plan that employs NR. This evaluation of site conditions is usually very extensive, involving many aspects of site history, geology, and hydrology. For example, surface contours can influence water infiltration, plume migration, and the ability to install monitoring wells or treatment systems. The soil type and nature of the aquifer material influences the practicality of treatment alternatives because contaminant distribution is directly related to subsurface geology (Charbeneau et al. 1992). Local aquifer use, potential exposure routes and receptors, and the location of drinking water, industrial, or agricultural wells should be identified as part of the overall site characterization. These parameters are instrumental for defining the relative health and environmental risk associated with contaminated groundwater.

The movement of groundwater directly affects the progress of NR in aquifers. Defining local aquifer characteristics contributes to an accurate prediction of the performance of NR. Hydraulic conductivity of the aquifer is determined by falling or rising head slug tests or pumping tests (Freeze and Cherry 1979). Thorough characterization of the aquifer is essential for predicting the long-term impact of the nonaqueous phase liquid (NAPL) product, residual product, and dissolved contaminant. Wells will be required for some tests, whereas simple monitoring points will be adequate for others.

The depth to groundwater and the groundwater gradient should also be determined in the contaminated area. If a significant change in gradient or depth is expected based on local topography and geology, these parameters should be determined in the downgradient direction. The groundwater flow rate and direction can be calculated based on hydraulic conductivity and groundwater gradient data

(Freeze and Cherry 1979). The rate of aquifer recharge and the areas where recharge from the surface and discharge to the surface are likely to occur are determined. These characteristics facilitate a more accurate determination of plume dilution. Discharge points are considered during risk evaluations, since they represent areas for potential contact with contaminated water.

Contaminant composition, location of NAPL, location of dissolved-phase contaminants, and the mass of contaminant present are critical aspects for understanding NR. Knowing the chemical composition of the contaminant permits changes in composition to be evaluated against time and distance migrated. This information is used to make predictions regarding the attenuation rate, the biodegradation rate, and the change in concentration due to physical mechanisms. The physical and chemical properties of the contaminant also affect the behavior of the contaminant in the subsurface. Once the composition of the contaminant is known, properties of individual components are usually available (Montgomery 1996). Particularly useful properties include aqueous solubility, the octanol–water partition coefficient, the soil adsorption coefficient, and the Henry's law constant.

During site characterization, contaminant chemical composition should be evaluated with 2 objectives. First, the concentration change in chemicals likely to biodegrade should be determined. Second, the change in chemicals that are not likely to biodegrade should be determined. Concentration changes that result from nonbiological processes can be used to correct for physical losses in the biodegradable compounds. Key target compounds for petroleum fuels that are very slow to biodegrade anaerobically include 2,3-dimethyl pentane, trimethyl benzenes, and tetramethyl benzenes (Cozzarelli et al. 1990, 1994; Wilson et al. 1993, 1994; Wiedemeier et al. 1994). These compounds have solubility and sorption characteristics similar to benzene, toluene, ethyl benzene, and xylenes (BTEX), and thus, they can serve as practical markers of contaminant movement and BTEX biodegradation. This approach is the simplest method for calculating dilution, dispersion, and sorption effects on contaminant concentration.

The standard procedure for quantitating remedial progress is measuring the reduction in contaminant concentration using USEPA-approved methods. The mass of contaminant in soil and dissolved in groundwater is also an important parameter for quantitating NR. Unfortunately, accurate measurement of contaminant mass is difficult and is usually not done accurately. Determination of contaminant mass involves estimation methods to extrapolate the amount of contaminant between monitoring points. Defining the mass of contaminant in a soil sample can be accomplished using several analytical methods. Defining the actual dimensions and concentrations of contaminant in 3 dimensions is challenging and expensive. One approach is to collect intact soil cores and analyze sections of each core to determine the depth distribution of contaminant and the mass of contaminant in each section (Figure 13-1) (Wilson 1993). Lateral contaminant distribution is estimated based on soil sampling at multiple locations.

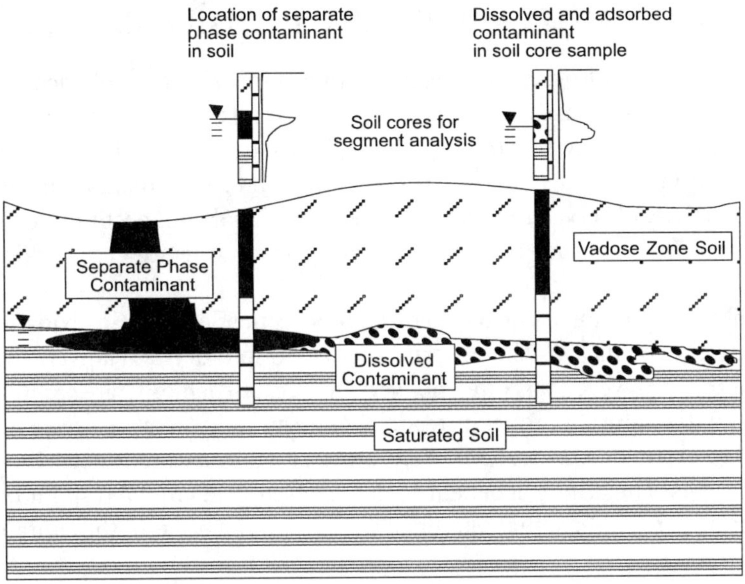

Figure 13-1 Analysis of soil core segments accurately defines location of separate phase and dissolved phase contaminants

The amount of contaminant present at a site can be estimated using a relatively simple technique known as the "Thiessen Method" (Fetter 1994). It was used originally to calculate the amount of rain that fell over an area by collecting data from rain gauges. Using monitoring well data, the method will yield an arbitrary and unbiased estimate of contaminant mass as well as the location of the mass. Results collected over time can provide useful insights into the NR process and the effect of a contaminant source such as NAPL on contaminant migration and dissolved concentration.

The accuracy of the Thiessen Method contaminant mass estimate is questionable, but the results are precise and can be compared over multiple time intervals. By combining the contaminant concentration and location data generated from soil core analysis with the Thiessen Method, the accuracy of the contaminant mass estimate can be improved. Geostatistical methods such as kriging can also provide reliable estimates of contaminant distribution and mass using either 2-dimensional or 3-dimensional data.

Physical and chemical properties of contaminants affect their distribution and movement in groundwater and soil gas. The octanol–water partition coefficient, the Henry's law constant, and the aqueous solubility of contaminants can be used to estimate the interaction of the contaminant with air, water, and soil. Retardation coefficients that reflect the relative mobility of a contaminant in an aquifer allow comparisons of the expected plume migration rate with observed plume movement.

BIOLOGICAL ACTIVITY MONITORING PARAMETERS

Biological activity monitoring parameters are indicators of microbiological activity in the subsurface. In all cases involving organic chemicals and some inorganic chemicals, microbial activity is the most likely, and in many cases the only, explanation for the observed changes in contaminant levels and geochemistry. Therefore, changes or activity observed in a contaminated area that do not occur or do not occur to the same magnitude in uncontaminated soil can often be attributed to biodegradation.

Microbial respiration is the biochemical process that leads to the oxidation of reduced organic chemicals. Demonstrating aerobic respiration is relatively simple in the vadose zone. However, it is often not very practical to demonstrate aerobic respiration in the saturated zone. Instead, laboratory studies may be used as surrogates to in situ demonstrations of biodegradation by naturally occurring microorganisms. The principal benefit of demonstrating microbial respiration in the laboratory is to confirm that microbial respiration can occur in the contaminated area.

The results of aerobic and anaerobic microbial respiration are often obvious in the subsurface when the concentrations of respiratory substrates (terminal electron acceptors [TEAs]) are compared within and outside of a contaminated area. The biodegradation potential of an aquifer (also known as the "expressed biodegradative capacity" [Weidemeier 1994]) can be estimated using the concentration of respiratory substrates and products.

In addition to estimating the biodegradation potential of the aquifer, respiratory substrate and product concentrations are critical data used to support the occurrence of intrinsic bioremediation. The observations described below are indicative of intrinsic biodegradation and are important aspects for evaluating the potential of successful intrinsic bioremediation at a site. Because of the energy derived from each substrate, they are preferentially used in the following order:

$$O_2 > NO_3^{-1} > Mn^{+4} \sim Fe^{+3} > SO_4^{-2} > CO_2$$

In addition to preferential utilization, respiration tends to be exclusive. For example, in the presence of oxygen, other respiratory substrates are not used. Similarly, when oxygen is depleted and nitrate is available, denitrification will prevail with little or no iron, manganese, and sulfate reduction or methanogenesis. This pattern of respiratory exclusion continues as high energy respiratory substrates are expended.

Groundwater that is undergoing intrinsic bioremediation will usually contain much less dissolved oxygen than a nearby noncontaminated groundwater. An area of oxygen-depleted groundwater usually coincides with the dissolved contaminant plume, and it often extends beyond the contaminant plume in the downgradient direction. The difference in dissolved oxygen between "clean" and contaminated groundwater is a useful indicator of intrinsic biodegradation.

The nitrate content of groundwater often varies with local land-use practices. Groundwater that is downgradient of agricultural areas, landfills, and industrial areas often has elevated nitrate concentrations. However, in areas with measurable nitrate, the groundwater within a contaminant plume will often have a much lower nitrate concentration than surrounding "uncontaminated" groundwater. This concentration change indicates nitrate-reducing activity. The absence or much lower concentration of nitrate in clean groundwater indicates that the presence of contaminant is required to support nitrate reduction (Ehrlich 1981; Kuhn et al. 1988; Major et al. 1988; Weidemeier et al. 1994).

Oxidized manganese (Mn^{+4}) can be reduced by microbial respiration (Ehrlich 1981; Lovely 1991; Weidemeier et al. 1994). Mn reduction occurs under conditions similar to those that support Fe reduction. There is some evidence that Mn reduction can support hydrocarbon biodegradation (Weidemeier et al. 1994); however, Mn does not seem to be a prominent respiratory substrate for anaerobic biodegradation at most sites. A practical approach to dealing with Mn reduction is to evaluate the other respiratory substrates or products. If contaminant degradation appears to be occurring but cannot be accounted for using more common respiratory substrates, Mn should be considered as a possible substrate supporting biodegradation.

Oxidized iron, Fe(III), is insoluble, and its presence is indicated by the red coloration of oxidized soil. Soil within contaminant plumes is often reduced. The biological conversion of oxidized iron, Fe(III), to reduced iron, Fe(II), can support hydrocarbon biodegradation. Iron reduction is not an efficient biochemical process, requiring about 42 pounds of Fe(III) per pound of hydrocarbon degraded. Nevertheless, Fe reduction is an important process because of the abundance of Fe in the soil. The laboratory speciation and analysis of Fe(III) and Fe(II) are not particularly accurate because of the reactivity of Fe. Samples are subject to corruption from the conversion of ferric iron to ferrous iron in acid-preserved samples and from the conversion of ferrous iron to ferric iron in unpreserved samples. Field measurement of Fe(II) in groundwater samples using colorimetric test kits usually yields satisfactory results with much less potential for the complications that plague laboratory measurements (Ehrlich 1981; Lovely 1991; Weidemeier et al. 1994).

Sulfate reduction can contribute to hydrocarbon biodegradation in the absence of oxygen, nitrate, and Fe(III). Sulfate-reducing conditions are characterized by a low oxidation–reduction (redox) potential and the production of sulfide (Ehrlich 1981; Edwards et al. 1992; Weidemeier et al. 1994). Under strictly anaerobic conditions and low redox, biodegradation of hydrocarbon leads to the reduction of carbon dioxide to methane (Grbic-Galic and Vogel 1987; Vogel et al. 1987; Wilson et al. 1993; Weidemeier et al. 1994).

Natural remediation rate calculations

The rate at which NR occurs often influences its acceptability as a strategy for restoring contaminated subsurface environments. Biodegradation rates may be determined using field data or laboratory data. Biodegradation rates are usually calculated with satisfactory results using a first-order rate equation. Laboratory data have the unfortunate disadvantage of usually overestimating biodegradation rates. Collection of rate data in the field with the accuracy to calculate a degradation rate in the absence of other physical loss mechanisms is challenging. Several different approaches have been applied to estimate the rate at which NR affects contaminant mass and concentration. The selection of an analytical approach is usually determined by the available data. Most approaches use an exponential decay equation (first-order rate equation) of the form:

$$C_i = C_o e^{-kt}$$

where

C_i = concentration at time "i"
C_o = initial concentration
t = time
k = rate constant

Five common approaches that incorporate this mathematical model are briefly described:

1) Timed flow vector approach
2) Steady-state aquifer approaches
3) Area, volume, and mass approaches
4) In-well calculation
5) Laboratory-based rate estimates.

Timed flow vector approach

The timed flow vector approach requires multiple monitoring wells that lie along the same groundwater flow path. Wells are sampled at time intervals that are approximately equal to the time required for the contaminant to migrate from one well to the next downgradient well. The data requirements for this approach include contaminant concentration at multiple time points, the groundwater seepage velocity, the distance between wells, and the time required for groundwater to migrate from one well to the next downgradient well. Simply solving for "k" yields the total attenuation rate, which includes the contribution of biological, physical, and chemical processes to contaminant concentration reduction. Biological losses can be estimated by comparing the change in concentration among biodegradable contaminants and poorly biodegradable contaminants that have similar chemical characteristics as the target compound. Examples of poorly degradable compounds that have been used to differentiate physical loss from biological loss of BTEX are 1,3,5- and 1,2,4-trimethyl benzene, which do not biodegrade very well under anaerobic conditions.

Steady-state aquifer approaches

Steady-state aquifer conditions refer to constant groundwater flow direction and velocity and constant input of contaminant from a source area or the absence of a continuing source of contamination. Under steady-state conditions, a single round of groundwater samples from wells that lie along a downgradient flow vector provides sufficient data to estimate attenuation rates. Groundwater seepage velocity and the concentration of contaminant in monitoring wells are required data. The time required for contaminant to migrate from one well to another and the concentration change from the upgradient well to downgradient wells are used to estimate attenuation rates. A least squares regression can be performed on datasets that contain 3 or more wells. The simple exponential decay equation provides only a total attenuation rate. Buscheck and Alcantar (1995) advocated an equation that incorporates a calculation of dispersion, advection, and retardation to estimate both attenuation and biodegradation rates.

Area, volume, and mass approaches

Changes in area, volume, or mass of contaminant in an aquifer or vadose zone can be used to estimate attenuation and biodegradation rates. Kriging is a useful tool for consistently defining the area or volume of contaminant in the subsurface in 2 or 3 dimensions. Kriging uses a mathematical algorithm to estimate the likely concentration of a chemical in the subsurface based on the concentration observed at nearby locations. Mass can also be approximated using kriging or methods such as the Theissen polygon method that defines the area represented by each well (Johnson et al. 1996). The use of a mathematical algorithm to calculate area, volume, or mass is very important so that fair calculations can be repeated for multiple sets of temporally separated data. The time difference between datasets and the change in area, volume, or mass is applied to the exponential decay equation to yield "k." This approach is probably the most reliable method for approximating the attenuation rate for vadose zones. Discerning the difference between biodegradation and nonbiological NR processes requires either the use of a compound that does not biodegrade but otherwise behaves similarly to the target compound or the measure of a degradation product that is directly correlated to the target compound.

In-well calculations

In-well calculations assume that the contaminant is not migrating. Over time, the change in concentration in single wells can be applied to the exponential decay equation to yield the attenuation rate. This approach is the least reliable because of the assumption that the contaminant is not migrating. Additionally, the very act of sampling the well has the potential to change the contaminant concentration in the immediate vicinity as a result of physical removal and aeration.

Laboratory-based rate estimates

Laboratory-based rate estimates are derived from microcosm studies that track contaminant biodegradation over time. This approach yields a true biodegradation rate; however, subsurface conditions are very difficult to reproduce in the lab. Laboratory-derived rates are almost always faster than in situ rates simply because laboratory conditions usually favor greater biological activity than do subsurface conditions.

The most reliable rate estimates are derived from in situ observations using multiple wells that lie along downgradient flow vectors and from plume area, volume, and mass calculations. In-well calculations and laboratory rate estimates are last-resort calculations that should be used only if other data are not available to support the use of another, more accurate rate estimation procedure.

PHYSICAL AND CHEMICAL MONITORING PARAMETERS

Physical and chemical monitoring parameters reveal abiotic processes that can result in contaminant dissipation. Unless these parameters are accounted for in the overall contaminant mass balance, biodegradation may be credited for more contaminant loss than actually occurs. Because most of the physical loss mechanisms do not result in contaminant degradation, these processes may not be acceptable approaches for site remediation. Chemical processes usually do not result in contaminant degradation; however, abiotic chemical reactions in soil and groundwater may convert a contaminant into another compound. Chemical transformations may be beneficial if they result in a compound that is less hazardous or more biodegradable.

In addition to accounting for abiotic contaminant loss mechanisms, chemical monitoring is the ultimate parameter that determines when a site is remediated. Therefore, monitoring to determine contaminant concentration reduction, contaminant mass loss, and possibly the appearance of degradation products is critical for defining the success of NR.

Dilution is an inevitable physical event. The significance of dilution is site dependent and difficult to predict with certainty; therefore, a measure of the dilution rate is useful for defining physical changes in contaminant concentration. Dilution affects the concentration of a contaminant, not the mass of contaminant. Diffusion and dispersion are the most common mechanisms of contaminant dilution; however, in shallow aquifers, infiltrated rainwater can also be a significant contributor.

Diffusion is a minor component of dissolved contaminant dilution because the tortuosity of soil greatly reduces the rate of diffusion. Although diffusion is usually insignificant in the dilution of dissolved contaminant (Freeze and Cherry 1979), it may be one of the principal factors in the dissolution of contaminant from NAPL into the groundwater. The rate of contaminant dissolution from an NAPL source is often the major rate-limiting component of NR.

The movement of a contaminant in a direction that is not the direction of groundwater flow is termed "dispersion." Dispersion is usually a significant physical process resulting in contaminant dilution. Dispersion commonly results in the side-to-side spread of a contaminant plume along the downgradient flow path. Like diffusion, dispersion does not result in a change of contaminant mass, although it does result in a concentration change. Dispersion is often estimated by examining plume dimensions in the downgradient and cross-gradient directions. The ratio of the cross-gradient movement to the downgradient movement represents the dispersion coefficient. Dispersion can be directly measured using an in situ tracer; however, this approach is time consuming and expensive. Dispersion can also be approximated using published values for defined aquifer conditions. The approximated value can be further refined by adjusting it iteratively in a groundwater transport model until the model matches the in situ condition. The net result of both diffusion and dispersion is a reduction in contaminant concentration. The combined effect of dilution and dispersion can be measured using the single procedure.

Volatilization results in a decrease in contaminant mass from soil or groundwater and an increase of mass in soil gas or atmosphere. Volatilization is controlled by Henry's law constant, octanol–water partition coefficient (K_{OW}), solubility, and density of the contaminant. Volatilization is most prominent in the capillary fringe and the groundwater fluctuation space when the contaminant is less dense than water or moves along the saturated–unsaturated soil boundary. The contribution of volatilization is difficult to predict during NR, since multiple conditions interact during volatilization. Volatilization is typically a minor contaminant loss mechanism, unless a soil gas or vadose zone soil treatment technology is used in conjunction with NR. An example is the operation of a bioventing or soil vapor extraction system in the vadose zone above a contaminated aquifer.

"Sorption" is a general term used to describe adsorptive and absorptive processes that result in the partitioning of contaminant from the aqueous or dissolved phase to the solid phase or soil (Freeze and Cherry 1979). In simple terms, contaminants stuck to soil are no longer dissolved in the groundwater. An adsorption coefficient (K_{OC}, the soil–water adsorption coefficient, describes the affinity of a chemical to partition between soil and water) can be estimated using the octanol–water partition coefficient and contaminant solubility. The adsorption coefficient can also be determined experimentally by observing the amount of contaminant in the soil and water. This type of experimental information usually fits a Freundlich adsorption isotherm, which can be used to calculate the adsorption coefficient.

The sorption and desorption of dissolved contaminant result in contaminant migration that is slower than groundwater velocity. The retardation coefficient describes the movement of dissolved contaminant relative to groundwater (Freeze and Cherry 1979). The retardation coefficient is important because it more accurately defines the retention time of a contaminant between 2 points along a flow vector. This information is central to the calculation of biodegradation rates.

SITE ASSESSMENT STRATEGIES

Quantifying the contribution of NR to subsurface restoration is the key to effectively and responsibly applying NR as a primary environmental management strategy. The same general approach to assessing the contribution of NR has been advanced by the U.S. Air Force, the U.S. Geological Survey (USGS), various USEPA regions, and state agencies. Three lines of evidence are generally recognized as useful for observing the occurrence of NR in the subsurface:

1) Evaluation of historical contaminant data to determine the change in concentration and mass of target compounds and the behavior of the dissolved contaminant plume
2) Evaluation of groundwater geochemistry, biodegradation intermediates, and respiratory substrates and products to demonstrate the contribution of naturally occurring biodegradation to subsurface restoration
3) Laboratory microcosm tests to demonstrate the biodegradation and the rate of biodegradation of target compounds under conditions likely to exist in the subsurface.

This same approach appears to be valid for any organic compound. The NR of metals relies on physical and chemical processes to a greater extent that organics. In either case, accurately quantifying the effect of physical and chemical processes is challenging but important. The change in concentration of poorly degradable co-contaminants, such as 1,3,5- and 1,2,4-trimethylbenzenes, pristane, and phytane, has been used to quantitate the contribution of physical and biological processes to the overall attenuation of target compound concentration and mass. Chloride has been used in some cases to contrast physical mechanisms with biological mechanisms at sites impacted with chlorinated aliphatics. In many cases, the change in concentration resulting from physical and chemical processes is approximated without particularly reliable field data.

In order to capture adequate evidence to support the use of NR, the following parameters are usually determined:

- Site geology and hydrogeology, including stratigraphy, depth to water, aquifer permeability, groundwater gradient, groundwater seepage velocity, porosity, and fraction organic C
- Property boundaries, exposure pathways, uses of groundwater, and location of production wells
- Contaminants in the subsurface, their source, the presence of lighter-than-water nonaqueous phase liquid (LNAPL) and/or denser-than-water nonaqueous phase liquid (DNAPL), the horizontal and vertical extent of contamination, and the presence of potential biodegradation or chemical transformation products
- Groundwater geochemistry, including as a minimum dissolved oxygen, redox potential, pH, conductivity, temperature, ferrous iron, nitrate and nitrite, sulfate, sulfide, dissolved methane, chloride, total organic C, alkalinity and/or dissolved carbon dioxide, dissolved H, and dissolved ethene and ethane at sites with chlorinated aliphatics
- Additional parameters that are recommended by some guidance documents: Mn, Ca, Mg, total kjeldhal nitrogen, ortho-phosphate, and ammonium.

In addition to the analytical parameters listed above, more complex sites may require advanced investigation methods. The use of geostatistical tools is particularly valuable for accurately defining 2-dimensional and especially 3-dimensional groundwater plumes. The use of statistical methods to define plumes insures objective, reproducible assessment of sites and contaminant distribution.

Calibrating the vertical velocity of groundwater has been performed by tracking the migration of tritium in the subsurface. The advent of the atomic age with the detonation of the first H bombs produced an elevated concentration of tritium in the atmosphere, which has subsequently been deposited in the subsurface. The depth of the elevated tritium signature and the time elapsed since the first testing of the H bomb provides a standard to quantify the downward groundwater migration rate.

At sites where multiple chlorinated compounds have been spilled, including compounds that may be either biodegradation products or spilled products, defining the contribution of reductive dechlorination for subsurface restoration can become so complicated that degradation products cannot be discerned from primary contaminants. Stable isotopes of carbon and chloride have been successfully used to demonstrate that biodegradation is affecting the isotope composition of groundwater impacted with chlorinated aliphatics (Weigand et al. 1997; Aravena et al. 1998; Sturchio et al. 1998). The preference of biological reactions to favor light isotopes can lead to a skewed distribution of stable isotopes in the subsurface, compared to background isotope ratios.

DEFINING A MONITORING PLAN FOR NATURAL REMEDIATION

Several approaches can be taken to increase the efficiency of long-term monitoring of NR-based corrective action plans. Once the plume has been delineated and its behavior has been evaluated, the following issues are important for long-term monitoring:

- "Sentry wells" may be installed in an intermediate point between the current downgradient edge of the plume and the nearest receptor, property boundary, or "point of compliance." Sentry wells are "clean" groundwater monitoring wells that are used to indicate movement of the contaminant plume. If contaminants are detected in sentry wells, more aggressive remedial actions are invoked to deal with the contamination.
- The contaminant concentration must be periodically evaluated; however, where plume migration is limited, where the sentry wells are protecting downgradient receptors, and where plume dimension and concentration are well documented, the frequency of contaminant analysis can be minimized. Similarly, the number of samples can be reduced to the minimum number that will confirm satisfactory performance of the NR remedial plan. There is no need to remap the entire contaminant plume at each sampling event. Sampling events should be used to confirm treatment with a negotiated confidence level.
- In addition to contaminant concentrations, groundwater parameters should be monitored. The choice of which parameters to monitor should be determined from the initial site assessment. For example, if no nitrate was detected in the groundwater during the site assessment, nitrate would not be a useful monitoring parameter. Similar assessments can be made for each groundwater chemistry parameter. If the parameter was not initially useful in defining the NR process, it need not continue to be analyzed. Field data on water table elevation are an important parameter that should always be included in the monitoring program. Seasonal or storm-specific changes in water table elevation can influence the dissolved concentration of contaminant. Understanding the normal concentration changes related to fluctuations in the water table elevation prevents rash decisions about the effectiveness of NR for groundwater restoration.

Modeling natural remediation

Predicting the long-term effectiveness of NR is often critical to its acceptance by regulatory agencies and the public. Modeling has proved a valuable tool for making such predictions. Two general approaches are usually taken to model the future performance of NR: analytical modeling and numerical modeling.

Analytical modeling relies on relatively simple calculations of groundwater flow, contaminant dispersion, adsorption, retardation, partitioning into groundwater, biodegradation, and abiotic transformations. Site-specific values for each of these parameters are used to predict the likely migration and attenuation of the contaminant. The accuracy of the prediction depends on the accuracy of the defining parameters. Attenuation rates are often modeled using chemical-specific data, one chemical at a time. Multivariate statistical analysis can be used to model intrinsic biodegradation (Tan 1994). The benefit of this approach is that it can evaluate all pertinent site data as a whole and derive an overall degradation rate for a multicomponent contaminant plume. Further refinements of this and other statistical approaches may lead to a new generation of analytical tools for evaluating and predicting NR.

Numerical modeling employs computer programs to simulate the behavior of an aquifer. One widely used numerical model for aquifer bioremediation is BIOPLUME II (Rifai et al. 1988). This model requires the input of site-specific parameters that define the hydrogeologic setting of the site; the behavior of the target contaminant in the aquifer; the influence of infiltration, pumping wells, and contaminant source areas; and the mechanisms by which the contaminant is remediated. In addition to simulating current conditions, the model also predicts what will happen to the contaminant plume in the future. This aspect of the numerical model is extremely valuable for weighing the benefits and risks of NR, defining the location to place sentry wells, and approximating the time required to complete the aquifer remediation. A major revision of BIOPLUME II has been completed recently; BIOPLUME III permits much more precise definition of anaerobic processes within the dissolved contaminant plume. Other aquifer bioremediation computer models include BIO1D (GeoTrans, Inc., Washington, DC) and BIOTRANS (Environmental Systems & Technologies, Inc., Blacksburg, VA) (Rifai 1993). Each model has useful features and may be pertinent for a given site. In addition to these models that are specifically designed for biodegradation, most groundwater fate and transport models incorporate an exponential decay constant that simulates biodegradation. Some models also accommodate the production and subsequent degradation of daughter products.

Risk assessments play a prominent role in the selection of suitable remediation strategies for contaminated subsurface sites. However, as currently practiced, risk assessments often overestimate risks (Milloy 1995; Brady et al. 1998). Risk-based corrective action (RBCA) is widely used in conjunction with underground storage tank (UST) replacement and remediation programs (American Society for Testing and Materials [ASTM] 1995). Risk-based, site-specific cleanup targets are gaining acceptance as the impracticability of actively restoring subsurface environments to pristine conditions or even reducing concentrations to maximum concentration limits (MCLs) becomes more apparent. Natural remediation is an important component of the environmental restoration and management strategy, especially for low-risk sites that pose minimal threats to human health and the environment. Natural remediation has the benefit of progressively, albeit slowly, reducing the risk associated with an impacted site.

ACCEPTANCE OF NATURAL REMEDIATION FOR SUBSURFACE ECOSYSTEMS

Natural remediation is a well-accepted remedial strategy for petroleum-contaminated subsurface environments. Although an appreciable level of site-specific information must usually be gathered and presented in a convincing manner before NR will be approved as the sole or partial solution for an impacted subsurface, the probability is high that approval will be granted if NR is occurring and provides acceptable reduction in risk.

The extension of NR to contaminants besides petroleum has met with varying degrees of acceptance. Chlorinated aliphatic compounds are generally considered susceptible to NR; however, the requirements to document and quantify the NR of these compounds are more rigorous than for petroleum. These sites are less likely to sustain environmentally acceptable levels of NR, compared to petroleum-contaminated sites. Suggestions have been made that as few as 15% of all sites contaminated with chlorinated aliphatics may be appropriate for NR.

Nitroaromatics, chlorinated pesticides and herbicides, polynuclear aromatic hydrocarbons (PAHs), and heavy metals and radionuclides are being considered as potentially susceptible to NR processes. However, data are sparse supporting the performance of NR processes for these common subsurface contaminants. Until precedents are set and additional site data become available, gaining approval for the use of NR for these contaminants will be difficult and should be expected to require an extremely thorough and complete site evaluation. "No action" decisions based on risk assessments and technical impracticability waivers actually are more likely for these sites than is the decision to approve NR.

Although NR will effectively reduce the risk associated with contaminated groundwater at a site, it may not be the best management alternative. Some of the suitable sites will also be easily and rapidly remediated using other technologies. When constraints such as remediation time or unacceptable additional plume migration do not favor NR, especially at small or geologically simple sites, other, more aggressive, technologies may be preferable to NR (Nyer et al. 1996). In such cases, NR is probably not the best environmental management choice.

Natural remediation is a powerful process that can, with time, result in site remediation. Natural remediation faces 3 obstacles that could hinder its use as a viable environmental risk management strategy:

1) Over use
2) Misapplication
3) Skepticism.

Recognition of these obstacles and conscious effort to avoid stumbling over them will assure responsible and reliable use of NR. Emerging environmental technologies are often indiscriminately applied to every site as though the technology can handle any problem. This unfortunate practice has compromised the perception of many useful technologies, including air sparging, soil vapor extraction, soil flushing, thermal desorption, and nearly every biological treatment technology. Skeptical, uninformed, or inflexible regulatory agencies can thwart the use of NR at ideal sites. Cooperation and flexibility among all involved parties and realization that even proven technologies were once innovative and unproven will help ease the way for applications of NR at suitable sites.

CONCLUSIONS

Natural remediation provides several measurable benefits as a strategy for subsurface restoration. These include reduced costs for remediation, performance that is often comparable to other more aggressive and costly technologies, and acceptance, under many circumstances, as an environmentally responsible action.

Natural remediation, also known in regulatory circles as "monitored natural attenuation," is still a new technology that will continue to grow in importance as a subsurface remediation strategy. The concepts will continue to be extended to more and more contaminants. Regulatory enforcement of environmental laws and the severity of environmental laws are sensitive to political whims; thus, the ease of achieving official recognition of the true contribution of NR is likely to vacillate. But regardless of the politics, NR will continue to silently contribute to environmental restoration. Taking full advantage of the earth's regenerative powers is only sensible. Environmental stewardship also requires conscientiousness and making every effort technically and economically possible to eliminate or at least minimize new abuses to the subsurface. Under these conditions, NR and aggressive environmental stewardship can address many of today's environmental issues.

REFERENCES

Aravena R, Beneteau K, Frape S, Butler B, Abrajano T, Major D, Cox E. 1998. Applications of isotopic finger-printing for biodegradation studies of chlorinated solvents in groundwater. In: Wickramanayake GB, Hinchee RE, editors. Risks, resources, and regulatory issues. Columbus OH: Battelle Pr. p 67–71.

[ASTM] American Society for Testing and Materials. 1995. ASTM Designation: E 1739-95, standard guide for risk-based corrective action applied at petroleum release sites. Conshohocken PA: ASTM.

Bradley PM, Chapelle FH, Landmeyer JE, Schumacher JG. 1997. Potential for intrinsic bioremediation of a DNT-contaminated aquifer. *Ground Water* 35(1):12–17.

Brady PV, Brady MV, Borns DJ. 1998. Natural attenuation CERCLA, RBCA's, and the future of environmental remediation. Boca Raton FL: Lewis.

Brady PV, Spalding BP, Krupka KM, Waters RD, Zhang P, Borns DJ, Brady WD. 1999. Site screening and technical guidance for monitored natural attenuation at DOE sites. Sandia National Laboratories. Sandia Report Sand 99-0464.

Buscheck TE, Alcantar CM. 1995. Regression techniques and analytical solutions to demonstrate intrinsic bioremediation. In: Hinchee RE, Wilson JT, Downey DC, editors. Intrinsic remediation. Columbus OH: Battelle Pr. p 109–116.

Cattaneo MV, Masson C, Hawari J, Sunhara G, Greer CW, Thiboutot S, Ampleman G. 1997. Natural attenuation of TNT in soil columns. In situ and on-site bioremediation: Volume 2. Fourth International In Situ and On-Site Bioremediation Symposium; 1997 April; New Orleans LA. Columbus OH: Battelle Pr. p 3–8.

Charbeneau RJ, Bedient PB, Loehr RC. 1992. Ground water remediation. Lancaster PA: Techomic.

Cozzarelli IM, Baedecker MJ, Eganhouse RP, Goerlitz DF. 1994. The geochemical evolution of low-molecular-weight organic acids derived from the degradation of petroleum contaminants in groundwater. *Geochim Cosmochim Acta* 58(2):863–877.

Cozzarelli IM, Eganhouse RP, Baedeker MJ. 1990. Transformation of monoaromatic hydrocarbons to organic acids in anoxic ground water environment. *Environ Geol Water Sci* 16.

Edwards EA, Wells LE, Reinhard M, Grbic-Galic D. 1992. Anaerobic degradation of toluene and xylene by aquifer microorganisms under sulfate-reducing conditions. *Appl Environ Microbiol* 58:794–800.

Ehrlich HL 1981. Geomicrobiology. New York NY: Marcel Dekker.

Fetter CW. 1994. Applied hydrogeology. Upper Saddle River: Prentice-Hall.

Freeze RA, Cherry JA. 1979. Groundwater. Englewood Cliffs NJ: Prentice-Hall.

Grbic-Galic D, Vogel TM. 1987. Transformation of toluene and benzene by mixed methanogenic cultures. *Appl Environ Microbiol* 53:254–260.

Johnson RL, Dupont RR, Graves DA. 1996. Assessing UST corrective action technologies: Diagnostic evaluation of in situ SVE-based system performance. Washington DC: U.S. Environmental Protection Agency (USEPA). EPA/600/R-96/041.

Krumholz LR, Li J, Clarkson WW, Wilber GG, Suflita JM. 1997. Transformations of TNT and related aminotoluenes in groundwater aquifer slurries under different electron-accepting conditions. *J Indust Microbiol Biotechnol* 18:161–169.

Kuhn EP, Zeyer J, Eicher P, Schwarzenbach RP. 1988. Anaerobic degradation of alkylated benzenes in denitrifying laboratory aquifer columns. *Appl Environ Microbiol* 54:490–496.

Lovely DR. 1991. Dissimilatory Fe(III) and Mn(IV) reduction. *Microbiol Rev* June 1991:259–287.

Major DW, Mayfield CI, Barker JF. 1988. Biotransformation of benzene by denitrification in aquifer sand. *Groundwater* 26:8–14.

McCormick NG, Feeherry FE, Levinson HS. 1976. Microbial transformation of 2,4,6-trinitrotoluene and other nitroaromatic compounds. *Appl Environ Microbiol* 31(6):949–958.

Milloy JS. 1995. Science-based risk assessment. National Environmental Policy Institute.

Montgomery JH. 1996. Groundwater chemicals desk reference. Boca Raton FL: Lewis.

Nyer EK, Kidd DF, Palmer PL, Crossman TL, Fam S, Johns II FJ, Boettcher G, Suthersan SS. 1996. In situ treatment technology. Boca Raton FL: Lewis.

Rifai HS. 1993. Mathematical modeling of intrinsic bioremediation at field sites. Proceedings of the Symposium on Intrinsic Bioremediation of Ground Water. EPA/540/R-94/515.

Rifai HS, Bedient PB, Borden RC, Haasbeck JF. 1988. Bioplume II. Houston TX: Rice Univ. National Center for Ground Water Research.

Spain JC. 1996. Future vision: Compounds with potential for natural attenuation. In: Proceedings of the Symposium on Natural Attenuation of Chlorinated Organics in Groundwater; 1996 September 11–13; Dallas TX. Washington DC: U.S. Environmental Protection Agency (USEPA). EPA/540/R-96/509.

Sturchio NC, Clausen JL, Heraty LJ, Huang L, Holt BD, Abrajano TA Jr. 1998. Chlorine isotope investigation of natural attenuation of trichloroethene in an aerobic aquifer. *Environ Sci Technol* 32(20):3037-3042.

Tan TC. 1994. Modeling of a natural bioremediation process using multivariate statistical approach [Master's thesis]. Knoxville TN: Univ Tenn.

Townsend DM, Meyers TE. 1996. Recent development in formulating model descriptions for subsurface transformations and sorption of TNT, RDX, and HMX. Vicksburg MS: U.S. Corps of Engineers, Waterways Experiment Station. Technical Report IRRP-96-1.

[USEPA] U.S. Environmental Protection Agency. 1996a. Overview of the technical protocol for natural attenuation of chlorinated aliphatic hydrocarbons in groundwater under development for the U.S. Air Force Center for Environmental Excellence. Symposium on Natural Attenuation of Chlorinated Organics in Ground Water. EPA/540/R-96/509.

[USEPA] U.S. Environmental Protection Agency. 1996b. Future vision: Compounds with potential for natural attenuation. Symposium on Natural Attenuation of Chlorinated Organics in Ground Water. EPA/540/R-96/509.

[USEPA] U.S. Environmental Protection Agency. 1997. Use of monitored natural attenuation at Superfund, RCRA corrective action, and underground storage tank sites. Washington DC: Office of Solid Waste and Emergency Response. Directive 9200.4-17.

[USEPA] U.S. Environmental Protection Agency Office of Solid Waste and Emergency Response. 1999. Use of monitored natural attenuation at Superfund, RCRA corrective action, and underground storage tank sites. OSWER Directive Number 9200.

Vance DB. 1998. MTBE: Character in question. *EnvironTechnol* Jan/Feb:16–17.

Vogel TM, Criddle CS, McCarty PL. 1987. Transformation of halogenated aliphatic compounds. *Environ Sci Technol* 21(8):722–736.

Weigand MA, Hecox G, Matherne C, Graves D. 1997. Investigation of the natural attenuation of chlorinated organics in groundwater. In situ and on-site bioremediation:

Volume 3, Fourth International In Situ and On-Site Bioremediation Symposium; 1997 April; New Orleans LA. Columbus OH: Battelle Pr. p 221–226.

Wiedemeier TH, Downey DC, Wilson JT, Kampbell DH, Miller RN, Hansen JE. 1994. Technical protocol for implementing the intrinsic remediation with long-term monitoring option for natural attenuation of dissolved-phase fuel contamination in ground water. Brooks Air Force Base, San Antonio TX: Air Force Center For Environmental Excellence.

Wiedemeier TH, Swanson MA, Moutoux DE, Gordon EK, Wilson JT, Wilson BH, Kampbell DH, Haas PE, Miller RN, Hansen JE, Chapelle FH. 1998. Technical protocol for evaluating natural attenuation of chlorinated solvents in ground water. USEPA EPA/600/R-98/128.

Wiedemeier TH, Swanson MA, Moutoux DE, Wilson JT, Kampbell DH, Hansen JE, Haas P. 1996. Overview of the technical protocol for natural attenuation of chlorinated aliphatic hydrocarbons in groundwater under development for the U.S. Air Force Center for Environmental Excellence. In: Proceedings of the Symposium on Natural Attenuation of Chlorinated Organics in Groundwater; 1996 September 11–13; Dallas TX. Washington DC: U.S. Environmental Protection Agency (USEPA). EPA/540/R-96/509.

Wilson JT, Cook M, Kampbell D. 1994. Intrinsic remediation of JP-4 jet fuel. Third Environmental Restoration Technology Transfer Symposium; 1994 August; San Antonio TX.

Wilson JT, Kampbell DH, Armstrong J. 1993. Natural bioreclamation of alkylbenzenes (BTEX) from a gasoline spill in methanogenic groundwater. Second International Symposium on In Situ and On Site Bioreclamation; San Diego CA.

Wilson JT. 1993. Testing bioremediation in the field. In situ bioremediation: When does it work? Washington DC: National Academy Pr.

14

Natural Remediation in the Great Lakes

Dora R. Passino-Reader, Michael A. Kamrin, James P. Hickey

THE VAST INLAND SEAS of the Laurentian Great Lakes span more than 1,200 km (750 mi) from west to east in North America. The 5 Great Lakes are Superior, Michigan, Huron, Erie, Ontario, and their notable connecting waters are the St. Marys River, the St. Clair and Detroit Rivers, and the Niagara River. The Great Lakes are the largest system of fresh, surface water on the earth (Government of Canada [GC] and U.S. Environmental Protection Agency [USEPA] 1995). The Great Lakes basin, which is home to 33 million people, provides water for consumption, transportation, power, recreation, extraction of natural resources, and other uses. Some of the world's largest concentrations of industrial capacity are in the Great Lakes region (GC, USEPA 1995): 25% of Canadian and 7% of U.S. agricultural production occurs in the basin.

The Great Lakes contain about 23,000 km^3 of water, with a surface area of 244,000 km^2. Physical characteristics such as climate, soils, and topography vary across the vast expanse of the basin. The major physical features and populations are shown in Table 14-1. The Great Lakes are largely ice covered in the winter, are stratified with a strong thermocline in the summer, and experience spring and fall vertical mixing of the water. This pattern determines the distribution of nutrients and biological species and the patterns for horizontal and vertical migration of fish and plankton.

Approximately 180 fish species were indigenous to the Great Lakes, although a number of extinctions and extirpations have occurred because of overfishing, pollution, exotic species, and loss of habitat. Indigenous fish species include lake

Natural Remediation of Environmental Contaminants. Michael Swindoll et al., editors.
 ISBN 1-880611-33-3.

Table 14-1 Physical features and populations of the Laurentian Great Lakes[a]

Feature		Lake					Total
		Superior	Michigan	Huron	Erie	Ontario	
Elevation	m	183	176	176	173	74	
Length	km	563	494	332	388	311	
Breadth	km	257	190	245	92	85	
Average depth	m	147	85	59	19	86	
Maximum depth	m	406	282	229	64	244	
Volume	km^3	12,100	4,920	3,540	484	1,640	22,684
Water (area)	km^2	82,100	57,800	59,600	25,700	18,960	244,160
Land drainage (area)	km^2	127,700	118,000	134,100	78,000	64,030	521,830
Total area	km^2	209,800	175,800	193,700	103,700	82,990	765,990
Shoreline length	km	4,385	2,633	6,157	1,402	1,146	17,017
Retention time	y	191	99	22	2.6	6	
Population: U.S. 1990		425,548	10,057,026	1,502,687	10,017,530	2,704,284	24,707,075
Population: Canada 1991		181,573		1,191,467	1,664,639	5,446,611	8,484,290
Total population		607,121	10,057,026	2,694,154	11,682,169	8,150,895	33,191,365
Outlet		St. Marys River	Straits of Mackinac	St. Clair River	Niagara River Welland Canal	St. Lawrence River	

[a] Source: GC, USEPA 1995.

trout, lake sturgeon, lake herring, walleye, and small- and largemouth bass. Preservation of existing fish and wildlife habitat and restoration of lost habitat are critical issues for sustainable use of the Great Lakes basin in the face of increasing human populations in the new millennium.

Contaminated sediments have been shown to have significant adverse impacts on human and ecological targets in the Great Lakes basin (Baumann et al. 1996; Lonky et al. 1996; Ludwig et al. 1996). The commercial and sport fishing industries can be negatively impacted by fish consumption advisories, the viability of commercial ports can be compromised by restrictions on dredging of navigational channels and disposal of dredged sediments, and municipalities can face greatly increased costs of water treatment. Ecological use impairments include degradation of fish and wildlife populations, degradation of phytoplankton or zooplankton populations, increased incidence of fish tumors and deformities, increased incidence of bird or other animal reproductive or developmental problems, degradation of the benthos, eutrophication or increases in undesirable water plants, degradation of esthetics, and loss of fish and wildlife habitat.

While the impacts are clear, progress in remediating sediment contamination has generally been slow. The reasons for this are varied and include lack of information on the sources and extent of the problems at particular sites, lack of standards by which to judge the severity of the problem and the success of cleanup efforts, lack of proven and cost-effective remediation techniques, funding limitations, and social and political factors. Another confounding factor is the uniqueness of each site within the Great Lakes and the resultant need to develop a different risk assessment and risk management strategy at each location.

Natural remediation (NR) is an attractive alternative at many Great Lakes sites. It is less costly than most other techniques and causes less disruption in the area during the remediation process. The primary issue faced by NR for a given site is whether it will be sufficiently effective to reduce use impairments to a degree acceptable to all of the stakeholders involved in that site. The case studies presented in this chapter provide illustrations of the types of problems that occur at sites with contaminated sediments and some of the remedial alternatives that have been tried or that are being considered at these sites. The third section of this chapter addresses the decision-making process at sites in the Great Lakes region and illustrates how the information collected at the sites is used to reach a resolution of the problem.

GREAT LAKES REGULATORY STATUS

Because the Great Lakes are shared by the U.S. and Canada, risk management requires both individual and joint actions by U.S. and Canadian governments, especially under the auspices of the International Joint Commission (IJC), established in 1909. These actions are undertaken by federal agencies on both sides of the border: state and local agencies in the U.S. and provincial agencies in Canada.

As the result of a number of studies on pollution of the Great Lakes, the U.S. and Canadian governments signed the Great Lakes Water Quality Agreement on 15 April 1972, which was revised and expanded in 1978 to address toxic chemical discharges into the Great Lakes (IJC 1978). In the 1978 agreement, there were stipulations that the nature and extent of sediment contamination be identified, that methods be developed to assess the impacts of such contamination, and that technologies that could be used to remediate sediment contamination be evaluated.

Under the 1987 Protocol (IJC 1987), an amendment to the Great Lakes Water Quality Agreement, a program was established to develop and implement remedial action plans (RAPs) to restore environmental quality at 42 sites in the Great Lakes basin that were designated as Areas of Concern (AOCs). As shown in Figure 14-1, the AOCs are located on the 5 Great Lakes and on channels that connect these lakes, generally in urban or industrial areas and often near the mouths of rivers that empty into the lakes. Pollution at these AOCs may result from local sources or from the accumulation of pollutants that entered the water upstream from both point and nonpoint sources. The nearshore sediments are very commonly contaminated with the highest levels of pollutants. The RAP process has 3 steps:

1) Identifying sources of contamination and defining the problem
2) Identifying actions that can be taken to restore the beneficial uses of the resource
3) Confirming that use restoration has occurred (IJC 1985).

Additional actions were taken by the U.S. government starting in 1987 when the U.S. Clean Water Act was amended to include authorization for the USEPA's Great Lakes National Program Office (GLNPO) to coordinate and conduct demonstration projects on the treatment of contaminated sediments. Five AOCs were chosen as sites where this program, known as the Assessment and Remediation of Contaminated Sediments (ARCS) Program (USEPA 1994c), was to be implemented. These AOCs are located at Saginaw Bay, Michigan; Buffalo River, New York; Grand Calumet River, Indiana; Ashtabula River, Ohio; and Sheboygan Harbor, Wisconsin.

In 1991, GLNPO was given responsibility for developing Lakewide Management Plans (LAMPs). More recently, under provisions of the U.S. Clean Water Act and the Great Lakes Critical Program Act, the USEPA developed and published a water quality guidance document known as the Great Lakes Initiative (GLI; USEPA 1995). While it might appear that the risk management measures so far described apply only to surface water quality, there are, in addition to Great Lakes-specific legislation, regulation, and policy, national risk management mandates that also have significant impacts on AOCs. In particular, risk assessment and risk management guidance issued under the U.S. Comprehensive Environmental Response, Compensation and Liability Act (CERCLA, also known as "Superfund") and the Resource

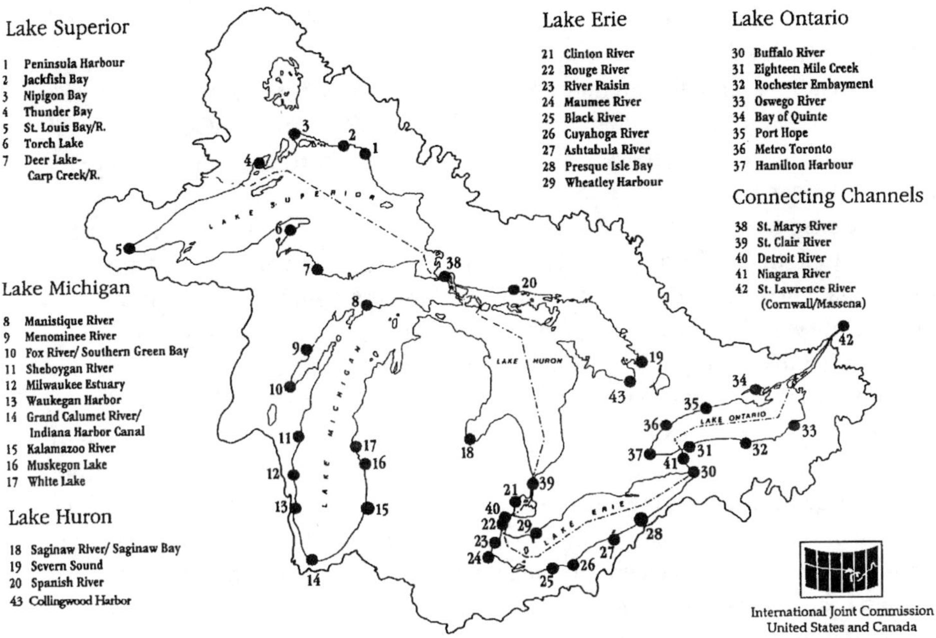

Figure 14-1 Outline map of the Laurentian Great Lakes, showing 42 Areas of Concern (AOCs) (Note: Collingwood Harbor, Lake Huron, has been delisted; Presque Isle Bay, Lake Erie, has been added. Adapted from Savagian and Uram 1995 and reprinted with permission from the International Joint Commission.)

Conservation and Recovery Act (RCRA) directly govern those U.S. AOCs or parts of AOCs that fall under these statutes. In addition, such federal guidance is used by the states to establish risk management approaches and goals for hazardous waste sites within their borders.

In addition to establishing guidelines for determining when hazardous wastes need to be cleaned up and to what level, the USEPA also provides guidance as to appropriate remedial action techniques (USEPA 1996). While NR is one of these alternatives, it has not yet been widely applied as an option in the U.S., probably because of the time required for NR to be effective. In some situations, levels of contaminants pose an immediate threat to ecological values and human health.

When the GLNPO selected preferred sediment remediation technologies for testing at the 5 AOCs in the ARCS program, 4 of the 5 remedies involved physical or chemical extraction or separation processes (USEPA 1991a).

The fifth demonstration technology involved active bioremediation; thus, NR was not included as a possibility.

The April 1998 revision of USEPA's Contaminated Sediment Management Strategy (USEPA 1998a) supports NR as the best strategy in certain circumstances, after reduction of pollution sources. Based on previous USEPA responses to the introduction of new remedial approaches, justification for the use of NR at a site will require more detailed information and analysis than is required for other options. Natural remediation may be the preferred alternative for those areas with less severe contamination, sites other than the "hot spots" in the AOCs. Natural remediation, or natural recovery, has been the management decision of choice for several AOCs in Canada (IJC 1997).

The past decade has been a period when management of risks from hazardous substances in the environment has moved from a command-and-control approach to one that encourages the use of consensus among stakeholders. In addition, it has been a time when the focus has gone from addressing each site in isolation to one where decisions as to whether and how vigorously to manage each site are based on risk comparisons. These include comparisons with risks that are qualitatively quite different (e.g., hazardous waste versus global warming) as well as comparisons of risks from the same contaminants in different media (e.g., sediments versus ambient air). In large part, this change has been due to the recognition that resources are limited and that priorities must be established among environmental contamination problems and also between environmental contaminants and other threats to human health and the environment.

In this atmosphere, the expectations that existed when AOCs were identified and when the ARCS program began have been tempered. Instead of rapid and extensive cleanups of contaminated sediments, risk managers are looking at cost-effective strategies that will substantially reduce immediate risks, and they also are looking at low-cost alternatives to provide long-term solutions. Thus, the critical issues are these:

1) How can we realistically assess the human and ecological risks of contaminated sediments at each site, and how do these risks compare with other risks competing for funding?
2) How can we realistically assess the effectiveness of various proposed remediation methods?
3) How can we assess the costs and benefits of each approach?
4) Perhaps most important, how can we generate the political will to implement solutions that will provide adequate protection of human health and the environment?

NATURAL REMEDIATION IN THE GREAT LAKES

The primary goal of remediation of contaminated sites in the Great Lakes is the restoration of healthy biological communities and ecosystems. The reestablishment of native species that have been extirpated and the restructuring of ecosystems that are structurally and functionally representative of pre-damage conditions are goals that are desirable but perhaps not attainable. Once contaminant levels have been reduced and other physical and chemical conditions of the site restored, native species will recolonize the site from healthier surrounding areas. In the case of AOCs, which generally are industrialized and heavily populated river mouths and harbors, the native species will be reestablished from downstream drift of invertebrates or movement of fish from the Great Lakes or from upstream areas.

The potential effectiveness of NR at various sites depends on a number of factors, including the degree of contamination, the types of contaminants, and the physical and chemical characteristics of the site. In some circumstances, these factors may preclude NR as a viable option for a site or for part of a site.

Documentation of successful remediation, whether from natural processes or active remediation, necessitates monitoring of biological species before and after remediation takes place. Biological monitoring should be designed to accurately assess the populations and community structures and to provide direct comparisons of the status of the populations and communities at different time periods. Unfortunately, because of the lack of institutional stability and the limitations of funds, very few consistent monitoring programs have been conducted at the AOCs. The near absence of consistent monitoring data makes documentation of NR difficult. One exception is the repeat sampling of fish communities by the same methods in the Collingwood Harbour AOC in 1986 and 1993 (Ferguson 1986; Collis and Proud 1994; Krantzberg and Houghton 1996). However, the lack of replicates for each year precludes statistical analyses of the data. Qualitative comparisons between the 2 years are instructive. For the other AOCs, we are forced to compare datasets collected in different locations, by different methods, and with different objectives. Nevertheless, we attempt to determine the trends with time in the biological communities so as to document possible progress in restoration of the AOCs.

CASE STUDIES OF AREAS OF CONCERN

The 5 ARCS priority AOCs (USEPA 1994a) have received considerable study to determine the extent and type of contamination at these sites as well as the most appropriate remediation options. These ARCS sites will be used to illustrate the status of Great Lakes remediation efforts (Jafvert and Rogers 1991). Data were compiled from numerous studies of these AOCs. The chemical data will be presented by site, as contaminant concentration over time from earliest to latest study, and within the context of the land usage and overall description of contaminant distribution and relative pollution (Table 14-2). The status and trends of the biological communities with time will be described to determine possible response to contaminant and other habitat changes. Remedial actions that have been taken and possible future remedial options will also be addressed. These options include methods that are direct (such as dredging) as well as indirect (such as prevention of additional contaminant input).

Table 14-2 USEPA pollution criteria[a]

Parameter	Classification (mg/kg dry wt)		
	Nonpolluted	Moderately polluted	Heavily polluted
Volatile solutions	< 5	5 to 8	> 8
COD	< 40,000	40,000 to 80,000	> 80,000
Nitrogen[b]	< 1,000	1,000 to 2,000	> 2,000
Oil and grease (hexane solubles)	< 1,000	1,000 to 2,000	> 2,000
Ammonia	< 75	75 to 200	> 200
Cyanide	< 0.10	0.10 to 0.25	> 0.25
Pb	< 40	40 to 60	> 60
Zn	< 90	90 to 200	> 200
P	< 420	420 to 650	> 650
Fe	< 17,000	17,000 to 25,000	> 25,000
Ni	< 20	20 to 50	> 50
Mn	< 300	300 to 500	> 500
As	< 3	3 to 8	> 8
Cd	—	—	> 6
Cr	< 25	25 to 50	> 50
Ba	< 20	20 to 60	> 60
Cu	< 25	25 to 50	> 50
Hg	—	—	≥ 1
PCBs (totals)	—	1 ≤ 10	≥ 10

[a] Source: USEPA 1977.
[b] Total kjeldahl nitrogen

While there is some continuity in sampling sites between study years, there is also variation in quality assurance/quality control (QA/QC) and collection parameters over time. In addition, data were often obtained for a particular parameter only once over the duration of the studies, making comparison difficult. Data (ppm as mg/kg in sediments or mg/L for solutions) are generally presented as averages over a series of depths. Decreases in concentrations may reflect either decreasing inputs or a net contaminant loss that may have occurred for a variety of reasons, including migration and biodegradation. Evidence for NR includes concentration reductions and the detection of degradation by-products (Neilson 1994; Larson and Weber 1994). In addition to contaminant data, ecosystem status and bioindicators provide support for improvements at each site. However, continuing contaminant inputs, even at lower levels, may still be a problem in some areas.

Buffalo River Area of Concern

The Buffalo River AOC includes the entire 12.5 km of the Buffalo River, as well as the Buffalo Ship Canal, Outer Harbor, Times Beach Confined Disposal Facility (CDF), and the Erie Basin Marina (New York Department of Environmental Conservation [NYSDEC] 1989; Lee et al. 1991; Passino-Reader et al. 1995). The Buffalo River and its sediments have been contaminated by past industrial and municipal discharges and disposal of waste. Sources of contamination include contaminated bottom sediments, inactive hazardous waste sites, combined sewer overflows (CSOs), and other point and nonpoint sources of pollution (Hartig and Law 1994). Steel production and oil refining were terminated in the early 1980s, thus alleviating some of the previous pollution pressures on the Buffalo River.

The Buffalo River is dredged for navigational purposes. The most recent navigational dredging was completed in 1992. Contaminant regulations prohibit open lake dumping of the dredged sediment, which must be placed in specifically confined sites.

Sediments are heavily polluted with metals, cyanide, polychlorinated biphenyl (PCBs), polynuclear aromatic hydrocarbon (PAHs), pesticides, oil, and grease (NYSDEC 1989). Studies conducted in 1981, 1983, and in 1985 found PCBs in 85 of 86 bottom sediment samples (NYSDEC 1989). PAHs were also found in most samples. In the 1981 study, chlordane and DDT were found in 16 and 15 samples, respectively. Concentrations of metals and cyanide exceeded USEPA criteria (Table 14-2). More recent studies reveal a range of values (Table 14-3). High values (Lee et al. 1991) and most recent sediment contaminant (1990) values (Ingersoll et al. 1993) include Pb 3,300 ppm/116 ppm, Zn 600 ppm/286 ppm, Hg 66.6 ppm/0.52 ppm, Cd 4.5 ppm/1.2 ppm, PCBs 3,238 ppm/ 3.2 ppm, and PAH 193.3 ppm/21.3 ppm. It is likely that very high values represent "hot spots." From the data in Table 14-3, it appears that there has been no real change in heavy metals and PCB contamination. Ni, Cd, and As levels have remained relatively steady; Cu, Cr, and Zn concentrations decreased and then leveled off; and Pb and Hg levels dropped. However, sediments containing PCBs, As, Cr, Cu, Mn, and Pb remain "heavily polluted," and those with Ni and Hg are "moderately polluted" (USEPA 1977).

The most recent sediment contaminant values show that there has been an apparent decrease in the volatile and high-molecular-weight PAHs, perhaps through evaporation, mobility, adsorption, or entrainment. Pesticide residue levels also appear to have dropped, possibly through weathering. Low levels of low-molecular-weight aromatics, aliphatics, and chlorinated alkanes measured in 1981 may indicate the presence of biodegradative processes, but better monitoring with time is needed for verification of this conclusion.

Table 14-3 Mean observed contaminant concentrations: Buffalo River, New York

	Sediment (mg/kg dry wt)								
	1979[a]	1981[a]				1983[a]	1985[a]	1989[b]	1990[c]
Contaminant	MOE	IWD-OR	USACE	USEPA	NYSDEC	RCK; NYSDEC	Erie Co.	RAP	ARCS
Total dichlorobenzenes		0.05		0.1		180			
1,4-dichlorobenzene				0.431					0.269
Total trichlorobenzenes		0.068				243			
Total tetrachlorobenzenes		0.001				510			
Pentachlorobenzenes		0.004				540			
Hexachlorobenzene	0.022	0.002	0.007			114			
Bis(2-ethylhexyl)phthalate				2.28					17.8
Butylbenzyl phthalate				0.12					2.69
Di-n-butyl phthalate			0.234	0.812					
PCBs (total)	0.71	0.48				0.99	2.71		3,240
Arochlor 1248			0.122	0.368					
Arochlor 1254			0.438	0.317					
PAH (total)						87.6			21.3
Naphthalene				11.6			4.44	0.31	0.505
Acenaphthene				5.3			1.17		
Fluorene				0.075	0.237	0.237	2.1	0.33	0.617
Phenanthrene				3.65	2.5	2.5	4.08	0.77	2.3
Methylphenanthrene					0.701	0.701			
Anthracene					0.855	0.855	4.09	0.18	0.847

Table 14-3 continued

Contaminant	Sediment (mg/kg dry wt)								
	1979[a]	1981[a]				1983[a]	1985[a]	1989[b]	1990[c]
	MOE	IWD-OR	USACE	USEPA	NYSDEC	RCK; NYSDEC	Erie Co.	RAP	ARCS
Methylanthracene					0.448	0.448			
Pyrene				4.62	5.48	5.48	3.17	0.93	2.17
Fluoroanthene				5.74	4.66	4.66	3.92	1.57	2.22
Benzofluorene					4.3	4.3			
Benzo(*a*)anthracene					1.34	1.34	2.18	0.61	0.868
Benzo(*a*)pyrene				4.53	1.23	1.23	2.06	0.95	1.12
Benzo(*e*)pyrene					3.01	3.01			
Benzo(*b*)fluoranthene				6.06	1.71	1.71	1.16	1.65	1.41
Benzo(*k*)fluoranthene					0.683	0.683	1.64		1.56
Chrysene				7.88	0.8	0.8	1.64	0.69	1.16
Perylene					5.75	5.75			
Benzo(*ghi*)perylene					1.36	1.36	1.73	0.45	0.566
Dibenzo(*a,h*)anthracene					0.869	0.869	1.54		
Indeno(1,2,3-*cd*)pyrene					1.54	1.54	2.07	0.45	0.666
Heptachlor epoxide		0.008		0.051			0.059	0.04	
Methoxychlor		0.001	0.01	0.102					
A-BHC	0.001	0.016					0.066		
B-BHC			0.008	0.02			0.119		
G-BHC	0.005		0.002	0.019					

Table 14-3 continued

	Sediment (mg/kg dry wt)								
	1979[a]	1981[a]				1983[a]	1985[a]	1989[b]	1990[c]
Contaminant	MOE	IWD-OR	USACE	USEPA	NYSDEC	RCK; NYSDEC	Erie Co.	RAP	ARCS
Chlordane			0.012	0.033		0.061			
Dieldrin	0.006			0.005					
Aldrin			0.002				0.045		
Alpha endosulfan		0.012	0.039	0.001					
Endrin		0.01		0.017			0.007		
p,p'-DDT		0.005	0.019	0.127			0.003		
o,p-DDT			0.017	0.003			0.015		
p,p'-DDE		0.008	0.002	0.034			0.025		
o,p-DDE	0.019			0.005					
p,p'-DDD	0.005	0.005		0.034			0.006		
o,p-DDD				0.005			0.017		
Mirex	0.012		0.012	0.026		0.036			
DCPA			0.096	0.001					
Ag				0.233		1.9	0.459		0.2
Al			9,260	11,300					
As	8.2	16	12.4						11.5
Cd	0.6	2	1.33	1.39		1.93	2.77	0.8	1.2
Cr	34	18	35.1	102		178	79.4	12	98.6
Cu	40	49	69.6	143		236	128	44	59.6

Table 14-3 continued

Contaminant	Sediment (mg/kg dry wt)								
	1979[a]	1981[a]				1983[a]	1985[a]	1989[b]	1990[c]
	MOE	IWD-OR	USACE	USEPA	NYSDEC	RCK; NYSDEC	Erie Co.	RAP	ARCS
Fe			27,700	27,000			40,700	28,800	
Hg	0.23	0.52	6.03	2		4.41	1.55	0.33	0.52
Mn			485	580			720	500	694
Ni	25	10	37.4	38.2		43	43.7	30	41.4
Pb	56	109	140	327		586	206	63	116
Zn	170	179	393	235		296	488	249	286

[a] Source: Lee et al. 1991 (Tables 13–23). MOE = Ontario Ministry of the Environment; IWD-OR = Inland Waters Directorate, Ontario Region, Environment Canada; USACE = U.S. Army Corps of Engineers; USEPA = U.S. Environmental Protection Agency; NYSDEC = New York State Department of Environmental Conservation; RCK = Rockwell, Claff, Kuehl 1983 in Lee et al. 1991.

[b] Source: Aqua-Tech 1989. RAP = remedial action plan.

[c] Source: USEPA 1993b (Chapter 4, Tables 4.12, 4.13, 4.15, 4.16, 4.37). ARCS = Assessment and Remediation of Contaminated Sediments program.

Ecosystem status and effects

A 1928 biological survey of the Buffalo River concluded that the lower Buffalo River was "obviously unfit for eggs or young of fish, and seemed to contain no form of fish life" (NYSDEC 1989). Low dissolved oxygen has been limiting for aquatic biota. Conditions have dramatically improved in recent years because of reductions in sources of contamination and increases in oxygen concentration. Fish surveys conducted in the 1980s and 1990s provide evidence of improved ecological conditions. Surveys of fish in 1981 and 1982 (Makarewicz et al. 1982) showed predominantly yellow perch, white sucker, rock bass, brown bullhead, pumpkinseed, spottail shiner, gizzard shad, emerald shiner, and carp (Table 14-4). Fish surveys (Kozuchowski et al. 1993) conducted in 1992 yielded from 340 to 1,335 total fish, depending upon the reach of the river. The predominant species were gizzard shad, emerald shiner, pumpkinseed, golden shiner, largemouth bass, and carp (Table 14-4). Using plankton net sampling (Table 14-4), they also found larval gizzard shad, *Morone* sp. (white perch or white bass), rainbow smelt, *Pomoxis* sp. (crappie sp.), *Lepomis* sp. (sunfish sp.), and yellow perch, for a total of 442 to 897 larvae per reach.

Conditions are also improving for benthic macroinvertebrates, which are often used for water quality monitoring. In 1976 and 1977, Bergantz (1977) found large numbers of pollution-tolerant tubificid worms in the Buffalo River. From 1969 to 1989, the number of major taxa of benthic macroinvertebrates increased (Passino-Reader et al. 1995). Whereas 1969 samples yielded primarily oligochaetes (tubificids) and chironomids, by 1989, fingernail clams, leeches, mayflies, caddisflies, and amphipods were also found (Table 14-5). Benthic macroinvertebrate sampling in the ARCS program (USEPA 1994c) showed that *Limnodrilus hoffmeisteri*, a pollution-tolerant worm, was the most abundant oligochaete in grab samples at all but 2 stations (USEPA 1994c). The chironomid community consisted primarily of genera known to be tolerant of organic enrichment. Of the mollusks present, *Musculium* and *Pisidium* have been reported from organically enriched environments. The unionids *Anodonta imbecilis* and *A. grandis* are considered mesosaprobic (Rosenberg and Resh 1993). As an indicator of invertebrate health, the ARCS study examined mouthpart deformities in chironomids (USEPA 1994c). The lower prevalence of mouthpart deformities in the Buffalo River AOC indicated that conditions there were less toxic compared with conditions in the Indiana Harbor–Grand Calumet River AOC or the Saginaw River AOC.

Overall, conditions for biota have improved at the Buffalo River AOC during the last 20 or more years, indicating a role for NR, especially the burial of highly toxic lower substrates by natural sedimentation with cleaner sediments flowing from upstream.

Table 14-4 Total number of fish collected in 1981–1982 by electrofishing or gill netting and in 1992 by electrofishing or plankton netting in the Buffalo River, New York

		1981–1982 fish collections[a]		1992 fish collections[b]	
Scientific name	Common name	Gill net	Electrofishing	Electrofishing	Plankton net
Lepisosteus osseus	Longnose gar	1			
Alosa pseudoharengus	Alewife	3	82	39	
Dorosoma cepedianum	Gizzard shad	64	51	691	1,302
Salvelinus namaycush	Lake trout		1		
Salmo trutta	Brown trout	1			
Oncorhynchus tshawytscha	Chinook salmon	1			
Oncorhynchus kisutch	Coho salmon	1			
Oncorhynchus mykiss	Rainbow trout	7	3	5	
Osmerus mordax	Rainbow smelt	2		351	
Esox lucius	Northern pike	17	1	19	
Esox masquinongy	Muskellunge	33	3		
Carassius auratus	Goldfish	13	36	87	2
Cyprinus carpio	Carp	276	46	119	13
	Carp x goldfish hybrid	12	3		
Notemigonus crysoleucas	Golden shiner	20	27	297	2
Hybopsis storeriana	Silver chub		2		
Cyprinella spiloptera	Spotfin shiner		1		
Notropis cornutus	Common shiner		4		

Table 14-4 continued

Scientific name	Common name	1981–1982 fish collections[a]		1992 fish collections[b]	
		Gill net	Electrofishing	Electrofishing	Plankton net
Notropis bifrenatus	Bridle shiner		1		
Notropis hudsonius	Spottail shiner		83	108	2
Notropis atherinoides	Emerald shiner		47	985	
Pimephales sp.				22	
Pimephales notatus	Bluntnose minnow		2	60	
Carpiodes cyprinus	Quillback	2	1		
Catostomus commersoni	White sucker	300	22	84	
Hypentelium nigricans	Northern hog sucker	2	1		
Moxostoma sp.	Redhorse species		65		
Moxostoma valenciennesi	Greater redhorse	1			
Moxostoma macrolepidotum	Shorthead redhorse	25	3		
Ameiurus nebulosus	Brown bullhead	92	15	236	
Noturus flavus	Stonecat	20			
Percopsis omiscomaycus	Trout-perch		1	1	10
Lota lota	Burbot			2	
Morone sp.					346
Morone americana	White perch	8	1	1	
Morone chrysops	White bass	2	1	4	
Ambloplites rupestris	Rock bass	108	34	46	

Table 14-4 continued

Scientific name	Common name	1981–1982 fish collections[a]		1992 fish collections[b]	
		Gill net	Electrofishing	Electrofishing	Plankton net
Lepomis sp.					296
Lepomis gibbosus	Pumpkinseed	15	172	340	
Lepomis macrochirus	Bluegill		2	8	
Lepomis gulosus	Warmouth		3		
Micropterus dolomieu	Smallmouth bass	28	23	53	
Micropterus salmoides	Largemouth bass	4	11	181	
Pomoxis sp.	Crappie species			628	
Pomoxis annularis	White crappie		5		
Pomoxis nigromaculatus	Black crappie	8	2	3	
Perca flavescens	Yellow perch	479	76	48	308
Percina caprodes	Logperch		2	1	
Stizostedion v. vitreum	Walleye	22	1	1	3
Aplodinotus grunniens	Freshwater drum	20	9	28	
Cottus cognatus	Slimy sculpin		2		
Total # of species		30	31	32	16
Total # of fish		1,586	685	3,566	3,327

[a] Source: Makarewicz et al. 1982.

[b] Source: Kozuchowski et al. 1993.

Table 14-5 Densities of major benthic invertebrate groups found in sediments of the channelized portion of the Buffalo River, New York, 1969–1989[a]

	Density (number/m^2)					
Group	1969	1970	1972	1977	1982	1989
Oligochaetes	2,530	1,890	2,100	39,000	38,840	8,730
Chironomids	80	18	2	38	240	430
Fingernail clams	3	0	6	312	94	70
Snails	14	10	18	40	460	59
Leeches	0	0	1	10	60	8
Flatworms	0	3	0	0	15	0
Mayflies	0	0	0	2	0	8
Caddisflies	0	0	0	0	0	3
Amphipods	0	0	0	0	0	9

[a] Source: Passino-Reader et al. 1995.

Saginaw River–Saginaw Bay Area of Concern

The Saginaw River–Saginaw Bay region contains 4 primary urban and industrial centers, with over half the land use in the region for agriculture (Bolattino and Tuchman 1993). Industrial and municipal discharges, CSOs, contaminated river and bay bottom sediments, urban and agricultural nonpoint source runoff, old waste disposal sites, and the atmosphere continue to contaminate the Saginaw River–Saginaw Bay AOC. Environmental problems are caused by cultural eutrophication (nutrients), toxic substances, bacterial contamination, sedimentation, and commercial and residential development (Hartig and Law 1994).

The sediment is characterized by "hot spots" (i.e., areas of high contaminant concentration) located below waste water treatment plants (WWTPs), automotive-industrial sites and CSO discharge points, as well as other uniform contamination throughout the river and bay. Sediment contamination in the river is most significant immediately downstream of Saginaw and Bay City, while the most contaminated bay sediments are north of the Saginaw River mouth (Michigan Department of Natural Resources [MDNR] 1988). High levels of PCBs are present near the outfalls of 2 industrial facilities that discharged directly into the river and to their WWTPs. The river is also contaminated by dioxins and furans (Bolattino and Tuchman 1993). Two Superfund sites, Velsicol and the Shiawassee River, contribute to the Saginaw system sediment PCBs.

Within the Saginaw River and Bay, a number of heavy metals have been detected. Cr and Pb were the most abundant metals in Saginaw Bay, followed by Cu, Ni, and Zn (MDNR 1988). Sediments within the AOC were classed as "heavily polluted" with 27 ppm PCB, 180 ppm Cr, 150 ppm Cu, 34,000 ppm Fe, 96 ppm Pb, 87 ppm Ni, and 56 ppm Zn. The highest 2,3,7,8-tetrachlorodibenzo-*p*-dioxin (TCDD) levels ranged from 1.2 to 1.5 ppb; more typical values were 0.11 to 0.29 ppb (Goudy 1991). Inner Bay sediments (1983) were "heavily polluted" with 16 ppm As, 422 ppm Ba, and "moderately polluted" with 63 ppm Cr, 25 ppm Cu, 45 ppm Pb, 32 ppm Ni, 96 ppm Zn. The Saginaw River mouth (ARCS 1989–1990 study) sediments were "heavily polluted" with 47.2 ppm As, 158.3 ppm Cr, 78.6 ppm Cu, 59.8 ppm Pb, 52.5 ppm Ni, and 239.9 ppm Zn. PCBs were present at a "nonpolluted" concentration of 0.309 ppm (Table 14-2).

Average surficial PCB concentrations decreased about an order of magnitude from 1980 to 1990 (Goudy 1991). Goudy's 1991 data also suggest that there has been no decrease in metal levels, if the values are assumed representative of the region. Continuous input may explain the lack of significant remediation of contaminants by any method. Cultural eutrophication is being offset by decreasing P input (Goudy 1991). Total P concentrations in the Saginaw River decreased 40% from 1974 to 1980, then 25% from 1980 to 1986. Orthophosphorus content (bioavailable P) declined 70% by 1980. Phosphorus concentrations have fallen less precipitously in the bay, possibly due to periodic sediment resuspension events.

Ecosystem status and effects

Populations of many important fish species have declined or disappeared because of overfishing and decreases in habitat quality. Historically, Saginaw Bay had a productive, commercial fishery with lake trout, lake herring, and walleye dominating the catch. Current commercial fishing is centered on carp, channel catfish, yellow perch, sucker, and freshwater drum. Fish consumption advisories because of toxic chemicals and conflicts with sport fishing adversely impact commercial fishing in Saginaw Bay (MDNR 1988). The economic value of the sport fishery in Saginaw is millions of dollars annually (MDNR 1988). Despite water quality problems, the Saginaw River continues to support recreational fishing for yellow perch, walleye, largemouth bass, smallmouth bass, northern pike, crappie, bluegill, and spawning runs of salmonids, white bass, and suckers (MDNR 1988). Subsistence fishing for carp and walleye occurs in the Saginaw River.

Saginaw Bay once supported a rich riverine invertebrate bottom fauna, but it underwent drastic changes in fauna in response to increased inputs of pollutants (MDNR 1988). High sediment oxygen demand eliminated many species of invertebrates (e.g., burrowing mayflies *Hexagenia* spp.), and these were replaced by pollution-tolerant oligochaete worms (e.g., tubificid [*Limnodrilus* spp.] and midges [*Chironomus* spp.]). Densities of macroinvertebrates in the inner bay increased from 1,756/m^2 in 1956 to 35,675/m^2 in 1978 (MDNR 1988). These changes and others point to decreasing sediment quality and a change from mesotrophic to highly eutrophic conditions in the inner bay.

After the invasion of Saginaw Bay by zebra mussels (*Dreissena polymorpha*), the density of adult zebra mussels on hard substrate in the inner bay increased to 75,296/m^2 by 1993 (Nalepa et al. 1995). Similar densities were not found for soft substrates. Although benthic macroinvertebrates are often used as indicators of water quality, with the establishment of *Dreissena* in Saginaw Bay, one can no longer assume the changes in benthic communities reflect changes in water quality (Nalepa and Fahnenstiel 1995). Hence, other indicators may be needed in Saginaw Bay to monitor the success of remediation, whether natural or intentional.

Extensive remediation is necessary to achieve a benthic community of pollution-intolerant organisms. The circumstances present in Saginaw Bay and River illustrate a range of conditions, indicating that NR may be appropriate for some areas but not for the heavily contaminated areas.

Since 1982 the Saginaw River has been dredged annually, and in 1983 the Bay was dredged. Overall, in a 10-year period beginning in 1982, 1,970,000 m^3 of sediment have been dredged from the river by the U.S. Army Corps of Engineers (USACE). When the Bay was dredged in 1983, 570,000 m^3 were removed. All sediment removed from the Saginaw River and Bay was contaminated and so was disposed of in a CDF (Bolattino and Tuchman 1993).

ARCS demonstrated bioremediation in Saginaw. Using native bacteria, the study team added nutrients to stimulate the growth of bacteria and varied the amount of oxygen available to the bacteria to increase the effectiveness of the bioremediation (USEPA 1994b). A bench-scale bioremediation study demonstrated that, while active PCB-dechlorinating populations were present, no appreciable PCB biodegradation occurs unless stimulated by amendment with higher Cl-containing congeners. Anaerobic reductive dechlorination of both historic and amended congeners then proceeded to mono- and dichlorospecies (Jones et al. 1996). Use of native bacteria demonstrated NR capacity. This site is not homogeneous, indicating that NR may be appropriate only at less contaminated parts of the AOC.

The colonization of Saginaw Bay by zebra mussels (*Dreissena*) has the potential to significantly impact biological communities and contaminant cycling in Saginaw Bay (Nalepa and Fahnenstiel 1995) and thus to impact possible remedial actions over time (Hartig and Law 1994).

Ashtabula River Area of Concern

Industrial development, including one of the largest and most diversified groups of chemical plants in Ohio (Sanders 1991), is heavily concentrated along Fields Brook, a major tributary of the Ashtabula River. The main section of Fields Brook is 5.6 km long and has 5 tributaries. It empties into the Ashtabula River about 2.4 km from the river mouth and Lake Erie. Fields Brook sediment has been classified as hazardous because of high chemical contamination, and it has been designated as a Superfund site. The AOC, especially Fields Brook, can be considered an "island" with high chemical contamination potential, surrounded by rural and agricultural land. Fields Brook contains about 30,000 m^3 of highly contaminated sediment (Bolattino and Tuchman 1993).

The Ashtabula River contains about 400,000 m^3 of contaminated sediment, with about half having a PCB concentration > 50 ppm. The outer harbor of the Ashtabula River contains about 100,000 m^3 of moderately contaminated sediment (Bolattino and Tuchman 1993), and 15 locations (in 1993) had PCB concentrations below 1 mg/kg (Hartig and Law 1994).

The horizontal distribution of the sediment constituents shows that the upper turning basin is a depositional sink for sediments, thus becoming a hot spot. Concentrations tend to decrease down river toward the harbor (Woodward-Clyde Consultants 1992). The upper turning basin is where the contaminated sediment being carried by Fields Brook is deposited into the Ashtabula River. No particular trend for vertical distribution was common to all parameters, but most concentrations increased with depth and then decreased as depth increased further. The sediment in the upper 2 ft (0.6 m) was found to be less impacted, compared to the overall average (Woodward-Clyde Consultants 1992). The less contaminated sediment in the upper layers provides evidence that NR is occurring through burial and deposition.

The major pollutants contaminating the sediments of the Ashtabula River are heavy metals, such as As, Ba, Cd, Cr, Pb, Hg, and Zn; PAHs; and chlorinated organics such as PCBs, hexachlorobutadiene, hexachlorobenzene, tetrachloroethene, and trichloroethylene. PCB concentrations throughout the AOC ranged from 0 to 660 ppm, with an average of 11.9 ppm. Most other organics were localized in hot spots (Ohio Environmental Protection Agency [OEPA] 1991; Sanders, 1991). The following maximum contaminant sediment concentrations found in the Ashtabula River AOC are from a variety of studies and are summarized in Tatem et al. (1990); the second figure is the most recent value: As 56 /1.8 ppm, PCBs 120/2.4 ppm, Cr 2,200/~18 ppm, hexachlorobutadiene 22/0.05 ppm, Pb 350/~26 ppm, hexachlorobenzene 32/0.13 ppm, Hg 4.7 to 14/~0.2 ppm, PAHs 188 /8.5 ppm, and Zn 830/ ~131 ppm.

Although the data from Tatem et al. (1990) are fragmentary, some tentative conclusions can be reached. Several metal loadings decreased; e.g., Cr and Ni dropped, and Hg and Pb dropped but then leveled off. However, loadings for As, Cu, and Zn remained level. As of 1988, sediments with As, Cr, Pb, Hg values were nonpolluting, but those for Cu, Ni, Zn remained moderately polluted (Table 14-2). At this time (Tatem et al. 1990), PCB loadings were decreasing but still at the moderately polluted level, with heavily polluting hot spots. Concentrations of both halogenated aromatics and aliphatics appeared to increase and then decline with time, possibly indicative of anaerobic PCB microbial degradation. Nitrogen levels were decreasing, nitrate concentrations were minuscule, and chemical oxygen demand (COD) was increasing, inhibiting potential microorganism activity. Volatile organic compound levels declined dramatically, possibly because of evaporation or movement through sediment. Phthalate concentrations seemed to decline, possibly from weathering or hydrolysis or microbial action, which are NR processes. PAH levels were stable, or increasing, indicating that conditions may not have been suitable for aerobic degradation or mineralization of PAHs.

Ecosystem status and effects

Fisheries collections from 1975 to 1977 and in 1984 in Ashtabula River and Harbor found species typical of the warmwater fish community in Lake Erie river mouths (Tatem et al. 1990). A fish tumor study indicated a higher probability of tumors in fish from the river, compared with the harbor or breakwater area (Mueller 1992). Mueller and Mac (1993) reported a 23% incidence of cancerous and precancerous lesions in livers of brown bullhead from Ashtabula River, compared with 4% at the reference site of the Huron River, Ohio.

Surveys of the benthic macroinvertebrates showed communities typical of polluted areas. Tubificids comprised the majority of the organisms, with a lesser number of chironomids (*Chironomus* and *Procladius*), gastropods (*Viviparus*), pelecypods (*Musculium, Sphaerium*), and leeches (*Hirudinea, Helobdella*) in some samples (Tatem et al. 1990). This state of affairs has existed since the early surveys in 1950 and 1951, when the sediments were described as "organically rich, dark colored muds" with high numbers of tubificids (Terlecky et al. 1975). Flint (1979) studied the responses of freshwater benthos to open-lake dredged spoils in Lake Erie. Dredged spoils from the Ashtabula Harbor and the Ashtabula River were deposited in Lake Erie near Ashtabula. Untreated sites nearby had 60% to 80% oligochaetes, compared with 98% oligochaetes at the disposal sites. Unstable communities were formed at disposal sites.

With regard to wildlife use of the Ashtabula River AOC, Bolsenga and Herdendorf (1993) describe Ashtabula River wetland as lying on nonforested landfill deposits inside the Ashtabula Harbor breakwalls. A lack of diversity of habitat accounts for a short list of bird species using the area. Improvement of wildlife habitat is needed at the Ashtabula AOC.

Remediation

Dredging and disposal of sediments in the Ashtabula River have posed a problem because of the highly toxic and polluted nature of the sediments. In 1986, the USEPA decided that incinerating, solidifying, and landfilling the contaminated sediment would be the final cleanup method (USEPA 1993a). However, remedial actions have been slow. No full-scale environmental cleanup has been undertaken to remediate contaminated sediments. But, sediment characterization work is at a stage where a number of options, including dredging only the hot spots, are being examined to determine the option that will pose the lowest risk for the lowest cost.

As for indirect remediation, several major dischargers to Fields Brook have ceased operations. A wastewater treatment project to collect and treat stormwater runoff and groundwater prior to discharge to Fields Brook was completed in 1992, at least 2 CSOs that discharge to Ashtabula Harbor were eliminated, and several connections to the sanitary sewer system were disconnected. This AOC demonstrates that stopping source inputs into the Great Lakes has contributed to improvement in quality and allowed NR to have a positive impact.

Sheboygan River Area of Concern

The Sheboygan River is one of the major tributaries to Lake Michigan. The Sheboygan River, Mullet River, and Onion River watersheds feed pollutants into the Sheboygan River and Harbor AOC. Within the AOC, industrial, agricultural, and residential areas line the rivers of the Sheboygan River basin, while agricultural lands border the tributaries west of the city. Agriculture accounts for 67% of area land use (Bolattino and Tuchman 1993; Hartig and Law 1994). Sediment input to the Sheboygan Harbor is estimated to be more than 20,000 m^3/y.

Pollutants of concern in the Sheboygan River and Harbor AOC include P, N, PCBs, PAHs, and heavy metals (Hartig and Law 1994). Sediments within the AOC are heavily polluted with PCBs and heavy metals. PCBs are the most significant problem, while Cr, Cu, and Pb have also contributed to the sediment pollution. The most severe sediment contamination occurs near the river mouth, exceeding 50 ppm PCBs at depths below 2 ft (0.6 m) from the sediment–water interface. PCB levels upstream of the Lower Kohler Dam in Kohler have generally ranged between 27 and 81 ppm, although concentrations as high as 4,500 ppm were found within the AOC before remedial dredging (Elder 1991; Bolattino and Tuchman 1993). There is heavy contamination of soil and groundwater with PAHs, cyanide, As, and Ni at a coal gasification plant site (Hartig and Law 1994). A significant landfill is also a probable source of pollutants, including heavy metals, to the Sheboygan River and Harbor. Recently, the maximum PCB concentration in river sediments was 503 ppm, but most of the river sediments contained PCBs in concentrations less than 20 ppm. Some floodplain soils had PCB concentrations up to 300 ppm (Bolattino and Tuchman 1993).

Based on the limited data available, PCB levels appear to be decreasing, in large part because of 2 courses of mechanical dredging. The PCBs are also being transformed by reductive dechlorination and oxidative degradation (Elder 1991). Low-molecular-weight chlorinated and oxygenated small chain organics appeared to increase, then decrease in the water column, which may indicate active biodegradation had occurred. An alternate explanation is that hot spots were sampled at some times and not others and that the decreases are not real. The results of the 1987 sampling show that the sediment was very heavily polluted with PCBs and moderately polluted with As; the 1988 sampling results indicate sediments were heavily polluted with Ba, Cu, and Pb and moderately polluted with As, Cr, Ni, and Zn (Table 14-2).

Ecosystem status and effects

The Sheboygan River is one of the major tributaries to Lake Michigan. The Sheboygan River AOC includes the lower 22.5 km of the Sheboygan River and the entire Sheboygan Harbor (Skogerboe et al. 1991). Historically, use of the Sheboygan River as a stocking site for trout and salmon contributed significantly to these fisheries in Lake Michigan. However, in 1987, the Wisconsin Department of Natural Resources (WDNR) banned the stocking of trout and salmon in the Sheboygan River, due in large part to high levels of PCBs in these fish (Eggold et al. 1996). A subsequent study was conducted by the WDNR to measure PCB accumulation in salmonine fish after stocking in the Sheboygan River (starting in 1990) and residence in Lake Michigan (Eggold et al. 1996). They determined that returning adult rainbow trout and coho salmon had levels that did not exceed the U.S. Food and Drug Administration tolerance level for PCBs of 2.0 mg/g wet weight. They recommended that the Sheboygan River be stocked with coho and chinook salmon and rainbow trout (Eggold et al. 1996). Active remediation by removal of PCB-contaminated sediments has been conducted as noted in the previous section.

In 1994, Schrank et al. (1997) sampled white suckers (*Catostomus commersoni*) from a contaminated site 1 to 2 km upstream from the mouth of the Sheboygan River and from a reference site 50 km further upstream. They selected white sucker as an indicator species for fish health because this species is common in the Sheboygan River and has been recommended as an environmental sentinel. Multiple indicators of fish health and of contaminant bioaccumulation were measured. They determined that white sucker in the lower site of the river absorbed significant amounts of PCBs and PAHs and exhibited hematological, biochemical, and histological alterations that suggested impaired fish condition. They provided these data as a baseline for future studies to evaluate cleanup efforts (Schrank et al. 1997). Unfortunately, due to lack of stability in institutions and funding, few attempts have been made to monitor remediation of contaminated sediments over periods of time using fish health indicators.

Remediation

In 1981 to 1989, the USACE dredged a total of 84,000 m^3 from the harbor mouth and reused some as beach nourishment material and industrial fill. No dredging of the navigation channel has been done since 1969 because of contaminated sediment disposal problems (Bolattino and Tuchman 1993). Natural remediation may be a suitable alternative in cases where dredging cannot be used as an active remediation method. Emergency removal of PCB-contaminated sediments in the upper portion of the Sheboygan River was completed in 1991. A total of 4,100 m^3 of PCB contaminated sediments were removed. Sediment was deposited in 2 CDFs, and other selected deposits were armored (Hartig and Law 1994). No estimates of the remaining volume of contaminated sediments are available.

A characterization of the PCB congener makeup of contaminated Sheboygan sediments indicated that concentrations of > 50 ppm were required to shift the congener makeup to lower chloro-ortho-enriched congeners. Anaerobic dechlorination was not observed in laboratory tests that used native microorganisms, but this was likely due to failure to reproduce the precise conditions found in nature (Sonzogni and David 1991). A bench-scale bioremediation study also showed that rates of anaerobic reductive dechlorination of PCBs are extremely slow, likely due to some limiting factor in the natural sediments. Native microorganisms were found capable of mediating reductive dechlorination of both historical and amended PCB congeners, but only after higher Cl-containing congeners were added. There was preferential dechlorination of the more highly chlorinated congeners and an accumulation of the ortho-substituted dichloro congeners (Jones et al. 1996).

Preliminary findings of a pilot bioremediation study in one confined treatment facility (CTF) indicated that the level of degradation found in the bench-scale studies was not observed in the pilot studies. Bioremediation was demonstrated (1994) on 2,100 m^3 of PCB-contaminated sediment from the river stored in a CTF. The contents of the CTF were manipulated to enhance naturally occurring biodegradation; nutrients were added to sediments already containing indigenous populations of microorganisms (bacteria and fungi), and the CTF was cycled between aerobic and anaerobic conditions (PCBs do not completely degrade either aerobically or anaerobically). Results showed that the PCBs present in the Sheboygan River sediments had already undergone a great deal of anaerobic dechlorination by NR. A properly engineered system is needed to deliver adequate amounts of oxygen to the sediments in order to break down the remaining partially dechlorinated PCB molecules (USEPA 1994a). These results suggest that NR has had a positive impact at this AOC and has the potential to lead to future reductions in contaminant levels. In addition, problems with disposal of contaminated sediments increase the desirability of NR.

Collingwood Harbour

Collingwood Harbour is situated on the southern shore of Nottawasaga Bay, which makes up the southern extension of Lake Huron's Georgian Bay (Figure 14-1). Three tributaries drain into Collingwood Harbour. On the south shore, Oak Street and Hickory Street Canals receive municipal storm sewer outflows and flow intermittently. The largest tributary, Black Ash Creek, drains an agricultural area as well as recreational in-stream and bypass ponds around the Town of Collingwood into the southeastern corner of the harbor. The harbor covers a 0.8-km^2 area, with a maximum depth (1989) of 6.4 m (21 ft). Use of the surrounding land has shifted from industrial to recreational and residential (Krantzberg 1995). The harbor includes a wetland complex, a WWTP outfall, a marina, a grain terminal, and former shipyards. In the past, high nutrient levels resulted in excessive aquatic plant growth (Hartig and Law 1994).

The data in Krantzberg and Houghton (1994) show that PCBs levels have been significantly reduced in Collingwood Harbor. In the 1993 sampling (Krantzberg and Houghton 1996), all metals were found at levels corresponding to the nonpolluted category (Table 14-2). However, while Cd levels had decreased, concentrations of other metals remained elevated at 2 sites of interest (CDF and shipyard). Hotter spots were sampled in 1991 than in 1989. The 1991 CDF sediment was heavily polluted with Cr and P and moderately polluted with Cu, Ni, and Zn. The 1991 shipyard and drydock area sediment was heavily polluted with Cu, P, Pb, and Zn and moderately polluted with Cr and Ni.

Removal and disposal of approximately 8,000 m^3 of contaminated sediment from shipyard slips was completed in 1994 with the capping of the CDF (Hartig and Law 1994). No navigational dredging will occur for at least 30 years. The concentrations of some metals still exceed open water disposal guidelines, but local uses and goals are not affected by these levels of metals (Krantzberg and Houghton 1996). A 35% drop in peak daily water pumpage from 1986 to 1990 assisted in reducing P loads to the harbor (Hartig and Law 1994). Mean harbor P concentrations have prevented nuisance algal growth since 1989.

Remediation and rehabilitation of fish and wildlife habitat have been ongoing within the harbor since 1991 (Krantzberg and Houghton 1996). One remedial objective was to ensure that fish species representative of mesotrophic environments (i.e., pike, bass, yellow perch, and walleye) are present. In particular, the target composition is as follows: 5% to 10% predators by number or 30% to 40% by weight, and 90% to 95% bottom feeders by number or 60% to 70% by weight (Krantzberg and Houghton 1996). After comparing the species and numbers of fish sampled in 1986 and 1993, Krantzberg and Houghton (1996) concluded that, with the exception of carp, brassy minnow, and gizzard shad, the objective of mesotrophic fish communities had been met for Collingwood Harbour.

A second remedial goal was to establish a benthic invertebrate community structure in the turning basin of Collingwood Harbour similar to that in reference sites with comparable physical and chemical characteristics. A benthic community survey conducted in 1991 revealed a species composition typified by an oligochaete–chironomid assemblage (Krantzberg 1995). Naid and tubificid oligochaetes were ubiquitous, as were the chironomini. Amphipods, gastropods, sphaerid mollusks and insect species included pollution-intolerant representatives. The benthic community structure in 1992 was shown to be comparable to the reference sites (Krantzberg 1995) and, as such, represents an important indicator of successful remediation. Furthermore, bioassays conducted on fathead minnows, mayflies, amphipods, and chironomids with sediments from the shipyards, harbor, and reference sites showed that there were no differences in mortality and growth between the contaminated and reference sites. However, reproduction of tubificid worms was inhibited by sediments from the shipyards. The above field surveys and bioassays showed that the goal had been reached and thus supported the delisting of Collingwood Harbour as an AOC (Krantzberg 1995). At this AOC, a combination of dredging, rehabilitation of physical habitat, and NR of sediment contaminants resulted in successful attainment of remedial goals.

RISK ASSESSMENT APPLICATIONS AT GREAT LAKES AREAS OF CONCERN

Resilience, stability, recovery, succession

Kraft (1993) evaluated available data on fish communities in the Green Bay AOC to "determine whether species-specific remediation goals are attainable or discernible, given background variability and trends in fish population abundance." Kraft reported that common fish species, such as yellow perch and spottail shiner, were abundant from 1978 to 1992, the period for which trawl data were collected. While there were increased numbers of white perch, gizzard shad, sheepshead, and young-of-the-year alewife in 1989 to 1991 compared to previous years, the opposite was true for young-of-the-year carp, burbot, bullhead, adult alewife, and common suckers (Kraft 1993). Dramatic changes occurred in the fish community in Lake Michigan during that period, with shifts from an alewife-dominated forage base to a bloater chub–dominated forage base (Passino-Reader et al. 1996). These changes may also have impacted the fish communities in Green Bay, independent from effects of sediment contaminants. Such changes in fish communities may result from introductions of exotic species, planting of predators (e.g., salmon and trout) from hatcheries, shifts in primary or secondary production, or eutrophication—causes that are unrelated to sources of contamination or remediation options.

Ecotoxicological models and role in predicting natural remediation

Velleux et al. (1993 [see 1995]) reported a mass balance model as part of the USEPA-supported Green Bay mass balance project. An objective of their model was to predict export of contaminants from the DePere dam in the lower Fox River to Green Bay, Lake Michigan. They calculated that most of the 25,000 to 40,000-kg reservoir of PCBs in the lower Fox River would remain in place, barring large storm events. Active sediment remediation efforts and flood control were predicted to reduce the current (1989) export of PCBs to Green Bay.

A mass balance model (GBTOX) was developed by DePinto (1994) for simulating transport and fate of PCBs in Green Bay. The model was used in a predictive mode to analyze the impact and response time of Fox River contaminant load reduction alternatives. The model was used to calculate that 92% of the contaminants coming into Green Bay during 1989 were from the Fox River discharge. Another result was that, while 304 kg of PCBs loaded into Green Bay during 1989, 745 kg were lost from the entire Green Bay system (water column and sediments). Of the 745 kg lost, 122 kg were exported to Lake Michigan, 158 kg were lost to the air by volatilization, and 465 kg were buried below the active sediment layer (upper 4 cm). Hence, these NR mechanisms included dispersal, volatilization, and burial.

Risk assessment at areas of concern

Risk reduction strategies have been proposed by the USEPA as the centerpiece of environmental protection efforts (Harris and Wenger 1993). This approach requires that an anthropogenic stressor (e.g., contaminant) can be linked to an impaired use criterion (e.g., fish consumption advisory). Harris and Wenger (1993) have developed a methodology that pairs the anthropogenic stressor with the impaired use and can be used to calculate the degree to which the stressor contributes to the ecosystem risk. Mathematical analyses are used to rank ecosystem risks with regard to remediation management options. Other methods for evaluating and prioritizing contaminated sites before remediation range from the baseline risk assessment (Crane 1995; Passino-Reader et al. 1995), which uses existing data, to hazard ranking of sites based on the principles of the Sediment Quality Triad (Chapman 1992; Wildhaber and Schmidt 1996).

Aquatic risk assessment

As part of the ARCS program of USEPA, Passino-Reader et al. (1995) developed and applied assessment techniques to perform a baseline hazard evaluation for aquatic life exposed to contaminated sediments at the Buffalo River AOC. This assessment followed, where possible, the approach derived for human health risk assessment at the Buffalo River AOC (Crane 1993). The assessment also followed the RAP for the site and provided a means of quantifying reference risk levels for chemicals determined to be of potential concern. The reference-level results reflected a conservative estimate of the potential exposures and risks experienced by Buffalo River aquatic species. The baseline or reference-level risk estimates were used in conjunction with other information to determine the need to remediate the site. If remediation strategies were developed, the baseline risk levels were then to be used as reference risk levels by which to judge the relative effectiveness of remediation alternatives.

Typically, the determination of which chemicals to include in an exposure and risk assessment is a 2-step procedure. First, a monitoring study is conducted and samples are analyzed for all chemicals included on a standard list of hazardous chemicals such as the USEPA priority pollutant list. Those contaminants that are frequently detected above recommended limits (e.g., sediment quality criteria

[SQC]) are included in the risk assessment. In the case of contaminated sediments, there are no SQC for the majority of the chemicals, only water quality criteria (WQC). However, aquatic toxicologists have developed the following approaches for estimating SQC: apparent effects threshold (AET), effects range-low (ER-L), sediment quality triad (SQT), screening-level concentration (SLC), and equilibrium partitioning (EqP).

Risk for noncarcinogenic chemicals was quantified using the hazard quotient approach (USEPA 1989): hazard quotient = exposure level / reference dose. Using this model, relative risks depend on both the aquatic receptor organism (fish or invertebrate) and the type of chemical. Applying this to the Buffalo River AOC as a typical exposure case, it was found that Cd, Cr, Cu, Fe, Pb, Hg, pyrene, and heptachlor epoxide presented a significant risk to most of the 8 receptor organisms. In the worst-case scenario, the above compounds plus Zn, indeno (1,2,3-cd) pyrene, endrin, and total PCBs represented a significant risk to aquatic receptors (Passino-Reader et al. 1995).

As a further step in prioritizing sites for remediation, Wildhaber and Schmidt (1996) ranked several sites at the Buffalo River, Grand Calumet River–Indiana Harbor (GC/IH), and Saginaw River AOCs using ARCS residue data on contaminants at the sites, laboratory bioassays exposing a suite of organisms to sediments from the sites, and benthic community information on relative pollution tolerance of macrobenthos at the sites. They ranked 2 sites at the GC–IH and Saginaw River as the most hazardous. On the average, the Buffalo River AOC ranked as the least hazardous of the 3 AOCs, although some sites (hot spots) at the Buffalo River ranked higher than particular sites at the other 2 AOCs.

Wildlife risk assessment

Several studies have reported the risks to wildlife of consuming Great Lakes fish. Heaton et al. (1995) fed carp from the Saginaw Bay AOC to adult mink (*Mustela vison*) prior to and throughout the reproductive period. The diets containing Saginaw Bay carp caused reproductive impairment and/or reduced survival of the kits. The lowest-observable-adverse-effect level (LOAEL) was 0.134 mg PCBs/kg body weight/d or 3.6 ng TEQs/kg body weight/d for adult female mink where toxic equivalents (TEQs) are the toxic equivalents of 2,3,7,8-TCDD (Heaton et al. 1995). They calculated the maximum allowable daily intake and hazard indices for PCBs for several species of Great Lakes fish. Shipp et al. (1998) conducted further investigations on multigenerational effects, specifically endocrine disruption, from consumption of PCB-contaminated carp from Saginaw Bay, on mink. For mink that were fed Saginaw Bay carp with 0.25 to 1.0 mg/kg PCBs, the PCBs exerted antiestrogenic activity by reducing estrogen binding sites in the liver, an important estrogen-responsive tissue.

Field surveys of fish-eating colonial birds have identified congenital malformations and reproductive failure in birds in several areas of the Great Lakes (Fox et al. 1991). Of the colonies of double-crested cormorants (*Phalacrocorax auritus*) surveyed, 73% of the chicks at the Green Bay AOC had bill defects, compared with 6% of the chicks in the reference areas. These defects are associated with elevated levels of dioxin-like compounds such as PCBs bioaccumulated in piscivorous birds (Tillitt et al. 1993). Epidemiological studies have pointed to PCBs and other halogenated organochlorines as the causative agents of these defects (Ludwig et al. 1996). This and similar studies indicate the need for remediation of the causative agents. Remediation of PCBs and similar teratogens at AOCs should lead not only to healthier fish and wildlife but also to healthier human populations, since reproductive effects in humans are implicated for populations that consume large quantities of Great Lakes fish from the same areas (Swain 1988).

Froese et al. (1998) examined sediments and 2 trophic levels (aquatic insects and insectivorous birds) to determine the trophic transfer of PCBs in the Saginaw Bay watershed. They analyzed sediments, emergent chironomid larvae, and tree swallow (*Tachycineta bicolor*) eggs and nestlings collected in 1992 for PCB congeners and determined the biota–sediment accumulation factor (BSAF). Based on the mono- and non-ortho PCB congeners in each matrix, they determined the TEQs of the PCBs in terms of 2,3,7,8-TCDD concentration. They determined a hazard quotient of 0.7 for threshold concentration TEQ in sediments that would protect the tree swallow compared with the BSAF calculated for tree swallow nestlings. Hence, the results of Froese et al. (1998) indicated that dioxin-like activity of PCB congeners in the sediments should not cause adverse effects on tree swallow at the locations studied in Saginaw Bay AOC. Given the historically high levels of PCBs in Saginaw Bay, these results suggest a degree of NR occurring with time in this AOC.

Crane (1995) conducted a comparative human health and wildlife risk assessment at the Buffalo River AOC as part of the ARCS program. She modeled 3 organic contaminants (total PCBs, benzo(*a*)anthracene, benzo(*a*)pyrene) and 2 heavy metals (Pb and Cu) that were significant contaminants at the Buffalo River AOC. Mink was the only piscivorous wildlife species considered because carp was the only forage fish for which contaminant concentrations were modeled for different remedial action scenarios. Mass balance modeling was applied to different remedial action scenarios to estimate changes in contaminant concentrations in sediment, water, and fish tissue (Atkinson et al. 1994). The remedial action scenarios modeled were as follows:

1) No action scenario: NR occurred with existing navigational dredging and external loadings.
2) Hamburg Cove scenario: Navigational dredging discontinued above Hamburg Cove (about halfway from the river mouth to the AOC upstream boundary).

3) Environmental dredging scenario: Several hot spots along the banks were removed by dredging along the entire river within the AOC.
4) No action: No loading scenario.
5) Hamburg Cove: No loading scenario.

For the risk assessment calculation, Crane (1995) used a simple hazard quotient method by which the forage contaminant concentration was compared to the no-observed-adverse-effect level (NOAEL). If the quotient was > 1, the mink were considered at risk. She found that the NOAEL was exceeded by 19 to 46 times for the various remedial alternatives. Environmental dredging resulted in the least risk to mink, whereas the other downstream remediation scenarios were almost identical in risk. Hence, NR, with or without existing loadings, was as satisfactory an alternative as all other scenarios with the exception of environmental dredging, which removes hot spots. Crane (1995) suggested additional work to obtain more accurate data on the diet of mink in the Buffalo River and to assess the potential exposure from all contaminants of concern to mink and other indicator species.

Human health baseline and comparative risk assessment

In 41 of the 42 AOCs in the Great Lakes, contaminated sediments are the major cause of impaired beneficial uses of these nearshore Great Lakes sites. As Crane (1996) stated, "The results of baseline risk assessments can be used by risk managers to make decisions regarding the need for immediate remediation, and to determine in the future whether risks have increased or decreased over time with changes in the levels of sediment contamination. Comparative risk assessments, which include a mass balance modeling approach, can be used to estimate and compare the risks that may be associated with different remedial alternatives (including the no action alternative)." Unfortunately, comparative risk assessments are seldom conducted because of the costs involved, so that a high degree of uncertainty is associated with decisions on remediation alternatives, including the efficacy of NR.

DECISION-MAKING AT AREAS OF CONCERN

While the RAP process provides a mechanism for addressing pollution concerns at AOCs, the process is constrained by legislation, regulations, and policies that are in effect. For example, if the AOC is or contains a hazardous waste site in the U.S., then remedies must meet the relevant criteria. In particular, if these areas are designated as Superfund sites, then the investigation, remediation, and evaluation must be consistent with USEPA Superfund guidance documents (USEPA 1989, 1991b). If they are designated as state hazardous waste sites, then RAP decisions must be in conformity with state rules. In Michigan, these rules include specific criteria that define: "How clean is clean?" based, among other factors, on future land use (Michigan Department of Environmental Quality [MDEQ] 1995).

Further, if these hazardous wastes move into U.S. surface water, remedial action decisions must consider the GLI numerical criteria, since these are also considered applicable or relevant and appropriate requirements (ARARs) that have to be met by any site cleanup under the U.S. Superfund Program, and these criteria are often used similarly by states in their own hazardous waste programs. Additionally, the antidegradation provisions of the GLI may limit remedial alternatives in certain areas of the Great Lakes, depending on individual state actions.

Risk management mandates such as Superfund, RCRA, GLI, and state and provincial laws and regulations not only provide criteria for deciding which remedial actions are chosen but also criteria for judging the success of remedial efforts. Since NR is one of the alternative cleanup techniques, the decision as to whether or not to apply this technology at a particular site is significantly impacted by the many and varied laws, rules, and agreements currently in force in the Great Lakes basin. In practice, remediation of contaminated sediments at U.S. AOCs has cost $362,322,405 as of March 1998 (USEPA 1998b). As stated in this report, "The majority of sites chose dredging as the environmentally and economically preferred option to remove contaminated sediments." Hence, active remediation has been applied at virtually all of the AOCs remediated thus far.

Even if this policy changes, as seems possible, the decision to recommend NR for at least some of the pollutants at an AOC depends on the stakeholders at the AOC. Thus, it is appropriate to examine in somewhat more detail the way that RAPs are carried out at AOCs. The key players at AOCs include the RAP teams, which generally consist of government agency scientists and technical experts who are responsible for problem identification and for providing technical information to another key actor, the public advisory committee (PAC). The RAP teams help in specifying appropriate restoration goals and identifying procedures needed to assess progress toward these goals. The PACs are the organizations responsible for deciding on restoration goals and remedial actions most appropriate for achieving these goals.

PACs are made up of representatives of a wide variety of stakeholder groups, and thus PAC decisions are supposed to reflect consensus judgements of all of these groups. Stakeholders may include representatives of agriculture, commerce, conservation organizations, environmental groups, health professionals, educators, land owners, local public officials, and citizens at large. In cases where AOCs include sites on both sides of the U.S.–Canadian border, representatives of interests in both nations serve on the advisory committee, and it is known as a binational public advisory committee (BPAC). The number of organizations represented on PACs or BPACs varies greatly, often depending on the size and complexity of the AOC.

Significance of risk communication in remedy selection

Because of the importance of the public advisory groups to the selection and carrying out of remedial alternatives, accurate and effective communication of the relevant scientific information to these groups is critical. Since these advisory groups often encourage public forums and other public input to the advisory committee process, communication to the public as well as to committee members is critical (Nuclear Regulatory Commission [NRC] 1989; Landre and Knuth 1993; Kellogg 1997). Effective communication with the public is also very important after remedial action decisions have been made because the public may be involved in facilitating some of these actions (e.g., recycling) and because continued public support is important to the success of the long-term monitoring that may be needed to assure that remedial actions meet the stated goals.

Communicating with the public must be done in a credible manner without any appearance that a particular agenda is being promoted. This, in turn, requires involvement of trusted members of the local, scientific, and regulatory communities in the risk communication process (Renn and Levine 1991; Slovic 1993; Peters et al. 1997). Their effectiveness will depend not only on their trustworthiness but also on the consistency of the messages provided by all.

Case study: The St. Clair River Area of Concern

This site is part of the connecting channel between Lake Huron and Lake St. Clair and runs between the U.S. (Michigan) and Canada (Ontario). It is used for shipping, boating, fishing, drinking water, power generation, and cooling and process water. It has been impacted by a variety of sources, including municipal sewage treatment plants, CSOs, industrial wastes, and other discharges. Chemicals found in the sediments include chlorinated organics, heavy metals, petroleum products, phenols, and suspended solids.

Because of this contamination, a number of uses are considered impaired: fish and wildlife consumption, beach access, fish and wildlife habitat, drinking water, shipping (through restrictions on dredging), and aesthetics. During the problem definition phase of the RAP process, highly contaminated sediment hot spots have been characterized, environmental quality measurements have been made, and land/use habitat information has been collected (Ontario Ministry of Environment and Energy [OMEE] 1995, 1996).

The large number of partners involved in the RAP process can complicate the risk management process; however, they also provide an opportunity to include a full range of inputs at an early stage of the RAP process. Such early inputs can help to avoid the all too common situation where decisions are proposed after much internal discussion among governmental organizations and are rejected by other groups that have not had a chance to have their opinions considered. In addition, incorporation of those who represent the various stakeholders in all aspects of the risk management process can help to facilitate the implementation of the planned remedial actions because they can carry back a message that they have had a part in shaping and thus has their support.

The risk management decisions that have been recommended at this AOC and, in some cases, carried out, are complex. They include changes in practices at a number of industrial and commercial facilities, upgrading of wastewater treatment facilities and infrastructure, dredging of limited areas of sediment, changes in agricultural practices, habitat modification such as constructing wetlands, and improvements in solid waste disposal. Identifying and implementing each of these remedial actions requires the participation of one or more of the stakeholders involved in the RAP process and reflects the importance of having continuous input from representatives of a very wide range of institutions and viewpoints (OMEE 1993).

At present, NR has not been chosen as an option at this AOC. Current USEPA sediment management strategy would support consideration of NR as an option (USEPA 1998a). For this remedial option to be chosen, it would need to be discussed as part of the RAP process, and any decision to utilize NR as one of the remedies could only be reached based on consensus discussions among all of the parties.

Current issues in remedy selection

Recently, a number of actions have been taken that are affecting the development and implementation of the RAPs. In general, these all reflect a decrease in the amount of funds devoted to AOCs. For example, the Province of Ontario no longer provides support for PACs, is no longer funding education programs as part of RAPs, and has eliminated some RAP coordinator positions (Wooster 1997). These actions indicate a reason to prioritize sites and use NR in appropriate areas. Economic realities will drive greater reliance on NR. In addition, support for monitoring, research, and rehabilitation activities in the Great Lakes has declined (Dolan and Hartig 1996). These actions come at a time when increased funding is needed, since the easier remediation steps have been undertaken and the more difficult and expensive ones remain. Thus, these funding declines are likely to have significant impacts. One will be a slowing down of both process and action at AOCs. A second will be an increase in the attractiveness of the less costly remediation options. Since NR is the least expensive option in most cases, it will become more attractive.

SUMMARY AND CONCLUSIONS

The summaries of what has been learned scientifically and what has been done to manage risks at AOCs, particularly those in the ARCS program, show that progress has been slow. In many cases, scientific data needed to characterize sediment contamination and adverse impacts of contamination are incomplete or missing.

The effectiveness of various remedial options is also unclear, especially with respect to experimental procedures. The most common direct remedial action has been the removal of contaminated sediments by dredging; the most common indirect approaches have been steps to limit or prevent additional contamination.

A recent report of the IJC Sediment Priority Action Committee (IJC 1997) examined all of the AOCs. It listed approximately 10 sites where NR is the preferred remedial alternative for at least a part of the AOC. Most of these are in Canada, but this alternative has also been chosen at a few sites in the U.S. This report also concluded that a major obstacle to cleanup in two-thirds of the AOCs is limitations on resources and funding. Other major obstacles that have been uncovered in one-third to one-half of the AOCs are regulatory complexity, lack of decision-making framework, limited corporate involvement, insufficient research and technology development, and limited public and local support.

Documentation of NR as well as verification of the efficacy of active remediation of contaminated sediments requires monitoring of contaminant levels and also the biological communities to determine their response to reductions in contamination. Measuring the bioavailability of remaining sediment contaminants is necessary to determine the potential for trophic transfer to top predators, including humans as well as salmonids and piscivorous wildlife. Examination of sentinel species and health indicators or biomarkers as well as structure and function of ecological systems is needed to verify progress. Unfortunately, these sorts of long-term studies are seldom conducted because of high costs and downsizing of staff and budgets of natural resource and environmental institutions, especially in the Great Lakes. The trend from a manufacturing-based economy in the Great Lakes to a service-based economy is resulting in decreased input of industrial chemicals to AOCs. However, the continuing urbanization and increased populations in AOCs preclude restoration of natural habitats in most of the nearshore areas, although these are valued for recreational and aesthetic enjoyment.

Natural remediation as a viable option

When we address the question of whether NR is a viable option, we must remember that more than one remedial approach may be appropriate at a particular AOC. The different techniques may be applied sequentially to certain parts of the AOC or may be applied simultaneously to various parts of the AOC. For example, hot spot areas may be dredged and less contaminated areas remediated by a combination of source reductions and NR. The remedial options may include both direct and indirect actions that, in combination, can reduce the levels of contaminants in sediments. At many of the ARCS sites, removal of sediments by dredging has occurred and has been accompanied by steps to limit inputs of pollutants. Exploration of other remedial techniques, including various forms of bioremediation, is also underway at many of these sites.

At many sites, a number of mechanical and chemical approaches to enhanced bioremediation are being studied (Schneider and Billingsley 1990; NRC 1993; Alexander 1994; Baker and Herson 1994; USEPA 1994b). These may be applied in situ by inoculation with exogenic organisms combined with infusion of oxidant (air or nitrates, etc.) and/or nutrients, or by use of bioreactors through which groundwater or sediment is passed, purified, and then replaced. Mechanical means, including pumping water and use of vacuum to reduce contaminant load, augment the processes. While the results are preliminary, some researchers are hopeful (Jafvert and Rogers 1991). However, other scientists believe that the results of bench-scale bioremediation studies at 4 ARCS priority AOCs indicate that in situ sediment treatments of contaminated sediments are not practical, and removal is required for proper treatment of PAH and PCB contaminants (Jones et al. 1996). Resolving these differences will require moving from the laboratory to the field, where bioremediation techniques can be studied under real conditions.

The effectiveness of intrinsic bioremediation, the "do-nothing and get results" solution to site contamination, can be very difficult to quantify conclusively unless the monitoring program is carefully planned and followed through (NRC 1997). Natural physical processes, such as volatilization, migration, and adsorption, that can cause contaminant concentration reductions over time are not easily distinguished from those caused by natural bioremediation. Simple disappearance of the contaminant does not allow us to distinguish between physical and biological processes. We must know the by-products of each and look for these to be more certain of the process. We may not know all the by-products to look for, or some of the other contaminants present may be the by-products. Sometimes we cannot tell one process from the other because we do not know enough (yet). In addition, NR may be hard to detect if there are continuing inputs so that contaminant concentration reductions are masked. One piece of evidence that suggests that NR may be a viable alternative is the observed reduction of pesticide residues around the Great Lakes in a remarkably short time period after the banning of the pesticides in the early 1970s (Hesselberg et al. 1990). This suggests that, in nature, time heals all wounds.

On the other hand, organisms capable of degrading or mineralizing the contaminants of concern may be present but limited by a site condition (or conditions), and engineered assistance will be required (NRC 1993; Alexander 1994; Baker and Herson 1994). In situ bioremediation can take considerable time and may present control problems to insure optimum operation, so in certain circumstances removal for treatment or storage in a CDF may be the most appropriate option (Schneider and Billingsley 1990).

Davenport and Spacie (1991) suggested that the phototoxicity of PAHs, together with their accumulation in sediments, poses a presently unrecognized environmental hazard from dredging and other disturbances. For the GC–IH AOC, phototoxicity was found in sediment elutriates that was not present in the absence of light. These results revealed hazards in shallow water environments near dredging sites (Davenport and Spacie 1991), and hence supported NR or lack of disturbance of these sediments.

Recommendations

Systematic programs are recommended to determine status and trends of contaminant levels in sediments and biota in the Great Lakes, including AOCs. The long-term contaminant monitoring program of lake trout and walleyes from the open waters of the Great Lakes required by the IJC should continue. Long-term monitoring of the status of fish and wildlife and their forage base, including plankton and benthos, should be supported. Monitoring of forage fish in the open waters of the Great Lakes has been ongoing for more than 2 decades by the Great Lakes Science Center (U.S. Geological Survey [USGS]). Similar programs are needed for nearshore fish and wildlife in AOCs. All of this must be supported by institutional stability and funding.

Because current trends in risk management suggest that fewer rather than more funds are likely to be available for cleanup of contaminated sediments, it is imperative that efforts be intensified to both better characterize and quantify human and ecological effects of such contaminants and to assess the effectiveness of lower-cost cleanup alternatives. Natural remediation is one of the lower-cost alternatives, but more research needs to be done to determine how effective it is as a function of the contaminant and the site characteristics. In addition, such research can help us to better understand how this approach can be used to complement other approaches so as to minimize overall remediation costs while maximizing benefits.

Summary of Great Lakes natural remediation

The Laurentian Great Lakes in North America are the world's largest system of fresh water, teeming with abundant natural resources. Because of the long history of settlement, development, and industrialization, a legacy of contaminated sediments now resides in many nearshore and tributary areas. These contaminated sediments impair ecological communities and impact human health and beneficial uses of these areas. The IJC has designated 42 AOCs that are impaired for use by humans, wildlife, and fish. RAPs to plan and implement remediation of contaminated sediments at all of the AOCs have been initiated by all involved parties, but many of these RAPs have yet to accomplish significant remediation because of insufficient funds. The only AOC to accomplish remediation of impaired uses to the extent of becoming de-listed is the Collingwood Harbour on northeastern Lake Huron in Ontario, Canada. A grassroots effort was coordinated by the OMEE and succeeded in restoring beneficial uses of the Collingwood Harbour.

The USEPA conducted an extensive demonstration project on ARCS in 5 of the AOCs in the U.S. The ARCS project described the current state of contamination, provided baseline risk assessments and one comprehensive risk assessment for humans and/or fish and wildlife, determined the mass balance of major contaminants, and conducted demonstration cleanup projects by several engineering technologies. Natural remediation or the "no action" alternative was considered at the ARCS sites.

Overall, the existence of stricter environmental laws during the last 30 years and a reduction in the manufacturing base in the Great Lakes has resulted in improvement in conditions in harbors, rivers, and nearshore waters. Problems remain, such as the inability to dredge certain harbors and remove sediments because of lack of disposal facilities for contaminated sediments. Because of the wide extent of contaminated sediments in the Great Lakes, much work remains to be done to document the condition of contaminated areas and the degree to which remediation of these areas is occurring from biotic and abiotic natural processes.

REFERENCES

Alexander M. 1994. Biodegradation and bioremediation. Chapter 4. San Diego CA: Academic Pr. p 11–13.

Aqua-Tech. 1989. Sediment analysis, Buffalo River and harbor, Buffalo, NY. Technical report No. G0193-06A, August. Prepared for Water Quality Section. Buffalo NY: U.S. Army Engineer Buffalo District.

Atkinson JF, Bajak T, Morgante M, Marshall S, DePinto JV. 1994. Assessment and Remediation of Contaminated Sediments (ARCS) Program: Model data requirements and mass loading estimates for the Buffalo River mass balance study. Chicago IL: Great Lakes National Program Office (GLNPO), U.S. Environmental Protection Agency (USEPA). EPA-905-R94-005.

Baker KH, Herson DS. 1994. Bioremediation. Chapters 2, 7, 9. New York NY: McGraw-Hill.

Baumann PC, Smith IR, Metcalfe CD. 1996. Linkages between chemical contaminants and tumors in benthic Great Lakes fish. *J Great Lakes Res* 22:131–152.

Bergantz J. 1977. Benthic survey: A comparison of various bio-indexing systems applied to benthic macroinvertebrate data. Buffalo NY: Great Lakes Laboratory, State Univ College Buffalo.

Bolattino C, Tuchman M. 1993. A summary of contaminated sediment activities within the United States Great Lakes areas of concern. Chicago IL: Great Lakes National Program Office (GLNPO), U.S. Environmental Protection Agency (USEPA). http://epaserver.ciesin.org/glreis/nonpo/nprog/aoc_rap/docs/aocsedtoc.html

Bolsenga SJ, Herdendorf CE. 1993. Lake Erie and Lake St. Clair handbook. Detroit MI: Wayne State Univ Pr.

Chapman PM. 1992. Sediment quality triad. In: Sediment classification methods compendium. Washington DC: Office of Science and Technology, U.S. Environmental Protection Agency (USEPA). [cited in Wildhaber and Schmidt 1996].

Collis J, Proud J. 1994. Collingwood Harbour habitat assessment and fish inventory. Toronto ON: Ontario Ministry of Environment and Energy (OMEE).

Crane JL. 1993. Baseline human health risk assessment: Buffalo River, New York, area of concern. Chicago IL: Great Lakes National Program Office (GLNPO), U.S. Environmental Protection Agency (USEPA). EPA-905-R93-008.

Crane JL. 1995. Comparative human health and wildlife risk assessment: Buffalo River, New York, area of concern. Chicago IL: Great Lakes National Program Office (GLNPO), U.S. Environmental Protection Agency (USEPA). EPA-905-R95-008.

Crane JL. 1996. Carcinogenic human health risks associated with consuming contaminated fish from five Great Lakes areas of concern. *J Great Lakes Res* 22:653–668.

Davenport R, Spacie A. 1991. Acute phototoxicity of harbor and tributary sediments from lower Lake Michigan. *J Great Lakes Res* 17:51–56.

DePinto JV. 1994. Application of a mass balance model, GBTOX, for diagnostic and remediation analysis of PCBs in Green Bay, Lake Michigan. *Lake Reservoir Manage* 9(2):68 (abstract only).

Dolan D, Hartig J. 1996. Reduction of funds for Great Lakes science: Results of November, 1995 research budget survey. *J Great Lakes Res* 22: 484–490.

Eggold BT, Amrhein JF, Coshun MA. 1996. PCB accumulation by salmonine smolts and adults in Lake Michigan and its tributaries and its effect on stocking policies. *J Great Lakes Res* 22:403–413.

Elder B. 1991. Sheboygan River and harbor. In: Jafvert CT, Rogers JE, editors. Biological remediation of contaminated sediments with special emphasis on the Great Lakes. Workshop Report; 1990 July 17–19; Manitowoc WS. Athens GA: Environmental Research Laboratory, U.S. Environmental Protection Agency (USEPA). EPA 600-9/91-001. p 50–54.

Ferguson MS. 1986. Collingwood Harbour fish habitat and inventory study 1986. Midburst ON: Huronia District, Ontario Ministry of Natural Resources (OMNR).

Flint RW. 1979. Responses of freshwater benthos to open-lake dredged spoils disposal in Lake Erie. *J Great Lakes Res* 5:264–275.

Fox GA, Collins B, Haykawa E, Weseloh DV, Ludwig JP, Kubiak TJ, Erdman TC. 1991. Reproductive outcomes in colonial fish-eating birds: A biomarker for developmental toxicants in Great Lakes food chains. II. Spatial variation in the occurrence and prevalence of bill defects in the young double-crested cormorants in the Great Lakes, 1979–1987. *J Great Lakes Res* 17:158–167.

Froese KL, Verbrugge DA, Ankley GT, Niemi GJ, Larsen CP, Giesy JP. 1998. Bioaccumulation of polychlorinated biphenyls from sediments to aquatic insects and tree swallow eggs and nestlings in Saginaw Bay, Michigan, USA. *Environ Toxicol Chem* 17:484–492.

[GC, USEPA] Government of Canada, U.S. Environmental Protection Agency. 1995. The Great Lakes: An environmental atlas and resource book. Toronto ON: EC and Chicago IL: Great Lakes National Program Office (GLNPO), USEPA. EPA 905-B-95-001.

Goudy G. 1991. Saginaw River/Bay AOC. In: Jafvert CT, Rogers JE, editors. Biological remediation of contaminated sediments with special emphasis on the Great Lakes. Workshop Report; 1990 July 17–19; Manitowoc, WS. Athens GA: Environmental Research Laboratory, U.S. Environmental Protection Agency (USEPA). EPA 600-9/91-001. p 42–49.

Harris JH, Wenger RB. 1993. A method for assessing environmental risk. 36th Conf. Intern. Assoc. Great Lakes Res., Program and Abstracts; June 4–10; DePere WI. Ann Arbor MI: International Association for Great Lakes Research. p 118.

Hartig JH, Law NL. 1994. Progress in Great Lakes remedial action plans: Implementing the ecosystem approach in Great Lakes areas of concern. Chicago IL: U.S. Environmental Protection Agency. [online]. University Center Mich.: Consortium for International Earth Science Information Network (CIESIN). http:epaserver.ciesin.org/glreis/nonpo/nprog/ aoc_rap/docs/progress/cover.html

Heaton SN, Bursian SJ, Giesy JP, Tillitt DE, Render JA, Jones PD, Verbrugge DA, Kubiak TJ, Aulerich RJ. 1995. Dietary exposure of mink to carp from Saginaw Bay, Michigan. 1. Effects on reproduction and survival, and potential risks to wild mink populations. *Arch Environ Contam Toxicol* 28:334–343.

Hesselberg RJ, Hickey JP, Nortrup DA, Willford WA. 1990. Contaminant residues in the bloater (*Coregonus hoyi*) of Lake Michigan, 1969–1986. *J Great Lakes Res* 16:121–130.

[IJC] International Joint Commission. 1978. An agreement between Canada and the United States of America on Great Lakes water quality. November 22, 1978. Windsor ON: IJC.

[IJC] International Joint Commission. 1985. Report on Great Lakes water quality. Windsor ON: Great Lakes Water Quality Board, IJC.

[IJC] International Joint Commission. 1987. Revised Great Lakes Water Quality Agreement of 1978 as amended by protocol signed November 18, 1987. Windsor ON: IJC.

[IJC] International Joint Commission. 1997. Overcoming obstacles to sediment remediation in the Great Lakes basin. Windsor ON: Sediment Priority Action Committee. Great Lakes Water Quality Board, IJC. http://www.ijc.org/boards/wqb/sediment.html

Ingersoll CG, Buckler DR, Crecelius EA, Lapoint TW. 1993. Assessment and Remediation of Contaminated Sediments (ARCS) Program: Biological and chemical assessment of contaminated Great Lakes sediment Chicago IL: Great Lakes National Program Office (GLNPO), U.S. Environmental Protection Agency (USEPA). EPA 905-R93-006.

Jafvert CT, Rogers JE, editors. 1991. Biological remediation of contaminated sediments with special emphasis on the Great Lakes. Workshop Report; 1990 July 17–19; Manitowoc WI. Athens GA: Environmental Research Laboratory, U.S. Environmental Protection Agency (USEPA). EPA 600-9/91-001.

Jones WJ, Araujo R, Rogers JE. 1996. Assessment and Remediation of Contaminated Sediments (ARCS) Program: Bench-scale evaluation of bioremediation for the treatment of sediments from the Ashtabula, Buffalo, Saginaw, and Sheboygan Rivers. Chicago IL: Great Lakes National Program Office (GLNPO), U.S. Environmental Protection Agency (USEPA). EPA-905-R96-012.

Kellogg W. 1997. Lessons from RAPs: Citizen participation and the ecology of community. *J Great Lakes Res* 23:227–229.

Kozuchowski E, Poole EA, Lowie C. 1993. Buffalo River fisheries assessment. Report on the results of the 1992 larval and adult fish survey. Amherst NY: Lower Great Lakes Fishery Resources Office, U.S. Fish and Wildlife Service. Admin. Rep. 93–03.

Kraft CE. 1993. Changes in the fish community of lower Green Bay, Lake Michigan. 36th Conf. Intern. Assoc. Great Lakes Res., Program and Abstracts; 1993 June 4–10; DePere WI. Ann Arbor MI: International Association for Great Lakes Research. p 116.

Krantzberg G. 1995. Using the burden of evidence approach for sediment management. Case study: Collingwood Harbour. In: Munawar M, Edsall T, Leach J, editors. The Lake Huron ecosystem: Ecology, fisheries, and management. Amsterdam, the Netherlands: SPB Academic. p 365–395.

Krantzberg G, Houghton E. 1994. Collingwood Harbour RAP Stage 3 document. Toronto, ON: Ontario Ministry of Environment and Energy (OMEE).

Krantzberg G, Houghton E. 1996. The remedial action plan that led to the cleanup and delisting of Collingwood Harbour as an area of concern. *J Great Lakes Res* 22:469–483.

Landre B, Knuth B. 1993. Success of citizen advisory committees in consensus-based water resources planning in the Great Lakes Basin. *Soc Nat Resour* 6(3):228–257.

Larson RA, Weber EJ. 1994. Reaction mechanisms in environmental chemistry. Chapters 1–4,6. Boca Raton FL: CRC Pr.

Lee CR, Brandon DL, Simmers JW, Tatem, HE, Skogerboe JG. 1991. Information summary, area of concern: Buffalo River, New York. Vicksburg MS: U.S. Army Engineer Waterways Experiment Station. Miscellaneous Paper EL-91-9.

Lonky E, Reihman J, Darvill T, Mather Sr J, Daly H. 1996. Neonatal behavioral assessment scale performance in humans influenced by maternal consumption of environmentally contaminated Lake Ontario fish. *J Great Lakes Res* 22:198–212.

Ludwig JP, Kurita-Matsuba H, Auman JH, Ludwig ME, Summer CL, Giesy JP, Tillitt DE, Jones PD. 1996. Deformities, PCBs, and TCBB-equivalents in double-crested cormorants (*Phalacrocorax auritus*) and Caspian terns (*Hydroprogne caspia*) of the upper Great Lakes 1986–1991: Testing a cause–effect hypothesis. *J Great Lakes Res* 22:172–197.

Makarewicz JC, Dilcher RC, Haynes JM, Shump K. 1982. Biological survey Buffalo River and Outer Harbor of Buffalo, N.Y. Volumes 1 and 2. Report to the U.S. Army Corps of Engineers, Buffalo District, Buffalo NY. Brockport NY: Biology Department, State Univ NY Brockport. Contract No. DACW49-81-0-0033.

[MDEQ] Michigan Department of Environmental Quality. 1995. Part 201 of the Natural Resources and Environmental Protection Act. Act 451 of 1994 as amended. Lansing MI: MDEQ.

[MDNR] Michigan Department of Natural Resources. 1988. Remedial action plan for Saginaw River and Saginaw Bay area of concern. Lansing MI: Great Lakes and Environmental Assessment Section, Surface Water Quality Division.

Mueller ME. 1992. Tumors and abnormalities in fish from the Ashtabula and Buffalo Rivers. Final report for ARCS. Chicago IL: Great Lakes Program Office, U.S. Environmental Protection Agency (USEPA).

Mueller ME, Mac MJ. 1993. Tumor surveys and 10-day bioaccumulation bioassays in the assessment and remediation of contaminated sediments. 36 Conf. Int. Assoc. Great Lakes Res; 1993 Jun 4–10; De Pere WI. Program and Abstracts, p 101.

Nalepa TF, Fahnenstiel GL. 1995. *Dreissena polymorpha* in Saginaw Bay, Lake Huron ecosystem: Overview and perspective. *J Great Lakes Res* 21:411–416.

Nalepa TF, Wojcik JA, Fanslow DL, Lang GA. 1995. Initial colonization of the zebra mussel (*Dreissena polymorpha*) in Saginaw Bay, Lake Huron: Population, recruitment, density, and size structure. *J Great Lakes Res* 21:417–434.

Neilson AH. 1994. Organic chemicals in the aquatic environment: Distribution, persistence, and toxicity. Boca Raton FL: CRC Pr. Chapters 4, 6, 7.

[NRC] National Research Council. 1989. Improving risk communication. Washington DC: National Academy Pr.

[NRC] National Research Council. 1993. In situ bioremediation. When does it work? Committee on In Situ Bioremediation, Water Science and Technology Board, Commission on Engineering and Technical Systems. Washington DC: National Academy Pr. Chapter 2.

[NRC] National Research Council. 1997. Contaminated sediments in ports and waterways: Cleanup strategies and technologies. Washington DC: Committee on Contaminated Marine Sediments, Marine Board, Commission on Engineering and Technical Systems.

[NYSDEC] New York State Department of Environmental Conservation. 1989. Buffalo River remedial action plan. Buffalo NY: NYSDEC.

[OEPA] Ohio Environmental Protection Agency. 1991. Ashtabula River Remedial Action Plan Stage I Investigation Report. Columbus OH: OEPA.

[OMEE] Ontario Ministry of Environment and Energy. 1993. St. Clair River RAP remedial options. Toronto ON: OMEE.

[OMEE] Ontario Ministry of Environment and Energy. 1995. Final Stage 2 RAP recommended plan and appendices. Toronto ON: OMEE.

[OMEE] Ontario Ministry of Environment and Energy. 1996. Environmental assessment of Upper St. Clair River sediments and benthic macroinvertebrate communities - 1994. Toronto ON: OMEE.

Passino-Reader DR, Hudson PL, Hickey JP. 1995. Baseline risk assessment for aquatic life for the Buffalo River, New York, area of concern. Chicago IL: Great Lakes National Program Office (GLNPO), U.S. Environmental Protection Agency (USEPA). EPA 905-R95-001.

Passino-Reader DR, Stedman RM, Madenjian CP. 1996. Status of forage fish stocks in Lake Michigan, 1995. Report to Lake Michigan Committee. Ann Arbor MI: Great Lakes Fishery Commission.

Peters RG, Covello VT, McCallum DB. 1997. The determinants of trust and credibility in environmental risk communication: An empirical study. *Risk Anal* 17:43–54.

Renn O, Levine D. 1991. Credibility and trust in risk communication. In: Kasperson RE, Stallen PJM, editors. Communicating risks to the public. Dordrecht, the Netherlands: Kluwer Academic. p 75–218.

Rosenberg DM, Resh VH, editors. 1993. Freshwater biomonitoring and benthic invertebrates. New York NY: Chapman and Hall.

Sanders P. 1991. Fields Brook Superfund site/Ashtabula River area of concern. In: Jafvert CT, Rogers JE, editors. Biological remediation of contaminated sediments with special emphasis on the Great Lakes. Workshop Report; 1990 July 17–19; Manitowoc WI. Athens GA: Environmental Research Laboratory, U.S. Environmental Protection Agency (USEPA). EPA 600-9/91-001. p 29–35.

Savagian A, Uram E. 1995. Lake Superior and the Great Lakes: A citizen's guide to economics and the environment. Madison WI: Sierra Club Great Lakes Ecoregion Program.

Schneider DR, Billingsley RJ. 1990. Bioremediation: A desk manual for the environmental professional. Des Plaines IL: Cahaers.

Schrank CS, Cormier SM, Blazer VS. 1997. Contaminant exposure, biochemical, and histopathological biomarkers in white suckers from contaminated and reference sites in the Sheboygan River, Wisconsin. *J Great Lakes Res* 23:119–130.

Shipp EB, Restum JC, Bursian SJ, Aulerich RJ, Helferich WG. 1998. Multigenerational study of the effects of consumption of PCB-contaminated carp from Saginaw Bay, Lake Huron, on mink. 3. Estrogen receptor and progesterone receptor concentrations, and potential correlation with dietary PCB consumption. *J Toxicol Environ Health, Pt. A* 54:403–420.

Skogerboe JG, Lee CR, Brandon DL, Simmers JW, Tatem HE. 1991. Information summary, area of concern: Sheboygan, Wisconsin. Vicksburg MS: Waterways Experiment Station, U.S. Army Corps of Engineers (USACE). Miscellaneous Paper EL-91-6.

Slovic P. 1993. Perceived risk, trust and democracy. *Risk Anal* 13:675–682.

Sonzogni WC, David MM. 1991. PCB dechlorination in the Sheboygan River, Wisconsin. In: Jafvert CT, Rogers JE, editors. Biological remediation of contaminated sediments with special emphasis on the Great Lakes. Workshop Report; 1990 July 17–19; Manitowoc WI. Athens GA: Environmental Research Laboratory, U.S. Environmental Protection Agency (USEPA). EPA 600-9/91-001. p 75–78.

Swain WR. 1988. Human health consequences of consumption of fish contaminated with organochlorine compounds. *Aquat Toxicol* 11:357–377.

Tatem HE, Brandon DL, Lee CR, Simmers JW, Skogerboe JG. 1990. Information summary, area of concern: Ashtabula River, Ohio. Vicksburg MS: Waterways Experiment Station, U.S. Army Corps of Engineers (USACE). Miscellaneous Paper EL-90-22.

Terlecky PM, Michalovic JG, Pek SL. 1975. Water pollution investigation: Ashtabula area. Great Lakes Initiative Contract Program. Chicago IL: Region V, U.S. Environmental Protection Agency (USEPA). EPA-905/9-74-008.

Tillitt DE, Kubiak TJ, Ankley GT, Giesy JP. 1993. Dioxin-like toxic potency in Forster's tern eggs from Green Bay, Lake Michigan, North America. *Chemosphere* 26:2079–2084.

[USEPA] U.S. Environmental Protection Agency. 1977. Guidelines for the pollutional classification of Great Lakes harbor sediments. Chicago IL: Region V, USEPA.

[USEPA] U.S. Environmental Protection Agency. 1989. Risk assessment guidance for Superfund. Volume 1, Human health evaluation manual (Part A). Interim Final. Washington DC: Office of Emergency and Remedial Response, USEPA. EPA/540/1-89/002.

[USEPA] U.S. Environmental Protection Agency. 1991a. Sediment remediation technologies selection. Chicago IL: Great Lakes National Program Office (GLNPO), USEPA. *Federal Register* 56(85):20222–20223.

[USEPA] U.S. Environmental Protection Agency. 1991b. Risk assessment guidance for Superfund. Volume 1, Part B. Washington DC: Office of Emergency and Remedial Response, USEPA. EPA 9285.7-01B.

[USEPA] U.S. Environmental Protection Agency. 1993a. Fields Brook Superfund site project update. Chicago IL: Region V, Office of Public Affairs, USEPA.

[USEPA] U.S. Environmental Protection Agency. 1993b. Assessment and remediation of contaminated sediments (ARCS) program: Biological and chemical assessment of contaminated Great Lakes sediment. Chicago IL: Great Lakes National Program Office (GLNPO), USEPA. EPA 905-R93-006.

[USEPA] U.S. Environmental Protection Agency. 1994a. Assessment and remediation of contaminated sediments (ARCS) program: Final summary report. Chicago IL: Great Lakes National Program Office (GLNPO), USEPA. EPA-905-S94-001. http://epaserver.ciesin.org/glreis/glnpo/data/ arcs/EPA-905-S94-001/EPA-905-S94-001.html

[USEPA] U.S. Environmental Protection Agency. 1994b. Report to congress on the Great Lakes ecosystem. Chicago IL: Great Lakes National Program Office (GLNPO), USEPA. EPA 905-R-94-004.

[USEPA] U.S. Environmental Protection Agency. 1994c. ARCS assessment guidance document. Chicago IL: Great Lakes National Program Office (GLNPO), USEPA. EPA 905-B94-002.

[USEPA] U.S. Environmental Protection Agency. 1995. Water quality guidance for the Great Lakes system. USEPA. *Federal Register* 60:15366–153425.

[USEPA] U.S. Environmental Protection Agency. 1996. Guidance for evaluation of federal agency demonstrations that remedial actions are operating properly and successfully under CERCLA Section 120(h)(3). Washington DC: Federal Facilities Restoration and Reuse Office, USEPA.

[USEPA] U.S. Environmental Protection Agency. 1998a. EPA's contaminated sediment management strategy. Washington DC: Office of Water, USEPA. EPA-823-R-98-001. http://www.epa.gov/OST/cs/manage/stratndx.html

[USEPA] U.S. Environmental Protection Agency. 1998b. Realizing remediation. A summary of contaminated sediment remediation activities in the Great Lakes Basin. Chicago IL: Great Lakes National Program Office (GLNPO), USEPA. http://www.epa.gov/grtlakes/sediments/realizing/realcover/html

Velleux M, Endicott D, Steuer J, Jaeger S, Patterson D. 1995. Long-term simulation of PCB export from the Fox River to Green Bay. *J Great Lakes Res* 21:359–372.

Wildhaber ML, Schmidt CJ. 1996. Hazard ranking of contaminated sediments based on chemical analysis, laboratory tests, and benthic community composition: Prioritizing sites for remedial action. *J Great Lakes Res* 22(3):639–652.

Woodward-Clyde Consultants. 1992. Ashtabula River investigation. Report. Ashtabula OH: Ashtabula River Group. Report 86C3609F-510.

Wooster M. 1997. Great Lakes United position statement on RAPs. *J Great Lakes Res* 23:224–225.

15

Metals Bioavailability as a Determining Factor in Natural Remediation at Milltown Reservoir Wetlands in Western Montana

Gary A. Pascoe and Greg Linder

ECOLOGICAL RISK ASSESSMENT (ERA) is one of the components of the remedial investigation/feasibility study (RI/FS) process under the Comprehensive Environmental Response, Compensation, and Liability Act (CERCLA) of 1980 that can drive remedial decisions at a chemically contaminated site. Ecological risk assessments (ERAs) are currently performed under a multi-step methodology (U.S. Environmental Protection Agency [USEPA] 1997a), consistent with the approach developed by USEPA's Risk Assessment Forum for assessing stressor (i.e., chemical and nonchemical) risks to the environment (USEPA 1992a–c, 1998) and adapted from the original risk assessment methodology developed by the National Academy of Sciences (National Research Council [NRC] 1983). The recommendations by Suter (1993) and earlier guidance (USEPA 1989) to include the 3 components of chemistry, ecology, and toxicology in the assessment of ecological risks are addressed within the Superfund ERA guidance.

Results of an ERA can help form the basis for setting goals to clean up chemical wastes to protect the nonhuman environment. Under CERCLA, the selected remediation for a chemically contaminated site must lower risks to levels that are protective of the ecological communities of the site. A high level of uncertainty in the results of an ERA will increase uncertainty in the subsequent remedial decisions and may lead to unnecessary effort and costs for active remediation of a site.

According to USEPA (1997b), the term "natural remediation" (NR) or "monitored natural attenuation" refers to the reliance on natural processes (within the

Natural Remediation of Environmental Contaminants. Michael Swindoll et al., editors.
 ISBN 1-880611-33-3.

context of a carefully controlled and monitored site cleanup approach) to achieve site-specific remedial objectives within a time frame that is reasonable compared to other methods. Natural remediation and natural attenuation refer to the variety of physical, chemical, or biological processes that, under favorable conditions, act without human intervention to reduce the mass, toxicity, mobility, volume, or concentration of contaminants in soil or groundwater. These in situ processes include biodegradation, dispersion, dilution, sorption, volatilization, and chemical or biological stabilization, transformation, or destruction of contaminants. Other terms associated with NR in the literature include "intrinsic remediation," "intrinsic bioremediation," "passive bioremediation," "natural recovery," and "natural assimilation." While some of these terms are synonymous with "NR," others refer strictly to biological processes, excluding chemical and physical processes.

This paper discusses chemical and biological factors that can influence the level of confidence, or uncertainty, in the ERA results. A case study of an ERA at the Milltown Reservoir Sediments Superfund Site in Montana is presented to illustrate how reducing chemical and biological uncertainties in the ERA affected remedial decisions about NR of metals in a wetland.

CHEMICAL AND BIOLOGICAL FACTORS IN NATURAL REMEDIATION

As described in early USEPA guidance for conducting ERAs at hazardous waste sites (USEPA 1989), 3 types of information are needed to establish a firm, causal relationship between toxic wastes and ecological effects:

1) Chemical analyses of the appropriate media are necessary to establish the presence, concentrations, and characteristics of hazardous chemicals.
2) Ecological surveys can establish that adverse ecological effects have occurred.
3) Toxicity tests are necessary to establish a link between the adverse ecological effects and the chemistry and toxicity of the wastes.

This triad of information is integrated in the characterization of ecological risks. Without all 3 types of data, other potential causes of the observed effects unrelated to the toxic effects of hazardous wastes, such as habitat alterations and natural variability, are difficult to determine. The degree of uncertainty in quantifying exposures and biological effects determines the degree of confidence in risk estimates. Higher uncertainty in the assessment of risks generally leads to more stringent remediation to protect the environment. Reducing uncertainty in the assessment of risks, by identifying and quantifying chemical and biological factors that govern chemical exposures and toxicology, leads to more realistic designs of remedial plans and associated costs. The decision to use NR as the preferred remedial alternative at a site must be based on a certain degree of site-specific knowledge about factors that affect chemical exposures and toxicity.

When potential biological effects are assessed at chemically contaminated sites, uncertainty can be minimized with increased site specificity of the study. For example, a benchmark-based assessment will be the least site specific because of the inherent design for wide applicability. Water quality criteria (WQC), as one such benchmark, were designed to be applicable nationwide. An ERA based on literature-derived values for biological effects is similarly limited but can be more readily applied to a given site, depending on the quality and specificity of the information.

To minimize uncertainty in the interpretation of an ERA, performance of site-specific field and laboratory studies may be necessary. For sites containing mining wastes, the bioavailability of the metals is a significant factor in predicting ecological risks and in subsequent decisions about the effectiveness of NR. Despite the relatively good understanding of key factors governing metals mobility and bioavailability in the environment (e.g., Bohn et al. 1979; Allen 1995), the use of this information to predict potential uptake and effects on biological organisms at mining waste sites has been difficult. The remainder of this chapter provides a synopsis of the technical findings of an evaluation of ecological risks associated with soils and sediments at Milltown Reservoir wetlands, and discusses the factors that govern metals bioavailability from soils and sediment and how they affected the subsequent remedial decisions for management of the wetlands.

CASE STUDY: MILLTOWN RESERVOIR WETLANDS

An ERA was developed for a metals-contaminated wetland in western Montana in the late 1980s and early 1990s (Pascoe and DalSoglio 1994; Pascoe, Blanchet, Linder, Brumbaugh et al. 1994). The wetlands at Milltown Reservoir are part of the Milltown Reservoir Sediments Superfund Site. The site consists of approximately 3 miles of riverine and palustrine wetlands formed by Milltown Dam on the Clark Fork River, Montana (Figure 15-1). The site is characterized by the accumulation of As- and Cu-enriched mine tailings in sediments and wetland soils. The mine tailings originated from mining activities more than 100 miles upstream in the Rocky Mountains and were transported downstream during floods in the early 1900s. Because of the potential threat of the metals to wetland biota, an initial remedial goal advocated by the public was removal of all metals-contaminated materials. The ERA was initiated under CERCLA to provide information for making subsequent decisions about the appropriate remediation for the site.

Risk assessment approach

The ERA for Milltown Reservoir followed Superfund methodology (USEPA 1989) that became available during the initial stages of the project. Because of the com-

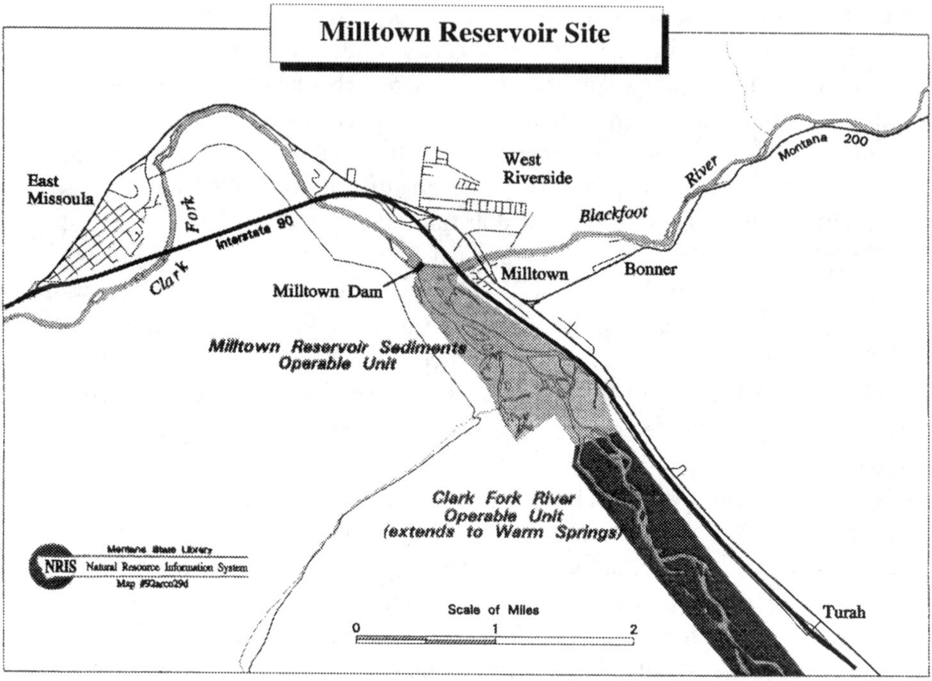

Figure 15-1 Milltown Reservoir wetlands, located in western Montana, 6 miles east of Missoula (reprinted from Lewis et al. 1999 with permission)

bined lack of applicable criteria for metal concentrations in sediments or wetland soils, the complexity of the site habitats, the likelihood of subtle ecological impacts resulting from seasonal influx of metals to the ecosystem, and the expected difficulty in predicting the bioavailability of the sediment and soil metals, an integrated field- and laboratory-based approach was developed for the risk assessment. This approach is consistent with Steps 3 (problem formulation) through 7 (risk characterization) of current Superfund guidance (USEPA 1997a) and was designed to address the 3 components of the triad of risk information mentioned above. It consisted of ecological surveys of terrestrial and aquatic habitats; quantitation of exposures of biota to metals and As in surface water, rooting zone water, soils, and sediments; and toxicological evaluations of these media both in the laboratory and in situ. The in situ studies were designed to complement the laboratory studies as well as to provide verification of conclusions based on laboratory results.

This approach was designed to reduce uncertainty in estimating exposures, bioavailability, and toxicity of the soil and sediment metals, and to provide reliable estimates of potential biological effects related to metals exposures. In order to address the site specificity of contaminant effects, the assessment focused on laboratory and field evaluations of lower trophic-level species and populations in the riverine, wetland, and terrestrial habitats.

Indicators of limited metals bioavailability at Milltown Reservoir Wetlands

Results of the integrated field and laboratory studies for the ERA at Milltown Reservoir have been summarized in detail elsewhere (Pascoe 1993; Linder et al. 1994, 1998; Pascoe, Blanchet, Linder, Brumbaugh et al. 1994; Pascoe and DalSoglio 1994). For the terrestrial habitat of the wetland, the studies found that the bioavailability of metals and As to small mammals was more limited than expected (Pascoe, Blanchet, Linder 1994, 1996). Table 15-1 compares the concentrations of various elements, estimated by food web modeling for small mammals at the wetland, with those concentrations found in body carcasses concurrently collected. The apparent bioavailability of soil elements to small mammals was less than 0.1%, much less than the typical assumption of 1% to 10% (Klaassen and Rozman 1991).

Table 15-1 Quantitation of the uptake of metals from wetland soils

		Body burden (μg/g)					
		Modeled[a]		Measured		Calculated mean bioavailable fraction (%)	
Metal	Avg soil conc (μg/g)	Meadow vole	Deer mice	Meadow vole	Deer mice	Meadow vole	Deer mice
As	61.5	2,208	4,215	0.2–0.3	< 0.02	0.01	< 0.0005
Cd	6.8	379	723	0.04–0.1	0.06–0.1	0.03	0.01
Cu	530	7,710	14,717	2.0–3.4	2.9–4.3	0.04	0.02
Zn	1,760	33,270	63,510	21.5–43.3	24.4–33.5	0.13	0.05

[a] In order to compare with measured body burdens, modeled values assume maximum absorption and no elimination.

The effect of the limited bioavailability of mining waste metals on toxicity tests is illustrated by results of earthworm (*Eisenia foetida*) bioassays (Linder et al. 1992). Table 15-2 compares responses expected for the concentrations of metals measured in surface soils at the wetland with the responses observed with in situ and laboratory bioassays of wetland soils. Based on the concentrations of total Cu in soil, substantial toxicity was predicted for earthworms (Hartenstein et al. 1981), yet only subtle effects were observed at the highest concentrations.

Table 15-2 Comparison of expected with observed endpoints of earthworm studies at Milltown Reservoir wetlands, Montana

Response	Expected[a]	Observed[b]
Decreased survival	+	–
Decreased growth	+	–
Reproductive effects	+	NT[c]
Morphological effects	NA[d]	+
Decreased population	+	–[e]

[a] Based on comparison of soil concentrations of metals with effects concentrations in literature, e.g., Hartenstein et al. 1981 (+ indicates expected endpoint response).
[b] Based on field and laboratory toxicity tests (Pascoe, Blanchet, Linder 1994) (+ indicates morphological effects observed as bioasay endpoints; – indicates expected response not observed).
[c] NT = not tested.
[d] NA = not available.
e Based on field observations at the site.

Comparison of sediment toxicity guidelines with no-effect-concentrations (NECs) calculated from sediment bioassays for the site illustrates the limitations of using only benchmark comparisons to characterize ecological risks (Ingersoll et al. 1997). Table 15-3 presents results of toxicity tests with whole sediments and sediment pore water, WQC for comparison with pore water test results, and sediment toxicity guidelines for comparison with whole-sediment test results. For Cu and Zn, the NECs were higher than the respective guidance values. These findings are indicative of the conservatism inherent in guidance values, as recently pointed out by O'Conner et al. (1998).

The subtle findings of toxicity tests performed with both terrestrial and aquatic exposure media at Milltown Reservoir generally were consistent with the lack of overt ecological impacts observed in terrestrial and aquatic habitats at the site (Linder et al. 1994; Pascoe and DalSoglio 1994; Pascoe, Blanchet, Linder 1994). An imbalance in community structure in sediment benthos (e.g., increased number of *Chironomidae* genera) was observed in sediments at stations with the highest extractable metals under oxidizing conditions (i.e., low flow and low acid-volatile sulfide [AVS] content) (Ingersoll et al. 1994). However, the ecological significance of the imbalance was considered of lesser importance to the ERA than was the potential transfer of metals from benthic invertebrates through the food chain to consumer fish (Pascoe, Blanchet, Linder, Brumbaugh et al. 1994).

Table 15-3 Indicators of bioavailability of sediment metals at Milltown Reservoir, Montana

	Sediment pore water (µg/L)		Whole sediment (µg/g)	
Metal	NOEC[a]	U.S. WQC[b]	NOEC[c]	SQG[d]
Cd	0.4	1.1	6.8	5 to 9
Cu	36	12	580	70 to 390
Zn	166	110	1,300	120 to 270

[a] NOEC = no-observed-effects concentration; sediment pore water toxicity tests with *Hyallela azteca*; length endpoint (Ingersoll et al. 1994; Pascoe, Blanchet, Linder, Brumbaugh et al. 1994).
[b] WQC = water quality criteria.
[c] Sediment toxicity tests with *Hyallela azteca*; length and sexual maturation endpoints (Ingersoll et al. 1994; Pascoe, Blanchet, Linder, Brumbaugh et al. 1994).
[d] SQG = sediment quality guideline; range of low and medium values from pool of national data from freshwater and marine studies (Long and Morgan 1991).

FACTORS GOVERNING METALS BIOAVAILABILITY

The following are physical and chemical factors that influence metal bioavailability. In general, the bioavailability of metals depends on the physical and chemical properties of both the metal wastes and the exposure medium. In the next section, those factors critical to the interpretation of metal bioavailability at Milltown wetlands are characterized.

Chemical activity

If a contaminant is introduced into a closed system that consists of 2 immiscible phases, it will be distributed between the 2 phases until equilibrium is reached. At equilibrium, there will be an escape of individual molecules from one phase to another, but the net exchange of chemical mass between the 2 phases will be equal to 0. This tendency of a chemical to escape from a phase is referred to as its "fugacity." Thus equilibrium exists when the chemical potential in all phases of a closed system is equal; equilibrium also exists when the fugacities of a chemical in the phases are equal.

At low concentrations typical of environmental contamination, fugacity and concentration can be directly related to a constant that quantifies the ability of a phase to contain the chemical. Partition coefficients express the concentration differential between 2 phases at equilibrium. The bioaccumulation potential of a chemical is dependent on the fugacity of chemicals and the containing abilities of the environmental medium and organism as the phases of concern. The chemical and physical factors that govern these containing abilities are numerous, as described below (Hamelink et al. 1994). More discussion on equilibrium partitioning (EqP) as it pertains to metals may be found in USEPA 1994.

Chemical stability

For chemicals to bioaccumulate, they must be stable and recalcitrant with respect to degradation. Metals are bioaccumulated by aquatic organisms either as ions in solution or as organometallic complexes (Hawker 1990). In soils, exposure occurs most frequently to metals either in solution in interstitial waters or through inorganic complexes that reflect the biogeochemical conditions of parent materials (Hawker 1990; Hamelink et al. 1994). Complexation of metals may facilitate or inhibit their bioaccumulation by affecting bioavailability. For example, organometalloids taken up by wetland biota may hydrolyze, thereby allowing the free metal ion to bind with sulfhydryl, amino, purine, and other reactive groups within the organism (Hawker 1990).

Chemical solubility

Generally, bioconcentration (defined as a net accumulation from water) increases as the water solubility of a chemical increases (Hamelink et al. 1994). Water solubility favors rapid uptake of a chemical by an organism but may also favor a rapid elimination of that chemical. Physical and chemical processes that increase the water solubility of a chemical decrease the tendency for that chemical to bioaccumulate, that is, reduce the exposure through dietary sources. However, the ionized forms of metals such as Hg, Cd, and Pb and the metalloid Se are soluble in water, which bind with tissues and thus are actively bioaccumulated by organisms (Hamelink et al. 1994). For example, Se has a sorptive affinity for hydrous metals and oxides, clays, and organic materials. Sorption by bed sediments or suspended solids can result in enrichment of Se concentrations in the bed sediments.

Acid-volatile sulfides

The identification of sulfides in sediment has become, along with oxidation–reduction (redox) potential, an important predictor of metals bioavailability for NR. Metals generally form sulfides of low solubility in sediments, especially under reducing conditions. Hong et al. (1995), for example, reported that adsorption of divalent metal cations to pyrite (iron sulfide) occurred as follows:

$$\mathrm{Co} < \mathrm{Cd} < \mathrm{Mn} < \mathrm{Ni} < \mathrm{Zn}$$

AVS is commonly used to estimate the influence of sulfide binding in sediments. AVS is that fraction of sulfide in sediments extracted by cold dilute HCl; it exists in natural sediments primarily as iron monosulfide complexes (Morse and Cornwell 1987; Allen et al. 1993; Ankley et al. 1994; USEPA 1994; Hare et al. 1994).

Under field conditions, metals will displace Fe from the monosulfide, and sulfide binding may therefore influence the bioavailability of various divalent metals in both marine and freshwater sediments and hence can play a major role in the NR of metals (Di Toro et al. 1990; Ankley et al. 1991; Carlson et al. 1991).

Recently, O'Connor et al. (1998) have suggested that the role of AVS must be considered within the context of other tools to evaluate metal bioavailability, since the complexities of chemical mixtures in sediments may not be adequately considered by an analysis of AVS and simultaneously extracted metal (SEM) concentrations alone. Nonetheless, AVS in sediments may be a useful predictor of metal bioavailability and toxicity to the benthos (Di Toro et al. 1990, 1991; Ankley et al. 1991; Carlson et al. 1991). For example, AVS and SEM have been found to predict the toxicity of some divalent metals (Cd, Pb, Zn, Cu, and Ni) in pore water (Ankley et al. 1994). That is, when the reactive metal-to-AVS molar ratio exceeds 1, the sediments may be toxic to benthic invertebrates. Metal toxicity to the benthos generally does not occur when the reactive metal-to-AVS molar ratio is less than 1 (Di Toro et al. 1990, 1991; Ankley et al. 1991; Carlson et al. 1991).

Eh and pH

Water or sediment pH (acidity or basicity) and sediment Eh (redox potential) affect the partitioning of metals in aquatic ecosystems (Allen 1995), while pH in soils exerts similar influences on metals in terrestrial habitats (Bohn et al. 1979). Most natural waters are weakly to strongly oxidized and are mildly acidic to mildly alkaline. In contrast, sediments are generally reduced and nearly neutral in pH. Both Fe and Mn are in the divalent, relatively soluble state in sediments. Formation of insoluble hydrous oxides that can coprecipitate or adsorb other soluble metallic species is favored by oxidation to the ferric and manganic forms. This process reduces the bioavailability of the metallic species and their potential for bioaccumulation. Insoluble sulfides of many metals, formed under reduced conditions, are rapidly oxidized when conditions become aerobic, first to elemental S and then to sulfate, releasing the soluble metal ion. Oxidation of Fe and Mn produces H ions, sulfide oxidation produces sulfate, and the result of these processes is an increase in acidity (Allen 1995). The solution of free metal ions and the subsequent formation of insoluble hydrous oxides, which tend to reduce the concentration of metal ions in solution by sorbing them, is favored by these acidic conditions. The increased solubility under acidic conditions and the sorption to hydrous oxides are competing processes.

Generally, free ions tend to be present in greater abundance and are thus more bioavailable at low pH and under oxidizing conditions. Under reducing conditions, metals are present largely as insoluble sulfides and are not bioavailable, which are favorable conditions for NR of metals. Sediment-associated metals that are not bound in the sediment crystal matrix are present either as ions, complexes, or precipitates (Allen 1995).

Hydrous ferric and manganese oxides

Hydrous ferric and Mn oxides form amorphous aggregates that contain large amounts of water (Boothman 1988). The surface areas of these aggregates are many times greater than those of clay minerals. Hydrated metal ions and soluble complexes are able to diffuse through the aggregate structures, in addition to being adsorbed on the surface. External concentration, pH and Eh, and to a limited extent ionic strength, affect the rates of adsorption and desorption processes. Hydrous oxides that form rapidly when reduced sediments are oxidized may scavenge soluble metalloids or metals, such as As, Cr, Cu, Ag, and Zn, from the water column by coprecipitation.

Cation exchange capacity and adsorption and desorption

Particle size and organic C content are major factors influencing rates of metal adsorption in soils and sediments. For example, about 10% to 60% of the sorption capacity of sediment particles typically appears to be accounted for in the labile fraction of sediment; that is, adsorption and desorption occur in a matter of minutes (Allen 1995). The sorption to or from the remaining sites may require days to weeks. Because of the length of time required for the second sorption phase, methods for estimating chemical concentrations that depend on equilibrium distribution of chemicals among environmental phases may overestimate the bioavailable fraction of a chemical in sediments. As a tool to predict the effectiveness of NR of soils or sediment, measurement of cation exchange capacity (CEC) can provide an indication of the capacity of the medium to maintain low bioavailability of sorbed metals.

Salinity

While of limited concern in freshwater habitats, soils potentially present a wide range of salinities across wetland types. While best described for marine and estuarine habitats, salinity affects bioaccumulation through physical or chemical processes, including desorption and solubility, as well as through physiological processes such as osmoregulation, membrane permeability, and respiration rate and volume (Hall and Anderson 1995). There is also competition among free ions for tissue-binding sites.

The effect of salinity on metal bioaccumulation is complex and element specific (Hall and Anderson 1995). In general, metals in solution have been reported to bioaccumulate to higher concentrations or increase in toxicity as salinity decreases, but the opposite may also occur because of the tendency for metals to remain in either solution or suspension in the fresh water (Hawker 1990). Euryhaline species are most resistant to toxic conditions at isosmotic salinities because of the minimization of osmotic stress (Sprague 1985).

The binding strength of Cd, Cu, Mn, and Zn to inorganic ligands decreases with increasing salinity, both by the competition of other major cations for binding sites and by favoring the formation of chloride complexes. The free ion form generally has the greatest bioavailability, but the variable amounts of dissolved and particulate C (related to salinity) confound the picture by providing sites for complexation. Generally, Cd and Se solubility and bioavailability are inversely related to salinity, and Zn uptake is unrelated to salinity. Cu is unpredictable and is especially affected by organic complexation, and Pb uptake increases with increasing salinity. Hg binds very tightly to particles and does not respond to salinity changes directly as other metals (USACE 1989).

Organic carbon

Total organic carbon (TOC) in soils and sediments consists primarily of humic matter and fulvic acids and may constitute a large portion of the sediments (Musani et al. 1980; Hawker 1990). The association of metals and metalloids, such as Cu, Pb, Se, and Ag, to sediment organic C is primarily by active bonding with functional groups. There is no simple relationship between TOC and bioavailability or bioaccumulation of metals (Hamelink et al. 1994).

Dissolved organic carbon

Dissolved organic carbon (DOC) in aquatic ecosystems or in soil interstitial water is composed primarily of humic substances produced by the degradation of dead plant material. Humic and fulvic acids make up 40% to 80% of DOC and are defined according to the effects of pH on their precipitation from aqueous solution. These organic acids are structurally complex colloidal and subcolloidal compounds containing large numbers of functional groups and straight and branched alkyl side-chains. The functional groups are what make these large molecules water soluble and also provide sites for metal ions in solution. The presence of uncontaminated humic acids as a factor in NR will reduce the uptake of metals by organisms when metals occur in water (USEPA 1994).

Sediment suspension and resuspension

Suspension of sediments can temporarily increase the concentration of associated chemical contaminants in the water column, as either dissolved or sorbed to suspended particles, or both. Suspension occurs as a result of the movement by benthic infauna (i.e., bioturbation), storms, dredging or disposal operations, and other activities that disturb the surface of the sediments. The increase in chemical concentrations in the water column is not a simple linear function of the mass of the sediment suspended because the contaminant-bearing organic content of the suspended sediment fraction is typically higher that the organic content of the consolidated deposited sediment (Allen 1995). For example, particulate organic matter can act as a scavenger of metals from solution, thus reducing the bioavailable fraction in the water column. The presence of divalent cations (Mg^{++}, Ca^{++}) in the water column can cause resuspended particulate, colloidal, and soluble organic matter to flocculate and settle from the water column. Lower-weight organic acids can be precipitated as metal fulvates and humates under these conditions. Trace elements may coprecipitate with flocculated material.

Suspension of uncontaminated sediments has been demonstrated to reduce the bioavailability of contaminants in overlying contaminated water by absorbing them from solution. Conversely, the suspension of contaminated sediments in clean water will result in greater bioaccumulation in exposed organisms in the water column. Fugacity favors desorption from particulates to the water, and metals such as Pb and Hg that are often bound with the particulate material may become bioavailable (Allen 1995). Thus, the potential for future resuspension of contaminated sediments can influence the predicted effectiveness of NR.

Sediment and soil particle size

The surface area of the particles per unit mass of matrix (sediment or soil) increases as particle size decreases (Allen 1995). As the surface area increases, the number of negatively charged sites for adsorption also increases, and therefore, the number of cations that can be carried on the sediment or soil increases. Depending upon the habitat setting, sediment and soil particles may have coatings of humic matter, with most of the organic C being associated with the fine-grained materials. Chemical contaminants in soils and sediments are generally associated with the fine-grained fractions. Biota dwelling in and ingesting fine-grained materials are potentially exposed to higher concentrations of contaminants than are organisms in coarse-grained sediments, although fine-grained or clay-like matrices may limit exposure because of a higher capacity for adsorption (Allen 1995) or because of ecological limitations associated with marginal habitats (Higgins and Thiel 1988).

FACTORS INFLUENCING THE BIOAVAILABILITY OF METALS AT MILLTOWN RESERVOIR WETLANDS

From the above understanding of potential attenuating factors on the bioavailability of metals in sediments and soils, certain existing conditions were identified at Milltown Reservoir wetlands that undoubtedly contributed to the relatively low occurrence of adverse effects (Pascoe, Blanchet, Linder, Brumbaugh et al. 1994). For example, the variability in terrestrial toxicity test results appears to be related to the heterogeneity of soil characteristics across the site (e.g., soil texture). As suggested by Figure 15-2, within–sample unit variability of the terrestrial habitat was relatively limited in comparison to variability among sample units, which supported the initial stratified design of the sampling and analysis. In sediments, adverse biological effects could be predicted for those sediments with high metals concentrations and subject to oxidizing conditions, as characterized by AVS content.

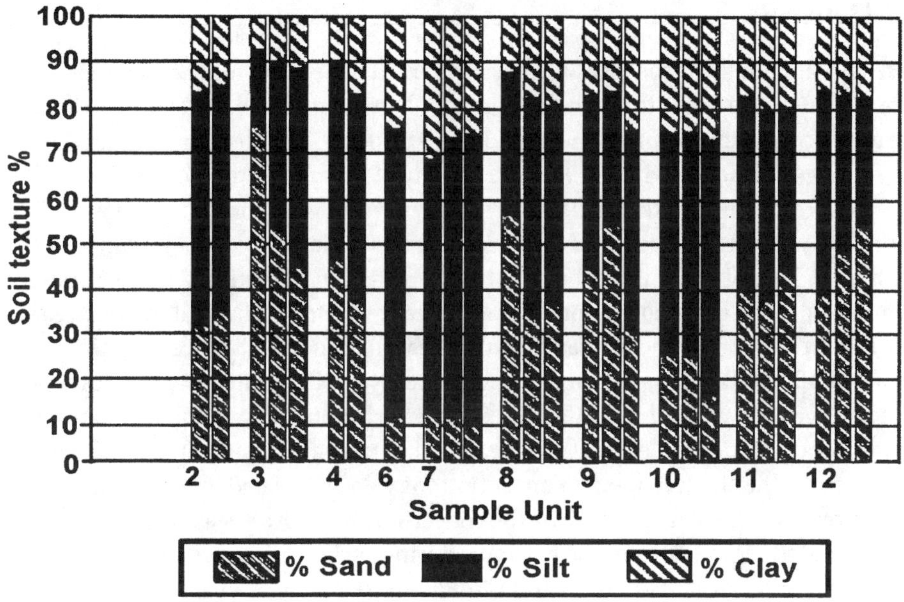

Figure 15-2 Sampling and testing units at Milltown Reservoir wetlands, Montana (reprinted from Lewis et al. 1999 with permission)

Similarly, although the distribution of selected target metals, both total and DTPA-extractable, in wetland soil varied spatially, within–sample unit variation appeared less than the variation among sample units. A nonparametric analysis of variance and multiple comparison test (Friedman's, P = 0.05) indicated that sample units could be ranked from high to low for selected target analytes (Cd, Cu, and Zn) as follows (Linder et al. 1994):

Sample Unit 6 > Sample Units 7, 10 > Sample Unit 2 > Sample Units 8, 9, 12 > Sample Units 3, 4, 11

By comparing and evaluating these target analyte rankings with results of the biological tests (e.g., earthworm and plant bioassays) performed on co-located soil samples, an integrated assessment of risk to biological and, by inference, ecological targets was developed (Linder et al. 1998). The integration was sufficient for making decisions about NR. The heterogeneity of the soils and sediments, and the large size of the wetlands, precluded identification of all of the factors that determined metals bioavailability throughout the site.

REMEDIAL DECISION FOR MILLTOWN RESERVOIR WETLANDS

Based on the results of the ERA, the current remedial decision for the wetland component of the site is for monitoring natural attenuation (i.e., NR) rather than the initial consideration of complete sediment removal. This decision was driven by the findings of lower bioavailability and toxicity of the metals than might be expected from benchmark and literature studies. Additional consideration was given to the likely impact that removal of the sediments would have on the wetland ecosystem, including the impact of increasing oxidation potential of newly exposed sediment and the subsequent potential for enhanced release of metals downstream of the reservoir. This concern was shown to be well founded when ice scour in the spring of 1996 (after completion of the ERA) resuspended contaminated sediment that had accumulated behind Milltown Dam, which resulted in marked elevations in metals concentrations in downstream waters that were above ambient WQC. More significantly, the elevated metals concentrations were associated with fish kills downstream of the dam. Because of the concerns over future releases of metal-contaminated sediment from the reservoir during such extreme events (see Pascoe 1994), the public oversight group and USEPA are creating plans for removal of depositional areas of sediment that have both high metals concentrations and a potential for oxidizing conditions, and hence a greater potential for dissolved metals releases during sediment-bed disturbances.

The differences in costs for the remedial decisions, that is, between the initial suggestion of complete sediment removal and the NR alternative, are easily apparent. Cost estimates for these remedial options for Milltown Reservoir are presented in Table 15-4. As can be seen, the difference in remediation costs when considering results of the site-specific ERA in the decision process can be substantial.

ROLE OF NATURAL REMEDIATION IN REDUCING ECOLOGICAL RISKS

Natural remediation processes typically occur in contaminated soil and sediment, but to varying degrees of effectiveness depending on the types of contaminants present and the physical, chemical, and biological characteristics of the medium.

Table 15-4 Comparison of estimated remedial costs for Milltown Reservoir wetlands, Montana

	Estimated cost	
Remedial option	Before ERA[a]	After ERA
Sediment removal	\$200 to \$500 million	
In-place controls		\$2 to \$5 million
Institutional controls/monitoring		< \$0.5 million

[a] ERA = ecological risk assessment

Natural remediation may reduce the potential risk posed by site contaminants in 3 ways:

1) The contaminant may be converted to a less toxic form through destructive processes such as biodegradation or abiotic transformations.
2) Potential exposure levels may be reduced by lowering total concentrations (through destructive processes, dilution, dispersion) or by lowering bioavailable concentrations (e.g., the factors discussed above for metals).
3) Contaminant mobility may be reduced by adsorption to the soil or rock matrix.

Natural remediation, under certain conditions (e.g., through sorption or redox reactions), may be an effective approach to reducing the levels of inorganic contaminants in soil and groundwater. Both metals and nonmetals may be attenuated by sorption reactions such as precipitation, adsorption on solid surfaces (e.g., soil) or into the solid matrix, or partitioning into organic matter. Sorption reactions are the dominant mechanism responsible for the reduction of mobility or change in toxicity of inorganic contaminants. In predicting the permanence of NR for metals, it is necessary to know what specific mechanism (type of sorption reaction) is responsible for their attenuation because some mechanisms are more desirable than others. For example, precipitation reactions and complexation in the solid structure of soil are generally stable, whereas surface sorption and liquid phase partitioning can be reversible.

As described earlier, changes in pH, redox potential, and chemical speciation may release chemicals into the environment. At the Milltown Reservoir site, understanding the changes in redox status of sediment metals under varying natural conditions was a key factor in predicting dissolved metals releases from sediments and consequent public health and ecological risks (Pascoe 1994). Inorganic contaminants typically cannot be degraded by NR processes (except for radioactive decay). Therefore, NR of inorganic contaminants is most applicable to sites where reducing mobility and/or toxicity, through lowering the chemical bioavailability, are determined to be sufficiently protective.

Some biological processes may yield transformation products more toxic than the parent contaminant (e.g., the methylation of inorganic Hg), which must be considered prior to implementation of any NR program, especially given the characteristics of some chemical mixtures released to the environment. Additionally, some NR processes may result in transfer of some contaminants from one medium to another (e.g., from soil to groundwater, from soil to the air, or from groundwater to surface water), which may be not acceptable within ecological contexts.

ROLE OF NATURAL REMEDIATION IN CLEANUP OF CONTAMINATED SOILS AND SEDIMENTS

Remedies selected for contaminated media must protect the environment; such protection may be achieved through a mix of treatment, containment, engineering controls, and other means identified as part of the remedy selection process. Natural remediation should not be considered a remedy at any contaminated site unless all remedial options have been closely considered in the selection process. Natural remediation is an appropriate alternative only where its use will be protective of the environment and when it will be capable of achieving site-specific remedial objectives within a reasonable time frame. The effectiveness of NR both in near-term and long-term time frames should be demonstrated in these ways:

- Sound technical analysis that provides confidence in the predictions for NR to achieve remedial objectives
- Performance monitoring
- Backup or contingency remedies, where appropriate.

Decisions to employ NR as a remedy or remedy component should be thoroughly and adequately supported with site-specific characterization data and analysis, which may be more detailed than that needed to support active remediation. In general, a greater amount of supporting information may be required to demonstrate the efficacy of NR at those sites with contaminants that do not readily degrade through biological processes (e.g., inorganic chemicals) or at sites where monitoring has been performed for a relatively short period of time. The amount and type of information needed for such a demonstration will depend upon a number of site-specific factors, such as the size and nature of the contamination problem and the proximity of receptors and the potential risk to those receptors. Where NR has been incorporated into remediation, a monitoring program must be developed.

SUMMARY AND CONCLUSIONS

Ecological risk assessment can be used as a tool under CERCLA to provide the needed supporting information for selecting an NR alternative for the cleanup of contaminated soils and sediments. Numerous factors govern the availability of soil metals for uptake by biological organisms and consequent biological and ecological effects. Simplistic ERAs may not account for all the potential factors and may inadequately estimate the limited bioavailability of metals at mining waste sites. The Milltown Reservoir Sediments Superfund Site risk assessment focused on significant biological parameters where limited metals bioavailability was a deciding factor in remedial decision-making. The decision to institute a long-term monitoring program for the wetlands rather than complete sediment removal was based largely on the results of the comprehensive ERA. The case study exemplifies how the inclusion of site-specific studies of chemical exposure and toxicology can provide the needed support for demonstrating the appropriateness of NR as a viable remedial alternative at metals-contaminated sites. Additionally, reducing the uncertainty in remedial options can result in lower remedial costs. Finally, the lessons learned from the ERA at Milltown Reservoir provide further support for the design of risk-based remedial goals that are reflected in realistic cleanup of environmental contamination.

ACKNOWLEDGMENTS

We would like to thank Julie DalSoglio (USEPA Region 8), Clarence Callahan (USEPA Region 9), and Bruce Duncan (USEPA Region 10) for providing background information on the regulatory applications of natural attenuation in remediation of contaminated soils and sediments, especially with reference to ecological risks.

REFERENCES

Allen HE. 1995. Metal contaminated aquatic sediments. Ann Arbor MI: Ann Arbor Pr.

Allen HE, Fu G, Deng B. 1993. Analysis of acid-volatile sulfide (AVS) and simultaneously extracted metals (SEM) for the estimation of potential toxicity in aquatic sediments. *Environ Toxicol Chem* 12:1441–1453.

Ankley GT, Pipps GL, Leonard EN, Benoit DA, Mattson VR, Kosian PA, Cotter AM, Dierkes JR, Hansen DJ, Mahony JD. 1991. Acid-volatile sulfide as a factor mediating cadmium and nickel bioavailability in contaminated sediments. *Environ Toxicol Chem* 10:1299–1307.

Ankley GT, Thomas NA, Di Toro DM, Hansen DJ, Mahony JD, Berry WJ, Swartz RC, Hoke RA, Garrison AW, Allen HE, Zarba CS. 1994. Assessing potential bioavailability of metals in sediments: A proposed approach. *Environ Manage* 18:331–337.

Bohn HL, McNeal BL, O'Connor GA. 1979. Soil chemistry. New York NY: Wiley.

Boothman WS. 1988. Characterization of trace metal associations with polluted marine sediments by selective extractions. In: Lichtenberg JJ, Winter JA, Weber CI, and Fradkin L, editors. Chemical and biological characterization of sludges, sediments, and drilling muds. Philadelphia PA: American Society for Testing and Materials, ASTM STP 976.

Carlson AR, Phipps GL, Mattson VR, Kosian PA, Cotter AM. 1991. The role of acid-volatile sulfide in determining cadmium bioavailability and toxicity in freshwater sediments. *Environ Toxicol Chem* 10:1309–1319.

Di Toro DM, Mahony JD, Hansen DJ, Scott KJ, Hicks MB, Mayr SM, Redmond MS. 1990. Toxicity of cadmium in sediments. The role of acid volatile sulfide. *Environ Toxicol Chem* 9:1487–1502.

Di Toro DM, Zarba CS, Hansen DJ, Berry WJ, Swartz RC, Cowan CE, Pavlou SP, Allen HE, Thomas NA, Paquin PR. 1991. Technical basis for establishing sediment quality criteria for non-ionic organic chemicals by using equilibrium partitioning. *Environ Toxicol Chem* 10:1541–1583.

Hall Jr LW, Anderson RD. 1995. The influence of salinity on the toxicity of various classes of chemicals to aquatic biota. *Crit Rev Toxicol* 25:281–346.

Hamelink JL Landrum PF, Bergman HL, Benson WH. 1994. Bioavailability: Physical, chemical, and biological Interactions. Boca Raton FL: Lewis.

Hare L, Carignan R, Huerta-Diaz MA. 1994. A field study of metal toxicity and accumulation by benthic invertebrates: Implications for the acid-volatile sulfide (AVS) model. *Limnol Oceanogr* 39:1653–1668.

Hartenstein R, Neuhauser EF, Narahara A. 1981. Effects of heavy metal and other elemental additives to activated sludge on growth of *Eisenia foetida*. *J Environ Qual* 10:372–376.

Hawker DW. 1990. Bioaccumulation of metallic substances and organometallic compounds. In: Connell DW, editor. Bioaccumulation of xenobiotic compounds. Boca Raton FL: CRC Pr.

Higgins RP, Thiel H. 1988. Introduction to the study of meiofauna. Washington DC: Smithsonian Institution Pr.

Hong J, Calmano W, Forstner U. 1995. Interstitial waters. In: Salbu B, Steinnes E, editors. Trace elements in natural waters. Boca Raton FL: CRC Pr.

Ingersoll CG, Brumbaugh WG, Dwyer FJ, Kemble NE. 1994. Bioaccumulation of metals by *Hyallela azteca* exposed to contaminated sediments from the Upper Clark Fork River, Montana. *Environ Toxicol Chem* 13:2013–2020.

Ingersoll CG, Dillon T, Biddinger GR. 1997. Ecological risk assessment of contaminated sediments. Pensacola FL: Society of Environmental Toxicology and Chemistry (SETAC).

Klaassen CD, Rozman K. 1991. Absorption, distribution and excretion of toxicants. In: Amdur MO, Doull J, Klaassen CD, editors. Casarett and Doull's toxicology: The basic science of poisons. 4th ed. New York NY: Pergamon Pr.

Lewis MA, Mayer FL, Powell RL, Nelson MK, Klaine SJ, Henry MG, Dickson GW. 1999. Ecotoxicology and risk assessment for wetlands. Pensacola FL: Society of Environmental Toxicology and Chemistry (SETAC). 375 p.

Linder G, Callahan C, Pascoe G. 1992. A strategy for ecological risk assessments for Superfund: Biological methods for evaluating soil contamination. In: Hoddinott KB,

Knowles GD, editors. Superfund risk assessment in soil contamination studies. Philadelphia PA: American Society for Testing and Materials. ASTM STP 1158.

Linder G, Hazelwood R, Palawski D, Bollman M, Wilborn D, Malloy J, DuBois K, Ott S, Pascoe GA, DalSoglio JA. 1994. Ecological assessment for the wetlands at Milltown Reservoir, Missoula, Montana: Characterization of emergent and upland habitats. *Environ Toxicol Chem* 13:1957-1970 (Annual review).

Linder G, Ingham E, Brandt CJ, Henderson G. 1992. Evaluation of terrestrial indicators for use in ecological assessments at hazardous waste sites. Corvallis OR: U.S. Environmental Protection Agency (USEPA), Environmental Research Laboratory. EPA/ 600/R-92/183.

Linder G, Pascoe G, DalSoglio J. 1998. An ecological risk assessment for the wetlands at Milltown Reservoir, Missoula, Montana: An overview. In: Lewis MA, Powell RL, Nelson MK, Henry MG, Klaine SJ, Dixon KB, Mayer FL, editors. Ecotoxicology and risk assessment for wetlands. Pensacola FL: Society of Environmental Toxicology and Chemistry (SETAC).

Long ER, Morgan LG. 1991. The potential for biological effects of sediment-sorbed contaminants tested in the national status and trends program. Seattle WA: National Oceanic and Atmospheric Administration (NOAA) Technical Memorandum NOS ORCA 52.

Morse JW, Cornwell JC. 1987. Analysis and distribution of iron sulfide minerals in recent anoxic marine sediments. *Marine Chem* 22:55–69.

Musani L, Valenta P, Nurnberg HW, Konrad K, Branica M. 1980. On the chelation of toxic trace metals by humic acid of marine origin. *Estuarine Coastal Marine Sci* 11:639–649.

[NRC] National Research Council. 1983. Risk assessment in the federal government: Managing the process. National Research Council, National Academy of Sciences. Washington DC: National Academy Pr.

O'Conner TP, Daskalakis KD, Hyland JL, Paul JF, Summers JK. 1998. Comparisons of sediment toxicity with predictions based on chemical guidelines. *Environ Toxicol Chem* 17: 468–471.

Pascoe GA 1993. Wetland risk assessment. *Environ Toxicol Chem* 12:2293–2307.

Pascoe GA 1994. Human health and ecological risk assessment of sediment releases ("continued releases"). Milltown Reservoir Operable Unit, Milltown Reservoir Sediments Site. Prepared for US Environmental Protection Agency, Region 8, Helena, MT. Seattle WA: Environmental Toxicology International, Inc.

Pascoe GA, Blanchet RJ, Linder G. 1994. Bioavailability of metals and arsenic to small mammals at a mining waste-contaminated wetland. *Arch Environ Contam Toxicol* 27:44–50.

Pascoe GA, Blanchet RJ, Linder G. 1996. Food-chain analysis of exposures and risks to wildlife at a metals-contaminated wetland. *Arch Environ Contam Toxicol* 30:306–318.

Pascoe GA, Blanchet RJ, Linder G, Brumbaugh W, Kemble N, Canfield T, Ingersoll C, Farag A, DalSoglio J. 1994. Characterization of ecological risks at the Milltown Reservoir-Clark Fork River Sediments Superfund Site, MT. *Environ Toxicol Chem* 13:2043–2058.

Pascoe GA, DalSoglio JA. 1994. Planning and implementation of a comprehensive ecological risk assessment at the Milltown Reservoir-Clark Fork River Sediments Superfund Site, Montana. *Environ Toxicol Chem* 13:1943–1956.

Sprague JB. 1985. Factors that modify toxicity. In: Rand GM, Petrocelli SR, editors, Fundamentals of aquatic toxicology. Washington DC: Hemisphere.

Suter GW. 1993. Ecological risk assessment. Chelsea MI: Lewis.

[USACE] U.S. Army Corps of Engineers. 1989. Factors influencing bioaccumulation of sediment-associated contaminants by aquatic organisms; factors related to sediment and water. Environmental effects of dredging technical notes, EEDP-01-18. Vicksburg MS: USACE.

[USEPA] U.S. Environmental Protection Agency. 1989. Ecological assessment of hazardous waste sites: A field and laboratory reference. In: Warren-Hicks W, Parkhurst BR, Baker SS, editors. Corvallis OR: Environmental Research Laboratory. EPA 600/3-89-013.

[USEPA] U.S. Environmental Protection Agency. 1992a. Framework for ecological risk assessment. Washington DC: Risk Assessment Forum. EPA 630/R-92-001.

[USEPA] US Environmental Protection Agency. 1992b. Report on the ecological risk assessment guidelines strategic planning workshop. Washington DC: Risk Assessment Forum. EPA 630/R-92-002.

[USEPA] US Environmental Protection Agency. 1992c. Peer review workshop report on a framework for ecological risk assessment. Washington DC: Risk Assessment Forum. EPA 625/3-91-022.

[USEPA] U.S. Environmental Protection Agency. 1994. Briefing report to the EPA Science Advisory Board on the equilibrium partitioning approach to predicting metals bioavailability in sediments and the derivation of sediment quality criteria for metals. Washington DC: Office of Water and Office of Research and Development. EPA 822-D-94-002.

[USEPA] U.S. Environmental Protection Agency. 1997a. Ecological risk assessment guidance for Superfund: Process for designing and conducting ecological risk assessments. Interim final. Washington DC: Solid Waste and Emergency Response. EPA 540-R-97-006.

[USEPA] U.S. Environmental Protection Agency. 1997b. Memorandum. Use of monitored natural attenuation at Superfund, RCRA corrective action, and underground storage tank sites. Timothy Fields, Jr., Acting Assistant Administrator. Washington DC: Office of Solid Waste and Emergency Response. June 9.

[USEPA] U.S. Environmental Protection Agency. 1998. Guidelines for ecological risk assessment. Washington DC: Risk Assessment Forum. EPA/630/R-95/002F.

16

The Future of Natural Remediation, Perspectives, and Research Needs

Ralph G. Stahl Jr., Michael Swindoll, Stephen J. Ells

THE PROTECTION OF ECOLOGICAL RESOURCES and the restoration of damaged ecosystems necessitates an understanding of both the risks posed to the environment by contaminants and the potential impact (i.e., positive or negative) of management decisions and actions. Because of the complexities inherent in ecosystems as well as societal and regulatory considerations, careful scrutiny of site conditions is typically required before decisions about risk management goals are made. Ecological risk assessment (ERA) is a tool designed to facilitate this understanding and to aid in the risk management decision process (Stahl et al. 2000). An ERA helps evaluate the risks posed by contaminants as well as the short- and long-term impacts associated with management actions. Based on information from the ERA and other inputs (Figure 16-1), risk managers decide which actions are appropriate.

In this book, we have identified natural remediation (NR) as a risk management alternative that may be applicable for some contaminated sites; therefore, it is an alternative that should be considered during an ERA (Stahl and Swindoll 1999). Natural remediation is the sum of naturally occurring processes that reduce potential risks to ecological resources posed by a wide array of environmental contaminants (NR is likely to reduce risks to humans as well, but that is not the focus of this book). The processes involved in NR and their significance and effectiveness at reducing ecological risks will vary among contaminants and sites. NR is not a

Natural Remediation of Environmental Contaminants. Michael Swindoll et al., editors.
 ISBN 1-880611-33-3.

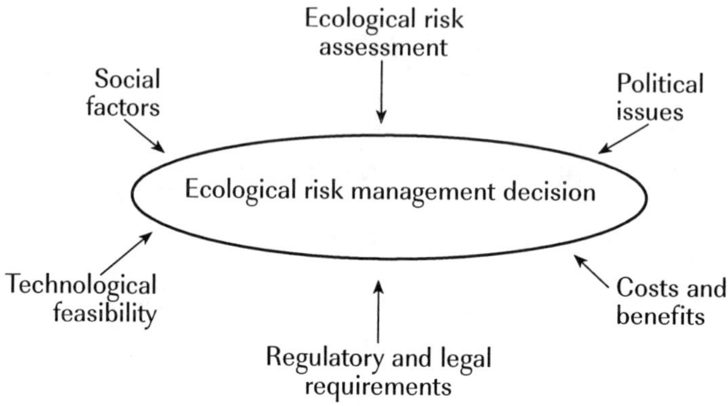

Figure 16-1 Inputs to the ecological risk management decision (from Stahl et al. 2000)

panacea for reducing ecological risks at all contaminated sites, as the effectiveness of NR at a particular site depends on a wide variety of factors.

Although the use and acceptance of NR as a remedial alternative has increased in recent years, it is clear from discussions in this book that NR is not appropriate for many sites (a summary of situations where NR would not be appropriate is provided at the end of this chapter). In addition, the perspectives of various stakeholders on the use of NR, including those of regulatory agencies, the regulated community, the public, and other groups, are presented in this chapter to provide the reader with an understanding of the many issues and considerations faced by risk managers. The chapter concludes with an assessment of the prospects for NR as a risk management alternative. Existing technical deficiencies and research needs are also identified.

INFLUENCES ON THE USE OF NATURAL REMEDIATION

Many influences affect ecological risk management decisions (Figure 16-1). Similar influences are likely when we consider actions to reduce risks to humans. These influences are diverse, and they present the risk manager with a substantial challenge when he or she attempts to synthesize the diverse inputs into a final decision. Similarly, when NR is being considered as a viable risk management action, the same or an enlarged set of influences is encountered.

Stakeholders

The term "stakeholder" can be defined in a number of ways, but in general, it means those individuals or groups who have some interest or connection to the decision being considered. The stakeholders can influence, either directly or indirectly, decisions concerning contaminated sites, and their involvement in decisions is becoming more commonplace (Yosie and Herbst 1998). Stakeholders vary with sites and can be involved with the risk management decision at various levels and at various times. The importance of this group lies with their ability to shape local and regional issues and decisions and with the fact that they may or may not be highly versed in the technical details of a proposed action. For example, stakeholders may or may not believe that NR is an appropriately "aggressive" approach to remediating a contaminated site. Further, given that NR may require a number of years to mitigate the toxicity and mobility of a contaminant, some stakeholders may be averse to such a lengthy process. On the other hand, having stakeholders involved in the early design of the ERA, particularly during the problem formulation phase, can provide them an opportunity and sufficient time to gain a better understanding of the role that NR can play in alleviating the risks associated with contaminants at a particular site.

Effectively communicating with stakeholders is essential to gaining their support for risk management decisions (Yosie and Herbst 1998). When presented with the context of what is known about a site and what actions are necessary to protect the environment, stakeholders can be expected to support decisions that they consider reasonable. In some ways, it is easier to explain NR to a nontechnical audience than to explain to them what an "assessment endpoint" means within the ERA context. Analogies for NR exist in everyday experience, including the biological transition of tree and lawn debris into useful compost that can be used to enhance vegetative growth.

Regulatory agencies

Individuals in regulatory agencies are charged with protecting the public's interests, including those natural resources that reside within the public trust. When managing contaminated sites, these same individuals are often challenged by the requests and requirements of a skeptical public, by the potentially dichotomous requirements of stakeholders, and by having to deal with responsible parties. Overlaying this complex situation is the regulatory and statutory diversity resulting from the inherent differences in federal versus state and state versus local jurisdictions.

With respect to NR, regulatory agencies may or may not have sufficient experience in applying it to contaminated sites, and for that reason, such agencies may not be eager to select it as a viable remedial action. On the other hand, stakeholders and others with influence on the decision may be averse to more aggressive remedial actions, including physical removal of the contaminated soils or sediments, and for that reason, they may lobby quite effectively for a less intrusive alternative such as NR.

Another important influence on whether or not NR is acceptable to a regulatory agency includes the requirement that the action comply with certain regulatory tenets, that it be protective (however defined), and that it be effective in mitigating or reducing exposure and risks over an extended period of time. In the final analysis, regulatory agencies attempt to balance the diverse needs of stakeholders and the regulated community, while complying with applicable regulations and policies to produce fair and equitable solutions to a problem.

Regulated community

The regulated community consists of the parties held accountable or responsible by environmental regulators for contaminated sites. The regulated community must bear the cost of risk management decisions and ultimately is held responsible for the liabilities associated with the contaminants. From the perspective of the regulated community, NR may be viewed as a viable risk management action if it has some of the same attributes that would make it attractive to regulators and other stakeholders.

Any action, including NR, must meet the key criterion of being compliant with regulatory requirements. From there, each potential action, including NR, can be screened against a number of other discriminating factors. First of all, will NR provide the same cost-effective risk reduction afforded by other actions, and if so, will it do so in a time frame that is acceptable to the company that funds the work as well as to the regulatory agency that will be asked to approve the timing. In this case, cost is a consideration, but it is not the single most important factor. In some cases, NR will require an extended period of monitoring and follow-up to ensure that the conditions favorable to risk reduction—whether biological, chemical, or physical—do not change. Such monitoring can lengthen the payout for NR and may not be financially attractive to the paying company. Undertaking a more aggressive action may be a more attractive alternative in this situation, but doing so could also have a much higher initial cost. The benefit to the company might be to discharge its liabilities quickly, thus freeing funds for other important projects.

Another key issue is whether NR is feasible and can be implemented at a specific site. The underlying premise of this book has been that NR should be evaluated as one of several potentially attractive risk management actions. Natural remediation can be applied successfully at sites where conditions are appropriate. Where they are not appropriate, for any number of reasons, then NR will not be feasible.

Even when NR is feasible at a particular site, it may not achieve the necessary remediation in a timely fashion. For example, a company may wish to keep a property in an industrial land-use application and may therefore move to remediate it more quickly so that it can be recycled into this new use.

TECHNICAL CHALLENGES AND RESEARCH NEEDS

Perhaps the greatest challenge faced by proponents of NR is the uncertainty associated with its long-term effectiveness or protectiveness. The implementation of long-term biological monitoring programs should help alleviate these concerns. In some cases, it may be appropriate to identify, up front, alternative courses of actions if monitoring efforts suggest that NR is not sufficiently protective. Financial assurance to support future actions may also be necessary, and companies may have to accrue funds for this potentiality when NR is the selected remedy.

In addition to the technical challenges faced by NR, there are a number of important research needs.

- Develop tools that allow quantification of the ability of a particular ecosystem to maintain itself. Today our knowledge in this area is rudimentary and dependent on a number of factors that themselves are subject to influence. Spatial and temporal factors will influence an ecosystem's ability to maintain itself, as will human activities. Sound information in this area will allow a more up-front evaluation of whether NR is feasible and may reduce the need for lengthy biological monitoring.
- Develop more robust data on the conditions under which chemical contaminants are likely to undergo NR, taking into account the diversity of sites and geographies likely to be encountered in the U.S. and elsewhere. Despite the fact that substantial information exists on the biodegradation of certain contaminants, the data are lacking for other processes such as photolysis, hydrolysis, and soil sorption and desorption kinetics for a great number of chemicals. More importantly, the soil conditions and types that would also play a major role in the degradation processes are sketchy and in need of further characterization. This information would also benefit the risk manager and risk assessor by allowing them to make an initial assessment of whether NR is feasible, given the particular contaminants at the site in question.
- Develop simple matrices of receptors and contaminants and reasonable exposure scenarios, which would allow an initial estimate of whether ecological receptors face an acutely toxic hazard. There are ongoing attempts by some groups to develop such matrices, which can provide risk assessors and risk managers with a "look-up table" of contaminants, exposure, and receptors. While this tool may be beneficial to a number of individuals assessing risks at contaminated sites, it would most certainly benefit those groups considering NR, since they would be able to determine whether a longer-term process to reduce risk, such as NR, is feasible.

Perhaps the most important consideration regarding the application of NR is the lack of a database that details the results of NR at various sites. Ultimately, the increased use of NR will provide this database and thus document those essential considerations for their applicability.

CONCLUSIONS

The underlying theme of this book is that NR should be considered as a viable risk management action within the context of other viable risk management actions that might be undertaken at a contaminated site. Based on the discussions in the book, it should be clear that NR is not a panacea for risk management at all sites. Each site should be evaluated individually and the conditions under which NR would be applicable explored and agreed to among risk assessors and risk managers. Where the proper conditions exist, NR has the potential to provide effective and inexpensive risk reduction, at potentially lower cost and with less direct impact on the environment than that of more aggressive remediation.

Natural remediation is a potentially viable risk management action where these conditions exist:

1) Physical, chemical, and biological conditions are appropriate.
2) The probability of acute hazard and exposure is low.
3) The regulatory climate does not mandate aggressive action.
4) Cost-effectiveness tests are met.

In conclusion, there are certainly situations where NR offers a less costly, less intrusive risk management option than is available with alternative actions.

REFERENCES

Stahl Jr RG, Bachman R, Barton A, Clark JR, DeFur P, Ells S, Pittinger CA, Slimak MW, Wentsel RS. 2000. Ecological risk management: A framework for and approaches to ecological risk-based decision making. Pensacola FL: Society of Environmental Toxicology and Chemistry (SETAC). In press.

Stahl Jr RG, Swindoll CM. 1999. The role of natural remediation in ecological risk assessment. *Human Ecol Risk Assess* 5(2):219-223.

Yosie TF, Herbst TD. 1998. Using stakeholder processes in environmental decision making. An evaluation of lessons learned, key issues and future challenges. Washington DC: Ruder Finn.

Abbreviations

1CP	1-chloropropane
1DCE	1,1-dichloroethene
1TCA	1,1,1-trichloroethane
2DCA	1,2-dichloroethane
2DCP	1,2-dichloropropane
2TCA	1,1,2-trichloroethane
2,4D	2,4-dichlorophenoxyacetic acid
ΔG	Gibbs free energy
$\rho\varepsilon$	electron activity
AE	assessment endpoint
AET	apparent effects threshold
AFCEE	Air Force Center for Environmental Excellence
AOC	area of concern
ARAR	applicable or relevant and appropriate requirement
ARCS	Assessment and Remediation of Contaminated Sediments program
ATSDR	Agency for Toxic Substances and Disease Registry
AUF	area use factor
AVS	acid-volatile sulfide
BAF	bioaccumulation factor
BCMOELP	British Columbia Ministry of the Environment, Lands and Parks
BPAC	binational public advisory committee
Bq	becquerel
BSAF	biota–sediment accumulation factor
BTEX	benzene, toluene, ethylbenzene, xylene
CA	chloroethane
CAM	crassulacean acid metabolism

 ISBN 1-880611-33-3.

CASM	Comprehensive Aquatic System Model
CCME	Canadian Council of Ministers of the Environment
cDCE	*cis*- dichloroethylene
CDF	confined disposal facility
CEAM	Center for Exposure Assessment Modeling
CEC	cation exchange capacity
CERCLA	Comprehensive Environmental Response, Compensation and Liability Act
CF	chloroform
CFC-11	trichlorofluoromethane
CFC-113	1,1,2-trichloro-1,2,2-trifluoroethane
CFR	Code of Federal Regulations
Ci	curie
CIESIN	Consortium of International Earth Science Information Network
CM	chloromethane
COD	chemical oxygen demand
CoM	cometabolic
CRREL	Cold Regions Research and Engineering Laboratory
CSO	combined sewer overflow
CT	carbon tetrachloride
CTF	confined treatment facility
DCA	1,2-dichloroethane
DCE	dichloroethylene
DCM	dichloromethane
DDE	dichlorodiphenyldichloroethylene
DDT	dichlorodiphenyltrichloroethane
DEQ	Oregon Department of Environmental Quality
DNA	deoxyribonucleic acid
DNAPL	denser-than-water nonaqueous phase liquid
DOC	dissolved organic carbon
DOM	dissolved organic matter
DQO	data quality objective
EBDRP	Elliott Bay/Duwamish Restoration Board
EDTA	ethylenediaminetetraacetic acid
Eh	oxidation-reduction (redox) potential
EqP	equilibrium partitioning
ERA	ecological risk assessment
ER-L	effects range-low
ETH	ethene

FCV	final chronic value
GAE	general assessment endpoint
GC/IH	Grand Calumet River–Indiana Harbor
GLI	Great Lakes Initiative
GLNPO	Great Lakes National Program Office
GUI	graphic user interface
ha	hectare
HCA	hexachloroethane
HCB	hexachlorobenzene
HCFC-21	dichlorofluoromethane
HCFC-123	1,1,1-trifluoro-2,2-dichloroethane
HCFC-123a	1,2-dichloro-1,2,2-trifluoroethane
HEP	habitat evaluation procedures
HPAH	high-molecular-weight polynuclear aromatic hydrocarbon
HQ	hazard quotient
IAEA	International Atomic Energy Agency
IBM	individual-based model
IJC	International Joint Commission
IUPAC	International Union of Pure and Applied Chemists
KCDNR	King County (Washington) Department of Natural Resources
K_d	distribution coefficient
K_H	Henry's law constant
K_{ie}	ion exchange constant
K_{oc}	organic C normalized partition coefficient
K_{ow}	octanol–water partition coefficient
LAMP	Lakewide Management Plan
LERAM	Littoral Ecosystem Risk Assessment Model
LNAPL	lighter-than-water nonaqueous phase liquid
LOAEL	lowest-observed-adverse-effect level
LPAH	low-molecular-weight PAH
LTRE	life table response experiment
MCL	maximum concentration limit (Chapter 13)
MCL	maximum contaminant level (Chapter 8)
MDCH	Michigan Department of Community Health
MDEQ	Michigan Department of Environmental Quality
MDNR	Michigan Department of Natural Resources
ME	measurement endpoint

MTBE	methyl tertiary butyl ether
MUSIC	MUlti-SIte Complexation model
NAPL	non-aqueous phase liquid
NAS	National Academy of Sciences
NCP	National Contingency Plan
NCRP	National Council on Radiation Protection and Measurements
NEC	no-effect concentration
NEPA	National Environmental Policy Act
NIST	National Institute of Standards and Technology
NMR	nuclear magnetic resonance
NOAA	National Oceanic and Atmospheric Administration
NOAEL	no-observed-adverse-effect level
NOM	naturally occurring organic matter, natural organic matter
NR	natural remediation
NRC	Nuclear Regulatory Commission
NRDA	Natural Resource Damage Assessment
NTA	nitrilotriacetic acid
NWQC	national water quality criteria
NYCDEP	New York City Department of Environmental Protection
NYSDEC	New York State Department of Environmental Conservation
OCP	organochlorine pesticide
OEPA	Ohio Environmental Protection Agency
OMEE	Ontario Ministry of Environment and Energy
OS	operator-split
OSWER	Office of Solid Waste and Emergency Response
p,p'-DDE	dichlorodiphenyldichloroethylene
PAC	public advisory committee
PAH	polynuclear aromatic hydrocarbon
PBT	persistent, bioaccumulative, toxic
PCA	pentachloroethane
PCB	polychlorinated biphenyl
PCDD	chlorodioxins
PCE	tetrachloroethylene
PEL	probable effect level
PH	petroleum hydrocarbon
PVC	polychlorinated vinyl chloride
QA/QC	quality assurance/quality control
QEA	quantitative environmental analysis

QSAR	quantitative structure-activity relationship
RA	risk assessor
RAP	remedial action plan
RBCA	risk-based corrective action
RCRA	Resource Conservation and Recovery Act
RI/FS	remedial investigation/feasibility study
RM	risk manager
RNA	ribonucleic acid
ROD	record of decision
SCS	Soil Conservation Service
SEM	simultaneously extracted metal (Chapters 9 and 15)
SEM	standard error of the mean (Chapter 6)
SI	Système International d/unités (international system of units)
SLC	screening-level concentration
SQC	sediment quality criteria
SQT	sediment quality triad
TCA	1,1,2-trichloroethane
TCDD	2,3,7,8-tetrachlorodibenzo-*p*-dioxin
TCE	trichloroethylene
TDCE	*trans*-1,2-dichloroethene
TEA	terminal electron acceptor
TeCA	1,1,2,2,-tetrachloroethane
TEF	toxic equivalency factor
TEL	threshold effect level
TEQ	toxic equivalent
TNT	2,2,6-trinitrotoluene
TOC	total organic carbon
TSS	total suspended solid
UMTRCA	Uranium Mill Tailings Radiation Control Act
UNESCO	United Nations Educational, Scientific, and Cultural Organization
UNSCEAR	United Nations Scientific Committee on the Effects of Atomic Radiation
USACE	U.S. Army Corps of Engineers
USCG	U.S. Coast Guard
USDA	U.S. Drug Administration
USDOI	U.S. Department of the Interior
USGS	U.S. Geological Survey

UST	underground storage tank
UV	ultraviolet
VC	vinyl chloride
VOC	volatile organic compound
WDNR	Wisconsin Department of Natural Resources
WDOE	Washington Department of Ecology
WER	water effect ratio
WET	wetlands evaluation technique
WQC	water quality criteria
WWTP	wastewater treatment plant

Index

A

B

Natural Remediation of Environmental Contaminants. Michael Swindoll et al., editors.

D

E

O

Q

R

T

U

V

W

X

Z